# Three-Dimensional Integration of Semiconductors

Kazuo Kondo • Morihiro Kada • Kenji Takahashi
Editors

# Three-Dimensional Integration of Semiconductors

## Processing, Materials, and Applications

Springer

*Editors*
Kazuo Kondo
Department of Chemical Engineering
Osaka Prefecture University
Osaka
Japan

Kenji Takahashi
Toshiba Corporation
Oita
Japan

Morihiro Kada
Osaka Prefecture University
Osaka
Japan

ISBN 978-3-319-18674-0 ISBN 978-3-319-18675-7 (eBook)
DOI 10.1007/978-3-319-18675-7

Library of Congress Control Number: 2015943688

Springer Cham Heidelberg New York Dordrecht London


Printed on acid-free paper

Springer International Publishing AG Switzerland is part of Springer Science+Business Media (www.springer.com)

# Preface

The concept of three-dimensional integration of semiconductors originated with the US patent by IBM in 1969. The patent is titled "Hourglass-shaped conductive connection through semiconductor structures" (http://www.google.com.mx/patents/US3648131). The original interconnect looks like an hourglass. Since 1969, the three-dimensional integration concept has spread out to semiconductor industries all over the world and more than 40 consortia and companies have been involved in this development. Forty-five years after its invention, three-dimensional integration of semiconductors is becoming very popular, and is about to be industrialized in advanced electronics in the very near future.

This book reviews the state of the art of three-dimensional semiconductor integration. Chapter 1 gives an overview of three-dimensional integration research and development history. Chapter 2 summarizes recent three-dimensional integration research and development activities and applications. Chapter 3 gives an explanation of through-silicon via (TSV) formation processes. Chapters 4 and 5 cover wafer handling, wafer thinning, and bonding of wafers and dies. Chapter 6 explains metrology and inspection. Chapter 7 discusses reliability and characterization issues. Chapter 8 covers trends in technology development of three-dimensional integration circuits testing. Finally, Chapter 9 summarizes research and development project results conducted by New Energy and Industrial Technology Development Organization (NEDO)/Association of Super-Advanced Electronics Technologies (ASET): Japan in 2008 to 2012.

We really hope that this book will help not only beginners in three-dimensional integration technology of semiconductors but also engineers who are already involved in this field, both in industry and academia. I was very much astonished when, in 2000, ASET members visited my university and asked for support to fill an interconnect via that was huge compared to the contact via created by the damascene process of copper electrodeposition. This introduction gave me the initial motivation to start my research on three-dimensional integration. Two editors,

Morihiro Kada and Kenji Takahasi, are former ASET leaders. The opportunity given by K. Howell of Springer to publish this book is very much appreciated.

Kazuo Kondo Osaka
Morihiro Kada Osaka
Kenji Takahasi Oita

# Contents

# Contributors

**Tsubasa Bandoh** Okamoto Machine Tool Works, Ltd., Annaka, Gunma, Japan

**Raleigh Estrada** Carl Zeiss X-ray Microscopy Inc., Pleasanton, CA, USA

**Michael Feser** Carl Zeiss X-ray Microscopy Inc., Pleasanton, CA, USA

**Gilles Fresquet** FOGALE nanotech, NÎMES, France

**Hiroaki Fusano** 3DI Dept. ATS BU, Tokyo Electron Ltd., Tokyo, Japan

**Allen Gu** Carl Zeiss X-ray Microscopy Inc., Pleasanton, CA, USA

**Takashi Haimoto** Dicer Grinder Marketing Team, Marketing Group, Sales Engineering Division, DISCO Corporation, Ota-ku, Tokyo, Japan

**Masaki Hashizume** The University of Tokushima, Tokushima, Japan

**Yoshinobu Higami** Ehime University, Ehime, Japan

**Yoshihito Inaba** Hitachi Chemical Co., Ltd., Ibaraki, Japan

**Toshihiro Ito** Okamoto Machine Tool Works, Ltd., Annaka, Gunma, Japan

**Bruce Johnson** Carl Zeiss X-ray Microscopy Inc., Pleasanton, CA, USA

**Morihiro Kada** Osaka Prefecture University, Osaka, Japan

The National Institution of Advanced Industrial Science and Technology (AIST), Ibaraki, Japan

**Tadashi Kamada** DENSO Corp., Aichi, Japan

**Shuichi Kameyama** Fujitsu Limited, Kanagawa, Japan

**Harufumi Kobayashi** Lapis Semiconductor, Kanagawa, Japan

**Kazuo Kondo** Osaka Prefecture University, Osaka, Japan

**Mitsuma Koyanagi** NICHe, Tohoku University, Miyagi, Japan

**Yutaka Kusuda** R&D Division, SAMCO Inc., Kyoto, Japan

**Kangwook Lee** NICHe, Tohoku University, Miyagi, Japan

**Shyue-Kung Lu** National Taiwan University of Science and Technology, Taipei, Taiwan (R.O.C.)

**Takahiko Mitsui** Okamoto Machine Tool Works, Ltd., Annaka, Gunma, Japan

**Fumihiko Nakazawa** FUJITSU LABORATORIES LTD., Kanagawa, Japan

**Tomoyuki Nonaka** R&D Division, SAMCO Inc., Kyoto, Japan

**Toshihisa Nonaka** Toray Research Center, Inc. (TRC), Shiga, Japan

**Kenichi Osada** Hitachi Ltd., Tokyo, Japan

**Sylvain Perrot** FOGALE nanotech, NÎMES, France

**Jean-Philippe Piel** FOGALE nanotech, NÎMES, France

**Zvi Roth** Florida Atlantic University, Boca Raton, FL, USA

**Kazuta Saito** 53M Japan Limited, Sagamihara-shi, Kanagawa, Japan

**Itsuko Sakai** Graduate School of Engineering, Nagoya University, Nagoya, Japan

**Osamu Sato** Lasertec Corporation, Yokohama, Japan

**Makoto Sekine** Graduate School of Engineering, Nagoya University, Nagoya, Japan

**Haruo Shimamoto** Advanced Industrial Science and Technology (AIST), Ibaraki, Japan

**Hiroshi Takahashi** Computer Science, Graduate School of Science and Engineering, Ehime University, Ehime, Japan

**Kenichi Takeda** Hitachi Ltd., Tokyo, Japan

**Hideo Takizawa** Lasertec Corporation, Yokohama, Japan

**Masahiko Tanaka** SPP Technologies Co., Ltd., Amagasaki, Hyogo, Japan

**Yoshitaka Tatsumoto** Lasertec Corporation, Yokohama, Japan

**Osamu Tsuji** R&D Division, SAMCO Inc., Kyoto, Japan

**Shiro Uchiyama** Micron Memory Japan, Kanagawa, Japan

**Senling Wang** Ehime University, Ehime, Japan

**Fumiaki Yamada** IBM Research, Tokyo, Japan

**Eiichi Yamamoto** Okamoto Machine Tool Works, Ltd., Annaka, Gunma, Japan

**Masahiro Yamamoto** 3DI Dept. ATS BU, Tokyo Electron Ltd., Minato-ku, Tokyo, Japan

**Tsuyoshi Yoshida** Okamoto Machine Tool Works, Ltd., Annaka, Gunma, Japan

**Hiroyuki Yotsuyanagi** The University of Tokushima, Tokushima, Japan

# About the Editors

**Dr. Kazuo Kondo** is a professor in the Department of Chemical Engineering, Osaka Prefecture University. He received his Ph.D. in chemical engineering from the University of Illinois in 1981. He has worked for Sumitomo Metal Industries, Hokkaido University, and Okayama University. He has 200 research publications and 100 patents. His major research is Copper Electrodeposition for TSV. His research extends to various fields, not only in electrodeposition but also in battery science and CVD. He is a member of the Electrochemical Society, IEEE, the Society of Chemical Engineering, Japan, the Japanese Institute of Electronics Packaging, the Surface Finishing Society of Japan, Material Japan, Electrochemistry Japan and the Japanese Society of Applied Physics.

**Morihiro Kada** is an invited researcher of the National Institute of Advanced Industrial Science and Technology (AIST) and a part-time researcher at Osaka Prefecture University. Prior to joining AIST and the university, he headed the Japanese national research and development project on semiconductor 3D-Integration technology in the Association of Super-Advanced Electronics Technologies (ASET). He has also been the general manager of the Advanced Packaging Development Department in Sharp Corporation. He received his B.E. in applied physics from Fukui University, Japan in 1970. He has more than 40 years of experience in semiconductor packaging engineering, with major emphasis on developing chip scale, chip stack package, and three-dimensional-system-in-package (3D-SiP), and is the global pioneer of 3D-Integration technology in the world.

**Dr. Kenji Takahashi** is a chief specialist at Memory Packaging Development Department, Memory Division, Semiconductor and Storage Company, Toshiba Corporation. He received an M.E. in chemical engineering from the University of Tokyo in 1984 and a Ph.D. in information science and electrical engineering from Kyushu University in 2010. His major research and development focuses on semiconductor packaging and chip package interaction, especially through-silicon via technology. He was the research manager of the Electronic System Integration Technology Research Department at the Association of Super-Advanced Electronics Technologies (ASET). He is a senior member of IEEE, and a member of the Society for Chemical Engineers, Japan; the Institute of Electronics, Information and Communication Engineers; and the Japanese Institute of Electronics Packaging.

# Chapter 1
# Research and Development History of Three-Dimensional Integration Technology

**Morihiro Kada**

## 1.1 Introduction

Semiconductor integrated circuits have been developed according to Moore's law; the conjecture made in 1965 was that the number of transistors in a dense integrated circuit (IC) will double every 2 years, and the industry has developed according to this trend [1]. Two different concepts have been proposed for future advancements. One is "More Moore," which suggests that technological progress will continue to follow scaling theory, and the other is termed "More than Moore," which emphasizes the evolution and diversification of function [2].

### *1.1.1 The International Technology Roadmap for Semiconductors*

The international semiconductor research community gathered in 2005, at the abovementioned meeting (the International Technology Roadmap for Semiconductors, ITRS), which led to the concept of "More than Moore." Two years later, at ITRS 2007, a number of such ideas were formally defined. We elaborate on two of these: scaling and functional diversification with reference to Fig. 1.1.

1. Scaling: Fig. 1.1, vertical axis of "More Moore"
   a. *Geometrical scaling:* Also referred to as constant field scaling, this design methodology involves reducing the horizontal and vertical dimensions of physical features of the on-chip logic and memory storage components to

M. Kada (✉)
Osaka Prefecture University, Osaka, Japan
e-mail: cmk26222@osakafu-u.ac.jp

The National Institution of Advanced Industrial Science and Technology (AIST), Ibaraki, Japan
e-mail: m-kada@aist.go.jp

 
K. Kondo et al. (eds.), *Three-Dimensional Integration of Semiconductors*,
DOI 10.1007/978-3-319-18675-7_1

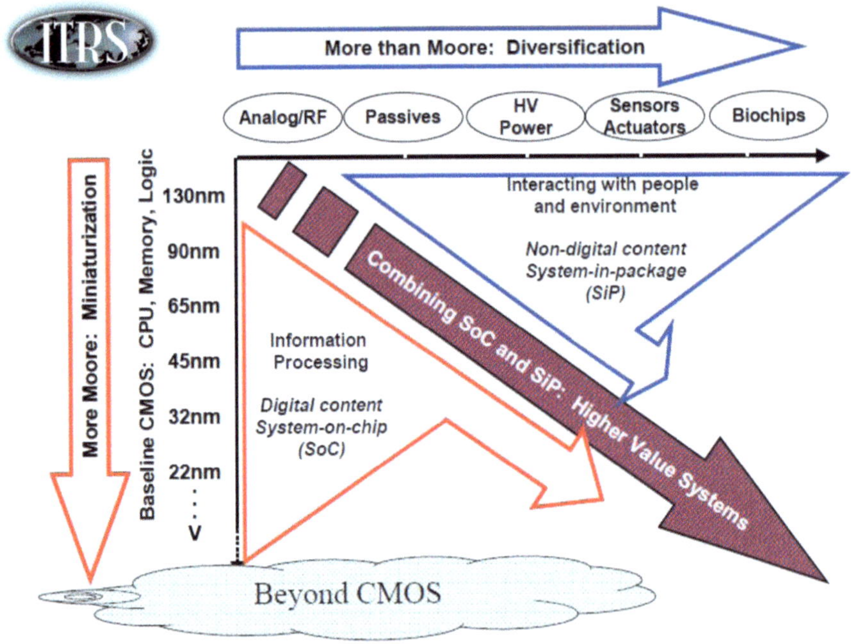


**Fig. 1.1** A diagrammatic representation of the concepts of "Moore's law" and "More than Moore". (Reproduced with permission from Ref. [3], Fig. 4). *RF* radio frequency, *CPU* central processing unit, *CMOS* complementary metal–oxide–semiconductor

improve density (cost per function reduction), performance (speed, power), and reliability (Fig. 1.1).

b. *Equivalent scaling:* This approach refers to (a) three-dimensional (3D) device structure ("design factor") improvements as well as other nongeometric processing techniques and the use of new materials that affect the performance of the chip; (b) novel design techniques and technologies, such as multi-core design. Equivalent scaling occurs in conjunction with geometric scaling and aims for the continuation of "Moore's law."

2. Functional Diversification: Fig. 1.1, horizontal axis of "More than Moore."

Moore's law is not the only way to provide additional value to the end user. A complementary approach is that of functional diversification, which refers to the incorporation of new functionalities into devices that are not necessarily scalings of existing hardware or software. Typical of this "More than Moore" approach is the migration of non-digital functionalities (e.g., radio frequency (RF) communication, power control, passive components, sensors, and actuators) from the system board level into a particular chip-level (system on a chip; SoC) or package-level (system in package; SiP) implementation. As the need increases for evermore complex software to be embedded into SoCs and SiPs, the role of the software itself in performance scaling may also need to be considered. The objective of the "More than Moore" design methodology is to incorporate digital and non-digital functionalities into compact systems.

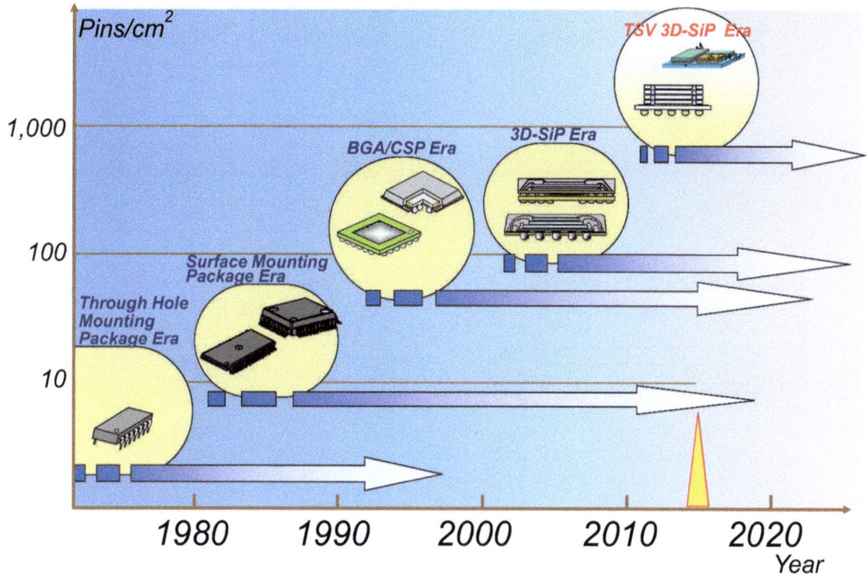


**Fig. 1.2** Toward the new TSV 3D-SiP Era. *3D-SiP* three-dimensional system in package, *TSV* through-Si via, *BGA/CSP* ball grid array/chip-scale package

### *1.1.2 3D Integration Technology*

Although 3D integration technology is not explicit in the definition of "More than Moore," it is generally considered to be one of the most important technology development strategies. The transistor scaling that has continued for more than 40 years is approaching the atomic level of silicon, and this physical limit will likely be reached in 10–15 years.

Entirely new device structures, such as carbon nanotubes, spintronics, and molecular switches are being developed to replace transistor technology. However, they will not be ready for 10–15 years. In the interim, 3D integration technology offers a viable solution for continued performance and economic advancement [4].

"More than Moore" is not just a solution to the limitation of "More Moore," it also recognizes the evolution and potential for improvements of packaging technology. Figure 1.2 illustrates the history of IC packaging technology. Every 10 years since the 1970s, packaging technology has undergone a technological revolution. The first decade of this century is the era of the 3D system in package (3D-SiP), and work has begun to develop new 3D technology termed through-Si via (TSV) that will define the present decade [5]. In TSV, the electrode passes completely through the silicon wafer (or chip). It represents the fusion of silicon wafer process technology (front end of line, FEOL) and semiconductor packaging technology (assembly/packaging).

3D integration using conventional technologies, such as with the wire bonding (WB) as shown in Fig. 1.3 (left), is referred to as 3D integration packaging technologies. In this book, we focus our attention on systems in which semiconductor chips are stacked and connected by TSV, as shown in Fig. 1.3 (right), which

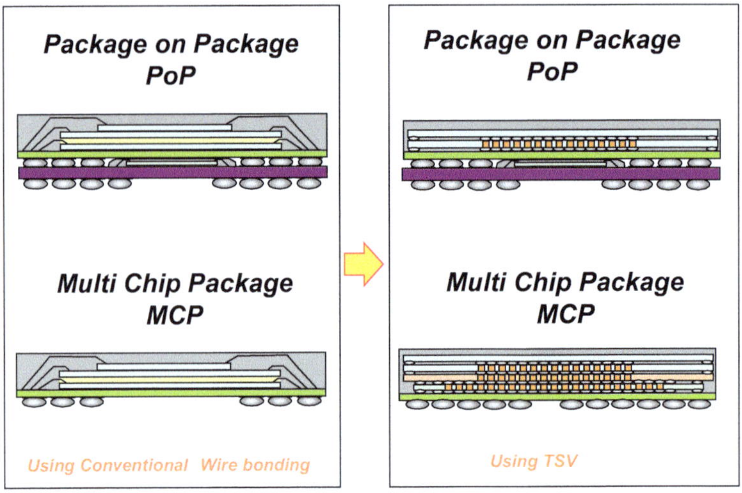


**Fig. 1.3** 3D integration packaging technology and 3D integration technology [6]

define 3D integration technology [6]. We do not discuss 3D integrated circuits (3D-IC) that use FEOL, such as 3D NAND in which transistors are stacked, nor the Intel tri-gate transistors that were introduced in the 22-nm generation ivy bridge CPU.

## 1.2 Motivation for 3D Integration Technology

The development impetus is accounted for in the following two points.

1. If semiconductor integrated circuit chips are connected using TSV, the interconnected distance is approximately 1/1000 of that using conventional WB (micrometers compared with millimeters). This results in dramatic reductions in electric resistance and capacitance, making possible high-speed operation and low power consumption.
2. It is difficult to make (wire) connections between the conventional packages on the mounting board on the order of thousands, but this task is straightforward, and on even grander scales, between Si chips using TSV. Thus, TSV-based systems that have several 1000 input/output (I/O) circuits are realizable, which also benefit from being lower power consumption devices with higher data transmission speeds.

3D integration need not be confined to like technologies. By combining semiconductor integrated circuits with, for example, micro-electro-mechanical systems (MEMS) devices, unique functionalities can be developed in what is termed heterogeneous 3D integration technology.

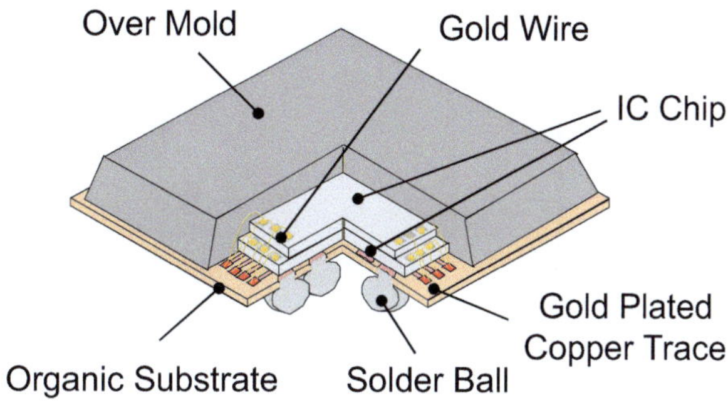


**Fig. 1.4** Typical construction of S-CSP (MCP). *IC* integrated circuit

## 1.3 Research and Development History of 3D Integration Technology

### *1.3.1 3D Packaging Technology*

Even as of 2015, the use of 3D-IC (TSV) is uncommon, with the exception of complementary metal–oxide–semiconductor imaging sensors (CIS). However, high-volume manufacturing of 3D integration packaging technology using WB continues.

In 1998, Sharp Corporation developed the world's first stacked two-chip chip-scale package (CSP) using WB [7]. Before that time, there was no notion of chip stacking in CSP. This led to its development for use in mobile phones, mostly by Japanese chip makers, such as Sharp, Mitsubishi, Hitachi, NEC, Toshiba, and Fujitsu. This technology was called stacked chip-scale (size) package (S-CSP) or multi-chip package (MCP). Figure 1.4 shows the typical construction of S-CSP (MCP).

S-CSP/MCP was first used to make combinations of NOR flash memory and of static random-access memory (SRAM), which are at the heart of all mobile phones. Consumer demand fuelled the development of smaller sizes and higher performance [8]. When Sharp developed the world's first stacked CSP, the combination memory development race was called the "East versus West War" over standardization by the Joint Electron Device Engineering Council. It became a demonstration of the strength of Japanese packaging technology.

Although, in the beginning, the interconnect technology was only WB, CSP stacking has given rise to the package on package (PoP) model, which also uses flip chip (FC) technology. Today, this approach is integral to modern smart phones and tablets; dynamic random access memory (DRAM) and application and/or base-band processors are stacked together using this technology. Upon these foundations, newer technologies continue to drive advances in telecommunications, such as through mold via (TMV) [9, 10]; see Fig. 1.5.

In the present-day flash memory, there are more than eight chips stacked into a single package [11]; see, for example, Fig. 1.6. This technology will likely continue being a mainstay of 3D integration packaging technology for some time yet. On the horizon are wireless interconnect technologies, such as capacitive and inductive 3D coupling [12].

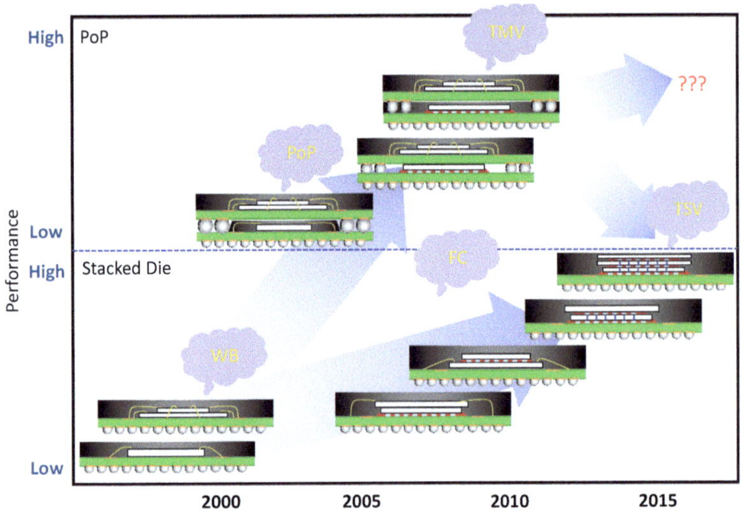


**Fig. 1.5** 3D integration packaging technology transition [10]. (With permission from Amkor Technology, Inc., Chandler AZ). *WB* wire bonding, *PoP* package on package, *FC* flip chip, *TMV* through-mold via, *TSV* through-Si via

**Fig. 1.6** World's first nine-chip stacked memory. (With permission from Toshiba Corporation)

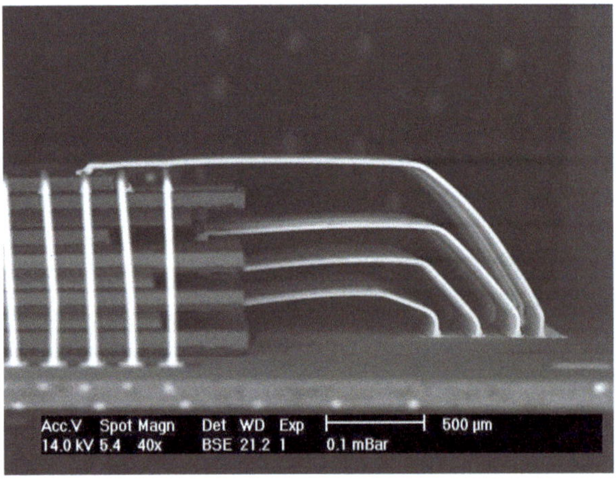


## *1.3.2 Origin of the TSV Concept*

The underlying concept of TSV technology is not new. International Business Machines Corporation filed the patent USP3,648,131 entitled, "Hourglass-shaped conductive connection through semiconductor structures" in November 1969, with the following abstract [13]:

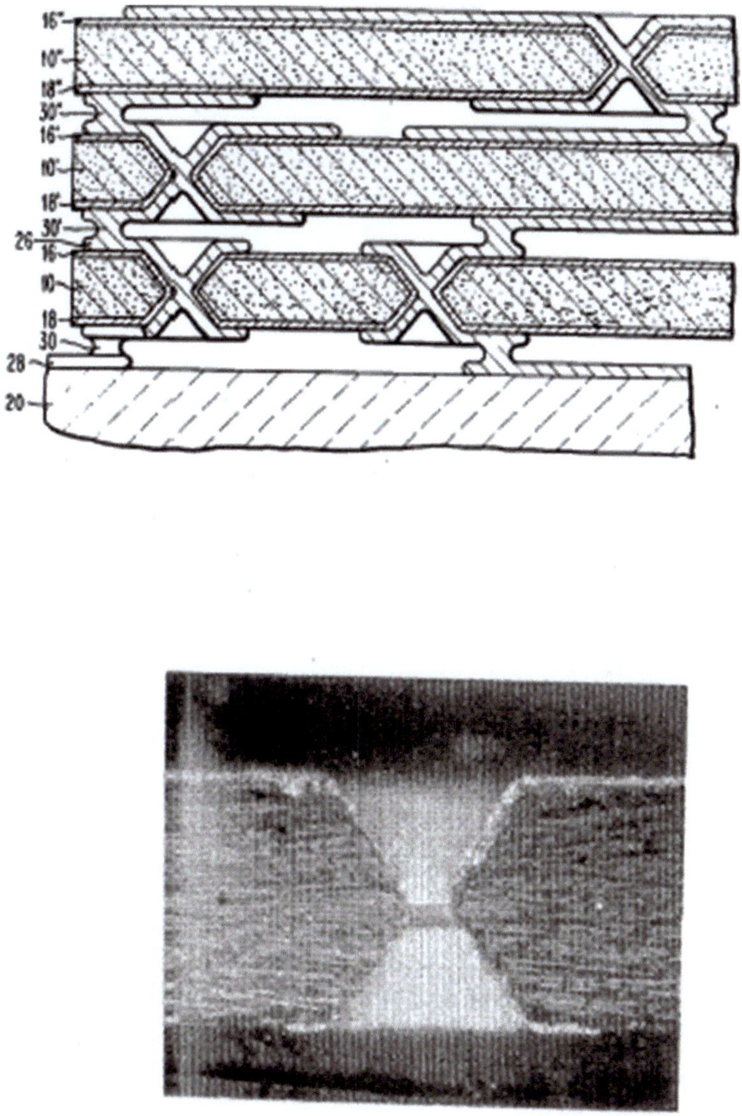


**Fig. 1.7** US Patent 3,648,131 A [13]

"An integrated semiconductor structure including the fabrication thereof, and more particularly, an improved means for interconnecting the two planar surfaces of a semiconductor wafer. To provide the electrically conductive interconnections through the wafer, a hole is etched, insulated, and metallized. Active or passive devices may be formed on either or both sides of the wafer and connected to a substrate by solder pads without the use of beam leads or flying lead bonding." The drawings are shown in Fig. 1.7.

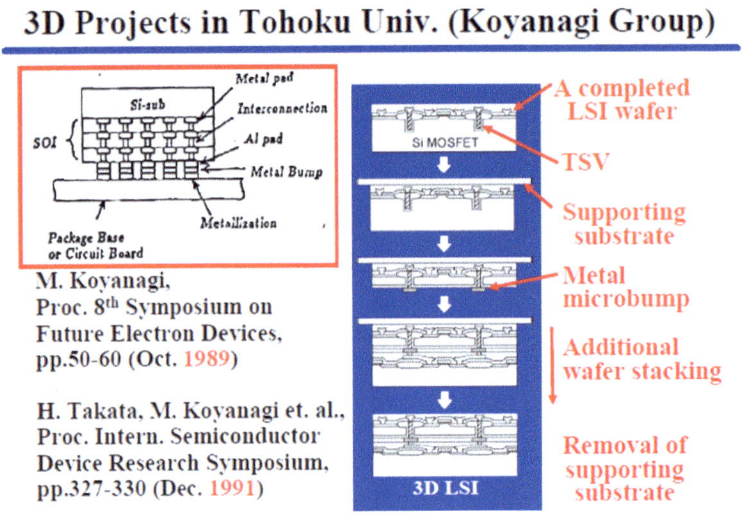


**Fig. 1.8** Extracts from presentations by Tohoku University researchers in 1989 [14] and 1991 [15]. (With permission from Tohoku University, Japan). *TSV* through-Si via, *LSI wafer* large-scale integration wafer

Later on, patents JP (S59)1984-22954 (June 1, 1983) and patent JP (S61)1986–88546 (October 5, 1984) were filed by Hitachi Ltd. and Fujitsu Ltd, respectively. The patent JP (S63)1988–156348 (December 19, 1986), by Fujitsu, describes a stacked chip structure. Figure 1.8 shows key schematics of chip stacking techniques sourced from 1989 and 1991 conference presentations by Tohoku University, Japan [14, 15].

### *1.3.3 Research and Development History of 3D Technology in Organizations*

Research and development of 3D integration technologies has been carried out through global efforts [16]. Some of the major contributions by region are summarized in Fig. 1.9 [17].

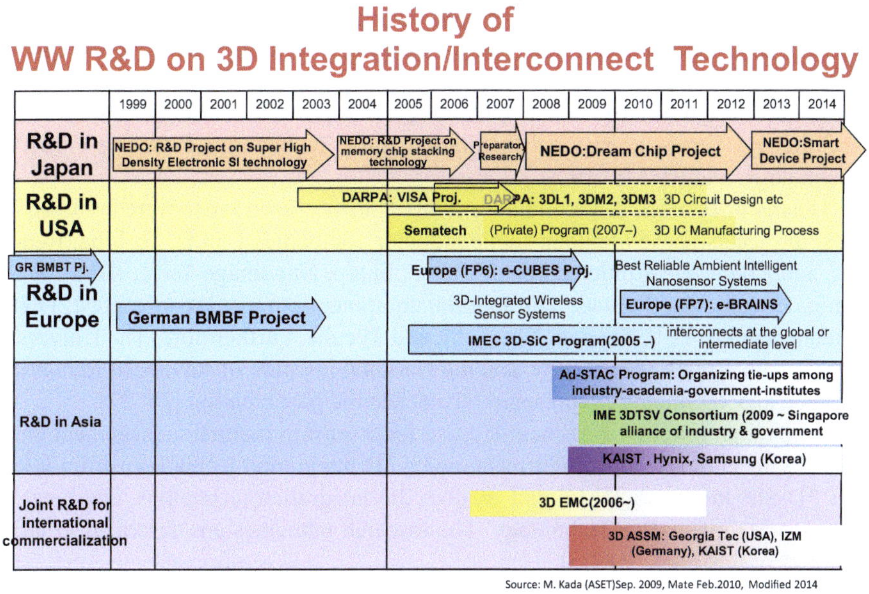


**Fig. 1.9** History of global research and development on 3D integration technology. (Adapted from Ref. [16]). *NEDO* New Energy and Industrial Technology Development Organization, *DARPA* Defense Advanced Research Projects Agency, VISA vertically interconnected sensor arrays, *AD-STAC* Advanced Stacked-System Technology and Application Consortium, *TSV* through-silicon via, *ASSM* All Silicon System Module, *EMC* Equipment and Materials Consortium, *SiC* stacked integrated circuit, *German BMFT* German Ministry of Research, *BMBF* Federal Ministry of Education and Research, *FP7* Seventh Framework Programme

#### 1.3.3.1 Japan

In Japan, research and development of the "Three-Dimensional Circuit Element R&D Project" by the Research and Development Association for Future (New) Electron Devices was conducted from 1981 to 1990, and the technology developed was termed "Cumulatively Bonded IC" (CUBIC): (in Japanese); TSV was not integral to the design. A thin film (approximately 2-μm thick) of electron channel metal–oxide–semiconductor field-effect transistor (nMOSFET) was laminated onto the bulk silicon device. The electrical interconnection of 1600 wiring contact arrays was checked, and the contact volume resistance of $5 \times 10^{-6}\ \Omega \cdot \text{cm}^2$ did not adversely affect the operation [18].

In Japan, the Association of Super-Advanced Electronics Technologies (ASET) carried out a research and development project of 3D integration technology using TSV during the 5-year period 1999–2003. The project was entitled, "R&D on High Density Electronic System Integration Technology" (in Japanese). Its execution was entrusted to the New Energy and Industrial Technology Development Organization (NEDO) organization of the Japanese government's Ministry of Economy, Trade, and Industry (METI) [19]. Following on were the "Stacked Memory Chip

Technology Development Project" (in Japanese), 2004–2006 [20] and the "Development on Functionally Innovative 3D-Integrated Circuit (Dream Chip) Technology Project" (in Japanese), 2008–2012. In 2010, research was conducted with a focus on "Design Environmental Technology, Interposer Technology, Chip-Testing Technology, Three-dimensional Integration Basic Technology, Flex chip (FPGA) Technology, and RF MEMS."

The majority of the semiconductor-related businesses in Japan were involved in these projects. These included semiconductor companies Elpida, Toshiba, Renesas, and Rohm; electronic companies NEC, Sharp, Nac Image Tech., IBM, Panasonic, Hitachi, and Fujitsu; and material/equipment companies Advantest, DNP, Ibiden, Shinko, TEL, Toppan, Yamaichi, and Zycube. Furthermore, The University of Tokyo, Tohoku University, and the National Institute of Advanced Industrial Science and Technology represented the academic participation [21–23].

In 2010, an interim assessment led to a focus shift to thermal management/chip stacking technology, thin wafer technology, 3D integration technology, ultra-wide bus 3D-SiP, mixed signal (digital–analog) 3D integration technology, and heterogeneous 3D integration technology. The research outcomes are described in later sections of this book [24, 25].

However, despite the sizeable investment by the Japanese government over these long periods of time, the national semiconductor industry is presently in decline and future research and development remains uncertain.

The WOW alliance based at the Tokyo Institute of Technology (based at The University of Tokyo until 2014) was founded in 2008 [26], and the "Three-Dimensional Semiconductor Investigation Center" (translated from the Japanese) in Kyushu commenced operations in 2011 [27].

#### 1.3.3.2 Japanese 3D Integration Technology Research and Development Project (Dream Chip)

The second full-scale national research and development (R&D) initiative of 3D integration technology using through-silicon via (TSV) was implemented over the 5-year period from 2008 to 2012. Super-Advanced Electronics Technologies (ASET) conducted the project "Development on Functionally Innovative 3D-Integrated Circuit (Dream Chip) Technology Project," and it was managed by the NEDO Organization that is based on "IT Innovation Program" of Japanese government's Ministry of Economy, Trade and Industry (METI). After the 2010 interim assessment, the two focus areas became 3D integration process basic technologies and application technologies using TSV. The former consisted of thermal management/chip-stacking technology, thin wafer technology, and 3D integration technology, while the latter focus area comprised ultra-wide bus 3D-SiP, mixed signal (digital–analog) 3D, and heterogeneous 3D; see Fig. 1.10. For details beyond research and development subjects and results, the reader is referred to Chapter 9 [17, 28].

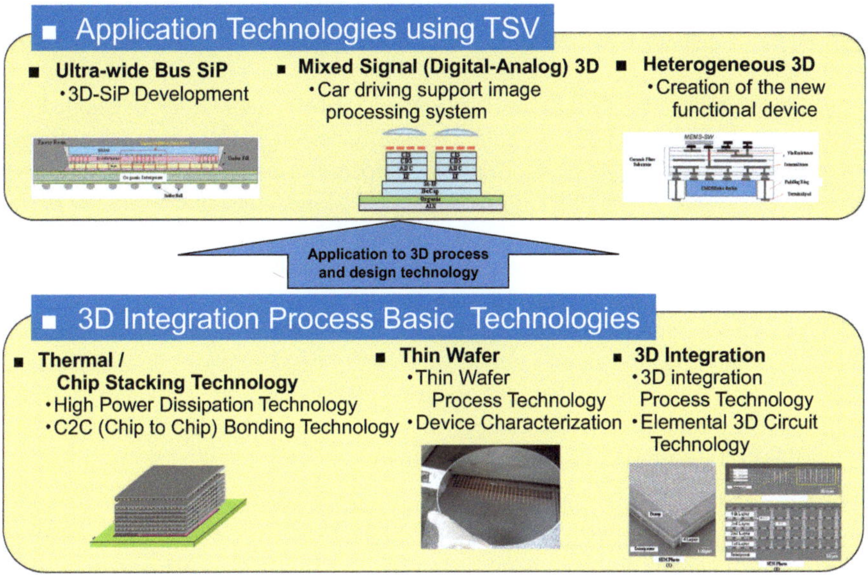


**Fig. 1.10** Research and development subject of the Dream Chip Project. (With permission from the Electrochemical Society: ECS) [17]. *TSV* through-Si via

#### 1.3.3.3 The USA

The US Defense Advanced Research Projects Agency (DARPA)'s work in microsystems technology has a long history. 3D-related research and development projects are controlled by the "Microsystems Technology Office (MTO)." These projects are:

1. Enhanced Digital (3D-IC Program)
   - Large amounts of cache memory
   - High memory bandwidth
2. Enhanced Analog (COSMOS Program)
   - Heterogeneous integration
   - Disparate process technologies (e.g., SiGe/Si, C.S./Si, SOI/Bulk)
3. Smart Focal Planes (Vertically Interconnected Sensor Arrays, VISA Program)
   - Processing at each pixel
   - High fill factors
4. Photonics (EPIC Program)
   - Optical and electronic tiers

DARPA funding and the Microelectronics Center of North Carolina Research and Development Institute (MCNC-RDI) supported a project that was started in 2003. The research and development device comprised VISA that implemented highly parallel and densely interconnected architectures with micron-sized vias penetrating stacks of detectors, which were made from analog, digital, and mixed signal circuits.

The Massachusetts Institute of Technology Lincoln Laboratory developed "A SOI-Based Wafer-scale 3-D Circuit Integration Technology." Integral to this project were precision wafer aligner-bonder, low temperature oxide–oxide bonding, and concentric 3D via, all 3D-enabling technologies. Major milestones were the delivery of the 3DL1, 3DM2, and 3DM3 chips in April 2006, and November 2007 and 2010, respectively. Participants of the 3DM3 chip project were from universities, commercial laboratories, and the business sector.

They are Arizona State University (ASU), North Carolina State University (NCSU), The Naval Research Laboratory (NRL), the US Department of Defense (DoD), Fermi National Accelerator Laboratory, IBM, The State University of New York (SUNY), MIT Lincoln Laboratory, and so on. As for third 3D-IC Multiproject Run (3DM3)3D Circuits, 39 designs were submitted as follows:

*3D Circuits* Anti-tamper authentication chip, stacked memory (SRAM & DRAM), stacked microprocessor, hi-speed transmit/receive, one-chip GPS, network-on-chip, reconfigurable neural network, SAR processor elements, RF-switching power converter, power management for 3D-IC's, Integrated RF MEMS, implantable biosensors, and bio lab on chip.

*3D Imaging Applications* International Linear Collider (ILC) pixel readout, low-power pattern recognition 3D vision chip, multi-core processor with image recognition, focal plane image processor, and sub-l- sized pixel imaging array

*3D Technology Characterization* 3D radiation test structures, jitter-clock skew-propagation delay, hi-speed I/O, RF building block, meta-material inductors, and stacked MOSFET.

3D design software were also developed by PTC, NTSU, R3Logic, and the University of Minnesota (UMN) [29–32].

Following are the overview of press releases by International SEMATEC from 2004 to 2006. "June 10, 2004 they released its several top technical challenges for 2005, one of several top technical challenges was 3-D interconnect on the list for the first time" [33]. "Then aiming to expand the range of potential solutions to the challenges of continued CMOS scaling, they launched a project to explore the feasibility of 3D interconnect technology for the semiconductor industry, in February 9, 2006" [34]. Also, "December 13, 2010, SEMATECH, the Semiconductor Industry Association (SIA), and Semiconductor Research Corporation (SRC) announced they had established a new 3D Enablement program to drive cohesive industry standardization efforts and technical specifications for heterogeneous 3D integration" [35].

In September 13, 2011, members of SEMATEC were Hynix, IBM, Intel, Samsung, ADI, and ON Semiconductor as the IDMS. As for foundries, Global Foundries, TSMC, and UMC. As for Fabless, HP, Altera, LSI, and Qualcomm. As for

OSATs, ASE. As for Suppliers Atotech, COSAR, NEXX, TEL, R&D Partners, CNSE/FRMC, NIST, and SRC [36].

#### 1.3.3.4 Europe

In Europe, 3D integration technology research has a long history, which was reviewed in the welcome presentation at the IEEE 3D System Integration Conference 2010 (3D-IC) in Munich, November 16–18 [37].

Siemens, AEG, Philips, and Fraunhofer IFT formed the 1987–1989 consortium for developing 3D integration technology. From the 1980s through to the early 2000s, projects were supported by the German Ministry of Research (German BMFT) and the Federal Ministry of Education and Research (BMBF). Subsequent projects were supported by the European Committee. From 2006 to 2009, the European ICT Project "e-CUBES (6th European Framework ICT)—3D-Integrated Wireless Sensor Systems, Technology Platform for 3D Heterogeneous Integration" was conducted. They said four optimized 3D integration technologies were successfully used in the development of three e-CUBES application demonstrators: thin-chip-integration technology (TCI/UTCS) for Philips' Health & Fitness demonstrator, TSV technology ICV-SLID, HoViGo for Infineon's Automotive demonstrator (TPMS) and Package-in-Package technology HiPPiP for Thales' Aeronautic demonstrator [38, 39].

And the latest was the Seventh Framework Programme (FP7: 2007–2013) of the European Union for research, technological R&D, and demonstration activities. "e-BRAINS"—Best Reliable Ambient Intelligent Nanosensor Systems for developing 3D Heterogeneous System Integration.

The project members were Infineon, Fraunhoer, Siemens, SINTEF, sensonor, imec, SORIN Group, CEA, IQE, FPFL, 3D plus, Tyndall, DMCE, TU, Vermon, ITE, MaganaDiagnositics, TECHNICHE UNIVERSITAT CHEMNITZ, and eesy-id. Applications are smart biosensor grain, Infrared imager, active medical implant, air quality system, and smart ultra-sound imaging probe [40, 41]. Following are the overview of press releases by IMEC from 2005 and CEA-Leti from 2011.

July 2005, "IMEC announced that IMEC launched advanced Packaging and Interconnect Center (APIC) and started to develop 3D Stacked IC "3D-SIC." APIC grouped more than 30 partners worldwide including integrated device manufacturers (IDMs), system houses, packaging, assembly and test houses and so on. The programs were based on interconnects at the global or intermediate level of the chip wiring hierarchy." [42].

Then "IMEC developed another R&D subject "3D-WLP," wafer level packaging, and "3D-SIP," traditional packaging interconnect technology. These were the part of IMEC's Industrial Affiliation Program (IIAP) on advanced interconnect technology for future technology node" [43].

And, "Qualcomm announced to participate in IMEC's industrial affiliation program (IIAP) on 3D-integration 2008 and also extends 3D research agreement with Qualcomm focusing on advanced technologies and devices 2011 [44, 45].

Also, "IMEC and TSMC announced 2009 that they have forged an Innovation Incubation Alliance to create a platform enabling the R&D of innovative product solutions using emerging More-than-Moore technology options [46].

Partners of IMEC 3D System Integration Program in 2011, were for Logic IDM, Panasonic, Intel, Fujitsu, Sony, As for Memory IDM Micorn, As for foundries TSMC, GlobalFanundries. As for fabless Quacomm, Xilinx, Nvidea, Altera. As for OSAT Amkor UTAC. As for EDA Synopsys, Acdnxe, Atrenta. As for Material suppliers Hitachi Chemical, ThinMaterials, Henkel, BASF. As for Equipment Suppliers Applied Material, Lam, TEL, Suss Screen, Ultratec, CascadeMicortech, Disco, Nanda Tech PVA Tepla, Smart Equipment Technology (Set) [47].

Followings are the announcement of CEA-Leti. "January 2011, they significantly expand its technology offering this month when it ramps up one of Europe's first 300 mm lines dedicated to 3D-integration applications." [48].

And "CEA-Leti announced January 2012 that the launch of a major new platform that provides industrial and academic partners with a global offer of mature 3D innovative technologies for their advanced products and research projects" [49].

#### 1.3.3.5 Asia

The research and development project entitled, "Advanced Stacked-System Technology and Application Consortium" (Ad-STAC), centering on ITRI, has been supported by the Taiwanese government since 2008. The initial announcement was as follows:

"Formed on July 23, 2008, Ad-STAC provides a unique platform for both technical exchanges and information sharing among the partners distributed worldwide." "The purpose of Ad-STAC is to unify companies in different field but not limited academia, government industries, and institute, to co-operate and improve the 3D IC technology [50].

Members were 22 companies in October 2010. Applied Materials Inc of Taiwan, Atotech, Deutschland GmbH, Brewer Science Inc., Cadence Design Systems, Inc. Graduate School of Engineering, the University of Tokyo, Hermes Epitek Corp., SÜSS MicroTec AG., Tazmo Co., Ltd, Unimicron Corporation, IV Technologies, Air Products, GPT, Disco, ASE, Leading Precision, Dupont, BASF, SPTS, UMC, SPIL, Cabot Microelectronics, and Cisco.

The full 300 mm line for 3D development was installed by the end of 2009 [51].

In South Korea, there has been speculation of a national project led by Hynix and Amkor but details presently remain confidential [52].

The Institute of Microelectronics (IME) Agency for Science, Technology, and Research (A*STAR) has been active since 2011. They announced, "Dec 6, 2011, IME and Tezzaron Semiconductor announced a research collaboration agreement to develop and exploit advanced Through Silicon Interposer (TSI) technology" [53].

June 5, 2012-IME and United Microelectronics Corporation agreed to develop Through-Silicon Via (TSV) technology for backside illuminated CMOS image sensors (CIS). [54]
Aug 17, 2012-IME and Huawei Technologies, signed MOU to develop and advance Through Silicon Interposer (TSI) technology. The two organizations will collaborate on advanced packaging with TSI, 2.5D/3D-IC research and development, and demonstrate heterogeneous 2.5D design and manufacturing flow. [55]

#### 1.3.3.6 International

Press releases of international 3D integration technology research and development consortia are as follows:

"Semiconductor 3D Equipment and Materials Consortium (EMC-3D) is a new consortium created to address the technical and cost issues of creating 3D interconnects using TSV technology for chip stacking and MEMS/sensors packaging. Equipment companies initiating the consortium are Alcatel, EV Group, Semitool and XSiL. Associate research members include Fraunhofer IZM, SAIT (Samsung Advanced Institute of Technology), KAIST (Korea Advanced Institute of Science and Technology) and TAMU (Texas A&M University). Material members include Rohm and Haas, Honeywell, Enthone, and AZ with wafer service support from Isonics [56].

Another international consortium's announcement was as follows: "The Microsystems Packaging Research Center (PRC) at Georgia Tech, in partnership with Fraunhofer IZM (Germany) and the Korea Advanced Institute of Science and Technology (KAIST) launched a global industry consortium titled"3D All Silicon System Module (3DASSM)" in October, 2008 [57].

## 1.4 Research and Development History of 3D Integration Technology for Applications

### *1.4.1 CMOS Image Sensor and MEMS*

Toshiba, Aptina, STMicroelectronics, and several other companies commercialized the complementary metal–oxide–semiconductor (CMOS) image sensor during 2007–2008, which used TSV as the target electrode formation technology. The advantage of using TSV was its compactness, permitting the design of miniature devices.

Toshiba named the new application of penetration electrode technology "Through Chip Via" (TCV). This space-saving technology made it possible to mount an assembly of camera module components in a wafer state. For example, the conventional substrates and wire bonding spaces were reduced by forming a solder ball in the back side of the chip.

STMicroelectronics' multi-chip MEMS devices, such as smart sensors and multi-axis inertial modules, enable a higher level of functional integration and performance in a smaller form factor [58–60].

### 1.4.2 DRAM

DRAM is anticipated to be the principle application driving volume manufacturing of full-scale 3D integration technology.

In June 2011, Elpida Memory shipped the world's first 8 GB DRAM sample. It consisted of four 2 GB (double data rate type 3; DDR3) modules stacked using TSV technology [61]. Then in August of the same year, Samsung Electronics announced that they had developed 30-nm-class 32 GB Green DDR3 DRAM using TSV package technology for next-generation servers [62].

Hynix became a member of SEMATECH's 3D Interconnect program at the College of Nanoscale Science and Engineering (CNSE) of the University at Albany [63].

Smart phone and tablet SoCs have reached impressive levels of performance in the past few years. However, limited memory bandwidth has become a bottleneck for further advancements, and the continued growth of the display resolution only exacerbates this problem. Despite their size difference, tablets have now overtaken laptop computers in display resolution.

JEDEC Solid State Technology Association, the global leader in the development of standards for the microelectronics industry, announced the availability of a new standard for wide I/O mobile DRAM: JESD229 Wide I/O Single Data Rate (SDR). They said "Widely anticipated by the industry, Wide I/O mobile DRAM is a breakthrough technology that will meet industry demands for increased levels of integration as well as improved bandwidth, latency, power, weight and form factor; providing the ultimate in performance, energy efficiency and small size for smartphones, tablets, handheld gaming consoles and other mobile devices." JESD229 using TSV was standardized in December 2011 in JEDEC [64, 65].

However, wide I/O memory, which has a bandwidth of 12.8 GB/s for mobile applications, has not been adopted because of the significant improvements of classical DDR memory. It is now possible to achieve the desired bandwidth using more traditional 2D-IC integration in both PC/server and mobile DRAM applications. JEDEC more recently moved to standardize Wide I/O2

One of the primary challenges facing DRAM engineers is the memory bandwidth required by high-performance computers and next-generation networking equipment. Conventional DDR is not suitable for these architectures. To address these needs, in October 2011, Samsung Electronics and Micron Technology announced a collaboration to implement as an open interface the Hybrid Memory Cube (HMC). They said "The Hybrid Memory Cube Consortium (HMCC) will work closely with fellow developers Altera, IBM (added by another announcement), Open-Silicon and Xilinx to collectively accelerate industry efforts in bringing to market a broad set of technologies. The consortium will initially define a specification to enable applications ranging from large-scale networking to industrial products and high-performance computing."

Then, in May 2012, HMCC announced that Microsoft Corp. has joined the consortium and in June 2012, they also announced that new members ARM, HP, and SK hynix, Inc. have joined the global effort to accelerate widespread industry adoption of HMC technology.

In August 2012, they announced that its developer members have released the initial draft of the HMC interface specification to a rapidly growing number of industry adopters [66–70].

### 1.4.3 2.5D with Interposer

The technology of mounting IC chips side by side, using a Si Interposer, rather than 3D stacking with TSV, has gained popularity. Because the Si interposer is technically a silicon chip with the wiring layer only, the IC is also referred to as being 2.5 dimensional.

Xilinx announcement in October 2010 for 2.5D is as follows. "The industry's first stacked silicon interconnect technology for delivering breakthrough capacity, bandwidth and power savings using multiple FPGA die in a single package for applications that require high-transistor and logic density, as well as tremendous levels of computational and bandwidth performance" [71, 72].

Then, just 1 year later, in 2011, Xilinx announced the progression. "First shipments of its Virtex®-72000T Field Programmable Gate Array (FPGA), the world's highest-capacity programmable logic device built using 6.8 billion transistors, providing customers access to an unprecedented 2 million logic cells, equivalent to 20 million ASIC gates, for system integration, ASIC replacement, and ASIC prototyping and emulation." [73].

One and a half years later, the competition started. Taiwan Semiconductor Manufacturing Co. Ltd. (TSMC) and Altera Corp, on March 22, 2012, announced "The joint development of a heterogeneous 3-D IC test vehicle using TSMC's chip-on-wafer-on-substrate (CoWoS) integration process. TSMC said the technology is an integrated process technology that attaches device silicon chips to a wafer through a chip-on-wafer bonding process [74] .

### 1.4.4 Others

Many other semiconductor companies, such as IBM [75] and Qualcomm [76], have also been developing 3D integration technology.

In September 2006, Intel announced at their developer forum, the promise of an 80-core chip within 5 years based on the innovation of connecting memory directly to processor cores. They also showcased TSV alongside the 80-core chip prototypes, which piggybacked 256 MB of SRAM directly to each chip's 80 cores [77].

But, in June 2010, they said as follows. "There are several problems with TSV technology: Lack of EDA design tools; complexity of designs; integration of

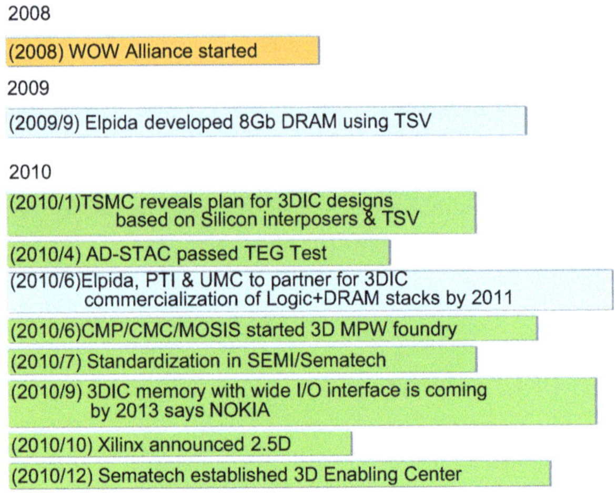


**Fig. 1.11** Worldwide research and development activities in 2008–2010. *DRAM* dynamic random access memory, *TSV* through-silicone via, *TSMC* Taiwan Semiconductor Manufacturing Co. Ltd, *AD-STAC* Advanced Stacked-System Technology and Application Consortium, *UMC* United Microelectronics Corporation, *MPW* multi-project wafer, *MOSIS* metal–oxide–silicon implementation service, *CMP* circuits multi-projects, *SEMI* semiconductor equipment and materials international, *I/O* input/output

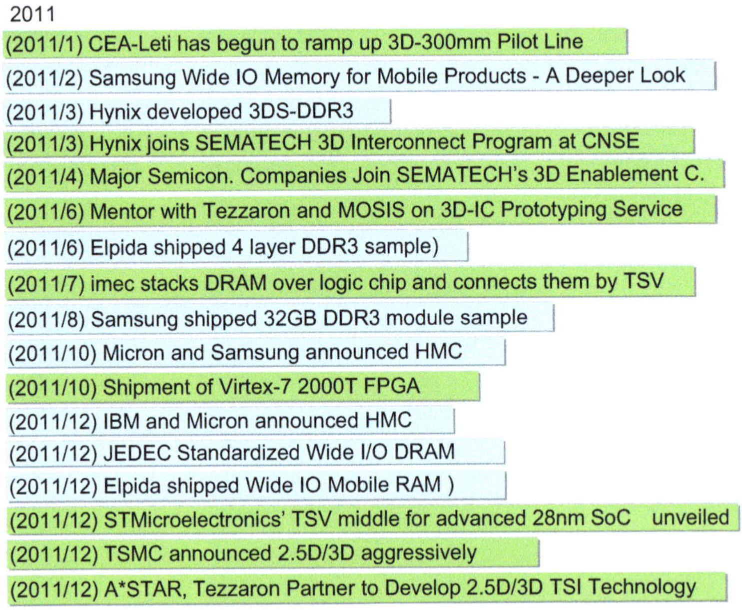


**Fig. 1.12** Worldwide research and development activities in 2011. *DDR3* double data rate type 3, *CNSE* College of Nanoscale Science and Engineering, *3D-IC* three-dimensional integrated circuit, MOSIS metal–oxide–silicon implementation service, *DRAM* dynamic random access memory, *TSV* through-silicone via, *HMC* hybrid memory cube, *FPGA* field-programmable gate array, *I/O* input/output, *RAM* random access memory, *SoC* system on chip, *TSMC* Taiwan Semiconductor Manufacturing Co. Ltd, *TSI* through silicon interposer, *A*STAR* Agency for Science, Technology, and Research

**Fig. 1.13** Worldwide research and development activities in 2012. *BIS* back-illuminated sensor, *CIS* complementary metal–oxide–semiconductor imaging sensors, *TSMC* Taiwan Semiconductor Manufacturing Co. Ltd, *JEITA* Japan Electronics and Information Technology Industries Association, *TSV* through-silicone via, *FPGA* field-programmable gate array, *A*STAR* Agency for Science, Technology, and Research, *UMC* United Microelectronics Corporation, *SiP* system in package

assembly and test; cost; and lack of standards. As previously reported, Intel Corp. is still searching for an application for TSVs. It does not make sense for Intel to go to 3D with CPU cache memory" [78].

A summary of worldwide research and development activities from 2008 to 2012 are shown in Figs. 1.11, 1.12, and 1.13.

The 3D integration technology using TSV has not yet reached high-volume manufacturing despite the sizeable global investment in research and development. There are several reasons for this. Standardization is difficult because of the diversity of architectures, such as Via Size and TSV, process order, for example, Via First/Last/Front/Back, and issues with the supply chain for high-volume manufacturing and high component costs.

However, the 3D integration technology in which TSV is used and the arguments surrounding its application and forecasted use is actively being discussed.

## References

1. Moore GE (1965) Cramming more components onto integrated circuits. Electronics 38(8):19
2. Dennard RH et al (1974) Design of ion-implanted MOSFETs with very small physical dimension. IEEE J Solid-State Circuits 9(5):256–268.
3. International Technology Roadmap for Semiconductors 2007 Edition, Executive Summary Interconnect
4. Garrou P (2007) Perspectives from the leading edge. The SIA meeting in NYC, mid Sept

5. Kada M (2005) Presentation slide toward the 3D-SiP era. SEMICON Japan
6. Kada M (2012) Presentation slide 3D-integrated circuits technologies—the history and future. ISSM Tokyo
7. Kada M (1999) Stacked CSP/a solution for system LSI. Chip Scale International Technical Symposium Semicon West, B1-B7, Sept13
8. Kada M, Smith L (2000) Advancements in Stacked Chip Scale Packaging (S-CSP), provides system-in-a-package functionality for wireless and handheld applications. Pan Pacific Microelectronics Symposium Conference, Jan. http://www.eet.bme.hu/~benedek/CAD_Methodology/Courses/packaging/PanPacific_StackedCSP_RevG1.pdf. Accessed 9 July 2014
9. Chen M et al (2011) Presentation slide TI OMAP4xxx POP SMT design guideline. http://www.ti.com/pdfs/wtbu/SWPA182C.pdf. Accessed 8 July 2014
10. Yoshida A (2013) Presentation slide bump & ball interconnect technology update. ICSJ 2013 Nov
11. Akejima S (2007) Hi-density flash memory packaging technology. J Jpn Inst Electron Packag 10(5):375–379 (in Japanese), https://www.jstage.jst.go.jp/article/jiep1998/10/5/10_5_375/_pdf. Accessed 8 July 2014
12. Niitsu K, Kuroda T (2010) An inductive-coupling inter-chip link for high-performance and low-power 3D system integration. Solid-State Circuits Technologies, ISBN: 978-953-307-045-2, INTECH, pp 281–306, Jan
13. International Business Machines Corporation (1969) US Patent US3648131 Hourglass-shaped conductive connection through semiconductor structures. Filed: Nov 7 1969
14. Fukushima T et al (2007) Presentation slide thermal issues of 3D ICs (M. Koyanagi, Proceeding of 8th symposium on future electron devices, pp 50–60 Oct 1989). http://www.sematech.org/meetings/archives/3d/8334/pres/Fukushima.pdf. Accessed 12 July 2014
15. Fukushima T et al (2007) Presentation slide thermal issues of 3D ICs (M. Koyanagi, Proc. 8th symposium on future electron devices, pp 50–60 Oct 1989). http://www.sematech.org/meetings/archives/3d/8334/pres/Fukushima.pdf. Accessed 12 July 2014
16. Kada M (2010) Prospect for development on 3D-integration technology and development of functionally innovative 3d-integrated circuit (dream chip) technology. 16th symposium on microjoining and assembly technology in electronics, P5-P12, Feb, (in Japanese)
17. Kada M (2014) R&D overview of 3D integration technology using TSV worldwide and in Japan. 2014 ECS and SMEQ joint international meeting (Oct 5–10)
18. Research & Development Association for Future Electron Devices (FED) Overview Report of R&D Result Spillover Effects and Prospect 3D Circuit Element R&D Project. 1981F-1990F (in Japanese)
19. Electronics and Information Technology Development Department (2004) Super high density electronic system integration technology. NEDO report of project assessment, Sept 30 (in Japanese)
20. NEDO Assessment Committee (2008) Stacked memory chip technology development project report of after project. Assessment Feb (in Japanese). http://www.nedo.go.jp/content/100096542.pdf. Accessed 25 June 2014
21. Kada M (2009) Development on functionally innovative 3D-integrated circuit (dream chip) technology. 3D system integration conference
22. Kada M (2009) Highly performance TSV is pursued towards "dream chip", simulator, peripheral technology, such as proving technology, are also developed. The semiconductor technology yearbook, Nikkei BP (in Japanese)
23. ASET (2009) R&D result of "dream chip project" ASET annual symposium 2010 (in Japanese)
24. NEDO R&D Assessment Committee (2010) Dream chip development project, report of interim project assessment, Nov (in Japanese). http://www.nedo.go.jp/content/100140983.pdf. Accessed 25 June 2014
25. NEDO R&D Assessment Committee (2013) Dream chip development project. Report of after project assessment Nov (in Japanese). http://www.nedo.go.jp/content/100545199.pdf. Accessed 25 June 2014

26. Electronic Journal (2009) Sept, pp 28–29 (in Japanese)
27. Home Page Research Center for Three Dimensional Semiconductors (2014) (in Japanese). http://www.tl.fukuoka-u.ac.jp/~tomokage/3dcenter/toppage.html. Accessed 22 April 2014
28. ASET (2013) Presentation slide dream-chip project by ASET (final result). March 8, http://aset.la.coocan.jp/english/e-kenkyu/Dream_Chip_Pj_Final-Results_ASET.pdf. Accessed 12 July 2014
29. Fritze M et al (2007) Presentation slide thermal challenges in DARPA's 3DIC Portfolio, Sematech workshop on "Thermal & Design Issues in 3D IC's" Albany, NY, Oct 11–12. http://www.sematech.org/meetings/archives/3d/8334/pres/Fritze-Steer.pdf. Accessed 19 July 2014
30. DA3RPA (2002) Fiscal Year (FY) 2003 budget estimates Feb pp 184–185, p 193, http://www.darpa.mil/WorkArea/DownloadAsset.aspx?id=1636. Accessed 22 April 2014
31. Research and Development Services in Support of the DARPA VISA Program Solicitation Number: DON-SNOTE-050228–001Agency: Department of the Navy Office: Space and Naval Warfare Systems Command Location: SPAWAR Systems Center Pacific (2005) Federal Business Opportunities, Research and Development Services in Support of the DARPA VISA Program. https://www.fbo.gov/index?s=opportunity&mode=form&tab=core&id=d0e6ddf71bfc03e9095d3b7b276d21b7. Accessed 3 March 2014
32. Keast C et al (2009) Presentation slide A SOI-based wafer-scale 3-D circuit integration technology. 3D architectures for semiconductor integration and packaging, Dec 11
33. Sematech (2004) Press release international SEMATECH identifies top technical challenges for 2005. http://www.sematech.org/corporate/news/releases/20040610a.htm. Accessed 17 April 2014
34. Sematech (2006) Press release SEMATECH launches 3D project to probe options for advanced interconnect. http://www.sematech.org/corporate/news/releases/20060209htm. Accessed 17 April 2014
35. Sematech (2010) Press release new 3D enablement program launched by SEMATECH, SIA and SRC. http://www.azonano.com/news.aspx?newsID=20908. Accessed 17 April 2014
36. Arkalgud S (2011) Presentation slide 3D interconnects 3D enablement center. Annual SEMATECH symposium Hsinchu, Sept 13. http://www.sematech.org/meetings/archives/symposia/10187/Session2/01Arkalgudl.pdf. Accessed 11 July 2014
37. Ramm P et al (2010) Presentation slide, welcome to the IEEE international 3D system integration conference (3DIC). Munich, Nov 16–18
38. Ramm P et al (2010) The European 3D technology platform (e-CUBES). IMAPS, http://www.sintef-norge.com/upload/IKT/9031/Ramm%20IMAPS%20Device%20Packaging%202010.pdf. Accessed 13 July 2014
39. Lietaer N et al (2009) Presentation slide 3D integration technologies for miniaturized tire pressure monitor system (TPMS). Lietaer09—IMAPS symposium Foredrage, http://sintef.org/upload/IKT/9031/Lietaer09%20-%20IMAPS%20Symposium%202009%20Foredrag.pdf. Accessed 25 June 2014
40. Ramm P et al (2013) Presentation slide the e-BRAINS project. ESSDERC/ESSCIRC, Bucharest Romania workshop: In the quest for zero power: enabling smart autonomous system applications. http://www.e-brains.org/data/events/uploads/Peter_Ramm_The_e-BRAINS_Project_ESSDERC_2013_WS_In_The_Quest_For_Zero_Power.pdf. Accessed 13 July 2014
41. http://www.e-brains.org/project/rtd/. Accessed 26 April 2014
42. IMEC (2005) Press release IMEC packaging research center attracts 30 companies. www.embedded.com/print/4054184. Accessed 2 March 2014
43. IMEC (2008) Brochure 3D @ IMEC, http://www2.imec.be/content/user/File/3D_brochure.pdf. Accessed 13 July 2014
44. Qualcomm (2008) Press release Qualcomm and IMEC collaborate on 3D integration research. http://www.cn-c114.net/577/a330029.html. Accessed 2 March 2014
45. IMEC (2011) Press release IMEC extends 3D research agreement with Qualcomm focusing on advanced technologies and devices. http://www2.imec.be/be_en/press/imec-news/imec-qualcomminsite.html. Accessed 2 March 2014

46. TSMC and IMEC (2009) Press release TSMC and IMEC join forces to bring novel technology solutions to emerging markets. http://www.leuveninc.com/event/36/784/TSMC_and_IMEC_join_forces_to_bring_novel_technology_solutions_to_emergin/. Accessed March 2, 2014
47. Beyne E (2011) Presentation slide 3D system integration technology convergence. Semicon Europe, Messe Dresden, Germany, Oct 10–13, http://semieurope.omnibooksonline.com/2011/semicon_europa/SEMI_TechARENA_presentations/3DICsession_02_Eric.Beyne_IMEC.pdf. Accessed 14 July 2014
48. CEA Leti (2011) Press release CEA-Leti Ramps up 300 mm line dedicated to 3D-integration applications. Accessed 18 Apr 2014
49. CEA Leti (2012) Press release CEA-Leti launches open 3D™ initiative. Accessed 18 April 2014
50. Ad-stac HP, http://ad-stac.itri.org.tw/memb/index_e.aspx. Accessed 26 April 2014
51. Tsai MJ (2011) Presentation slide overview of ITRI's TSV Technology, 2011-06-22, http://www.sematech.org/meetings/archives/symposia/9237/Session%205%203D%20interconnect/1%20MJ_Tsai_ITRI.pdf. Accessed 9 July 2014
52. Kim G (2009) Presentation slide TSV based 3D technologies in Korea. TSV technology conference, NIKKEI MICRODEVICES, 2009-04-16
53. A*STAR (2011) Press release A *STAR Institute of Microelectronics and Tezzaron Team Up to Develop 2.5D3D through-silicon interposer technology. http://www.bizwireexpress.com/showstoryACN.php?storyid=26505538. Accessed 26 April 2014
54. A*STAR (2012) Press release A-STAR institute of microelectronics and UMC to develop TSV technology for BSl image sensor used in mobile applications. http://www.advfn.com/news_A-STAR-Institute-of-Microelectronics-and-UMC-to-De_52659319.html. Accessed 26 April 2014
55. A*STAR (2012) Press release A*STAT of microelectronics and Huawei announced joint effort to develop 25D/3D through-silicon interposer technology. https://www.astar.edu.sg/Portals/30/news/IMEFuturewei%20Press%20Release%20Final.pdf. Accessed 26 April 2014
56. Siblerud P, Kim B (2007) Presentation slide EMC-3D consortium overview and COC Model". Jan 22–26, http://atlas-old.lal.in2p3.fr/elec/EMC3DEu/documents/Semitool-Consortium%20Overview.pdf. Accessed 18 July 2014
57. 3DASMM Consortium (2008) News release new consortium formed focusing on Si interposer technologies. http://www.i-micronews.com/news/3D-consortiumof-formed-focusing-Si-interposer-technologies,1464.html. Accessed 26 April 2014
58. Toshiba (2007) Press release strengthening of CMOS image sensor business by in-house production of COMS camera module for mobile phone. (in Japanese), http://www.toshiba.co.jp/about/press/2007_10/pr_j0102.htm. Accessed 26 April 2014
59. Micron (2008) News release micron introduces wafer level camera technology with TSV interconnects. http://www.i-micronews.com/news/Micron-wafer-level-camera-TSV-interconnects,1025.html. Accessed 26 April 2014
60. ST Micro (2011) Press release ST microelectronics first to use through-silicon vias for smaller and smarter MEMS chips. http://www.bizjournals.com/prnewswire/press_releases/2011/10/11/NY84151. Accessed 26 April 2014
61. Elpida (2011) Press release Elpida to start to ship sample of 8 Gbit DDR3 SDRAM of X32 using TSV. (in Japanese). http://techon.nikkeibp.co.jp/article/NEWS/20110627/192909/. Accessed 27 April 2014
62. Samsung (2011) Press release Samsung develops 30 nm-class 32 GB green DDR3 for next-generation servers using TSV package technology. http://www.samsung.com/global/business/semiconductor/news-events/press-releases/detail?newsId=4014. Accessed 27 April 2014
63. Hynix (2011) Press release Hynix semiconductor joins SEMATECH 3D interconnect program at UAibany Nano College. http://electroiq.com/blog/2011/03/hynix-semiconductor/. Accessed 27 April 2014
64. JEDEC (2012) Press release JEDEC publishes breakthrough standard for wide I/O mobile DRAM. http://www.jedec.org/news/pressreleases/jedec-publishes-breakthrough-standard-wide-io-mobile-dram. Accessed 27 April 2014

65. JEDEC (2011) STANDAD JESD229 wide I/O single data rate (Wide I/O SDR) wide I/O single data rate. December
66. Samsung and Micron (2011) Press release micron and samsung launch consortium to break down the memory wall. http://investors.micron.com/releasedetail.cfm?releaseid=611879. Oct 6, Accessed 3 July 2014
67. IBM and Micron (2011) Press release IBM to produce Micron's hybrid memory cube in debut of first commercial, 3D chip-making capability. IBM news room. https://www-03.ibm.com/press/us/en/pressrelease/36125.wss. Accessed 15 Aug 2014
68. HMCC (2012) Press release microsoft joins hybrid memory cube consortium. http://investors.micron.com/releasedetail.cfm?releaseid=671388. Accessed 3 July 2014 (May 8)
69. HMCC (2012) Press release consortium to accelerate dramatic advances in memory technology announces new members. http://investors.micron.com/releasedetail.cfm?ReleaseID=686974. Accessed 3 July 2014 (June 27)
70. HMCC (2012) Press release first draft of hybrid memory cube interface specification released. http://news.micron.com/releasedetail.cfm?ReleaseID=700331. Accessed 3 July 2014 (August 14)
71. Xilinx (2010) Press release Xllinx stacked silicon interconnect extends FPGA technology to deliver 'More than Moore' density, bandwidth and power efficiency. http://press.xilinx.com/2010–10-26-Xilinx-Stacked-Silicon-Interconnect-Extends-FPGA-Technology-to-Deliver-More-than-Moore-Density-Bandwidth-and-Power-Efficiency. Accessed 27 April 2014
72. Xilinx (2012) White paper: Virtex-7 FPGAs, Xilinx stacked silicon interconnect technology delivers breakthrough FPGA capacity, bandwidth, and power efficiency Virtex-7 FPGAs. http://www.xilinx.com/support/documentation/white_papers/wp380_Stacked_Silicon_Interconnect_Technology.pdf. Accessed 11 July 2014 (December 11)
73. Xilinx (2011) Press release Xllinx ships world's highest capacity FPGA and shatters industry record for number of transistors by 2X. http://press.xilinx.com/2011–10-25-Xilinx-Ships-Worlds-Highest-Capacity-FPGA-and-Shatters-Industry-Record-for-Number-of-Transistors-by-2X. Accessed 27 April 2014
74. TSMC and Altera (2010) Press release TSMC, Altera team on 3-D IC test vehicle. http://www.eetimes.com/document.asp?doc_id=1261410. Accessed 12 June 2014
75. Knickerbocker JU (2012) IBM presentation slide 3D integration & packaging challenges with through-silicon-vias (TSV). USA NSF Workshop—2/02/2012, http://weti.cs.ohiou.edu/john_weti.pdf. Accessed 3 July 2014
76. I-Micronews (2012) Qualcomm integrates Wide IO Memory onto 28 nm logic chip. http://www.i-micronews.com/news/Qualcomm-integrates-Wide-IO-Memory-onto-28-nm-logic-chip,9605.html. Accessed 27 April 2014 (Oct 3)
77. TechFreep (2006) Hardware news Intel's TSV connects processors to memory. http://techfreep.com/intels-tsv-connects-processors-to-memory.htm. Accessed 27 April 2014 (Sept 28)
78. EETimes (2010) India news 3D TSV chips still pre-mature. http://www.eetindia.co.in/ART_8800610003_1800007_NT_192ccb4b.HTM. Accessed 6 Aug 2011 (18 Jun)

# Chapter 2
# Recent Research and Development Activities of Three-Dimensional Integration Technology

**Morihiro Kada**

## 2.1 Recent Announcement of Research and Development Activities

Since 2013, research and development activities in three-dimensional (3D) integration technology have been accelerating. The themes announced in 2013 and 2014 are described below and shown in chronological order in Figs. 2.1, 2.2, 2.3, 2.4.

(2013/Jan) Novati Technologies Licenses Ziptronix's Direct Oxide Bonding and Direct Bond Interconnect Patented Technologies [1].

"Ziptronix, Inc. has signed a licensing agreement with Novati Technologies, Inc. for the use of its patents covering direct bonding technology, ZiBond and DBI. We believe that Ziptronix's patented direct bonding technology enables the industry's best performance for applications such as 3D memory, BSI image sensors and a developing host of other applications."

(2013/Jan) TORKI K., Presentation Slide (2013), 3D-IC Integration, CMP annual users meeting, 17 January 2013, PARIS, "A very collaborative work has been achieved and still ongoing between the partening CMC, CMP, MOSIS, FermiLab, Tezzaron, HEP Labs, NCSU" [2].

(2013/Jan) STATS ChipPAC and UMC Unveil World's First 3D IC Developed under an Open Ecosystem Model [3].

"STATS ChipPAC and UMC announced the world's first demonstration of TSV-enabled 3D IC chip stacking technology developed under an open ecosystem collaboration. The 3D chip stack, consisting of a Wide I/O memory test chip stacked upon a TSV-embedded 28 nm processor test chip, successfully reached a major milestone on package-level reliability assessment."

M. Kada (✉)
Osaka Prefecture University, Osaka, Japan
e-mail: cmk26222@osakafu-u.ac.jp; m-kada@aist.go.jp

The National Institution of Advanced Industrial Science and Technology (AIST), Ibaraki, Japan



K. Kondo et al. (eds.), *Three-Dimensional Integration of Semiconductors*,
DOI 10.1007/978-3-319-18675-7_2

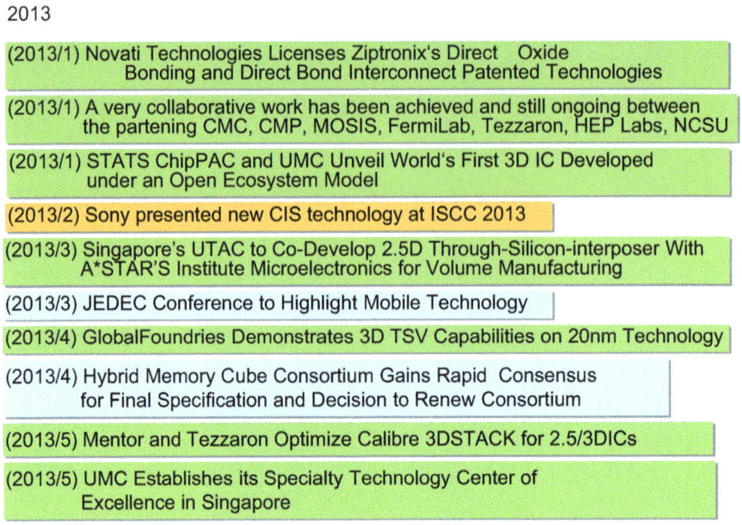


**Fig. 2.1** Worldwide research and development activities in January–May 2013. CMP chemical–mechanical polishing, *MOSIS* metal oxide semiconductor implementation service, *CMC* commercial metal company, *HEP* high-energy physics, *NCSU* North Carolina State University, *UMC* United Microelectronics Corporation, *CIS CMOS* image sensor, *UTAC* United Test and Assembly Center Ltd, *JEDEC* Joint Electron Device Engineering Council, *TSV* through-silicon via, *IC* integrated circuit

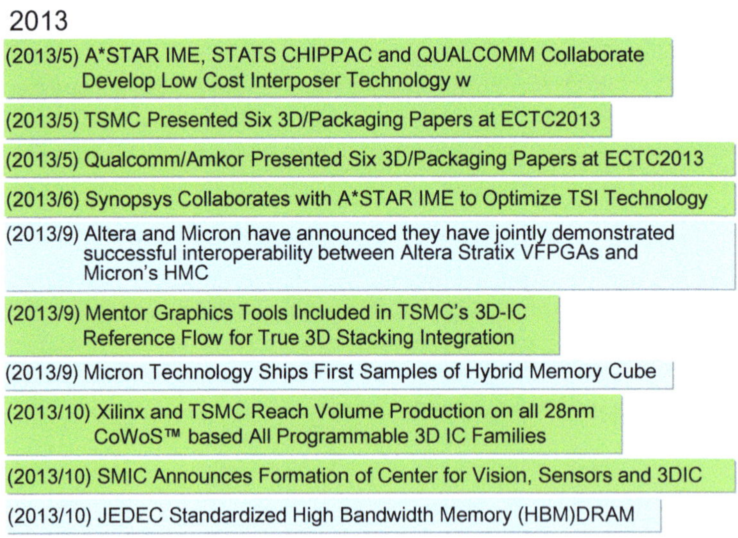


**Fig. 2.2** Worldwide research and development activities in June–October 2013. *IME* Institute of Microelectronics, *TSI* top surface imaging, *VFPGA* virtual field-programmable gate array, *HMC* hybrid memory cube, *TSMC* Taiwan Semiconductor Manufacturing Company, *JEDEC* Joint Electron Device Engineering Council, *HBM* high bandwidth memory, *DRAM* dynamic random-access memory

2013

(2013/11) Micron Technology announced today it is planning to adapt its HMC for petascale supercomputer systems

(2013/12) Pico Computing Delivers the First PCI Express Card to Integrate Micron's Hybrid Memory Cube and Multiple Altera Stratix V FPGAs

((2013/12) CEA-Leti Signs Agreement with Qualcomm to Assess Sequential 3D Technology

(2013/12) Xilinx announced a new record breaking 4.4M logic cell device,,,,, using second generation SSI technology

(2013/12) AMD and Hynix announce joint development of HBM memory stacks

(2013/12) SK Hynix to start production of HBM DRAM chip modules using 3D TSV chip-packaging technology

**Fig. 2.3** Worldwide research and development activities in November–December 2013. *IC* integrated circuit, *PCI* peripheral component interconnect, *VFPGA* virtual field-programmable gate array, *AMD* advanced micro device, *HBM* high bandwidth memory, *DRAM* dynamic random-access memory, *TSV* through-silicon via, SMIC Semiconductor Manufacturing International Corporation, *SSI* stacked silicon integration

2014

(2014/2) Hybrid Memory Cube Consortium Continues to Drive HMC Industry Adoption With Release of Second-Generation Specification

(2014/3) Altera and Intel Extend Manufacturing Partnership to Include Development of Multi-Die Devices

(2014/3) Nvidia's Pascal to use stacked memory

(2014/4) ASE expands System-in-Package business model through industry partnership with Inotera

(2014/6) Micron Collaborates With Intel to Enhance Knights Landing With a High Performance, On-Package Memory Solution

(2014/8) Samsung Starts Mass Producing Industry's First 3D TSV Technology Based DDR4 Modules for Enterprise Servers

(2014/10) AMD R9 380X in February & R9 390X & 370X Announced ….. planning on using 3D stacked HBM

**Fig. 2.4** Worldwide research and development activities in 2014. *HMC* hybrid memory cube, *ASE* Advanced Semiconductor Engineering, *TSV* through-silicon via, *AMD* advanced micro device, *HBM* high bandwidth memory, *DDR* double data rate

(2013/Feb) Sony and Olympus presented new CIS technology at ISSCC 2013 [4].

"The Sony ISX014 8MP sensor features 1.12 μm pixels and integrated high speed ISP. The pixel layer and logic layer part are manufactured as separate chips and stacked by using TSVs. Previously the pixel and logic circuit of Sony's back side illuminated (BSI) CMOS image sensor were formed during the same fabrication process."

(2013/Mar) Singapore's UTAC to Co-Develop 2.5D Through-Silicon-Interposer With A*STAR'S Institute of Microelectronics for Volume Manufacturing [5].

"A*STAR (IME) and UTAC announced a collaboration to develop a 2.5D Through-Silicon-Interposer (TSI) platform which will enable UTAC to join the scarce list of suppliers in offering fine-pitch 2.5D TSI packaging solutions."

(2013/Apr) GlobalFoundries Demonstrates 3D TSV Capabilities on 20 nm Technology [6].

"GLOBALFOUNDRIES announced the accomplishment of a key milestone in its strategy to enable 3D stacking of chips for next-generation mobile and consumer applications. At its Fab 8 campus in Saratoga County, N.Y., the company has demonstrated its first functional 20 nm silicon wafers with integrated Through-Silicon Vias (TSVs)."

(2013/May) Mentor and Tezzaron Optimize Calibre 3DSTACK for 2.5/3DICs [7].

"Mentor Graphics and Tezzaron Semiconductor announced they are collaborating to integrate the Mentor® Calibre® 3DSTACK product into Tezzaron's 3D-IC offerings. The new integration will focus on fast, automated verification of die-to-die interactions in 2.5D and 3D stacked die configurations by verifying individual dies in the usual manner, while verifying die-to-die interfaces in a separate procedure with specialized automation features."

(2013/May) UMC Establishes its Specialty Technology Center of Excellence in Singapore [8].

"UMC announced that the company has established Fab 12i in Singapore as its 'Center of Excellence' to spearhead the company's R&D and manufacturing for advanced specialty process technologies. The Center of Excellence was set up with an initial investment of US $ 110 million, and will undertake R&D collaborations with local research institutes such as Singapore's Institute of Microelectronics. Technologies being developed include CMOS image sensor backside illumination, embedded memory, high voltage applications and TSV (through silicon via) connections."

(2013/May) A*STAR IME, STATS CHIPPAC and QUALCOMM Collaborate Develop Low Cost Interposer Technology [9].

"A*STAR Institute of Microelectronics (IME), Qualcomm Technologies Inc., and STATS ChipPAC announced a collaboration to develop technology building blocks for Low Cost Interposers (LCI) for 2.5D ICs in May 2013."

(2013/May) TSMC Presented 3D technology at ECTC2013 "Reliability Characterization of Chip-on-Wafer-on-Substrate (CoWoS) 3D IC Integration Technology" [10].

"TSMC reported 'CoWoS' 3D IC integration technology has been developed with comprehensive reliability characterization. The copper interconnect reliability of silicon interposer are not impacted by the TSV insertion, in terms of EM, SM, IMD TDDB, and Vbd/TDDB of MiM de-cap. μBump EM and TSV EM are characterized to provide a design guideline for maximum current carrying capability."

(2013/May) Qualcomm/Amkor Presented Six 3D/Packaging Papers at ECTC2013 [11, 12].

"The trial production evaluation towards utilization of TSV base three-dimensional structure package of Wide I/O memory and logic chip is progressing steadily at Qualcomm. 3D-TSV structure which stacked 28 nm generation logic device and maximum of four chips Wide I/O memory were bonded using TSV was reported for the first time in the world. Qualcomm made the device in cooperation with U.S. Amkor Technology as an experiment."

(2013/Jun) Synopsys Collaborates with A*STAR IME to Optimize TSI Technology [13].

"Synopsys announced that it joined Singapore's A*STAR Institute of Microelectronics (IME)-led 2.5D TSI Consortium to provide the framework for heterogeneous 3D-IC systems using through-silicon interposer (TSI) technology."

(2013/Sep) TSMC and Cadence Deliver 3D-IC Reference Flow for True 3D Stacking [14].

"Cadence Design Systems announced that TSMC has collaborated with Cadence to develop a 3D-IC reference flow which features innovative true 3D stacking. The flow, validated on a memory-on-logic design with a 3D stack based on a Wide I/O interface, enables multiple die integration. It incorporates TSMC 3D stacking technology and Cadence® solutions for 3D-IC, including integrated planning tools, a flexible implementation platform, and signoff and electrical/thermal analysis."

(2013/Sep) Mentor Graphics Tools Included in TSMC's 3D-IC Reference Flow for True 3D Stacking Integration [15].

"Mentor Graphics Corp. announced that its solutions have been validated by TSMC with a true 3D stacking test vehicle for TSMC's 3D-IC Reference Flow. The flow expands support from silicon interposer offerings to include TSV-based, stacked die designs. Specific Mentor® offerings include capabilities for metal routing and bump implementation, multi-chip physical verification and connectivity checking, chip interface and TSV parasitics extraction, thermal simulation, and comprehensive pre- and post-package testing."

(2013/Oct) Xilinx and TSMC Reach Volume Production on all 28 nm CoWoS™ based All Programmable 3D IC Families [16].

"Xilinx and TSMC announced production release of the Virtex®-7 HT family, the industry's first heterogeneous 3D ICs in production. With this milestone, all Xilinx 28 nm 3D IC families are now in volume production. These 28 nm devices were developed on TSMC's Chip-on-Wafer-on-Substrate (CoWoS™) 3D IC process that produces significant silicon scaling, power and performance benefits by integrating multiple components on a single device."

(2013/Oct) SMIC Announces Formation of Center for Vision, Sensors and 3DIC [17].

"Semiconductor Manufacturing International Corporation (SMIC) announced today the formation of SMIC's Center for Vision, Sensors and 3DIC (CVS3D). CVS3D consolidates and strengthens SMIC's R&D and manufacturing capabilities for silicon-based sensors, thru-silicon-via (TSV) technology and other middle-end wafer process (MEWP) technologies."

(2013/Dec) CEA-Leti Signs Agreement with Qualcomm to Assess Sequential 3D Technology [18].

"CEA-Leti announced an agreement with Qualcomm Technologies, Inc. to assess the feasibility and the value of Leti's sequential 3D technology. In recent years, Leti has been actively working on a new 3D integration technology process called sequential 3D integration that enables the stacking of active layers of transistors in the third dimension. In comparison with 3D-TSV technologies, advantageously used to stack separate die, sequential 3D technology is anticipated to process all the functions in a single semiconductor manufacturing flow."

(2013/Dec) Xilinx will be the industry's largest capacity doubled to 4.4 million devices logical unit [19].

"Xilinx announced owns 4.4 million logic cells Record product, its density is the industry's highest density product Virtex ® -7 2000T more than twice, the device to make it successful in the market for two generations of high-end devices maintain a leading edge, and provide customers with a value advantage beyond process nodes .... VU440 using advanced 3D IC technology, the capacity on the 20 nm process node has exceeded all previous competitive 14/16 nm technology plan publicly."

(2014/Mar) Altera and Intel Extend Manufacturing Partnership to Include Development of Multi-Die Devices [20].

"Altera Corporation and Intel Corporation today announced their collaboration on the development of multi-die devices that leverage Intel's world-class package and assembly capabilities and Altera's leading-edge programmable logic technology. The collaboration is an extension of the foundry relationship between Altera and Intel, in which Intel is manufacturing Altera's Stratix® 10 FPGAs and SoC s using the 14 nm Tri-Gate process."

(2014/Apr) ASE expands System-in-Package business model through industry partnership with Inotera [21].

"Advanced Semiconductor Engineering, Inc. announced a joint development with Inotera Memories, in a move to further strengthen ASE's System-in-Package (SiP) capabilities. Complementing ASE's established portfolio, Inotera will provide manufacturing services for silicon interposer, an interconnect device on silicon wafer for 2.5D IC solutions. This collaborative business model, combining Inotera's strong front-end wafer processing capability with ASE's advanced IC packaging and testing technology, will serve to deliver solutions featuring high quality, stable yield and an efficient cost structure, to a broader customer base and market."

We now describe research and development activities that are expected to lead to high-volume manufacturing.

## 2.2 Dynamic Random-Access Memory

### *2.2.1 Through-Silicon Via Technology for Dynamic Random-Access Memory*

Joint Electron Device Engineering Council (JEDEC) Solid State Technology Association has heavily promoted the adoption of through-silicon via (TSV) for dynamic random-access memory (DRAM) applications based on the improvements in speed-per-power consumption (lower pj/bit). Notable technologies include wide I/O DRAM for mobile, high bandwidth memory (HBM) for high-performance computing (HPC), and double data rate (DDR).

3D stacking (3DS) technology can increase memory capacity by stacking multiple DRAM chips into a single package using TSV. The suffixes in DDR4_2H,

DDR4_4H, and DDR4_8H DRAM denote two (2-high), four, and eight stacking layers, respectively. For example, a 16 GB DRAM can be constructed from stacking eight layers of 16-Gb chips. This allows a master-/slave-type architecture with many improvements over traditional DDR3 stacks, such as increased memory capacity, lower power consumption, and higher data transmission rates.

However, JEDEC is divided on the matter of whether DRAM is ready for TSV volume manufacturing. Samsung, the largest DRAM manufacturer, has been aggressively developing TSV technology. Although there was a delay since their original announcement at the *Institute of Electrical and Electronics Engineers'* (IEEE) Symposium on Computers and Communications—ISCC 2011, Samsung began mass production of DDR4 in August 2014.

The following is the announcement of Samsung [22].

"Samsung Electronics, Ltd. announced today that it has started mass producing the industry's first 64 gigabyte (GB), DDR4, registered dual Inline memory modules (RDIMMs) that use three dimensional (3D) 'through silicon via' (TSV) package technology."

It took 45 years from the beginning of TSV research and development for industry to begin high-volume manufacturing devices with this technology. Although some challenges related to reliability and cost remain, it would seem that TSV production will continue to gain momentum.

### 2.2.2 Wide I/O and Wide I/O2 Mobile Dynamic Random-Access Memory

JEDEC released standard JESD229 in December 2011 for the new Wide I/O mobile DRAM. However, Wide I/O mobile DRAM did not take off in smartphone industries, which had experts perplexed. In hindsight, we see that the supply chain was complex and bandwidth requirements have increased beyond expectations; there was also a thermal issue. Consumer demand instead led to the development of LPDDR3 (DDR3 for mobile computers) that provided an economical solution to achieving higher bandwidths.

DRAM bandwidth requirements continue to increase rapidly, but for the near future, the LPDDR4 technology will satisfy smartphone requirements. After abandoning Wide I/O, JEDEC introduced the Wide I/O2 task group (TG) standard in September 2011, planned to freeze the specification at the end of 2013, and estimated device production at the beginning of 2015. Wide I/O2 is expected to have more appealing performance, including a bandwidth greater than 50 MB/s and lower power consumption [23].

Following is the JEDEC conference news on March 16, 2013. "JEDEC to Discuss Wide I/O 2, LPDDR4 Standards at Conference, mobile technology will be the focus of a conference hosted by JEDEC standard-setting organization in early. Industry leaders will discuss various aspects of mobile chip standards in general as well as will cover the next-generation memory standards for smartphones and tablets, including LPDDR4 and Wide IO 2" [24]. Table 2.1 shows properties of mobile DRAM [25, 26, 27, 28].

**Table 2.1** Properties of mobile DRAM

| | LPDDR2 | LPDDR3 | LPDDR4 | WideI/O | WideI/O2 |
|---|---|---|---|---|---|
| *Standardization JEDEC* | *JESD209-2F* | *JESD209-3B* | | *JESD229* | |
| Published | Rev. June 2013 (April 2011) | August 2013 | Under developing | December 2011 | Under developing |
| Data rate per pin (Mbit/sec) | 1066 | 1600 | 3200 | 266 | 800 |
| I/O count per channel | 32 | 32 | 32 | 512 | 512 |
| Channel | 1 or 2 | 1 or 2 | 1 or 2 | – | – |
| Bandwidth (GB/s) | 4.3/8.5 | 6.4/12.8 | 12.8/25.6 | 17 | 51.2 |
| VDD | 1.2 | 1.2 | 1.1 | 1.2 | 1.1 |
| Power consumption (W) | | 2.3 | 1.5 | 0.8–1.0 | |
| Power efficiency (pJ/bit) | 9.8 | 8.4 | 5.5 | 5.3 | 2.9 |
| High Volume Manufacturing (HVM) | 2010 | 2013 | 2015 | Dropped | 2015 |

No 3D products are included

## 2.3 Hybrid Memory Cube and High Bandwidth Memory Dynamic Random-Access Memory

Demand for bandwidth and power efficiency in digital applications has no foreseeable limit. HMC and HBM development will be driven by networking/server applications, for which the evolution of DDR technology will reach its limit. HMC precedes HBM; the announcements made in 2013–2014 were as follows.

### *2.3.1 Hybrid Memory Cube*

(2013/Apr) Hybrid Memory Cube Consortium Gains Rapid Consensus for Final Specification and Decision to Renew Consortium [29].

"More than 100 developer and adopter members of the Hybrid Memory Cube Consortium (HMCC) announced they've reached consensus for the global standard that will deliver a much-anticipated, disruptive memory computing solution."

(2013/Sep) Altera and Micron have announced that they have jointly demonstrated successful interoperability between Altera Stratix V FPGAs and Micron's HMC [30].

"Altera Corporation and Micron Technology announced they have jointly demonstrated successful interoperability between Altera Stratix® V FPGAs and Micron's Hybrid Memory Cube (HMC). This technology achievement enables system designers to evaluate today the benefits of HMC with FPGAs and SoCs for next-generation communications and high-performance computing designs."

(2013/Sep) Micron Technology Ships First Samples of Hybrid Memory Cube Volume Production Planned for 2014 [31].

"Micron Technology Inc. announced today that it is shipping 2 GB Hybrid Memory Cube (HMC) engineering samples. HMC represents a dramatic step forward in memory technology, and these engineering samples are the world's first HMC devices to be shared broadly with lead customers."

(2013/Nov) Micron Technology announced today it is planning to adapt its HMC for petascale supercomputer systems, representing a dramatic step forward in memory technology [32].

"Micron Technology announced it is planning to adapt its Hybrid Memory Cube (HMC) for petascale supercomputer systems, representing a dramatic step forward in memory technology. HMC is designed for applications requiring low-energy, high-bandwidth access to memory, which is the most important requirement for supercomputers."

(2013/Dec) Pico Computing Delivers the First PCI Express Card to Integrate Micron's Hybrid Memory Cube and Multiple Altera Stratix V FPGAs [33].

"Pico Computing announced they have created the world's most powerful blade server. The EX800 delivers compute density never before realized in a single PCI Express card and features Micron's groundbreaking Hybrid Memory Cube (HMC) technology which provides unprecedented levels of highbandwidth, lowpower, random access memory performance."

(2014/Feb) Hybrid Memory Cube Consortium Continues to Drive HMC Industry Adoption With Release of Second-Generation Specification [34].

"The Hybrid Memory Cube Consortium (HMCC), dedicated to the development and establishment of an industry-standard interface specification for the Hybrid Memory Cube (HMC) technology, today announced its continued work to build the HMC ecosystem and support for the industry adoption of this groundbreaking technology through the development of a new interface specification."

(2014/May) Fujitsu exhibited the main board of the next supercomputer first time in Japan and the performance per size is 22 times of "KEI": (in Japanese) [35].

"Eight DRAM modules per MPU are carried in a main board. This DRAM module is 'Hybrid Memory Cube (HMC)' of U.S. Micron Technology."

(2014/June) Micron Collaborates With Intel to Enhance Knights Landing With a High Performance, On-Package Memory Solution [36].

"Micron Technology announced an ongoing collaboration with Intel to deliver an on-package memory solution for Intel's next-generation Xeon Phi™ processor, codenamed Knights Landing."

Also Micron reported at 2012 Symposium on VLSI Technology Digest of Technical Papers as follows. The HMC is a 3D DRAM architecture that reduces the access latency between logic and DRAM, improves the bandwidth and reduces power consumption. The main application targets are servers, graphics, and networking systems. The HMC is a heterogeneous stack that is internally connected using TSVs. For example, a standard 1-Gb 50-nm DRAM building block (DRAM layer) can be combined with various versions of application-specific logic (logic layers) using TSVs. In the HMC prototype, 1866 TSVs on 60-μm pitch were used, which had an energy consumption of 10.48 pj/bit—much smaller than the 65 pj/bit of DDR3 modules. In terms of architecture, the DRAM layer is a slave to the logic layer timing control. The logic layer also contains adaptive timing, calibration, refresh, and thermal management capabilities that are hidden from the host. Thus, HMC can use a simple abstracted protocol and high-speed links, such as a serializer/deserializer (SerDes), to connect a host or another cube. By increasing the number of TSVs (i.e., the number of links), the total bandwidth can reach 320 GB/s and beyond [37].

HMC Consortium developers are Altera, ARM, IBM, Micron, Open-Silicon, Samsung, SK hynix, and Xilinx. And adopter members are more than 100 [38].

### *2.3.2 High Bandwidth Memory Dynamic Random-Access Memory*

(2013/Dec) SK Hynix to start production of HBM DRAM chip modules using 3D TSV chip-packaging technology [39].

"SK Hynix said today that it has developed a next generation of HBM, or high bandwidth memory DRAM chips using a 3D TSV, or through silicon via chip packaging technology. The HBM is a new breed of memory chip standard that standard-setting body JEDEC has defined as a next generation of graphics applications for super computers and servers."

(2013/Dec) AMD and Hynix announce joint development of HBM memory stacks [40].

"We're finally reaching a breakthough in consumer-level memory technology, with AMD and Hynix nearing mass production quantities by next year, according to articles at ElectroIQ and SemiWiki."

(2014/Oct) AMD R9 380X in February & R9 390X & 370X Announced [41].

"The announcement from AMD of the R9 3xx series being worked on is hardly a surprising turn of events. The R9 380X (based on the Pirate Islands architecture) is said to be aimed squarely at Nvidia's recently released Nvidia Maxwell GTX970 and 980."

JEDEC standardized "HBM DRAM" (denoted ESD235) for the next-generation high-performance memory in October 2013. It is the stacking DRAM standard aimed at super-wide bandwidths on the order of 1 TB/s. A stack of multiple DRAM devices communicate across independent interfaces called channels. Individual dies contribute additional capacity and channels to the stack (up to a maximum of eight channels per stack) [42].

SK Hynix reported that HBC DRAM only draws 1.2 V of power per 1 GB of data processed per second. Through its 1024 I/O gateways, 128 GB of data can be transferred per second, approximately four times faster than DDR type-5 graphics (GDDR5 synchronous random-access memory) with 40 % less power required [43]. High-volume production started in late 2014, and the first products using HBM are anticipated in 2015. Unlike mobile-oriented wide I/O using TSV, HBM is not intended for stacking DRAM on graphics processor units (GPUs) or central processing units (CPUs) directly. Rather, HBM connects DRAM and logic devices using interposers. The interface width is 1024 bits, and it can realize 128 GB/s for one stack and 512 GB/s to 1 TB/s for two to four stacks.

Industry may want to use HBM DRAM as the inheritor of GDDR5 of a broadband DRAM. This would include video card GPUs for HPC, memory for throughput processors, cache memory for CPUs, and memory for network processors. HBM not only has higher bandwidth but also has lower power consumption. Dense and fine pitch interconnect enables simple low-power interfaces as well as fine control of the DRAM. This leads to low-power consumption in the physical layer (PHY), improving performance threefold; a four-stack HBM only consumes 30 W at 1 Gbps compared with 85 W at 8 Gbps for GDDR5 [44, 45]. Table 2.2 shows properties of DRAM for high-performance applications [28, 37, 39, 44, 45, 46, 47, 48, 49, 50, 51].

**Table 2.2** Properties of DRAM for high-performance applications

| | High bandwidth memory (HBM) | Hybrid memory cube (HMC) |
|---|---|---|
| *Standardization* | *JEDEC JESD235* | *HMC consortium* |
| Published | Oct 2013 | Initial draft<br>aug 2012 |
| Data rate per pin (Mbit/sec) | 1024 (1st gen.)<br>2048 (2nd gen.) | 10,000 |
| I/O count per channel | 1024 | 16(Tx) + 16(Rx) |
| Channel | 1/2/4 | 2/4 |
| Bandwidth (GByte/s) | 128 (1st gen.)<br>256 (2nd gen.) | 80/160 |
| VDD | 1.2 | 1.2 |
| Power consumption(W) | 3.3(@128 GB/s) | 8.2(@128 GB/s) |
| Power efficiency (pJ/bit) | 3.2 | 8 |
| HVM | 2014 | 2014 |

## 2.4 FPGA and 2.5D

While volume manufacturing of 3D integration technology was still in development, Xilinx, Inc. and Altera Corporation, two FPGA manufacturers introduced TSV as 2.5D technology. Xilinx put multiple FPGA chips side by side on the interposer using a TSMC 65 nano process with TSV and then began shipping their Virtex-72000T programmable logic devices in October 2011. Some 6 months later, Altera, in a joint project with Taiwan Semiconductor Manufacturing Company (TSMC) developed a heterogeneous 3D integrated circuit (3DIC) test device using Chip-on-Wafer-on-Substrate (CoWoS) technology.

At present, it is costly to manufacture a Si interposer supporting TSV [52]. Glass may prove to be an alternative economical interposer material.

Another competing technology is fan-out wafer level CSP (FO-WLP), which performs wiring and packaging within the wafer without the need for an interposer. These approaches are distinct from but complementary to 3D integration technology.

## 2.5 Others

Toshiba Corporation has mentioned the possibility to using TSV in their multichannel NAND devices. If this were realized, then it would expand the TSV market considerably. Sony announced at the International Solid-State Circuits Conference ISSCC 2013 that they will use TSV for their back side illumination (BSI) chip, which is based on CIS + Logic. The anticipated adoption of TSV in the Logic + Memory and Logic + Logic (analog) industries is further behind.

## 2.6 New Energy and Industrial Technology Development Organization Japan

### *2.6.1 Next-Generation "Smart Device" Project [53]*

In Japan, New Energy and Industrial Technology Development Organization (NEDO) commenced the research and development of the "Next-Generation 'Smart Device' Project" in November 2013, which succeeds the "Dream Chip Project." It is envisaged to be the electronic technology that makes possible energy saving and advanced safety driving technology for collision evasion. The 5-year project is scheduled for completion in March 2018 (the end of the Japanese financial year). The key milestones are as follows:

1. Development of the in-vehicle obstacle-sensing device:

The sensing device technology that simultaneously measures the positions and distances of multiple obstacles in real time under all weather and visibility conditions.

1. Development of the obstacle detection and risk cognition processor:

The application processor that recognizes multiple obstacles from the sensor data, predicts motion, and distinguishes the likelihood of near collisions.

1. Development of the probe data processor:

The low-power consumption processor for high-speed analysis of the circumference information collected from multiple cars.

### *2.6.2 Background, Purpose, and Target of "Smart Device" Project*

Energy saving and the improvement of driver safety are goals of the next-generation traffic society. Environmentally, such a system must take into account the automotive combustion system; for example, the inefficient fuel consumption during battery jump starts, hard breaking, and slow driving in traffic congestion. And plainly, collision reduction will save both driver and pedestrian lives. The construction of a system that collects vehicle proximity data and effects immediate action is highly desirable.

In Europe, much emphasis is already placed on collision evasion technologies for safety improvements. The development of the autonomous run technology in the USA anticipates deployment around 2020. In Japan, collision evasion technology based on vehicle-to-vehicle communications and traffic congestion relief technology using road-to-vehicle communication are in development.

The Japanese market for total obstacle sensing devices, including the materials and manufacturing, is currently estimated to be worth 1 trillion Yen by the 2020 fiscal year.

At the heart of the "Smart Device" will be the application processor that recognizes an obstacle from the information retrieved from in-vehicle sensors and distinguishes danger. By reducing the number of accidents, this device will contribute to reducing traffic congestion, thereby lowering carbon emissions. It will also contribute to strengthen the global competitiveness of the Japanese automotive industry. The deployment target of the "Smart Device" is 2018, which will need to realize the following objectives.

1. Development of an in-vehicle obstacle-sensing device

This device will simultaneously measure in real time the position and distance of an obstacle, be it vehicle or pedestrian, under all weather and visibility conditions. The research and development will include the miniaturization technology of the device.

1. Development of the obstacle detection system and the risk cognition application processor

The hardware must have a high-speed and low-power consumption architecture to enable real-time processing of sensor data. The software will predict the motion of obstacles, such as pedestrians and other vehicles, and will quantitatively estimate the danger of a collision.

1. Development of the probe data processor

Driving support on a per car basis will be provided by high-end server systems that analyze real-time vehicle data and retrospective traffic congestion models. The Exabyte scale telematics presents a substantial technological milestone in this project, which is aiming for deployment in the 2020 financial year.

Realization of the next-generation traffic society requires an information system for each car that constantly transmits its measurements and assessments of the local transportation network to a central server. Likewise, the server's responsibility is to provide client cars with traffic diagnostics and a stochastic map of accident occurrence.

## References

1. Novat and Ziptronix (2013) Press release Novati Technologies Licenses Ziptronix's Direct Oxide Bonding (ZiBond) and Direct Bond Interconnect (DBI) patented technologies for advanced 3D integrated assemblies. http://mobile.reuters.com/article/pressRelease/idUSnBw-1CjnM9a + 124+BSW20130116?irpc = 942. Accessed 29 April 2014
2. TORKI K (17 January 2013) Presentation slide 3D-IC integration. CMP annual users meeting Paris. http://www.monolithic3d.com/uploads/6/0/5/5/6055488/09_3d_integration_2013.pdf. Accessed 29 April 2014

3. STATS ChiPAC and UMC (2013) Press release STATS ChipPAC and UMC unveil world's first 3D IC developed under an open ecosystem model. http://www.statschippac.com/news/newscenter/2013/news01292013b.aspx. Accessed 29 April 2014
4. Insights From Leading Edge News (2013) IFTLE 172 Sony TSV stacked CMOS image sensors finally arrive in 2013. http://electroiq.com/insights-from-leading-edge/2013/12/iftle-172-sony-tsv-stacked-cmos-image-sensors-finally-arrive-in-2013/. Accessed 4 March 2014
5. IME and UTAC (2013) Press release Singapore's UTAC to co-develop 2.5D through-silicon-interposer with A*STAR's Institute of Microelectronics for volume manufacturing. http://www.noodls.com/view/432D47164CD4DB4ED000D09B71C9D3C6B634E85F?3477xxx1364464809. Accessed 29 April 2014
6. GlobalFoundries (2013) Press release GLOBALFOUNDRIES demonstrates 3D TSV capabilities on 20 nm technology. http://www.techpowerup.com/182292/globalfoundries-demonstrates-3d-tsv-capabilities-on-20nm-technology.html. Accessed 29 April 2014
7. Mentor and Tezzaron (2013) Press release mentor and Tezzaron Optimize Calibre 3DSTACK for 2.5/3D-ICs. http://www.mentor.com/company/news/mentor-tezaron-calibre-3dstack. Accessed 29 April 2014
8. UMC (2013) Press release UMC establishes its Specialty Technology Center of Excellence in Singapore. http://www.umc.com/english/news/2013/20130522.asp. Accessed 29 April 2014
9. A*STAR (IME), Qualcomm and STATS ChipPAC (2013) Press release, A*STAR IME, Stats Chippac and Qualcomm collaborate to develop low cost interposer technology. http://www.noodls.com/view/822844F17FDB7A4F108A8AFEBBB5EA9EDB1A48AF. Accessed 28 April 2014
10. Lin L et al (2013) Reliability characterization of Chip-On-Wafer-On-Substrate (CoWoS) 3D IC integration technology. Sixty third electronic components & technology conference, May 28–31, P366–P371
11. Kim DW et al (2013) Development of 3D Through Silicon Stack (TSS) assembly for wide IO memory to logic devices integration. Sixty third electronic components & technology conference, May 28–31, P77–P80
12. Nikkei (30 October 2013) Technology online news Qualcomm and Amkor 3D stacked 28 nm generation logic and maximum of four Wide I/O memories [ECTC]. http://techon.nikkeibp.co.jp/article/EVENT/20130530/284551/. Accessed 27 April 2014
13. Synopsys (2013) Press release synopsys collaborates with A*STAR IME to optimize TSI technology. http://news.synopsys.com/2013-06-24-Synopsys-Collaborates-with-A-STAR-IME-to-Optimize-TSI-Technology. Accessed 11 July 2014
14. Cadence (2013) Press release TSMC and Cadence deliver 3D-IC reference flow for true 3D stacking. http://www.cadence.com/cadence/newsroom/press_releases/Pages/pr.aspx?xml = 091913_3dic&CMP=home. Accessed 28 Oct 2014
15. Mentor Graphics Corp (2013) Press release mentor graphics tools included in TSMC's 3D-IC reference flow for true 3D stacking integration. http://www.mentor.com/products/mechanical/news/mentor-tsmc-3d-ic-reference-flow. Accessed 15 April 2014
16. Xilinx (2013) Press release Xilinx and TSMC reach volume production on all 28 nm CoWoS™-based all programmable 3D IC families. http://press.xilinx.com/2013-10-20-Xilinx-and-TSMC-Reach-Volume-Production-on-all-28nm-CoWoS-based-All-Programmable-3D-IC-Families. Accessed 9 March 2014
17. SMIC (2013) Press release SMIC announces formation of center for vision, sensors and 3DIC. http://www.prnewswire.com/news-releases/smic-announces-formation-of-center-for-vision-sensors-and-3dic-228617611.html. Accessed 28 Sept 2014
18. CEA-Leti (2013) Press release CEA-Leti signs agreement with Qualcomm to assess sequential 3D technology. http://www.businesswire.com/news/home/20131208005051/en/CEA-Leti-Signs-Agreement-Qualcomm-Assess-Sequential-3D#.U19t4_l_vX1. Accessed 29 April 2014

19. Xilinx (2013) Press release Xilinx doubles industry's highest capacity device to 4.4 M logic cells, delivering density advantage that is a full generation ahead. http://www.wantinews.com/news-5975003-Xilinx-will-be-the-industry-39s-largest-capacity-doubled-to-44-million-devices-logical-unit.html. Accessed 30 Sept 2014
20. Altera and Intel (2014) Press release Altera and Intel extend manufacturing partnership to include development of multi-die devices. http://newsroom.altera.com/press-releases/nr-intel-packaging.htm. Accessed 31 March 2014
21. ASE and Inotera (2013) Press release ASE expands System-in-Package business model through industry partnership with Inotera. http://www.inotera.com/English/Press_Center/Press_Releases/2014-04-07+Inotera+and+ASE+Joint+Press+Release.htm. Accessed 7 April 2014
22. 22ï¼Samsung (28 August 2014) Press release Samsung starts mass producing industry's first 3D TSV technology based DDR4 modules for enterprise servers. http://www.samsung.com/global/business/semiconductor/news-events/press-releases/detail?cateSearchParam=&searchTextParam=&startYyyyParam=&startMmParam=&endYyyyParam=&endMmParam=&newsId = 13602&page=&searchType=&rdoPeriod=A#none. Accessed 2 Sep 2014
23. Choi JY (2012) Presentation slide beyond LPDDR3 directions. LPDDR3 symposium 2012. http://www.jedec.org/sites/default/files/JY_Choi_Final_LPDDR3.pdf. Accessed 10 July 2014
24. X-bit labs News (2013) JEDEC to discuss wide I/O 2, LPDDR4 standards at conference. http://www.xbitlabs.com/news/memory/display/20130316080548_JEDEC_to_Discuss_Wide_I_O_2_LPDDR4_Standards_at_Conference.html. Accessed 27 April 2014
25. JEDEC (2011) STANDARD JESD209-2F June 2013, Low Power Double Data Rate 2 (LPDDR2) (Revision of JESD209-2E, April 2011), October 2013
26. JEDEC (2013) STANDARD JESD209-3B, Low Power Double Data Rate 3 (LPDDR3) (Revision of JESD209-3A, August 2013)
27. Kim M (2013) Presentation slide high performance & low power memory trend. http://www.semiconwest.org/sites/semiconwest.org/files/docs/SW2013_Minho%20Kim_SK%20Hynix.pdf. Accessed 30 July 2014
28. SK Hynix (2014) Presentation slide memories in 2014 & introducing the new generation memories. https://intel.activeevents.com/sz14/connect/fileDownload/session/F5F7AD6840DE76F60EF6D7B8B8A91EA0/SZ14_GSSS006_100_ENGf.pdf. Accessed 30 July 2014
29. HMCC (2013) Press release hybrid memory cube consortium heralds 2013 as turning point for high performance memory ICs, gains rapid consensus for final specification and decision to renew consortium. http://investors.micron.com/releasedetail.cfm?ReleaseID = 753753. Accessed 29 April 2014
30. Altera and Micron (2013) Press release Altera and Micron lead industry with FPGA and hybrid memory cube interoperability. http://investors.micron.com/releasedetail.cfm?ReleaseID = 788635. Accessed 28 April 2014
31. Micron (2013) Press release Micron technology ships first samples of hybrid memory cube. http://investors.micron.com/releasedetail.cfm?releaseid = 793156. Accessed 24 Jan 2015
32. Micron (2013) Press release Micron's hybrid memory cube earns high praise in next-generation supercomputer. http://investors.micron.com/releasedetail.cfm?ReleaseID = 805283. Accessed 29 April 2014
33. PRWEB News (2013) Pico computing delivers the first PCI express card to integrate Micron's hybrid memory cube and multiple Altera Stratix V FPGAs. Accessed 24 Jan 2015
34. HMCC (2014) Press release hybrid memory cube consortium continues to drive HMC industry adoption with release of second-generation specification. http://investors.micron.com/releasedetail.cfm?ReleaseID = 828028. Accessed 19 March 2014
35. Nikkei (2014) Technology on line Fujitsu exhibited the main board of the next supercomputer first time in Japan and the performance per size is 22 times of "KEI" (in Japanese). http://techon.nikkeibp.co.jp/article/NEWS/20140513/351641/, 2014/05/14 08:00. Accessed 2 July 2014

36. Micron (2014) Press release Micron collaborates with Intel to enhance knights landing with a high performance, on-package memory solution. http://investors.micron.com/releasedetail.cfm?ReleaseID = 856057. Accessed 2 July 2014
37. Jeddeloh J, Keeth B (2012) Hybrid memory cube new DRAM architecture increases density and performance. Proceedings of 2012 symposium on VLSI technology digest of technical papers, pp 87–88
38. Hybrid Memory Cube Consortium (2014) http://hybridmemorycube.org/about.html. Accessed 15 Aug 2014
39. SK Hynix (2013) Press release SK Hynix to start production of HBM DRAM chip modules using 3D TSV chip-packaging technology. http://itersnews.com/?p = 62940. Accessed 29 April 2014
40. Tech Soda News (2013) AMD and Hynix announce joint development of HBM memory stacks. http://techsoda.com/amd-hynix-hbm-memory/. Accessed 29 April 2014
41. RedGamingTech News (2014) AMD R9 380X in February & R9 390X & 370X Announced. http://www.redgamingtech.com/amdr9380xfebruaryr9390×370xannounced/1/. Accessed 24 Jan 2015
42. JEDEC (2013) STANDARD JESD235 High Bandwidth Memory (HBM) DRAM. October 2013
43. Business Korea (2013) News release SK Hynix develops 4x faster DRAM. http://www.busin9esskorea.co.kr/article/2766/moore%E2 % 80 % 99s-law-sk-hynix-develops-4x-faster-dram. Accessed 28 March 2014
44. Goto H (28 April 2014) PC watch news HBMDRAM for the next generation GPU which realizes 1TB/sec is in to volume production stage (in Japanese). http://pc.watch.impress.co.jp/docs/column/kaigai/20140428_646233.html. Accessed 20 May 2014
45. Goto H (1 May 2014) PC watch news details of "HBM" stacked with TSV technology which succeed GDDR5 memory (in Japanese). http://pc.watch.impress.co.jp/docs/column/kaigai/20140501_646660.html. Accessed 20 May 2014
46. BLACK B (2013) Presentation slide DIE STACKING IS HAPPENING!. 12-9-2013. http://www.microarch.org/micro46/files/keynote1.pdf. Accessed 10 Aug 2014
47. My Navi News (31 January 2014) Advanced technology seen at the largest society of supercomputer "SC13" (in Japanese). http://news.mynavi.jp/column/sc13/020. Accessed 20 May 2014
48. My Navi News (31 January 2014) Advanced technology seen at the largest society of supercomputer "SC13" (in Japanese). http://news.mynavi.jp/column/sc13/021. Accessed 20 May 2014
49. Borkar S (7 October 2012) Presentation slide ubiquitous computing in the coming years—technology challenges and opportunities. Semicon West
50. Kimmich G (2013) Presentation slide 3D—what's next, D43D. Workshop June 2013, Grenoble. http://www.leti-innovationdays.com/presentations/D43DWorkshop/Session3-StateOfTheArts/D43D13_Session_3_1_GeorgKimmich.pdf?PHPSESSID=f7b640026ea14ad29a39a245618e010d. Accessed 3 July 2014
51. Kim J, Kim Y (2014) Presentation slide HBM: memory solution for bandwidth-hungry processors. http://www.setphaserstostun.org/hc26/HC26-11-day1-epub/HC26.11-3-Technology-epub/HC26.11.310-HBM-Bandwidth-Kim-Hynix-Hot%20Chips%20HBM%202014 % 20v7.pdf. Accessed 27 Oct 2014
52. Xie JY (7 October 2013) Presentation slide developing high density packaging with low cost stacking solution. Semicon West
53. NEDO (July 2013) Basic plan next generation smart device project (in Japanese). http://www.nedo.go.jp/content/100542603.pdf. Accessed 20 July 2014

# Chapter 3
# TSV Processes

**Masahiko Tanaka, Makoto Sekine, Itsuko Sakai, Yutaka Kusuda, Tomoyuki Nonaka, Osamu Tsuji, Kazuo Kondo**

## 3.1 Deep Silicon Etching by Bosch Process

**Masahiko Tanaka**

### *3.1.1 Introduction*

The silicon etching process which is the so-called Bosch process was invented in 1992 by Robert Bosch GmbH [1]. This very special etching process with cyclic etching step and deposition step enables an extremely deep Si etching, namely several tens to hundreds microns. There is no doubt that this innovative process technology made Si-microelectromechanical systems (MEMS) realized and most of the Si-MEMS device manufacturers are now using this technology. The development of application of this Bosch process to through-silicon via (TSV) etching was started around 2000, and it is well established now [2–4].

In this chapter, the technological basics of the Bosch process, features of the process including TSV etching characteristics are described. The features of Bosch process etching equipment are also described.

M. Tanaka (✉)
SPP Technologies Co., Ltd., 1–10, Fuso-cho, Amagasaki, Hyogo 660-0891, Japan
e-mail: masahiko.tanaka@spp-technologies.co.jp

M. Sekine · I. Sakai
Graduate School of Engineering, Nagoya University, C3-1(631) Chikusa, Nagoya 464-8603, Japan
e-mail: sekine@plasma.engg.nagoya-u.ac.jp

Y. Kusuda · T. Nonaka · O. Tsuji
R&D Division, SAMCO Inc., 36 Waraya-cho, Takeda, Fushimi-ku, Kyoto 612-8443, Japan
e-mail: p-cvd@samco.co.jp

K. Kondo
Osaka Prefecture University, Osaka, Japan
e-mail: kkondo@chemeng.osakafu-u.ac.jp



K. Kondo et al. (eds.), *Three-Dimensional Integration of Semiconductors*,
DOI 10.1007/978-3-319-18675-7_3

### 3.1.2 *Basic Characteristics of the Bosch Process*

For plasma etching of Si in mainstream semiconductor manufacturing, $Cl_2$-or HBr-based process has been used. These two basic chemistries provide the etching sidewall protection with by-products which has relatively low vapor pressure and tends to deposit on the sidewall where ion bombardment from the plasma is limited. These processes were very well established and widely used in semiconductor industry in the etching rate range of several 100 nm/min and the photoresist selectivity range of 2–5. For Si etching in semiconductor, as there was no serious requirement of etching rate and selectivity improvements when Si was highlighted as micromechanical material [5], these usual Si etching processes are not enough for this new application especially in the aspect of photoresist selectivity. The Bosch process was invented in such a circumstance where extremely high selectivity and high etching rate had become inevitable for Si-MEMS [6–8]. Although fluorinated radicals were known to provide the highest Si etching rate among the halogen group, the reaction by-product $SiF_x$ is very much volatile and not easy to be deposited on the sidewall. The innovation of Bosch process is to implement short chemical vapor deposition (CVD) step to prevent the sidewall erosion by F radicals. Figure 3.1 shows a schematic diagram of cross section of samples after each step of Bosch etching cyclic process. It is a combination of $SF_6$ plasma etching and $C_4F_8$ plasma deposition. (a) After the etching step, the Si etching profile by F radicals is isotropic, and all the surface of the wafer including the top of photoresist, sidewall of both of photoresist and etched Si, and bottom of etched Si is covered by $CF_x$ polymer after the CVD step with $C_4F_8$ chemistry. (b) The ion bombardment from plasma removes the polymers on the flat area: top of the photoresist and the bottom of the etched Si. (c) Then, the high-rate F-radical etching of exposed Si at the bottom is performed. A (a)–(c) cycle is done to the depth required. The photoresist is being eroded only during the step (b) after the deposited polymer is removed, thus the photoresist selectivity of this process is quite high, namely 30–100:1, so that Si-MEMS, which needs hundreds of micron etching, is enabled by this Bosch process.

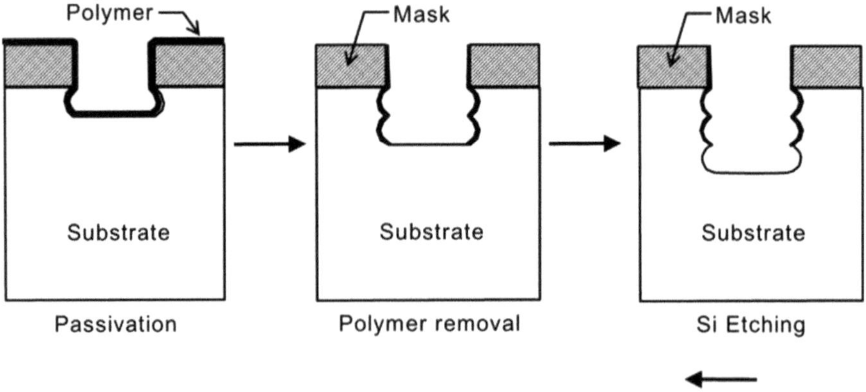


**Fig. 3.1** Concept of Bosch process

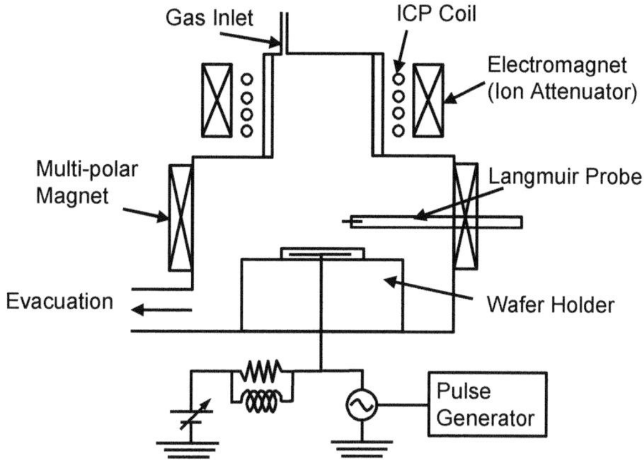


**Fig. 3.2** Schematic diagram of experimental etching process chamber. *ICP* inductively coupled plasma

An experimental typical etching process chamber for Bosch process is shown in Fig. 3.2. Usually inductively coupled plasma (ICP) is used for Bosch process by the features of high-density remote plasma source which provides a high rate and good controllability of ion energy by applying the radio frequency (RF) bias on wafer holder. ICP was a well-established technology from the late 1980s to early 1990s. It was, in early days, applied on photoresist removal process and then diverted to Al-alloy etching and Si etching. Several trials of applying this technology on $SiO_2$ etching were conducted. But none of them succeeded as $SiO_2$ etching prefers relatively lower density plasma and very high energy ion bombardment. It also needs sensitive control of carbon-rich (C-rich) polymer deposition on the under-layer, namely poly-Si or metal interconnect. ICP tends to decompose $C_xF_y$ too much to fail into less controllable C-rich polymer deposition. So, for the $SiO_2$ etching in mainstream semiconductor parallel plate, capacitively coupled plasma (CCP) is still being used.

In Bosch process, the Si etching rate depends on the amount of F radicals and the ion bombardment is only necessary in the polymer removal step, which is (b) in Fig. 3.1. During the main etching step, which is (c) in Fig. 3.1, if the ion density and/or ion energy is high, the photoresist etching rate becomes high so that photoresist selectivity becomes low. In the commercial Bosch process etching system, there are several kinds of methods to keep F radical amount high but make ion bombardment minimal. Figure 3.3 shows the ion attenuation effect of the magnetic field induced by the electromagnet surrounding ICP source in Fig. 3.2, which is one of the methods to reduce ions keeping radical amount high. Figure 3.4 shows the effect of ion attenuation on the etching rate and photoresist selectivity. The ion density was measured by Langmuir probe in the chamber. By applying the magnetic field, the selectivity improves several times while the etching rate remains almost same. Another way of controlling ions/radicals is the optimization of the process condition in each step which is one of the key know-hows of each equipment supplier. Some supplier divides (b) and/or (c) into more steps to realize more precise and stable control of ion/radical amount.

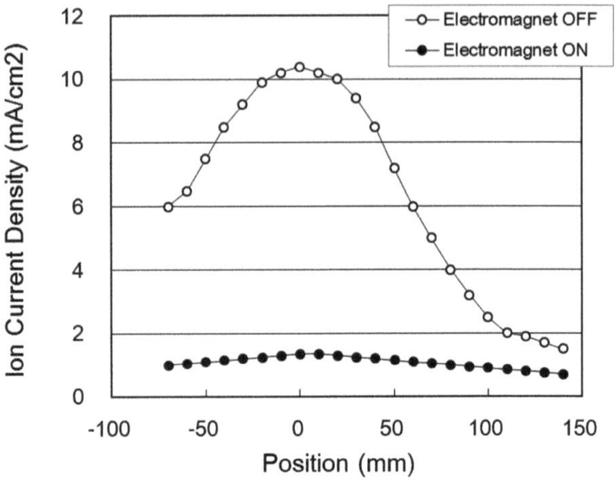


**Fig. 3.3** Ion attenuation by magnetic field

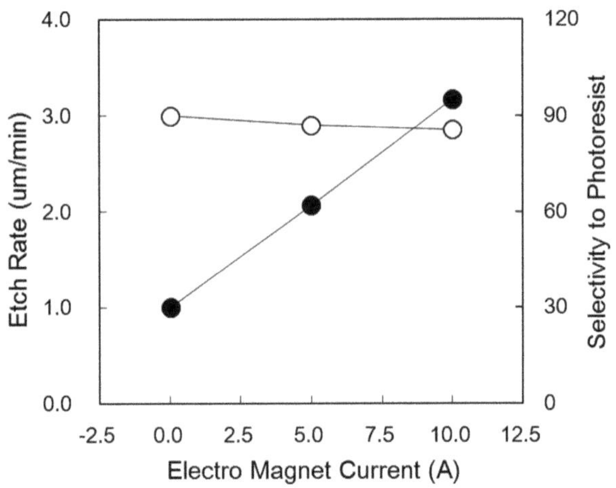


**Fig. 3.4** Effect of ion attenuation in etching step

When Si etching is performed down to the dielectric under-layer, lateral erosion of Si at the boundary between Si and under-layer tends to occur, which is so called "notching". This effect was found in poly-Si gate electrode etching in mainstream semiconductor in mid 1990s, and the mechanism is called "electron shading effect" [9]. A schematic diagram of the mechanism of notching formation is shown in Fig. 3.5. In the case of mainstream gate electrode etching, the issue is not only notching but also the gate dielectric charges up itself which induces the degradation of metal–oxide–semiconductor field-effect transistor (MOS FET) characteristics. Fortunately, the aspect ratio (AR) of poly-Si gate electrode was not so high so that the optimization of process condition worked well to prevent this issue, while Si etching in MEMS the AR is very high, namely 10–100, so that only process optimization is not enough to prevent notching. One commercialized solution of this issue is pulsed biasing. When RF bias is off, the self-bias voltage drops to the

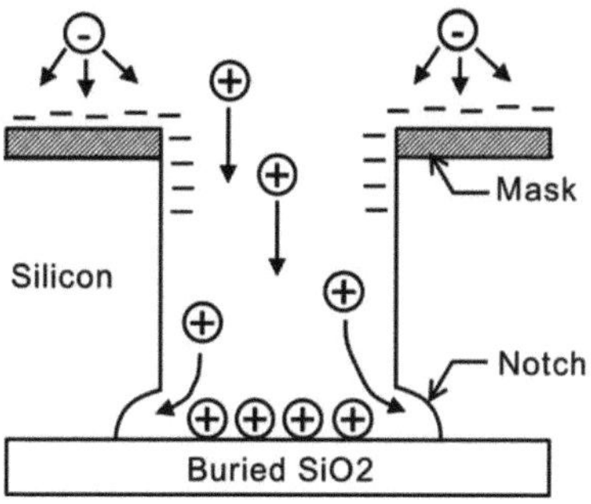


**Fig. 3.5** Mechanism of notch formation

plasma sheath level and electrons with higher energy can reach onto the under-layer which results in the charge being neutralized to the level by which ions cannot be redirected. Figure 3.6 shows an example of notch prevention by pulsed bias [10]. This countermeasure for notching is very important for Si etching in via-last TSV which has an AR of 5–10 and the dielectric layers at the bottom of via hole. The knowhow in MEMS is one of the obvious advantages of the application of Bosch process on TSV.

Another challenge of Si etching in TSV is that etching rate decreases as the AR increases. Figure 3.7 shows an example of AR dependence (ARD) on etching rate. This ARD is caused by the decrease of the supply of etchant at etching-front Si surface and the decrease of evacuation of by-products. In the case of blind via hole, the constant etching rate provides the easier control of via depth. There is a solution which is the so-called parameter ramping technique in which the etching parameters are being changed during the process. Figure 3.8 shows an example of the improved ARD by parameter ramping.

In Bosch process, the periodic sidewall roughness is generated with its cyclic process steps, which is so-called "scallops". Generally, the lateral depth of each scallop is enlarged when the etching rate becomes higher but that can be reduced in several methods. The simplest way is to reduce the step time. Figure 3.9 shows the typical relationship between the etching rate and the depth of scallops.

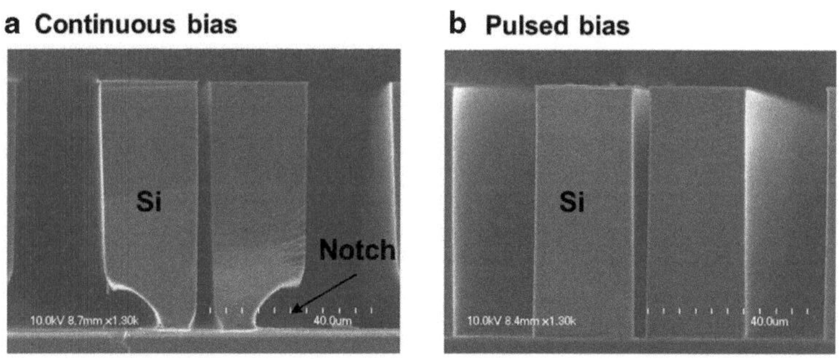


**Fig. 3.6** Notch prevention by pulsed bias

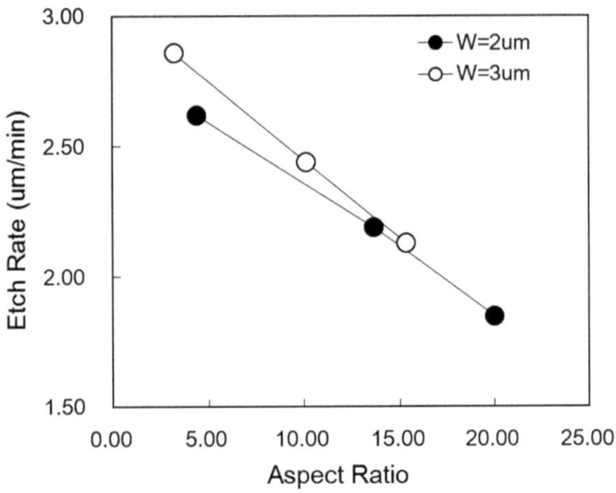


**Fig. 3.7** Aspect ratio dependence on etching rate

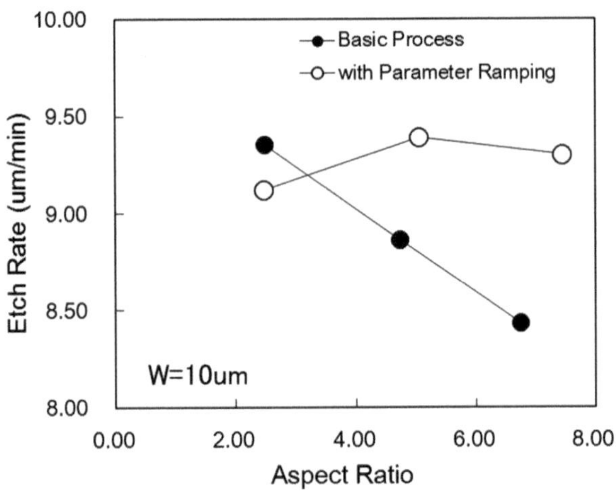


**Fig. 3.8** Improved ARD by parameter ramping

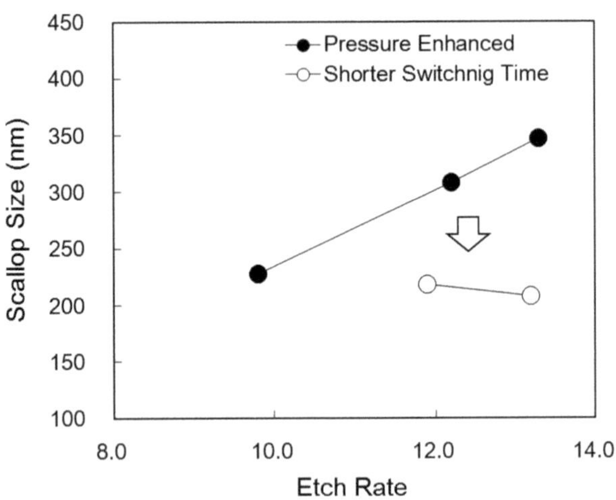


**Fig. 3.9** Scallops versus etching rate

When the gas pressure in etching step increases, the etching rate increases but by reducing the step time the scallops can be reduced. In this case, the gas exchange efficiency is important. Equipment suppliers pay attention on that and put several techniques which are not being used for mainstream semiconductor equipment, for instance using high-speed mass flow controller (MFC) directly mounted on the process chamber, quick valve operation, and so on. Another way of reducing the depth of scallops is the optimization of additive chemistry in $SF_6$ etching step. By implementing these kinds of techniques the depth of scallop has been reduced down to several nanometers in leading-edge systems.

### *3.1.3 Bosch Etching Equipment for TSV*

In the case of Bosch-type etching equipment, because they had been developed in MEMS field, there was no urgent requirement for 300 mm wafer processing. While the TSV applications are mainly with 300 mm wafers so that equipment suppliers needed to improve the plasma uniformity. On the other hand, the mainstream semiconductors' suppliers who started to apply their established front-end wafer-processing technology to Bosch-type switching process needed to establish more robust hardware for very quick chemistries exchange and very quick plasma impedance matching for these different chemistries.

"Pegasus 300" is a leading-edge commercial Bosch-type etcher developed by SPP Technologies Co. Ltd. in the former category, which is shown in Fig. 3.10. It equips a unique coaxial plasma source to enhance the capability to larger wafers such as 300 mm or more. Figure 3.11 shows a schematic diagram of coaxial dual

**Fig. 3.10** Outlook of SPP Technologies "Pegasus 300"

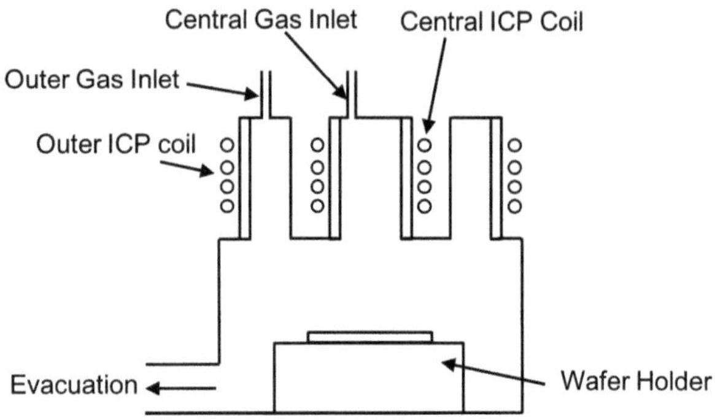


**Fig. 3.11** Schematic diagram of coaxial dual inductively coupled plasma (ICP) source

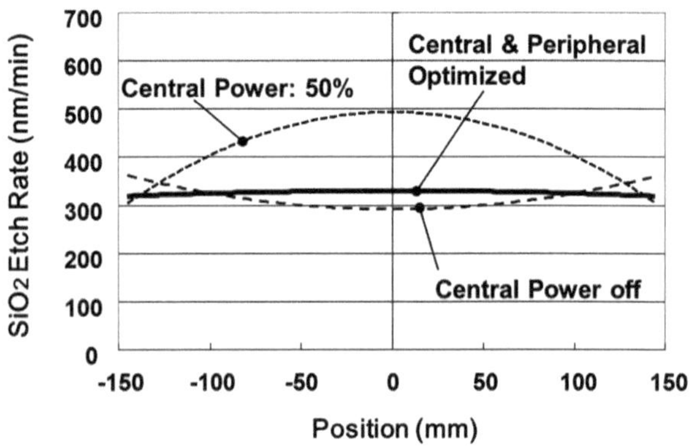


**Fig. 3.12** Plasma density control

ICP source [11]. The purpose of this structure is obviously to control the distribution of plasma density by tuning the power balance between central area and peripheral area of the plasma source. Figure 3.12 shows an example of $SiO_2$ etching rate distribution which is reflected by ion density distribution controlled by the balance of introduced powers.

In early stage, the depth of scallops was several hundred nanometers but by implementing several techniques mentioned before, the current practical depth of scallops is below 100 nm. Figure 3.13 shows a typical etching shape of via-middle TSV after etching. The diameter is 5 μm and the depth is 65 μm. Si etching rate is 5 μm/min, the nonuniformity in 300-mm wafer is ± 1.5 %, selectivity to photoresist is 60 and the depth of scalloping is 80 nm.

Another preferable performance for TSV etching is in situ $SiO_2$–Si–$SiO_2$ etching especially in via-last process. There are several potential process schemes for via last, and in some cases, the TSV bottom oxide on the metal interconnect should better be removed before the next process step. In the case of mainstream Si etcher, the wafer-stage biasing power is relatively small to control the ion energy low enough to prevent damage on the gate dielectric layer so that it cannot etch $SiO_2$. Also for the usual Bosch-type etching system, the wafer-stage biasing power is not high because the high ion energy is useful only in passivation removal step. A recently

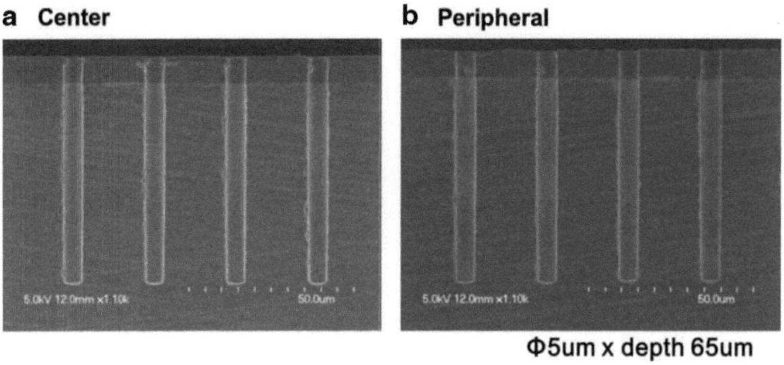


**Fig. 3.13** Typical etching profile for via middle

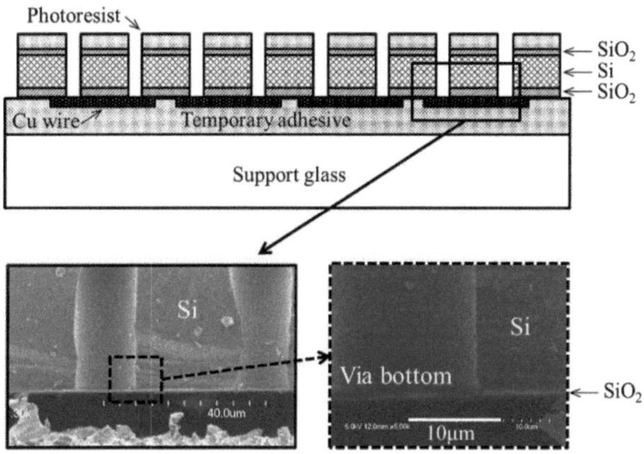


**Fig. 3.14** In situ etching for top $SiO_2$, Si, and bottom $SiO_2$

developed Bosch-type etcher for TSV Pegasus 300 equips the optimized higher power-biasing system so that it can etch all of $SiO_2$ mask, Si, and bottom $SiO_2$ layer successively [12]. Figure 3.14 shows an example of in situ top $SiO_2$, Si, and bottom $SiO_2$ etch. It can provide a simpler manufacturing line configuration.

## 3.1.4 Conclusions

The basics of Bosch process concept are described. This unique technology, which was invented in 1992, is the key enabler of Si-MEMS and now widely being used in volume production of Si-MEMS. Leading-edge Bosch etching equipment for TSV is also described. This is one of the key solutions for 3D integration of complementary metal–oxide–semiconductor (CMOS) imaging sensors or memories especially for via-last scheme which needs relatively deep via hole without the notching problem at the bottom. The basic technology is ready for the coming 3D era and will be improved along with the requirements which will be raised in volume production.

## 3.2 High-Rate Silicon Via Etching and Basics of Sidewall Etch Reaction by Steady-State Etch Process

**Makoto Sekine, Itsuko Sakai**

### *3.2.1 Introduction*

Etch processes to form through-silicon via (TSV) structures, where plural chips are electrically connected by vertical interconnects through stacks, have provided an incentive for further studies of the deep silicon etching technology. Although TSV etch is used for making stacked 3D chips, completely thorough etching of the Si substrate is not necessary. Because the chips are thinned in packaging processes, usually a 100-μm depth is enough. For high productivity, TSV etch requires high rate to enable high throughput and smooth sidewalls to enable the filling of conducting materials optimally in subsequent processes. These requirements make the TSV etch very challenging. These TSVs are typically large features with diameters of 10–50 μm and up to 150 μm depth depending on the applications and integration scheme.

The success of high-AR Si etch depends on the controlling of the lateral etch rate as well as enhancing the vertical etch rate. There are three types of technologies to potentially attain the requirements: time-multiplexed alternating (TMA) process, steady-state etch processes with cryogenically cooled wafer (cryogenic process), and around near-room temperature process.

The TMA process for high-AR Si etch, which alternates etching and polymerization steps known as Bosch process [13] , allowed a higher AR to be obtained with very high etch selectivity over mask materials, such as photoresist and $SiO_2$. The TMA etch process has proven effective for high AR trenches especially in MEMS applications. Each polymerization step is followed by an etch step which rapidly removes the polymer layer on the bottom of the feature while partially removing the polymer on the sidewall. The partially remained polymer protects the sidewall during the etch step. Etch and passivation steps are alternated until the specific etch depth. The polymer is similar to polytetrafluoroethylene (PTFE) and approximately 50 nm in thickness on the sidewall and bottom after the deposition step [14]. The advantages of TMA processes are straight profile, high etch selectivity, and relatively high rate. Disadvantages include a scallop-shaped sidewall surface profile due to the alternating etch and deposition processes. This scallop shape may be responsible for causing some failures in the subsequent via-filling process.

As a steady-state process, cryogenic etching attempts to suppress the etch reaction at sidewall, while enhance the reaction at the bottom. It could achieve high etch rate and high selectivity over photoresist by keeping the wafer at very low temperature [15, 16]. The cryogenic etching uses $SF_6$ and $O_2$ gases to passivate and etch simultaneously. The component of passivation film should be $SiO_xF_y$. Compared with fluorocarbon polymer passivation for the TMA process, $SiO_xF_y$ passivation film might be more difficult to be etched with neutral species, F radical. Ion bombardment with relatively high energy could clear the passivation from the bottom surface and allow vertical silicon etch without the lateral etching and the sidewall surface roughness that is observed for the TMA process. Temperature and oxygen flow rate ratio are the main parameters for controlled sidewall surface reaction to achieve fine profile. Higher oxygen addition may cause a narrowing of the trench width and a tapered profile would appear. On the other hand, lowering temperature tends to enhance the passivation with thin surface layer and improve the profile verticality from bowing profile. Increasing the bias power and reducing the pressure both enhance the ion bombardments and make the sidewall profile more vertical. Cryogenic etch uses a hardware with a chiller system by utilizing a liquid nitrogen to keep the wafer temperature below − 100 °C. The cryogenic system is not popular nowadays due to the practical difficulty of maintaining a very low wafer temperature. The structure of wafer electrode must be very complex, and the time consumed in increasing and decreasing the wafer temperature results in a low productivity.

High-density plasma etching at near-room temperature is a promising technology to overcome these problems seen in TMA and cryogenic processes by using conventional high-density plasma etch tools with fine plasma chemistry control. In this chapter, an ultrahigh rate Si etch process using capacitively coupled plasma (CCP) generated in a magnetically enhanced reactive ion etching (MERIE) system with $SF_6$ gas-based chemistry [17, 18] is introduced and the key factors for obtaining the very high etch rate and the profile control are discussed.

The approach taken for obtaining an ultrahigh etch rate was to increase the supply of fluorine radicals to the Si surface as much as possible. The means suited for generating high-density plasma at high pressures; therefore, the MERIE was chosen for study. In MERIE, the magnetic field enhances the plasma density because the electrons are inhibited to cross the magnetic field lines, which keeps the power-loss rate low, and also the electrons are confined to a small volume above the cathode (wafer).

For understanding the sidewall reaction mechanism where the fluorine and oxygen atoms competitively react to the sidewall of TSV, Si surface placed in the downflow region of $SF_6$–$O_2$ gas plasma was analyzed by in situ X-ray photoelectron spectroscopy (XPS). This in situ system is inevitable, because once the etched surface with any reaction layer is exposed to the atmosphere, they are immediately oxidized by water vapor and oxygen in the air, and usually only $SiO_2$-like film containing small amount of fluorine is detected. Effects of relative densities of radicals, substrate temperature, and $SiF_4$ as addition were examined and discussed.

### *3.2.2 MERIE Process for TSV Application*

The etch processes using MERIE [19–23] realized a significant increase of etch rates for Si and $SiO_2$ when the magnetic field is applied. Just as important is the characteristic property of MERIE that the high-density plasma is confined close to the cathode where the wafer is placed [24]. Because of this property, it was expected that a high flux of fluorine radicals as well as ions can be supplied to the wafer surface, especially at high-pressure conditions such as several tens of pascals, where the diffusion of the plasma would be small. Although the MERIE plasma is inherently nonuniform due to the $\mathbf{E} \times \mathbf{B}$ drift of electrons near the cathode sheath

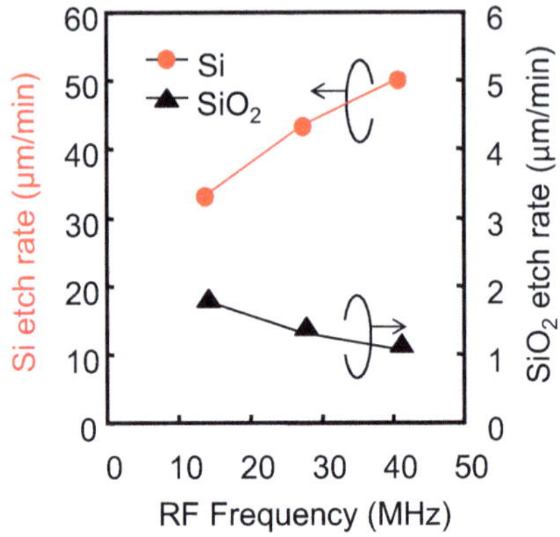


**Fig. 3.15** Si and $SiO_2$ etch rates as a function of RF

edge, by using the dipole ring magnet (DRM) the plasma distribution can be controlled to be relatively uniform by the optimized magnetic field configuration [23]. In this study, the extremely high-rate processes required for TSV were studied using CCP MERIE with DRM, from the view of process parameters of RF , gas pressure, and additional amount of oxygen gas.

#### 3.2.2.1 Effect of RF

The most important requirement for large TSV etching is ultrahigh etch rate as well as the precise profile control due to the large depth. The sidewall should be straight or slightly tapered because the requirement for the AR is not so high and small amount of undercut and bowing shape is acceptable for realistic manufacturing. The high etch rate process was explored using MERIE system with DRM and $SF_6$ and $O_2$ gas chemistry at relatively high pressures [19].

Figure 3.15 shows the etch rates of Si and $SiO_2$ (mask material) for the RFs of 13.56, 27.12, and 40.68 MHz. The RF power was 2200 W; the $SF_6$ and $O_2$ flow rates were 200 and 40 sccm; the pressure was maintained at 33 Pa. Samples were Si substrates with $SiO_2$ mask patterns of 40 μm size holes. The Si etch rate increased when the RFwas increased from 33.2 μm/min at 13.56 MHz to 50.3 μm/min at 40.68 MHz. The $SiO_2$ etch rate decreased when the RF was increased. It was found that the ultrahigh Si etch rate of 50 μm/min was achieved at a high RF.

The impact of RF on etch profile is shown in Fig. 3.16 for the 1-min etch time. The hole etched at 13.56 MHz, Fig. 3.16a, showed a typical bowing profile. It changed from bowing to straight figure for higher frequency as seen in Fig. 3.16b and c. The bowing profile was caused by isotropic reaction induced by neutral F radicals. It has been reported that the variation of Si etch rate is related to the density of atomic fluorine when using $SF_6$ gas chemistry [24, 25]. Also, the addition of $O_2$ resulted in reducing isotropic etch reaction by forming an oxidation layer on the sidewall, that is, a sidewall protection mechanism [26]. Therefore, for realizing ultrahigh etch rates and etch profile control simultaneously, the effective plasma

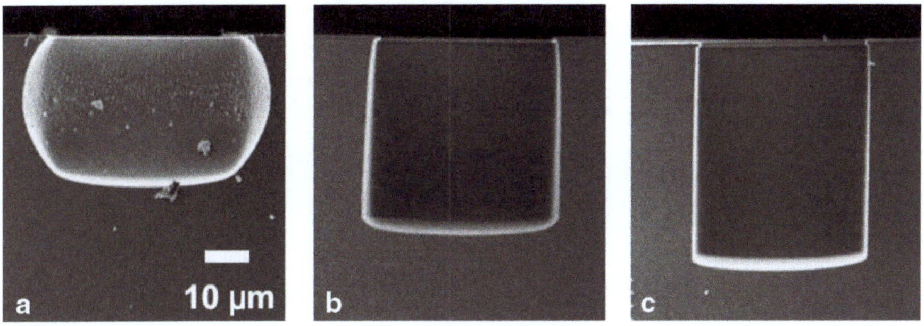


**Fig. 3.16** Etch profile dependence on RF. **a** 13.56 MHz. **b** 27.12 MHz. **c** 40.68 MHz

parameters were investigated on the density of F and O as an indicator of Si etch rate and profile control. The relative densities of F and O were estimated using an optical emission spectroscopy combined with actinometric method [27], where Ar gas was introduced at the 10 % of total gas flow rate. During the measurements, the cathode was covered by an Si wafer with a thermally grown $SiO_2$ film.

Relative emission intensities from F (704 nm) and O (845 nm) over Ar (750 nm) are shown in Fig. 3.17. They are normalized by the intensity ratio of the experiment for 13. 56 MHz. The F density increased when the RF was increased from 13.56 to 40.68 MHz and so was seen in the Si etch rate. It means that Si etch rate is closely correlated with F density. When F density and Si etch rate increased for the frequency, the profile changed to straight shape, although the isotropic reaction should be enhanced with increasing F. It is noticeable that a more profound increase of O density compared to the F density was observed and it suggests that the sidewall passivation reaction, that is, the oxidation of sidewall surface, was much enhanced at higher frequency. This indicates that certain amount of O radicals form an oxidized layer on the sidewall to suppress the lateral etch rate, but it does not suppress Si etch reaction at the hole bottom where bombarded by energetic ions.

**Fig. 3.17** F and O emission intensities and Si etch rate as a function of RF, normalized at 13.56 MHz

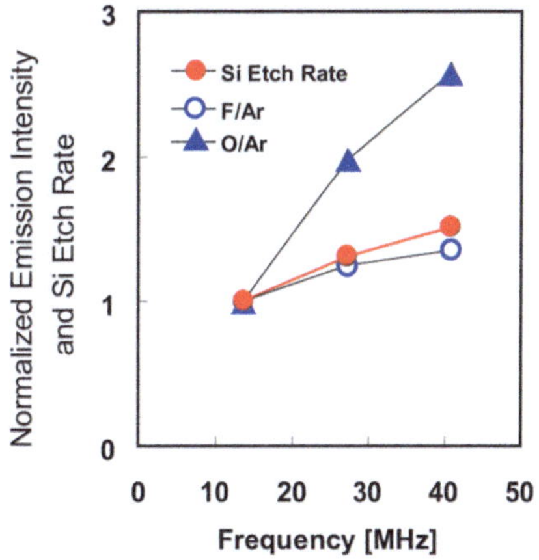

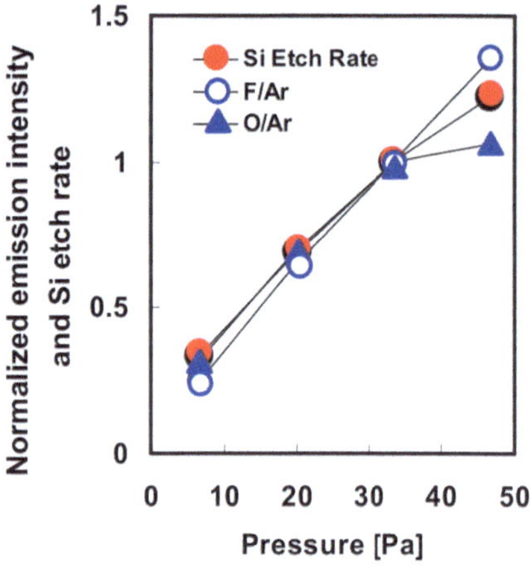


**Fig. 3.18** F and O emission intensities and Si etch rate as a function of pressure, normalized at 33 Pa

Furthermore, the Si selectivity over $SiO_2$ mask increased at a higher frequency. This is because the $SiO_2$ etch rate mainly depends on ion energy, as is typically represented by averaged DC potential at wafer surface ($V_{dc}$). The $V_{dc}$ decreased with the increase in plasma density at higher frequency. In CCP MERIE with $SF_6$ and $O_2$, the F and O emission intensity increased when the RF was increased from 13.56 to 40.68 MHz. The Si etch rate increased and the profile became anisotropic at the high RF of 40.68 MHz.

#### 3.2.2.2 Effect of Pressure

The pressure dependence of the Si etch rate and the relative intensities for F and O emission to Ar emission at 40.68 MHz RF was investigated as shown in Fig. 3.18. The etch rate and relative optical emission intensities were normalized by the values at 33 Pa. The Si etch rate and the relative signal intensities of F and O emissions increased with the increase in pressure from 6.7 to 47 Pa, and here again the Si etch rate correlated very well. The F density and Si etch rate increased as the pressure. It has an ultrahigh etch rate, more than 50 μm/min, which is one of the key etch requirements for the etch process of deep holes more than 100 μm. Consequently, the Si etching using CCP MERIE with $SF_6$ gas chemistry at high RF power and high pressures is an effective process to the TSV etching.

#### 3.2.2.3 Effect of Oxygen Addition

For the profile control as well as high etch rate, $O_2$ addition using the ultrahigh rate process of CCP MERIE with $SF_6$ gas chemistry at high pressure and high RF was investigated [3.17]. The etched profiles (nominal hole diameter of mask was 8 μm) are shown in Fig. 3.19 as the flow rates of additive $O_2$ gas were changed from 0 to

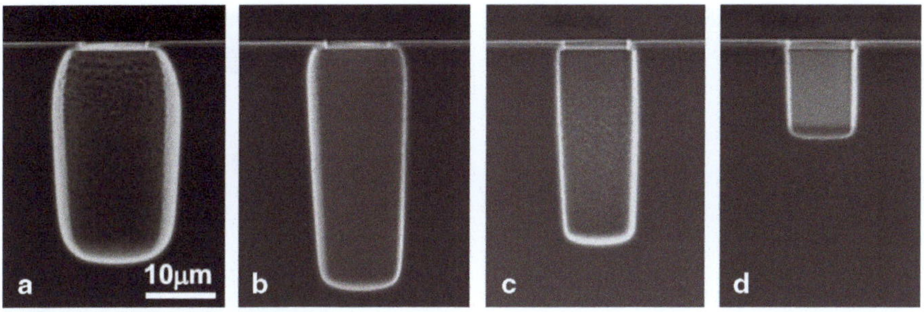


**Fig. 3.19** Etched profiles of 8 m holes with **a** 0 sccm, **b** 40 sccm, **c** 80 sccm, and **d** 160 sccm $O_2$ addition to 200 sccm of $SF_6$

160 sccm for the etch conditions, $SF_6$ flow rate of 200 sccm, total pressure of 47 Pa, RF powers of 1500 and 200 W for 60 and 3.2 MHz biases, respectively. The etch time was 1 min. Samples were Si substrates with $SiO_2$ mask patterns.

With no $O_2$ addition, bowing profile was observed as shown in Fig. 3.19a. The amount of bowing in the etch profile decreased at 40-sccm $O_2$ addition and the etch depth increased (Fig. 3.19b). By increasing $O_2$ addition to 80 sccm, there was no more bearing of isotropic reaction and the profile became almost straight, but the etch depth decreased slightly. Finally, the profile was quite straight but the etch depth decreased even more for 160-sccm $O_2$ condition as seen in Fig. 3.19d. The profile clearly changed with the increase of $O_2$ addition. When the $O_2$ addition was 0–40 sccm, it was a bowing profile due to the isotropic Si etching by F radicals, while at the $O_2$ addition of 80 and 160 sccm the profiles were much straight because of the suppression of Si etch reaction by the sidewall protective oxidized layer by O radicals. However, even at the $O_2$ addition of 160 sccm, the protection layer itself was not clearly identified by the scanning electron microscope (SEM) observation at the sidewalls.

Then, the behavior of F and O radicals was compared with the tendency of etch rate variation for $O_2$ addition. Figure 3.20 shows the relative emission intensities of F and O atoms to the Ar atom as a function of $O_2$ flow rate. The intensities were

**Fig. 3.20** Si etch rate, and F and O emission intensities normalized by the $O_2$ flow rate of 160 sccm, as a function of $O_2$ flow rate

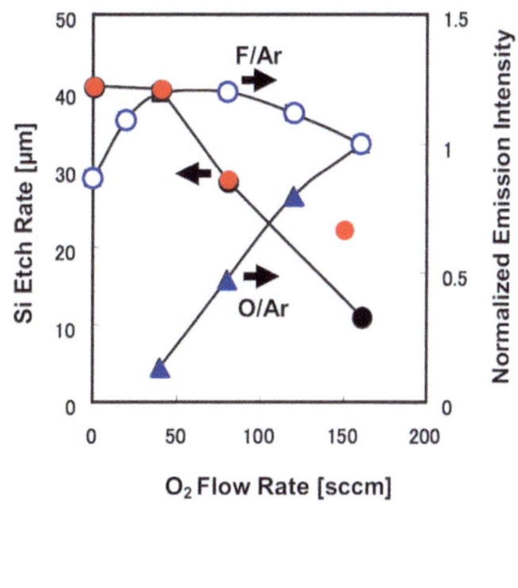

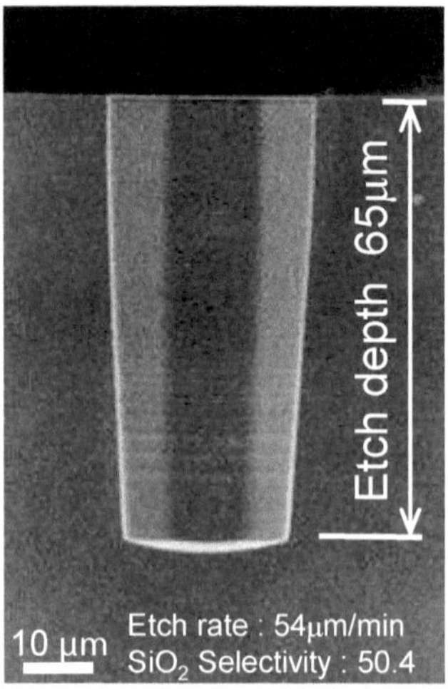


**Fig. 3.21** Typical etch profile of TSV using CCP MERIE reactor

normalized by the value at 160-sccm $O_2$. Closed circles are for the Si etch rate of 95-μm holes. The relative F emission increased slightly first with the $O_2$ flow rate increase up to 75 sccm, and then decreased with further $O_2$ addition. It was reported that the $O_2$ addition to $SF_6$ plasma provides O radicals that first increase the F atom density by preventing the recombination reaction of SFx and F, then decrease it, because of the dilution effect. The relative O emission increased drastically with $O_2$ flow rate increase according to its partial pressure. This is the reason why the profiles come closer to one with straight sidewall. Contrarily, the Si etch rate did not correspond to the F density change. Rather, the Si etch rate was inversely correlated to the O radical density. Thus in this case, with large amounts of $O_2$ addition, the excessive O radicals suppressed Si etching at the hole bottom. This is the reason why the profile became straight and the Si etch rate decreased when the $O_2$ flow rate was increased.

Based on the investigation of process parameters for optimal TSV etching, 54 μm/min with $SiO_2$ mask selectivity of 50.4 was successfully achieved as shown in Fig. 3.21. The sidewall is slightly tapered profile that is ideal to be filled with dielectrics and conductors for the subsequent electrical wiring in the via holes. Actually, this profile was observed after the mask removal and some amount of undercutting still occurred in this process. In order to minimize the undercut at the sidewall, surface reaction mechanism on the sidewall is examined in the next section.

### 3.2.3 *Investigation of Sidewall Etch Reaction Induced by $SF_6$–$O_2$ Plasma*

As performed in Sect. 3.2.2, the high etch rate of Si is realized by supplying a large amount F radical, while the etching profile is controllable with adding $O_2$.

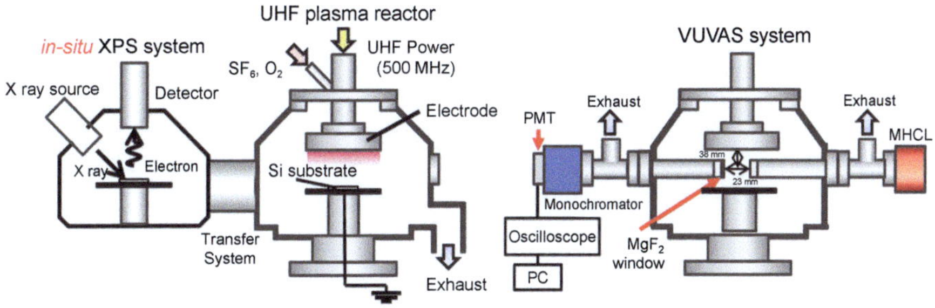


**Fig. 3.22** 500-MHz UHF $SF_6$–$O_2$ plasma reactor and in situ XPS analysis system (**a**), setup for VUVAS (**b**)

However, this competition reaction of etch, oxidation, and redeposition is very complicated and not easy to understand. Furthermore, even for large amount of oxygen addition, sidewall etching is still observed, while the bottom etch rate decreased rapidly in such high oxygen composition. In this section, the etch reaction is examined by diagnosing the $SF_6$–$O_2$ plasma and Si surface exposed to the plasma. Since it is not easy to analyze the sidewall of TSV directly, Si surface exposed to a downflow plasma of $SF_6$ and $O_2$ gas, which is the Si surface representing the sidewall of TSV, was examined [28]. The effects of wafer temperature and addition gas ($SiF_4$) are also investigated for establishing well-understood processes [29].

Si substrate ($1 \times 1$ cm$^2$) with $SiO_2$ mask ($8 \times 8$ μm square opening) was etched by a 500-MHz ultrahigh frequency (UHF) $SF_6$–$O_2$ plasma reactor shown in Fig. 3.22. The source power to the upper electrode was 500 W. The masked Si wafer was placed on the bottom grounded electrode. The electrodes gap was 40 mm. $SF_6$ and $O_2$ gases were introduced with the total flow rate of 350 sccm and the pressure was kept at 50 Pa. The stage temperature was changed from 45 to 7 °C. After etching, the Si sample was transformed in vacuum to in situ X-ray photoelectron spectroscopy (XPS) analysis system, namely in situ XPS, where the surface atomic composition and chemical bindings were analyzed. Vacuum ultraviolet absorption spectroscopy (VUVAS) [30] and optical emission spectroscopy (OES) were used for the plasma diagnosis. Absorption of VUV lines at 130.22, 130.49, and 130.60 nm for the ground level oxygen atom was measured by the VUVAS system shown in Fig. 3.22b and we can estimate the O radical absolute density [31]. OES was used for monitoring the relative change of F radial density by actinometry.

In this etch reactor, the plasma density just above the bottom (sample) electrode was substantially low, and the ion bombardment effect on the Si substrate could be negligible. Furthermore, the etch profiles were isotropic and the radical reaction was dominant, thus the Si wafer on the bottom electrode was exposed to a kind of downflow plasma and we could assume the Si surface as representing the sidewall of TSV in this experiment.

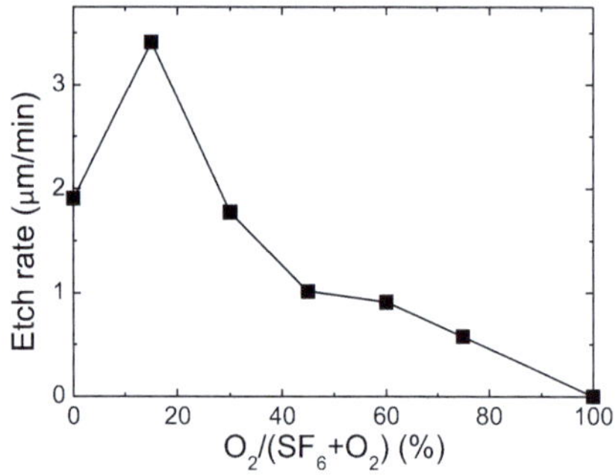


**Fig. 3.23** $O_2$ flow rate ratio dependence of Si etch rate in downflow plasma etching

### 3.2.3.1 Effect of Oxygen Addition

Si etch rate as a function of the gas flow rate ratio is shown in Fig. 3.23. On increasing $O_2$ content up to 30 %, Si etch rate was increased, then it was decreased. However, even for 75 % $O_2$ condition the etching reaction progressed. In a real TSV etch process explored in Sect. 3.2.2 [16], the $O_2$ content for obtaining the directional profile was less than 20 %. It was suggested that the realistic amount of O radical supply to the surface during F radical etching might not be enough to form a side-wall protection film by oxidizing the surface reaction layer. To confirm this, we need the absolute density for F radicals.

After a 1-min plasma exposure, the Si surface was analyzed by in situ XPS. Figure 3.24 shows Si2p spectra for the conditions for $O_2$ flow rate ratio of 0–90 %. By the plasma exposure, silicon oxyfluoride (SiOF) layer was formed on the Si surface. The peak position shifted to higher binding energy as increasing the $O_2$ flow rate ratio. These shifted peaks attributed to the Si bonded to O and F, that is, SiOF. Increase in the number of peaks means the increase of the SiOF layer thickness. The small gradual energy shift of SiOF peak toward higher binding energy indicated that the amount of F atoms binding to Si atoms increased.

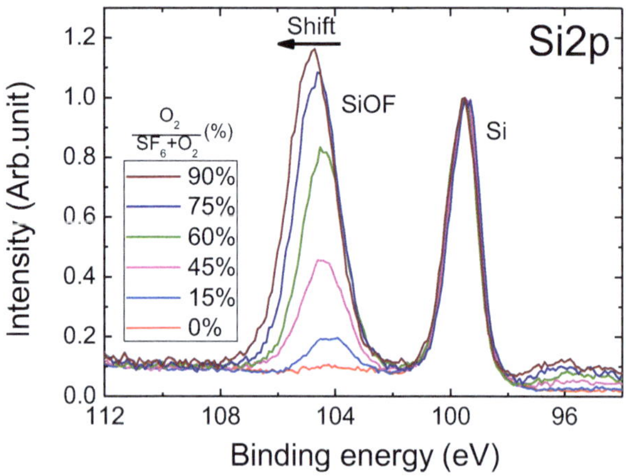


**Fig. 3.24** Si2p XPS spectra for the conditions for $O_2$ flow rate ratio of 0–90 % after a 1-min plasma exposure

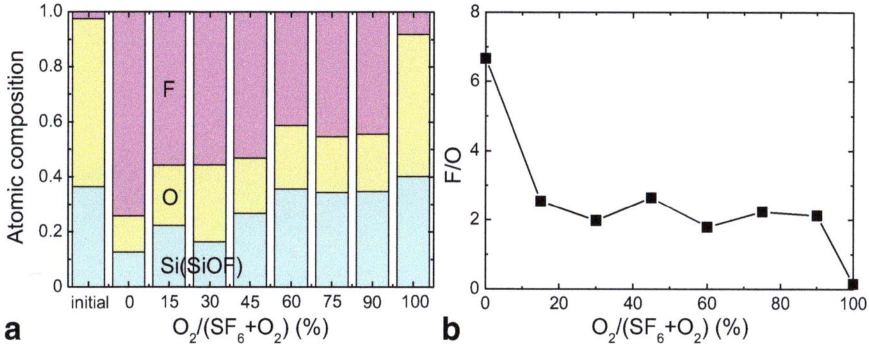


**Fig. 3.25** **a** Atomic composition of the SiOF layer observed in Fig. 3.19 as a function of the $O_2$ gas flow rate ratio. **b** The ratio of fluorine to oxygen (F–O) as $O_2$ flow ratio

**Fig. 3.26** SiOF layer thickness estimated from the XPS data

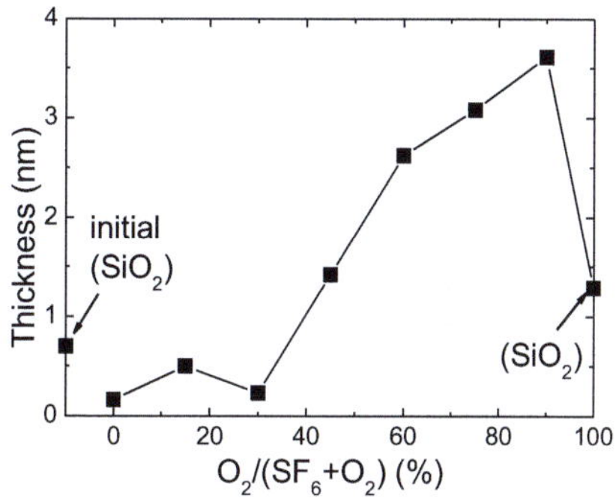


Figure 3.25a shows the atomic composition of the SiOF layer observed in Fig. 3.24 as a function of the $O_2$ gas flow rate ratio. The ratio of fluorine to oxygen (F–O) is also plotted. As $O_2$ content was increased to 15 %, the O composition increased. It did not change much for further increase of $O_2$. On the other hand, the F–O ratio is almost same value, about 2 for 15–90 % of $O_2$. From the data above it was speculated that the surface was covered with $(SiOF_2)n$, the –Si–O– chain structure with the F terminations for the remaining Si bonds.

The SiOF thickness was calculated from the XPS peak ratio of SiOF and Si, as shown in Fig. 3.26. We estimated the reaction layer thickness assuming that the inelastic mean free path of electron in reaction layer has a similar tendency to $SiO_2$ [32]. In $O_2$ flow rate ratio from 0 to 30 %, the SiOF thickness was low and less than initial $SiO_2$ thickness. On increasing $O_2$ from 45 to 90 %, the SiOF thickness was increased. It was 3.6 nm at 90 % $O_2$. For the oxygen plasma, the $SiO_2$ thickness was less than the thickness of SiOF layer for 60–90 % $O_2$ content. In the low oxygen condition, the thickness did not depend on O radical density. When $O_2$ content exceeded 30 %, thicker SiOF layer was formed with increasing $O_2$ flow rate where the thickness increased monotonically with the $O_2$ flow rate ratio from 45 to 90 %. Thickest one was observed for 90 % $O_2$. It means that a small amount of F radical enhanced SiOF formation.

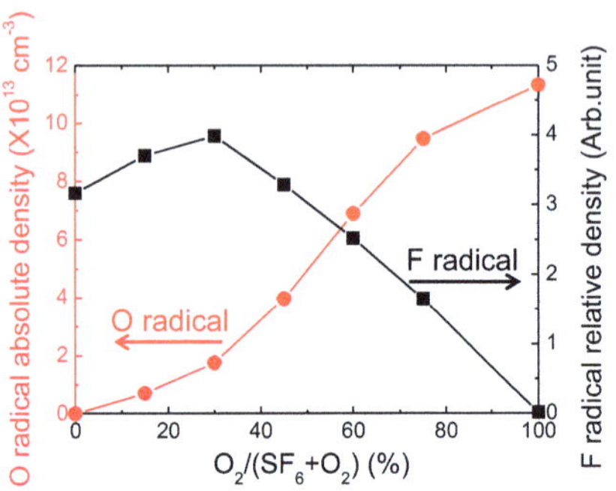


**Fig. 3.27** O radical absolute density and F radical relative density plasma as changing $O_2$ flow rate in $SF_6$–$O_2$ plasma

Figure 3.27 shows the O radical absolute density and the F radical relative density in the $SF_6$–$O_2$ plasma. On increasing $O_2$ from 15 to 100 %, the O radical density was increased from $7.01 \times 10^{12}$ cm$^{-3}$ to $1.13 \times 10^{14}$ cm$^{-3}$ almost monotonically. The F radical relative density had maximum value at 30 %. It was considered that the F radical density was increased though the $SF_6$ gas flow ratio was decreased on increasing $O_2$ content percentage from 0 to 30 %. The reaction of O and $SF_x$ might generate much F radicals and inhibit the recombination of F radicals [24]. On comparing with the dependence of the etch rate on $O_2$ flow rate ratio, where the etch rate showed the peak at 15 % $O_2$ addition and drastically decreased for further $O_2$ addition, oxygen radicals were very effective in reducing the sidewall (lateral) etch rate. However, even for 75 % $O_2$ condition, lateral etch reaction continued.

#### 3.2.3.2 Effect of Substrate Temperature

Figure 3.28a shows Si etch depth after a 3-min etching as a function of temperature with the $O_2$ flow rate ratio of 0, 15, and 60 %. It shows that Si etch rate decreased as the temperature decreased. In the range of low temperature, the decreased tendency

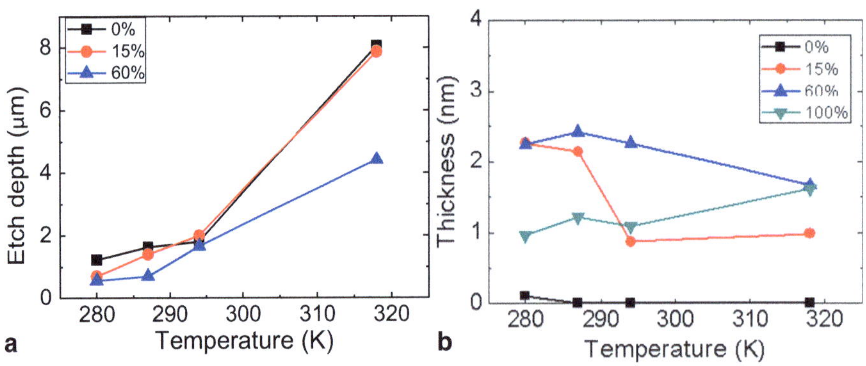


**Fig. 3.28** Si etch rate as a function of temperature with the $O_2$ flow rate ratio of 0, 15, and 60 % (**a**), estimation of film thickness (**b**)

of Si etch rate became saturated with the addition of $O_2$ gas. After a 1-min plasma exposure, the Si surface was analyzed to estimate the SiOF layer thickness by XPS as shown in Fig. 3.28b. The reaction layer was not formed at 0% $O_2$. At 15 and 60% $O_2$ flow rate ratio, the reaction layer thickness was high at low temperature. The thickness decreased at high temperature. Composition ratios for F–O for these SiOF films were also around 2, regardless of the temperatures in this range. It was considered that the atomic composition of reaction layer was not changed so much and similar bonding such as $SiOF_2$ was formed regardless of the temperature and $O_2$ flow rate ratio. It was considered that the etch products including Si, F, and O became difficult to volatilize at lower temperature. However, it looks like there is no strong correlation between the etch rate and the reaction layer thickness.

The F radical relative density was maximum at 30% of $O_2$ as shown in Fig. 3.27. On increasing $O_2$ from 15 to 100%, the O radical density was increased from $7.01 \times 10^{12}$ to $1.13 \times 10^{14}$ cm$^{-3}$. It was suggested that such large amount of O radical supply to the surface during F radical etching might be not enough to stop the etch reaction (as shown in Fig. 3.23) by oxidizing the surface reaction layer. On the other hand, in a real TSV etch process, the $O_2$ content for obtaining the directional profile was less than 20%. The other elements such as a redeposition of etch products containing Si could contribute to reduce the F radicals that diffuse in the sidewall film toward the Si bulk; then the net etching could be stopped at the sidewall.

### 3.2.3.3 Effect of $SiF_4$ Addition

In order to reduce the lateral etching while keeping the etch rate for the vertical direction at via bottom, the addition of $SiF_4$ gas was examined because it contains both of the etch species, F and deposition species, Si. Figure 3.29a and b shows reaction layer thickness and etch depth measured after a 3-min plasma exposure on changing the $O_2$ gas addition with and without $SiF_4$ addition. The wafer temperature was kept at 280 K. As observed in Fig. 3.29b, the etch depth decreased with $SiF_4$ addition. Both etch depth with and without $SiF_4$ gas tend to decrease as $O_2$ increases. At 75% $O_2$, SiOF thickness jumped up to around 9 nm (Fig. 3.29a) and

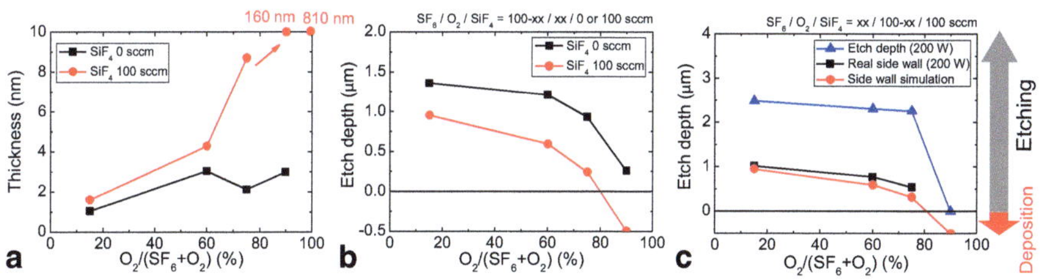


**Fig. 3.29** **a** Thickness of surface SiOF layer and **b** Si etch depth with and without $SiF_4$ gas addition as a function of $O_2$ flow rate ratio. **c** Etched depths for via bottom (*blue triangle*), sidewall (*black square*) for real via-hole etching with substrate bias (200 W), and Si etch depth without bias, (*red diamond*), i.e., identical with (**b**). All etching time was 3 min

the etch reaction almost stopped. Thick deposition film was clearly observed on the Si exposed to the plasma with more than 90 % $O_2$ and $SiF_4$ gas. Further, $O_2$ or $SiF_4$ gas addition causes thicker film deposition.

Etched depth for a real via hole for the plasma chemistry with $SiF_4$ gas was measured with a 400-kHz bias power of 200 W. The hole mask was a square with 8 μm on each side. The etched depth with the bias power for the hole bottom and its sidewall is shown in Fig. 3.29c as a function of gas flow rate ratio, $O_2$–($O_2$ + $SF_6$). Plots for the "side wall simulation" are identical to the etch depth of Si without the bias power in Fig. 3.29b. The etched depths for the real sidewall corresponded well with that of simulated ones and the experimental method which used to simulate the sidewall reaction on a flat Si substrate used here is valid. The etched depth of the hole bottom was around 2.5 μm up to 75 % $O_2$. Then the etching stopped with 90 % $O_2$ condition. It seems that optimum condition to etch the bottom while stopping the lateral etch reaction should be around 80 % $O_2$. Since the plasma conditions are so different from the real etch system used in Sect. 3.2.2, the densities of each radical are so different and so are the absolute etch rate values. However, the basic trends and SiOF layer formation on the sidewall must be informative to understand the mechanism, to build and control the process in optimum.

Based on the results shown above, reaction probability on the sidewall as function of SiOF layer thickness using an empirical formula for the pure chemical F atom etching [33, 34] is,

$$\mathrm{ER}_{\mathrm{Si/F}}(\mathrm{nm/min}) = 2.86\times10^{-13} n_{\mathrm{FS}} T \exp(-1248/T),$$

where $\mathrm{ER}_{\mathrm{Si/F}}$ is the etch rate with no reduction by the SiOF layer on surface, $n_{\mathrm{FS}}$ is F atom density near the surface, and $T$ (K) is the temperature. Since F atom density was just estimated relatively for each condition here, relative reaction probability Pr was defined as a function of SiOF layer thickness, $\theta$,

$$\mathrm{ER}_{\mathrm{Si/FO}} = \mathrm{Pr}(\theta)\,\mathrm{ER}_{\mathrm{Si/F}},$$

$\mathrm{ER}_{\mathrm{Si/FO}}$ is the reduced etch rate by the SiOF layer and the values of $\mathrm{ER}_{\mathrm{Si/FO}}$ were normalized by the maximum etch depth data. Figure 3.30 shows the relative reaction probability as a function of SiOF layer thickness. It indicates that SiOF layer of

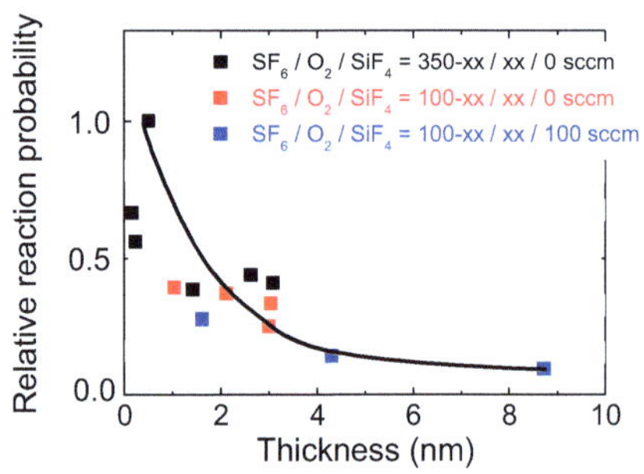


**Fig. 3.30** Relationship of relative reaction probability and SiOF layer thickness

9 nm is still not enough to stop the etch reaction. By simple extrapolation, approximately 50-nm thickness of SiOF layer is necessary to keep the etch rate negligible. This value is similar to the PTFE-like protection film thickness reported in TMA etching (Bosch) process. Although the composition of these films is different, ca. 50-nm thickness might be needed to reduce the amount of F diffusion to an adequate level at the interface of the film and Si bulk.

### *3.2.4 Conclusion*

Ultrahigh rate etching process was explored for TSV applications by using CCP MERIE with $SF_6$–$O_2$ gas chemistry in the viewpoints of $O_2$ content, RF, and gas pressure. Extremely high etch rates were obtained with higher RF of 40.68 MHz and higher pressure up to 47 Pa. Higher the fluorine atom density generated in the gas phase, higher was the etch rate obtained. Oxygen atom effectively reduced the sidewall lateral etch reaction. Finally, a 40-µm diameter TSV hole was etched at 54 µm/min with the selectivity of 50 over $SiO_2$ mask.

To investigate the reaction mechanism on the sidewall where the fluorine and oxygen atoms competitively react to Si surface, the effects of relative densities of radicals, substrate temperature, and $SiF_4$ gas addition were examined. Si surface exposed to the downflow plasma of $SF_6/O_2$ gas was analyzed to understand the reaction at the Si surface representing the sidewall of TSV. In the competitive reaction for etch and oxidization, the reaction layer including Si, F, and O was formed and the thickness was changed by O radical density and temperature. The F–O ratio in the reaction layer was kept around 2. Thicker the SiOF layer, lower the etch rate by the neutral fluorine. It was estimated that about 50-nm-thick SiOF layer is necessary to stop the etch reaction at the TSV sidewall.

## 3.3 Low-Temperature CVD Technology

**Yutaka Kusuda, Tomoyuki Nonaka, Osamu Tsuji**

### *3.3.1 Introduction*

Recently, the progress of information technology has brought with it the rise of mobile technology (smartphones, tablets, etc.). The functions and capabilities of this technology are expected to advance even further. Due to this, further miniaturization and high-density packaging of electronic components used in mobile technology is also anticipated. By using TSV technology and stacking etched Si substrates on top of each other (instead of lining them up side by side), miniaturization for three-dimensional large-scale integration (3D-LSI) packaging is possible.

The integrated circuit (IC) manufacturing process comprises three subprocesses: (1) transistor process (front end of line (FEOL)), (2) wiring process (back end of line (BEOL)), and (3) packaging. The TSV process varies depending on where it is situated with respect to these subprocesses. It can come before the transistor process (via-first process), between the transistor and wiring processes (via-middle process), or between the wiring and packaging processes (via-last process).

The advantage of via-last process is that it will not complicate device formation during the transistor and wiring subprocesses. However, the challenge for via-last processing is the necessity to deposit insulation film at low temperatures. After device formation, the Si wafer is bonded to a glass or silicon carrier using an adhesive that is heat resistant up to 150 °C. Backgrinding and chemical mechanical polishing (CMP) are used to thin the Si wafer backside and then the vias are etched through. Due to the temperature limit of the adhesive, insulation film deposition within the vias must be implemented at 150 °C or lower temperature. Because the silicon via has high AR, deposition of insulation film with high step coverage is necessary to coat the via sidewalls. SAMCO's cathode-coupled liquid source chemical vapor deposition (LS-CVD) systems enable deposition of insulation film with excellent step coverage within high-aspect vias at 150 °C or lower, and also allow for film stress control [35–38].

### *3.3.2 Cathode-Coupled Plasma-Enhanced CVD (LS-CVD)*

Cathode-coupled plasma-enhanced (PE) CVD (LS-CVD) involves the introduction of a liquid source tetraethyl orthosilicate (TEOS) + $O_2$ into the reaction chamber under vacuum through a showerhead type inlet. The reaction chamber has a parallel plate electrode configuration. Wafers are loaded onto the powered lower electrode. Plasma discharge occurs when RF power (13.56 MHz, max. 1 kW) is introduced into the lower electrode after chamber pumpdown and TEOS–$O_2$ introduction. A negative self bias is generated on the lower electrode simultaneously with plasma discharge. The LS-CVD mechanism can be seen in Fig. 3.31.

Figure 3.32 shows the relationship between power density and self bias at different pressures. The trend shows that higher RF power and lower pressure result in more negative self bias. A more negative self bias causes higher TEOS and $O_2$ dissociation as well as increased reaction between the Si and $O_2$ species in the plasma. This allows for deposition of high-density $SiO_2$ films.

Figure 3.33 shows the tendency in film stress and film density with respect to RF power density. The trend shows that higher RF power density yields higher film density and more compressive films. Compressive film stress levels off at approximately 0.8 W/cm$^2$. Figure 3.34 shows the correlation between stress and film density (the more compressive the film, the higher the film density). Because cathode-coupled CVD enables the deposition of compressive films with larger than 300 MPa stress, high-density films can be obtained.

TEOS is safer than $SiH_4$ and results in lower utilities costs. Furthermore, the precursor formed by TEOS that results in growth of film has high mobility and

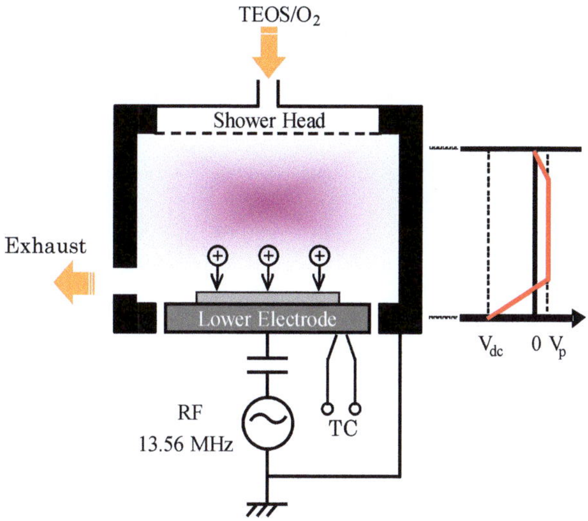


**Fig. 3.31** The image shows a cathode-coupled PECVD. RF power is introduced to the lower electrode and this generates a large DC potential at wafer surface ($V_{dc}$). TEOS–$SiO_2$ is deposited by ion bombardment

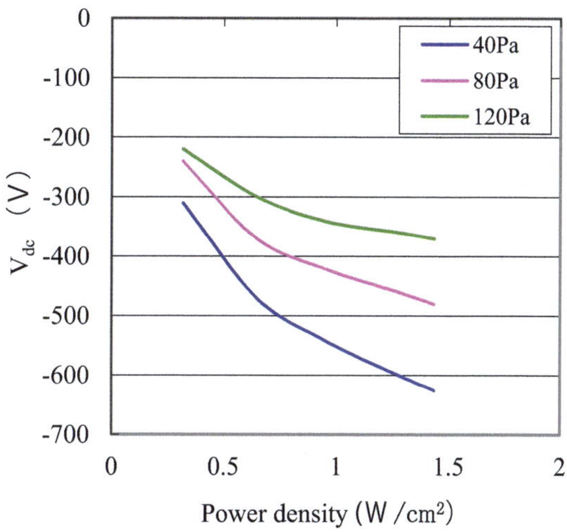


**Fig. 3.32** Power density versus $V_{dc}$ (self bias)

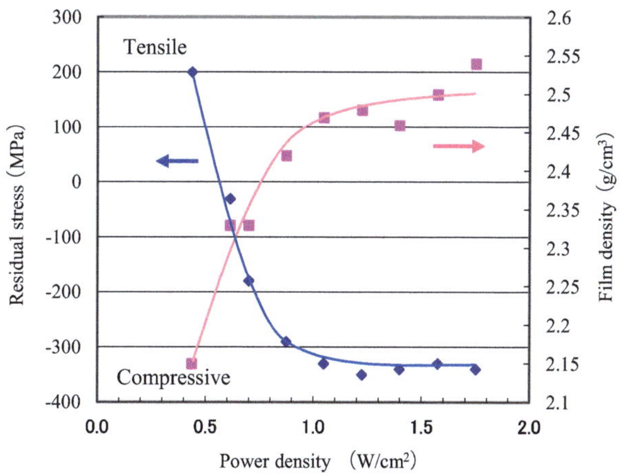


**Fig. 3.33** Power density versus residual stress

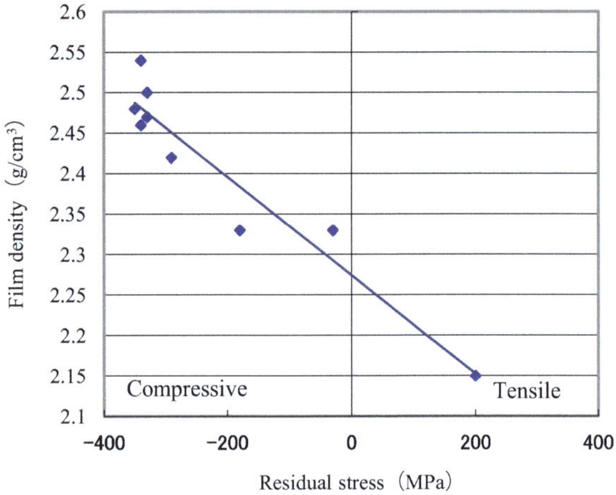


**Fig. 3.34** The correlation between residual stress and film density

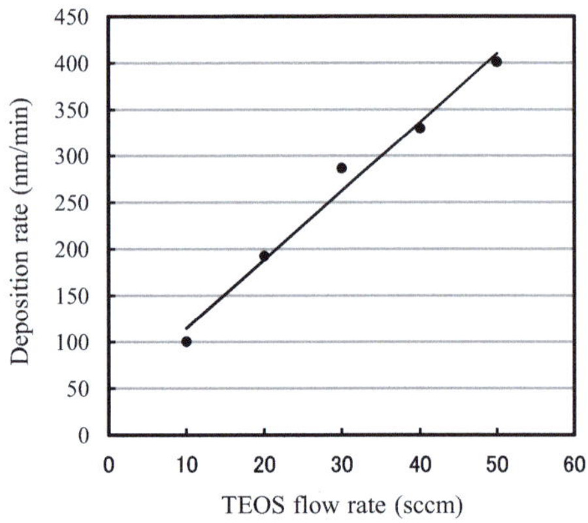


**Fig. 3.35** TEOS flow rate versus $SiO_2$ deposition rate

allows for excellent step coverage. By utilizing cathode coupling CVD technology, high-density films with excellent step coverage can be obtained.

$SiO_2$ films that resemble quartz (high-quality film with low absorbency index and low refractive index (RI)) were deposited at high temperatures (~400 °C). These films are especially useful for optical waveguide applications. Figure 3.35 shows the relationship between TEOS flow rate and deposition rate. Higher TEOS flow rates yield higher deposition rates. Normally, flow rates of 100–200 nm/min are used. Figure 3.36 shows the film thickness, thickness uniformity, RI, and RI uniformity of P, B, and Ge doped wafers. RI uniformity was ±0.01–0.02% regardless of dopant type, which can reduce the loss of light in optical waveguides. Furthermore, when P and B are doped together, reduction in anneal temperature is possible without change in RI.

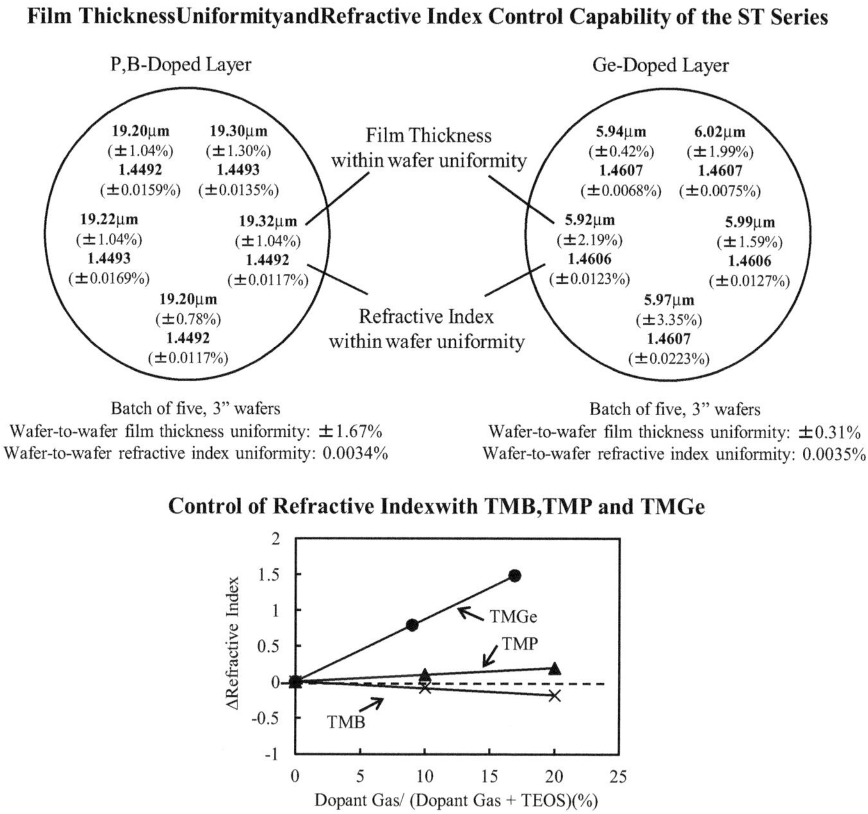


**Fig. 3.36** TEOS–$SiO_2$ films doped P, B, and Ge for optical waveguide

## *3.3.3 Low-Temperature $SiO_2$ Deposition*

### 3.3.3.1 Wafer Temperature During Low-Temperature Deposition

For the via-last process, the Si substrate is fixed to a glass or silicon carrier using an adhesive and then thinned to approximately 100 μm or thinner using backgrinding and CMP. The adhesive contains epoxy or acrylic resin that can endure temperatures up to 150 °C. Due to the temperature constraints of the adhesive, low-temperature CVD is necessary.

The ion incidence provided by cathode-coupled CVD systems enables the deposition of high-density films at low temperatures. However, ion incidence causes both lower electrode temperature and wafer temperature to rise. Wafer temperature is affected by both the lower electrode heater as well as heat from plasma discharge. Plasma heat derives from ion energy, and ion energy is correlated with power density. In order to keep wafer temperature under 150 °C, a power density of 0.8 W/

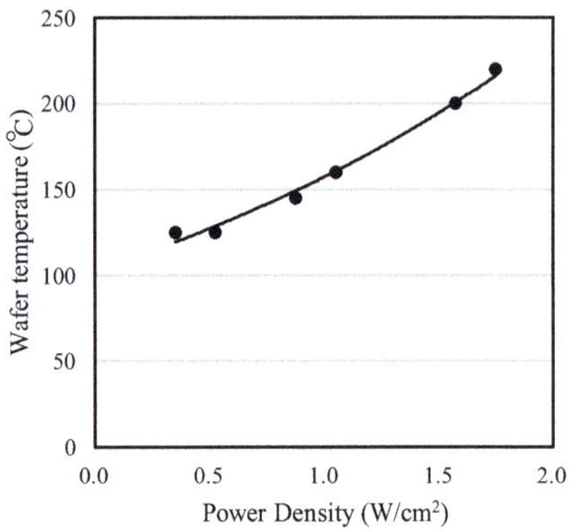


**Fig. 3.37** Power density versus wafer temperature

**Fig. 3.38** An example of 1-μm-thick deposition of $SiO_2$ on a PET film at 150 °C

cm² or lower is required. Figure 3.37 shows the relationship between power density and wafer temperature. Substrate temperature was set at 80 °C. Figure 3.38 shows low-temperature deposition (150 °C) of a 1-μm-thick $SiO_2$ film over a polyethylene terephthalate (PET) film that has low heat resistance. Although there was a bow in the $SiO_2$ film, there were no cracks observed in the film.

### 3.3.3.2 Step Coverage in Si-Via Holes

Figure 3.39 shows the difference in step coverage between TEOS- and $SiH_4$-based $SiO_2$ insulation film deposition in vias with 10:1 AR. Wafer temperature in both TEOS and $SiH_4$ processes was constant at 150 °C. As seen in the SEM micrographs, the TEOS-based process resulted in a uniform, crack-free film that carpeted the entire via profile (opening, sidewalls, and bottom). In contrast, the $SiH_4$-based process resulted in uniform deposition in the profile opening, but deposition on the profile bottom was hardly observed. With these results, it is clear that TEOS is superior to $SiH_4$ in terms of step coverage.

In the via-last process, $SiO_2$ film deposited within the via hole (upper and lower sidewalls, bottom) is not as thick as that on top of the via hole. Figure 3.40 shows

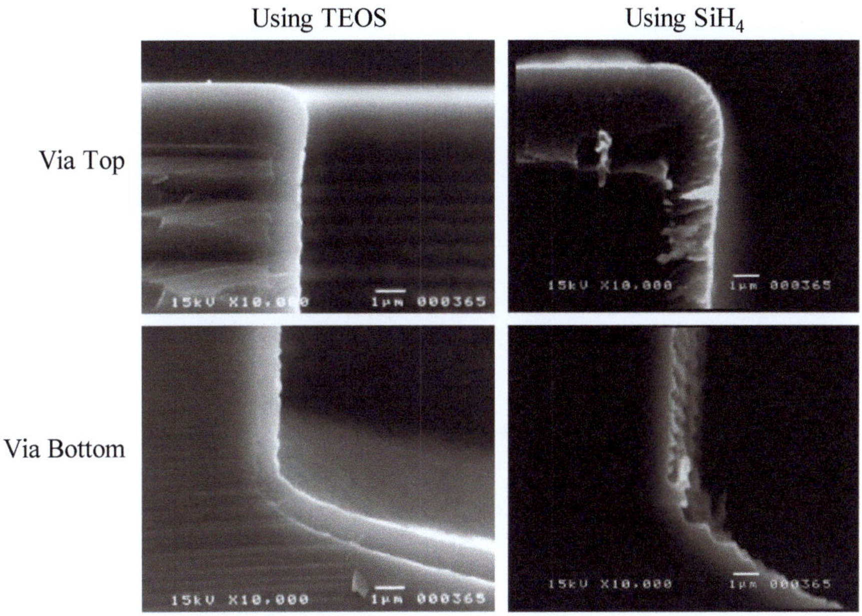


**Fig. 3.39** Comparison between TEOS and $SiH_4$ step coverage. Normally, it is difficult for the insulation film to embed the bottom of the hole, but using TEOS-based cathode-coupled PECVD, the insulation film easily embeds bottom of the hole through migration

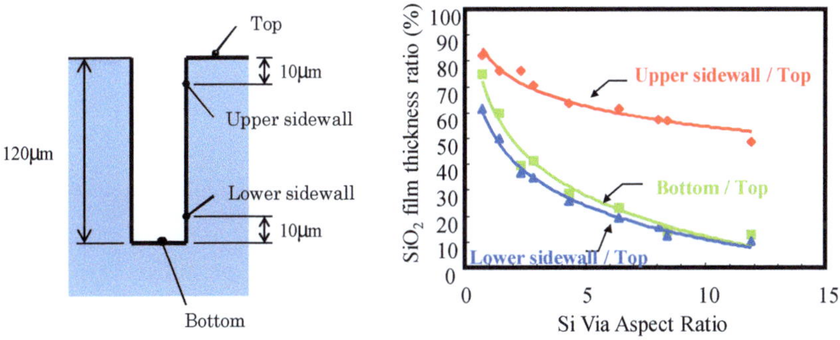


**Fig. 3.40** The relationship between step coverage and aspect ratio

the thickness ratio of $SiO_2$ film between the upper sidewall, lower sidewall, and bottom, with the top of the via. As AR increases, the thickness ratio between the upper sidewall, lower sidewall, and bottom, with the top of the via decreases. Even with an AR of 10:1, a thickness ratio of 10 % is achieved.

Normally, the Bosch process [39] is used for Si-via fabrication [40]. However, etching with the Bosch process results in a via-hole profile with the following characteristics:

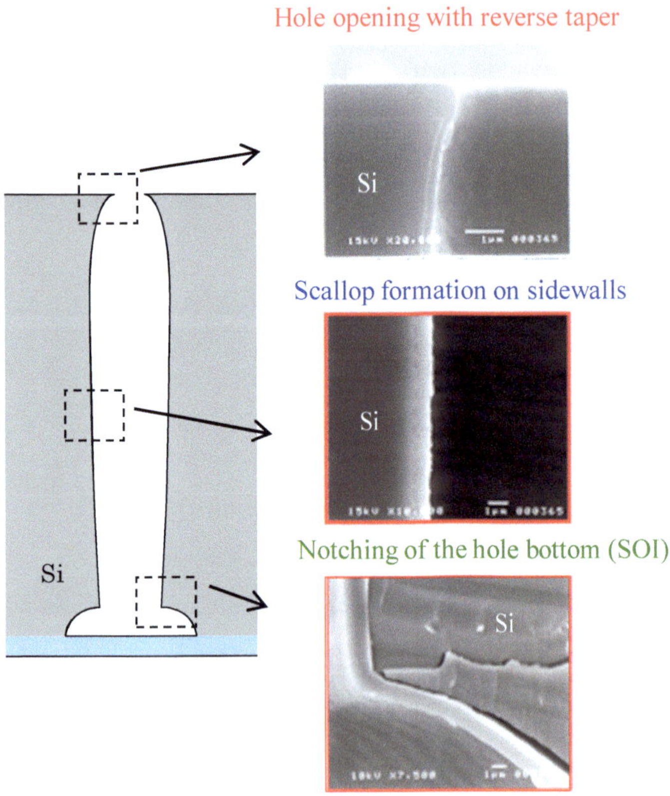


**Fig. 3.41** Step coverage of Si-via hole with a challenging profile

1. Via opening with reverse taper
2. Scallop formation on sidewalls
3. Notching on the via bottom (for via-last process and silicon on insulator (SOI) applications)

These profile characteristics make it challenging to achieve full coverage in the holes during $SiO_2$ insulation film deposition. However, as seen in Fig. 3.41, step coverage of TEOS-based $SiO_2$ film was allowed to cover all surfaces of a Bosch-type via, in spite of the challenges presented by the via profile.

#### 3.3.3.3 Electrical Characteristics of $SiO_2$ Film Deposited at Low Temperature

We measured breakdown voltage of a comb drive unit with 0.1-μm-thick sidewall $SiO_2$. The wafer temperature during the CVD process was 150 °C. The test resulted in a breakdown voltage of 76 V (7.6 MV/cm). The result proved the effectiveness in terms of electrical characteristics of $SiO_2$ film deposited by cathode-coupled PECVD. Figure 3.42 shows the measurement of the leakage current in a quasi via. The via was 30 μm wide and 30 μm deep. Leakage current density was $10^{-5}$ A/cm$^2$.

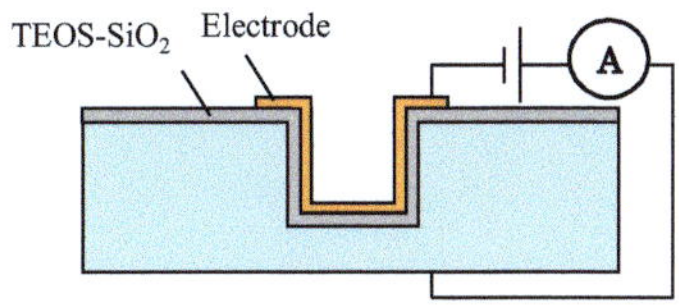


**Fig. 3.42** A test sample of Si-via hole picture for measurement of leakage current

#### 3.3.3.4 Stress Control of $SiO_2$ Film Deposited Using LS-CVD

High compressive stress is preferred to obtain high-density films. However, when film thickness exceeds 3 μm, a bow in the wafer is usually observed. There are cases where this bow can be an issue for other steps in the semiconductor process, such as photolithography. Figure 3.43 shows the change in film stress with respect to power density as well as the 24-h time-lapse effect on film stress. RF power density can be used to control film stress, as seen in Fig. 3.43. However, after 24 h in an atmospheric environment, films that initially had stress $>-70$ MPa became compressive films. Stress for these films tended to shift to approximately $-70$ MPa regardless of initial stress level. In other words, regardless of the initial stress target, it was not possible to obtain completely stress-free $SiO_2$ films.

Tensile and compressive multilayer films have behaved differently from single layer films. Compressive films with "A" thickness served as barrier films and were deposited on top of tensile films with "B" thickness. Figure 3.44 shows the relationship between thickness ratio of tensile ($t_1$) and compressive ($t_2$) films and film stress. Stress could be controlled within the range of $-170$ to 200 MPa, and films that were essentially stress-free were obtained simply by adjusting the tensile-compressive film thickness ratio. Furthermore, minimal shift in stress was observed after 24 h.

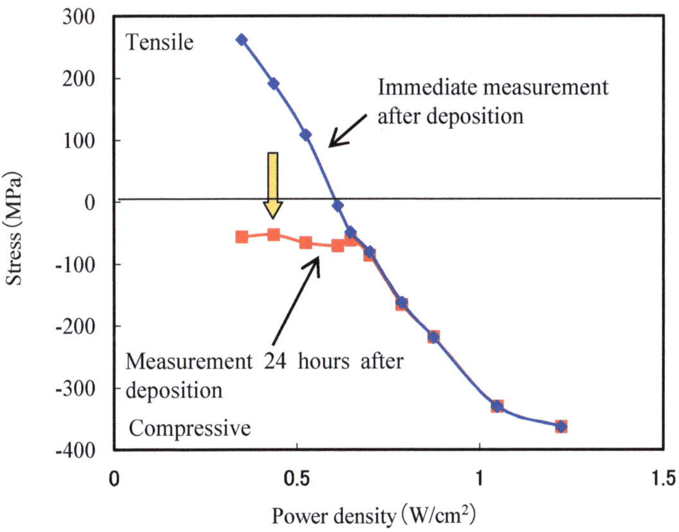


**Fig. 3.43** Film stress shift in 24 h (immediate measurement after deposition and 24 h after deposition.)

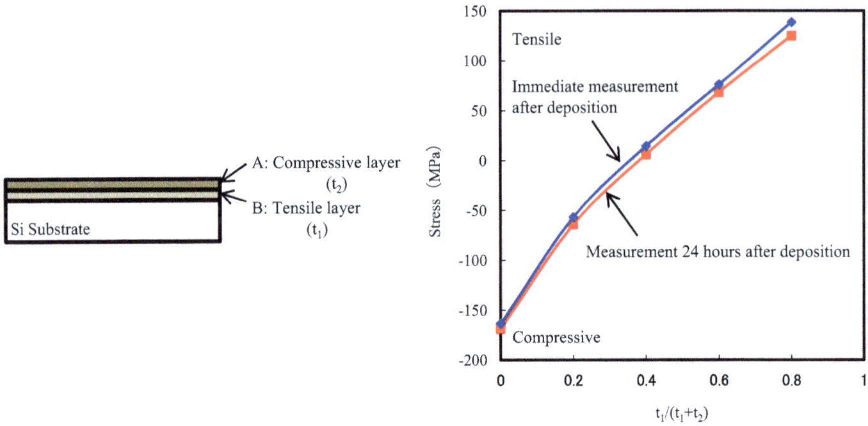


**Fig. 3.44** Dependence of film stress changing the film thickness for tensile and compressive layer

### *3.3.4 Conclusion*

This chapter has detailed the mechanism behind our cathode-coupled LS-CVD technology for low-temperature $SiO_2$ deposition as well as its advantages and applications (such as MEMS, and etc.). In the near future, LS-CVD is expected to further impact the miniaturization and functionalization of mobile technology by enhancing the high-density packaging capability of electronic components.

In addition to the PD–270STL, SAMCO also offers a cassette-to-cassette type LS-CVD system (PD–270STLC) for mass production applications. LS-CVD technology will likely contribute to the progress of 3D-LSI and MEMS applications in the near future.

## 3.4 Electrodeposition for Via Filling

**Kazuo Kondo**

### *3.4.1 $Cu^+$ Ion as an Accelerant Additive of Copper Electrodeposition*

Copper has lower resistivity than aluminum and can be electrodeposited easily. Hence, copper can be used for the chip wiring and TSV. Copper electrodeposition to fill the via is an indispensable process for TSV. Additives are necessary to fill the deep via.

The role of the additive is classified as an inhibition effect outside the trenches or vias and an acceleration effect inside the trenches and vias. The inhibition effect consists of the combination of PEG (polyethylene glycol) and $Cl^-$ [41]. The acceleration effect consists of bis (3-sulfopropyl) disulfide (SPS). More recent studies have focused on this acceleration effect, and the influence of SPS has been discussed. Moffat and West have proposed the absorption of an accelerator, and a curvature-enhanced mathematical model has been proposed which assumes adsorption of an accelerator at the curvature in the via bottoms [42, 43].

On the other hand, the formation of $Cu^+$ is a crucial intermediate step and the $Cu^+$ is always produced during the electrodeposition process.

$$Cu^{2+} + e^- \underset{k_{-1}}{\overset{k_1}{\rightleftarrows}} Cu^+ \tag{3.1}$$

$$Cu^+ + e^- \underset{k_{-2}}{\overset{k_2}{\rightleftarrows}} Cu \tag{3.2}$$

These reactions are reversible processes. The reaction constant $k_1$ for the $Cu^{++}$ to $Cu^+$ is $2 \times 10^{-4}\,\text{mol m}^{-2}\text{s}^{-1}$ and $k_{-1}$ for the $Cu^+$ to $Cu^{++}$ is $8 \times 10^{-3}\,\text{mol m}^{-2}\text{s}^{-1}$. $k_2$ for the $Cu^+$ to metallic copper is 130 mol $m^{-2}\,s^{-1}$ and $k_{-2}$ for the metallic copper to $Cu^+$ is $3.9 \times 10^{-7}\,\text{mol m}^{-2}\text{s}^{-1}$. Large value of $k_2$ for 130 mol $m^{-2}\,s^{-1}$ means that once the $Cu^+$ is formed, the reaction to reduce it to metallic copper is extremely rapid [44]. If $k_1$ becomes a large value as equivalent to $k_2$ with an additive, the reduction of $Cu^{++}$ to metallic copper will be rapid and the additive and $Cu^+$ are the accelerators. Furthermore, $Cu^+$ is transparent and odorless. Hence, it is extremely difficult to detect $Cu^+$ and accordingly a few studies exist which are related to the acceleration effect of $Cu^+$.

K. Kondo, the author, has used a trench bottom electrode to measure the acceleration effect. I have added SPS, PEG, and $Cl^-$ ions in addition to the basic bath of copper sulfate and sulfuric acid. The potential at the trench bottom electrodes was swept in the negative direction and the narrower the trench bottom electrode, the greater is the current [45]. Kondo considered that the acceleration effect with the narrower electrode is due to the free accelerant of the intermediate with $Cu^+$ and SPS, however, further studies are necessary for the detailed discussion about the role of $Cu^+$ by forming the $Cu^+$ within these trenches.

A rotating ring-disk electrode (RRDE) has been used to detect the $Cu^+$. The RRDE has two electrodes, disk and ring electrodes (Fig. 3.45). As shown in Eqs. (3.1) and (3.2), $Cu^+$ always forms during the process of electrodeposition. So, if we electrodeposit on the disk electrode, the $Cu^+$ intermediate forms above this disk electrode. With spinning the RRDE, this intermediate is thrown out with the centrifugal force. The intermediate always passes across the ring electrode. If the ring potential is set equal to the positive potential such as +300 mV, the oxidation reaction of intermediate occurs. Then all you have to do is count the number of electrons, current, to detect the amount of $Cu^+$ formation.

$$Cu^+ \rightarrow e^- + Cu^{2+} \tag{3.3}$$

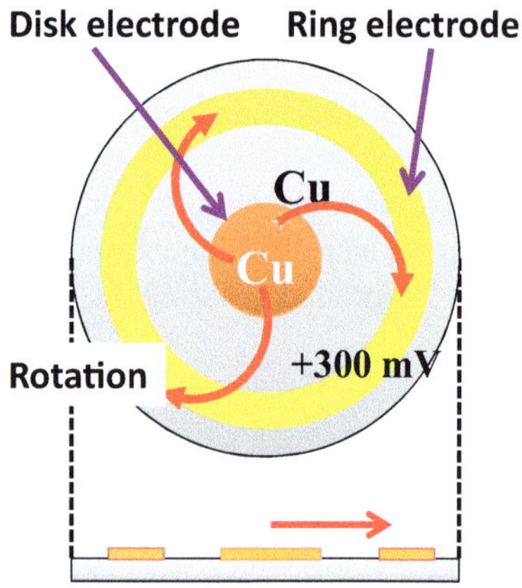


$$Cu^{+} \rightarrow e^{-} + Cu^{2+}$$

**Fig. 3.45** Schematic illustration of rotation-ring disk electrode and reaction occurring on the ring electrode with the positive ring potential

Figure 3.46 shows the relation between this ring current and disk overpotential. If the disk overpotential is negative (electrodeposit copper on the disk), the ring current shows a very small value of 0.1 μA. If the disk overpotential is positive (dissolution of copper), there is a drastic increase in the ring current of 100 μA. The $Cu^{+}$ detected at the ring of the RRDE is 1000 times higher for the copper dissolution, if it is compared to the $Cu^{+}$ for the electrodeposition [46]. Hence, we formed a large amount of $Cu^{+}$ within the confined area of the trench in the trench bottom electrodes and experimentally verified the relation between the $Cu^{+}$ and acceleration.

The effect of $O_2$ bubbling and $N_2$ bubbling has been initially tested in the non-stirred bath. The bath consists of the basic bath and 1 ppm SPS, 400 ppm PEG, and

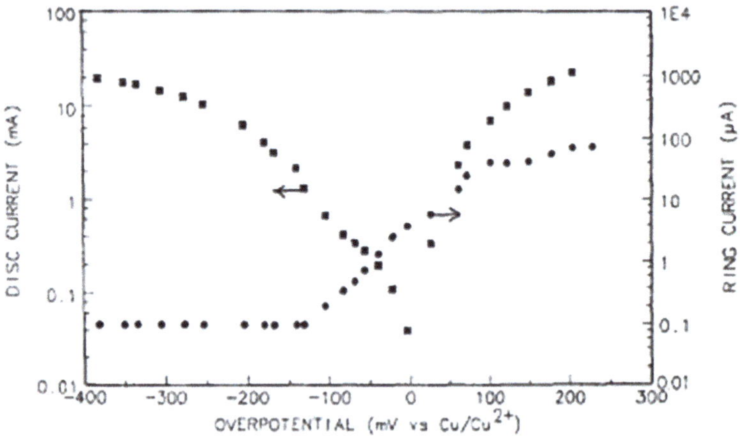


**Fig. 3.46** Relation between ring current and overpotential by using the rotation disk electrode

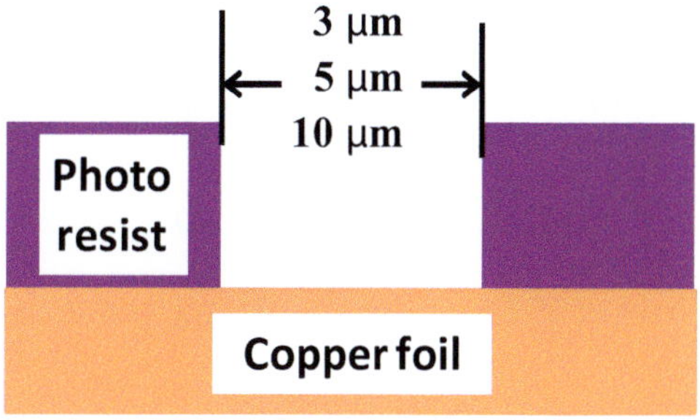


**Fig. 3.47** Schematic illustration of trench bottom electrode

50 ppm $Cl^-$. The trench bottom electrode of widths of 3, 5, and 10 μm is illustrated in Fig. 3.47. Prior to the linear sweep voltammetry (LSV), the $Cu^+$ was formed by dissolving the copper electrode at the trench bottom for 12 s at 10 mA/cm². Figure 3.48 shows the LSV result with $N_2$ bubbling. The potential and current density is illustrated for the trench bottom electrode widths of 3, 5, and 10 μm. Drastic increases in the current densities due to the acceleration effect are observed and the current density increased up to −40 mA/cm² for the 3-μm trench bottom electrode width at −0.05 V versus saturated calomel electrode (SCE).

Furthermore, the current densities increase with the narrower trench bottom electrode width. On the contrary, the currents show a very low value of −1.0 mA/cm² for the LSV result with $O_2$ bubbling (Fig. 3.49). The current densities are same with the trench bottom electrode widths of 3, 5, and 10 μm. The drastic difference in the current densities with the $O_2$ gas concentration in the electrolyte must be caused by the $Cu^+$ intermediate formed in the trench of the trench bottom electrode. The increase in the current densities for the narrower trench of the trench bottom electrode must be caused by the accumulation of a $Cu^+$ intermediate in the trench.

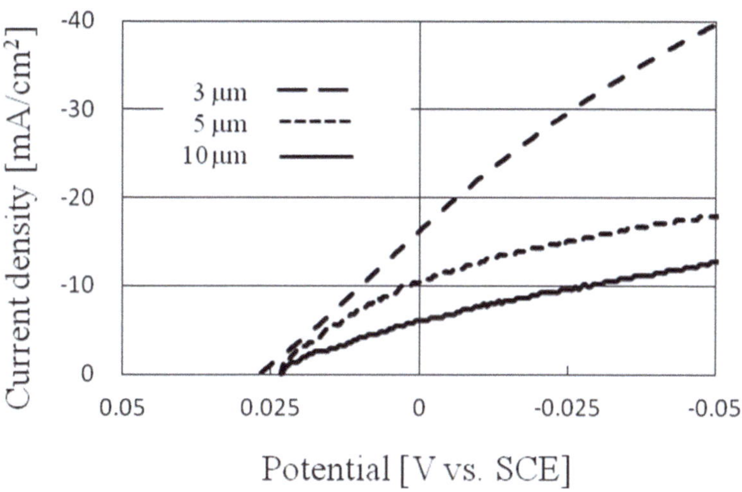


**Fig. 3.48** Result of LSV measurement with $N_2$ bubbling (SPS: 1 ppm, $Cl^-$: 50 ppm, PEG: 400 ppm)

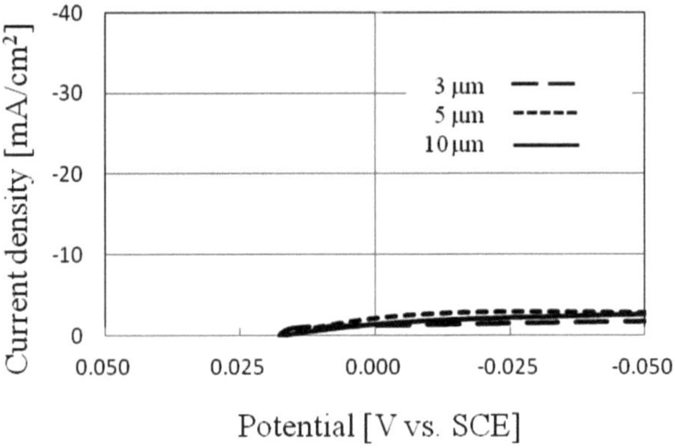


**Fig. 3.49** Result of LSV measurements with $O_2$ bubbling (SPS: 1 ppm, $Cl^-$: 50 ppm, PEG: 400 ppm)

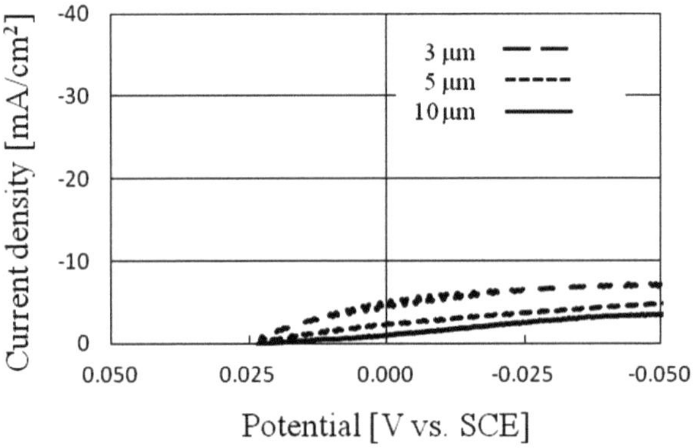


**Fig. 3.50** Result of LSV measurements with $N_2$ (stirring rate of stirrer is 600 rpm, SPS: 1 ppm, $Cl^-$: 50 ppm, PEG: 400 ppm)

In order to prove that this $Cu^+$ intermediate is free, floating in the electrolyte, and not adsorbing on the electrode, forced convection has been applied by the stirrer at 600 rpm. The current densities have been then measured by LSV by initially dissolving the copper electrode at the trench bottom. The LSV measurements are shown in Fig. 3.50. The current densities decrease to a few mA/cm$^2$ and there is almost no difference in the current densities for the trench widths of 3, 5, and 10 μm. The acceleration effect has deceased. This is because the free $Cu^+$ intermediate flows out of the trenches due to the stirring at 600 rpm.

Next, in order to investigate the $Cu^+$ and the additives, we have eliminated the additives one by one and the LSVs have been measured. We then have eliminated SPS and PEG and only the $Cl^-$ has been added. The current shows a marked increase for −70 mA/cm$^2$ for 3 μm at −0.05 V versus SCE (Fig. 3.51). The narrower the trench, the more the increase in current densities. $Cl^-$ is an important additive for the acceleration, and the acceleration must be related to the free $Cu^+$ intermediate and electron bridge formation of $Cl^-$ [47]; however, these details are currently under investigation.

Without dissolving the copper forming at the trench bottom electrodes, the current densities have been measured at the constant potential of −0.15 V versus SCE. This

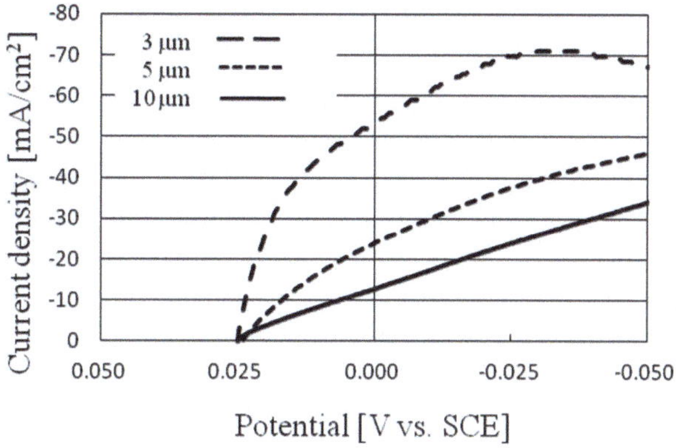


**Fig. 3.51** Result of LSV measurements with $N_2$ bubbling ($Cl^-$: 50 ppm)

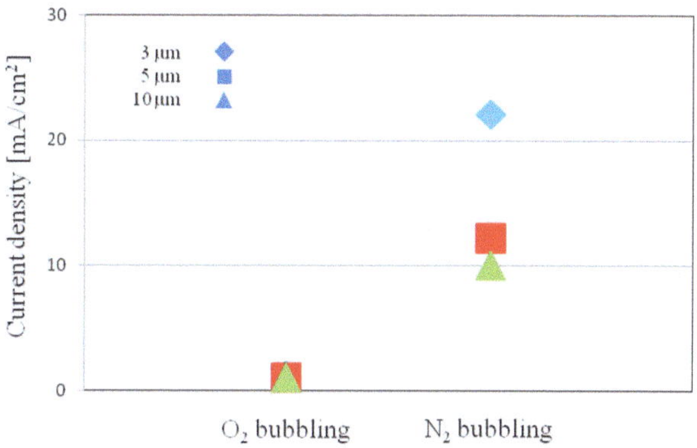


**Fig. 3.52** Result of constant potential measurements with $N_2$ bubbling and $O_2$ bubbling. Potential at −0.15 mV (SPS: 1 ppm, $Cl^-$: 50 ppm, PEG: 400 ppm)

is because we want to prove that the $Cu^+$ forming through the electrodeposition process is an accelerator. The $O_2$ and $N_2$ bubbling results are shown in Fig. 3.52. The difference in the $O_2$ and $N_2$ bubbling is the *x*-axis and the current densities are the *y*-axis. With $O_2$ bubbling, the current densities are low values of about 1.0 mA/cm$^2$. However, the current densities markedly increase with the $N_2$ bubbling. With the narrower trenches, the current densities increase and −23 mA/cm$^2$ has been measured for the 3-µm trench width. These constant potential measurements without dissolving the copper electrode also show that the free $Cu^+$ intermediate formed through electrodeposition is the accelerant [48].

### *3.4.2 Relation Between Via Filling and $Cu^+$ Ion by Periodical Reverse Current Waveform*

We have already reported that a 10-µm-diameter via with an AR of 7.0 is perfectly filled within 37 min by using diallylamine leveler and oxygen gas bubbling [49–52]. By using a number of additives, such as chloride, SPS as accelerator PEG

**Table 3.1** Bath composition

| *Basic bath composition* | |
|---|---|
| $CuSO_4 \cdot 5H_2O$ | 200 g/L |
| $H_2SO_4$ | 25 g/L |
| *Additives* | |
| $Cl^-$ | 70 ppm |
| SPS | 2 ppm |
| PEG (M.W. 10,000) | 25 ppm |
| SDDACC | 1.5 ppm |

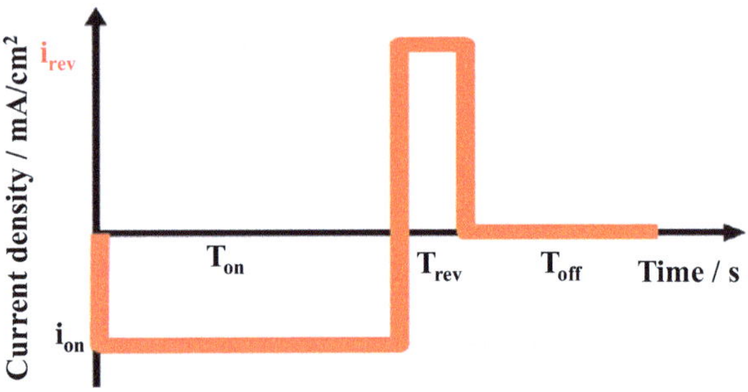


**Fig. 3.53** The schematic illustration of periodic reverse pulse current waveform

as inhibiter, and quaternary diallylamine as leveler, the electrodeposition time has been shortened from 60 to 35 min [53].

In this study, we studied smaller via. Via diameter is 4 μm and AR is 7.5. The same bath composition that was used for a 10-μm-diameter via filling in Table 3.1 [53] has been adopted for the 4-μm-diameter via filling. Furthermore, by the RRDE, we have evaluated produced $Cu^+$ ion concentration during reverse pulse current waveform (Fig. 3.53, $T_{rev}, i_{rev}$) of copper electrodeposition. This reverse current dissolves copper and hence forms $Cu^+$ inside the via.

Figure 3.54 shows via cross sections of electrodeposits for 20 min with changing $i_{rev}/|i_{on}|$ ratios. Figure 3.54a is $i_{rev}/|i_{on}| = 0$, Fig. 3.54b is $i_{rev}/|i_{on}| = 2.0$ and Fig. 3.54c is $i_{rev}/|i_{on}| = 6.0$. In Fig. 3.54a and b, the voids exist at via top or via bottom (the arrow points of Fig. 3.54a and b). However, in Fig. 3.54c, the void does not exist.

Next, we have measured the ring current by changing the periodic reverse pulse current waveform. The result is shown in Fig. 3.55. The $x$-axis is the time and the $y$-axis is the ring current. In this measurement, $i_{rev}/|i_{on}|$ ratios have been changed to 0, 2.0, 4.0, and 6.0. Figure 3.55a is $i_{rev}/|i_{on}| = 0$, Fig. 3.55b is $i_{rev}/|i_{on}| = 2.0$, Fig. 3.55c is $i_{rev}/|i_{on}| = 4.0$, and Fig. 3.55d is $i_{rev}/|i_{on}| = 6.0$. From Fig. 3.55, the peak ring current at the on time is about 0 μA for $i_{rev}/|i_{on}| = 0$, the peak ring current at the on time is about 40 μA for $i_{rev}/|i_{on}| = 2.0$, the peak ring current at the on time is about 150 μA for $i_{rev}/|i_{on}| = 4.0$, and the peak ring current at the on time is about 230 μA for $i_{rev}/|i_{on}| = 6.0$. Ring current increases with the increasing $i_{rev}/|i_{on}|$ ratios. This means that produced $Cu^+$ ion concentration on the disk electrode is markedly increasing with the increasing periodic reverse pulse current waveform

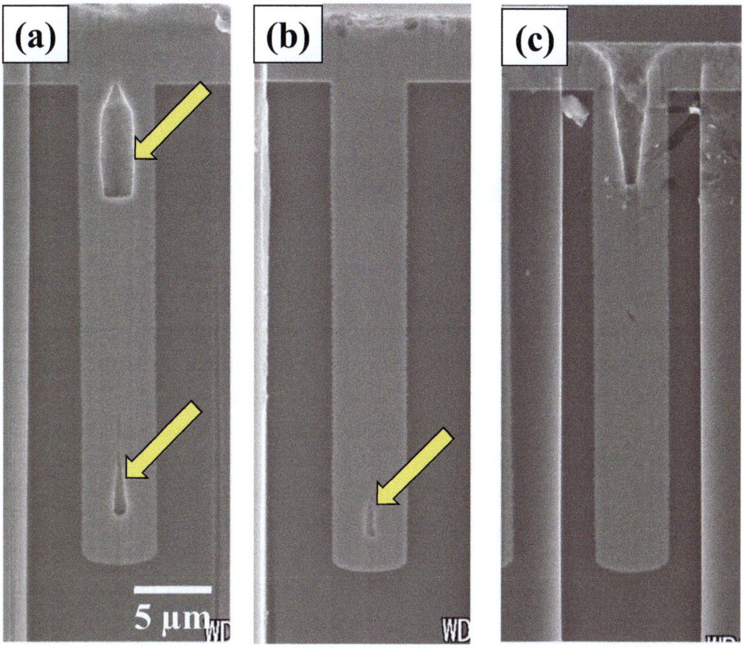


**Fig. 3.54** The via cross sections of electrodeposits for 20 min with changing $i_{rev}/|i_{on}|$ ratios. **a** $i_{rev}/|i_{on}| = 0$. **b** $i_{rev}/|i_{on}| = 2.0$. **c** $i_{rev}/|i_{on}| = 6.0$

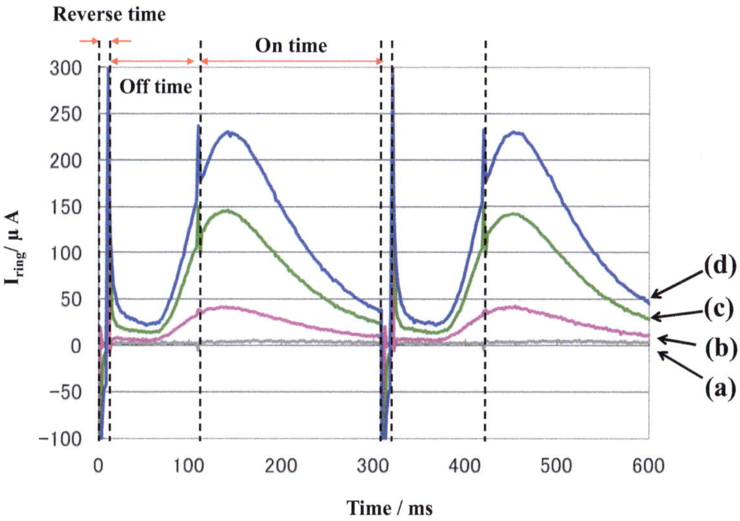


**Fig. 3.55** Comparison of the ring current by various $i_{rev}/|i_{on}|$ ratios using RRDE. **a** $i_{rev}/|i_{on}| = 0$. **b** $i_{rev}/|i_{on}| = 2.0$. **c** $i_{rev}/|i_{on}| = 4.0$. **d** $i_{rev}/|i_{on}| = 6.0$

$i_{rev}/|i_{on}|$ ratios. Hence, we reasonably conclude that $Cu^+$ ion concentration on the disk electrode is markedly increasing during reverse current of periodic reverse pulse current waveform.

### 3.4.3 Simulation of $Cu^+$ Ion Distribution Inside the Via

R. Akolkar analyzed the mass transportation of $Cu^{++}$ ion within the via by adapting the 1D ionic diffusion and electrode kinetics [54]. We have confirmed the increase in ring current, that is $Cu^+$ ion concentration, by increasing the reverse current which has been measured by the RRDE (Fig. 3.55). However, the direct measurement of $Cu^+$ ion concentration profile within the via is impossible. We have adopted the simulation model of $Cu^+$ ion 1D diffusion and electrode kinetics [54]. These electrode kinetics parameters ($k_m$) have been determined by comparing the current distribution both from experiment and simulation model. The $Cu^+$ ion concentration profiles in the via have been calculated by using these $k_m$.

We have calculated the via inside $Cu^+$ ionic concentration distribution by a 1D diffusion model for depth direction. This is because radial concentration gradient is ignorable when it is smaller than the depth concentration gradient. We also assumed that $Cu^+$ ionic transportation inside the via is mainly transported with the ionic diffusion. Figure 3.56 shows the mass balance of $Cu^{++}$ ion and $Cu^+$

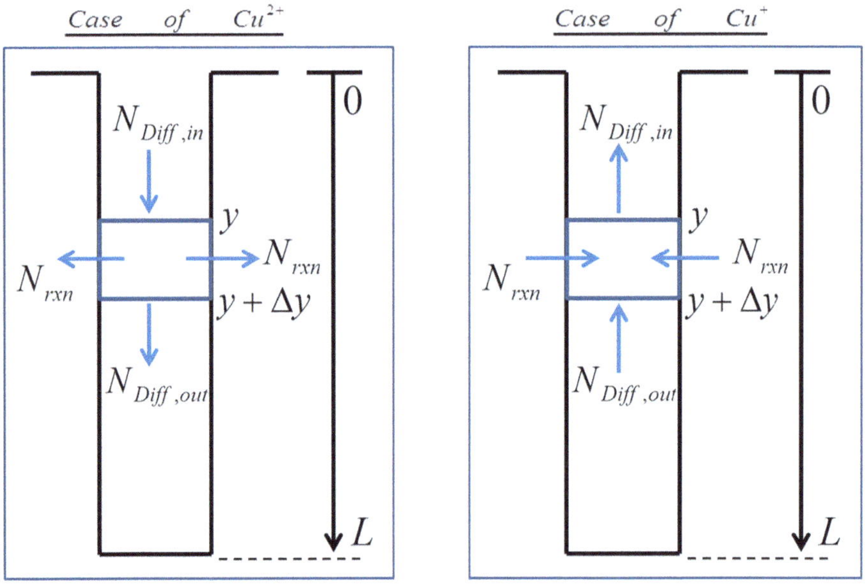


**Fig. 3.56** Schematic representation of $Cu^{++}$ ion and $Cu^+$ ion diffusion at via inside. $N_{Diff}$ and $N_{rxn}$ are the diffusion and reaction fluxes

ion. The ionic diffusion flux in the depth direction, according to Fick's law, is expressed as:

$$N_d = -D_i \frac{\mathrm{d}C_i}{\mathrm{d}y}, \tag{3.4}$$

where $D_i$ is the ionic diffusion coefficient, $C_i$ is its concentration, and $y$ is the depth coordinate. The current distribution on the 1D TSV sidewall due to Cu electrodeposition of a current distribution $i$ is expressed by the Butler–Volmer equation as (Eq. 3.5):

$$N_{rxn} = \frac{i}{nF} = \frac{i_0}{nF}\left(\frac{C_i}{C_{i,bulk}}\right)e^{b\eta}, \tag{3.5}$$

$C_{i,bulk}$ is the bulk ion concentration, $i_0$ is the exchange current density, $b$ is the Tafel slope, and $\eta$ is the overpotential. Using the reaction kinetic parameters, we can also expand on Eq. 3.3 as Eq. 3.6.

$$N_{rxn} = k_m C_i, \tag{3.6}$$

where $k_m = i_0 e^{b\eta/nFC_{i,bulk}}$ is the first-order rate constant. Furthermore, the $i$ th ionic time-dependent diffusion equation is expressed as Eq. 3.7.

$$D_i \frac{\mathrm{d}^2 C_i}{\mathrm{d}y^2} \pm \frac{i}{n\mathrm{F}}\frac{2}{r} = \frac{\mathrm{d}C_i}{\mathrm{d}t}, \tag{3.7}$$

where $n$ is the number of electrons transferred, $F$ is the Faraday constant, and $r$ is the via radius. The boundary conditions are shown as follows:

$$\text{At } y = 0, C_{\mathrm{Cu}^{2+}} = C_{\mathrm{Cu}^{2+},bulk}, C_{\mathrm{Cu}^{+}} = C_{\mathrm{Cu}^{+},bulk}, C_{\mathrm{O}_2} = C_{\mathrm{O}_2,bulk} \tag{3.8}$$

$$\text{At } y = L, \frac{\mathrm{d}C_{\mathrm{Cu}^{2+}}}{\mathrm{d}y} = 0, \frac{\mathrm{d}C_{\mathrm{Cu}^{+}}}{\mathrm{d}y} = 0, \frac{\mathrm{d}C_{\mathrm{O}_2}}{\mathrm{d}y} = 0, \tag{3.9}$$

where $L$ is the via depth. The bulk $Cu^+$ ion concentration is calculated by the bulk $Cu^{++}$ ion concentration from Eq. 3.10. If the bulk $Cu^{++}$ ion concentration is 0.8 M, the bulk $Cu^+$ ion concentration becomes $4.81 \times 10{-}^4$ M. The maximum bulk $O_2$ concentration is 20 ppm.

$$K_{\mathrm{Cu}} = C_{\mathrm{Cu}^{+}}{}^2 / C_{\mathrm{Cu}^{2+}} = 5.8 \times 10^{-7} \tag{3.10}$$

Figure 3.57 shows the calculated $Cu^+$ ionic concentration distributions inside the via produced with $i_{\mathrm{rev}} / |i_{\mathrm{on}}| = 6.0$ of the periodic reverse pulse current waveform. Figure 3.57a shows the result without $O_2$ in the bulk electrolyte and Fig. 3.57b

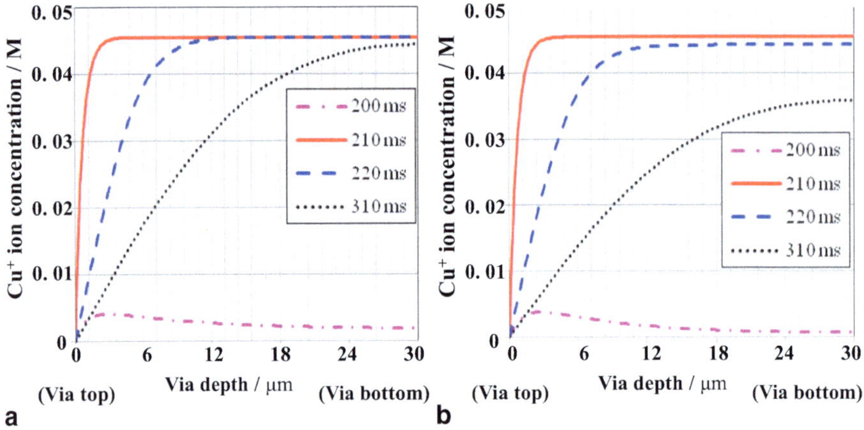


**Fig. 3.57** The $Cu^+$ ion concentration distribution during electrodeposition (at $t=200$ ms), dissolution (at $t=210$ ms), and off-time (from $t=220$ to 310 ms) with $i_{rev}/|i_{on}| = 6.0$. **a** With nitrogen gas purging. **b** With oxygen gas purging

shows the result with $O_2$. The $x$-axis shows the via depth and the $y$-axis shows the $Cu^+$ ion concentration inside the via and zero $x$-axis is the via top. In these graphs, the result from $t=0$–200 ms represents the $Cu^+$ ionic concentration distribution during electrodeposition, the result from $t=200$–210 ms represents that during dissolution, and the result from $t=220$–310 ms represents that during off-time. First, from Fig. 3.57a, the $Cu^+$ ion concentration is several $10^{-3}$ M inside the via at the end of electrodeposition ($t=200$ ms) without $O_2$ in the bulk electrolyte, however, which is twice as much as in the bulk $Cu^+$ ion concentration, this concentration is very low as compared to the other phases of the periodic reverse pulse current waveform, such as the dissolution or the off-time. Second, the $Cu^+$ ion concentration markedly increases at the end of dissolution ($t=210$ ms). The $Cu^+$ ion concentration increases to about 0.045 M which is about 450 times higher than that at the end of electrodeposition ($t=200$ ms). This means that the $Cu^+$ ions form drastically during dissolution with the reverse current of the periodic reverse pulse current waveform. Moreover, the $Cu^+$ ion concentration is constant from the via middle to the via bottom.

Third, the $Cu^+$ ion concentration is significantly decreased near the via top during off-time from $t=220$ ms to 310 ms. For example, the $Cu^+$ ion concentration at 6 μm depth decreased about 60% from 0.045 M at the end of dissolution ($t=210$ ms) to 0.018 M at the end of off-time ($t=310$ ms). This is due to the produced $Cu^+$ ion diffusing in the bulk electrolyte from the via top because of the bulk $Cu^+$ ion concentration being lower than that inside the via during off-time. From these results, we conclude that the $Cu^+$ ions are produced during dissolution. The $Cu^+$ ion concentration gradient occurs from higher concentration at the via bottom to lower concentration at the via top due to the produced $Cu^+$ ion diffusion in the bulk electrolyte from the via top during off-time. The via-bottom electrodeposits are thick with $i_{rev}/|i_{on}| = 6.0$ if compared with the electrodeposits with $i_{rev}/|i_{on}| = 0$. This is

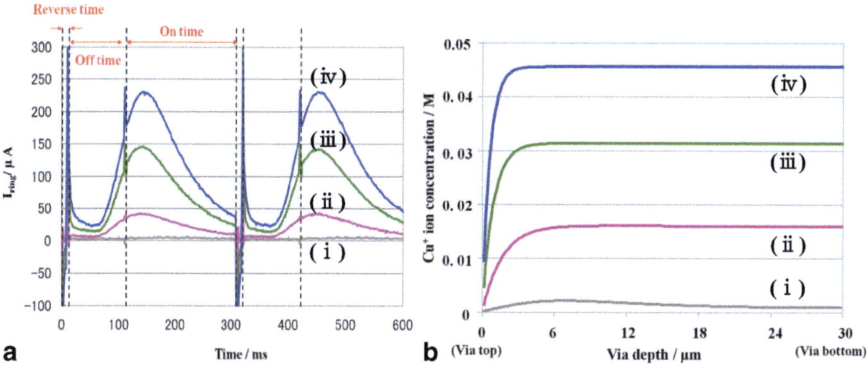


**Fig. 3.58** Comparison of $Cu^+$ ion concentration with various $i_{rev}/|i_{on}|$ during dissolution. **a** The electrochemical measurement results, *x*-axis is the time and *y*-axis is the ring current [6]. **b** The calculated results during dissolution, *x*-axis is the via depth and *y*-axis is the $Cu^+$ ion concentration. (i) $i_{rev}/|i_{on}| = 0$, (ii) $i_{rev}/|i_{on}| = 2.0$, (iii) $i_{rev}/|i_{on}| = 4.0$, and (iv) $i_{rev}/|i_{on}| = 6.0$

due to the high $Cu^+$ ion concentration inside the via if compared with low $Cu^+$ ion concentration with $i_{rev}/|i_{on}| = 0$.

Additionally, with $O_2$ in the bulk electrolyte, the $Cu^+$ ionic concentration distributions at electrodeposition and dissolution are similar to that without $O_2$ in the bulk electrolyte ($t$=200 and 210 ms in Fig. 3.57a and b). However, the $Cu^+$ ion concentrations with $O_2$ in the bulk electrolyte during off-time decrease whole via inside as compared to that without $O_2$ in the bulk electrolyte ($t$=220 and 310 ms in Fig. 3.57a and b). At the end of off-time, the via bottom $Cu^+$ ion concentration with $O_2$ in the bulk electrolyte becomes 0.038 M which is about 20% lower than that without $O_2$ in the bulk electrolyte ($t$=310 ms in Fig. 3.57a and b). This means that the dissolved $O_2$ has diffused inside the via and oxidized the produced $Cu^+$ ion to the $Cu^{++}$ ion.

Lastly, we have calculated the $Cu^+$ ionic concentration distributions inside the via by changing $i_{rev}/|i_{on}|$ of the periodic reverse pulse current waveforms as 0, 2.0, 4.0, and 6.0. Figure 3.58a shows the result of the electrochemical measurements by using RRDE that is reported in our previous study [53, 55]. The *x*-axis is the measuring time and the *y*-axis is the ring current. The *x*-axis is equal to zero in the via top. These ring currents correspond to the produced $Cu^+$ ion concentration on the electrode surface. Figure 3.58b shows the calculated the $Cu^+$ ionic concentration distributions inside the via during dissolution. The *x*-axis is the via depth and the *y*-axis is the $Cu^+$ ion concentration. In these graphs, $i_{rev}/|i_{on}|$ ratios are changed as 0, 2.0, 4.0, and 6.0, (i) is the result of $i_{rev}/|i_{on}| = 0$, (ii) is the result of $i_{rev}/|i_{on}| = 2.0$, (iii) is the result of $i_{rev}/|i_{on}| = 4.0$, and (iv) is the result of $i_{rev}/|i_{on}| = 6.0$.

From the electrochemical measurement result shown in Fig. 3.58a, the peak ring current at the on-time is about 0 μA for $i_{rev}/|i_{on}| = 0$. The peak ring current at the on-time is about 40 μA for $i_{rev}/|i_{on}| = 2.0$. The peak ring current at the on-time is about 150 μA for $i_{rev}/|i_{on}| = 4.0$. The peak ring current at the on-time is about 230 μA for $i_{rev}/|i_{on}| = 6.0$. The ring currents increase with the increasing $i_{rev}/|i_{on}|$ ratios of the periodic reverse pulse current waveforms. This means that produced $Cu^{+}$ ion concentration on the reactive surface is markedly increasing with the increasing periodic reverse pulse current waveform $i_{rev}/|i_{on}|$ ratios.

From the calculated result as shown in Fig. 3.58b, the $Cu^{+}$ ion concentration inside the via during dissolution is about 0.002 M for $i_{rev}/|i_{on}| = 0$. The $Cu^{+}$ ion concentration inside the via during dissolution is about 0.016 M for $i_{rev}/|i_{on}| = 2.0$. The $Cu^{+}$ ion concentration inside the via during dissolution is about 0.031 M for $i_{rev}/|i_{on}| = 4.0$. The $Cu^{+}$ ion concentration inside the via during dissolution is about 0.045 M for $i_{rev}/|i_{on}| = 6.0$. The $Cu^{+}$ ion concentrations inside the via during dissolution increase with the increasing $i_{rev}/|i_{on}|$ ratios of the periodic reverse pulse current waveforms. This result is in accordance with the electrochemical measurement results using RRDE (Fig. 3.58a).

Now I would briefly like to introduce our latest TSV via filling data. We can fill the 5-μm-diameter and 25-μm-deep via within 5 min. Available data nowadays is about 30 min. This will be described in the next chapter in detail. The via has a slight tapered shape (Fig. 3.59). The bath is shown in Table 3.1and pulse reverse current has been applied.

**Fig. 3.59** Five-minute electrodeposition time $|i_{on}| = -90\,\text{mA/cm}^2$

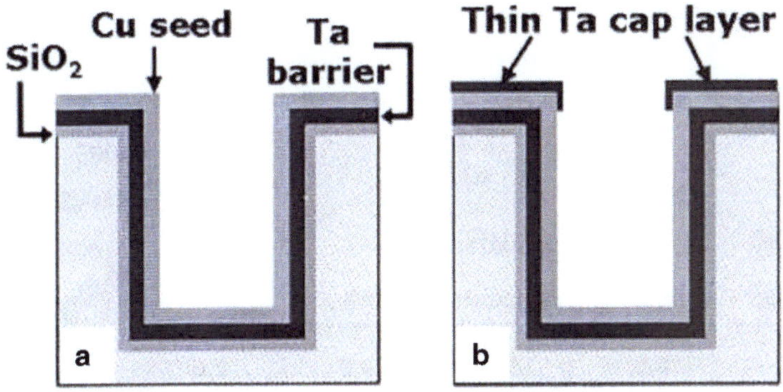


**Fig. 3.60** Illustration of cross section of Ta barrier, copper seed, and thin Ta cap layers. **a** The silicon surface is fully covered by Ta barrier and Cu seed layer. **b** After the silicon surface is fully covered by Ta and Cu layers, a thin Ta cap layer is applied on the top surface of the Cu seed layer

### *3.4.4 High-Speed Via Filling Electrodeposition by Other Organizations*

O. Lhun, A. ARadisic, and P. M. Verecken at Interuniversity Microelectronics Center (IMEC) have done considerable work on TSV copper electrodeposition [56, 57]. Their initial work addressed filling of 5-µm-diameter and 25-µm-deep vias. Additives PEG, SPS, Janus Green B (JGB), and $Cl^-$ were used. Figure 3.60 illustrates their invention. In Fig. 3.60a, the silicon surface is fully covered by a Ta barrier and Cu seed layer. In Fig. 3.60b, after the silicon surface is fully covered by Ta and Cu layers, a thin Ta cap layer is partially covered just on the top surface of the Cu seed layer. This partially covered Ta cap layer is their invention. Figure 3.61 shows the filled copper center height and filling time. With fully covered substrate, the copper center height increases rapidly at the initial time, and 2800 s were required to fill the via. On the other hand, with the partially covered substrate, the center height growth has an incubation period, and from 450 s the center height increases rapidly. A total of 865 s were required to fill the via, which is one third of the fully covered substrate.

JGB was used as a leveler in combination with the additives above. With a larger amount of JGB, more inhibition was observed in current–voltage curves. At the bottom and outside of the via, electrodeposition is inhibited due to a higher JGB concentration. This inhibition may be related to the bottom-up mechanism, however, a detailed mechanism has not been clearly described [58]. A refinement in the crystal size at the via mouth has been observed by focused ion beam (FIB) examination of cross sections. By use of the same commercial additives, 2–8 AR vias were filled successfully [58]. Both coupon level and wafer level have been tested. The 5-µm-diameter and 40-µm-deep via was filled in 1.5 h.

H. Kadota of Hitachi Kyowa [59] invented a vertically configured electrodeposition cell with a high electrolyte flow velocity of 5 m/s. Pulse current was applied with a long off-time of 1.0 s at a current density of 10 mA/cm2. It took 90 min to fill the 10-µm-diameter and 70-µm-deep via (Fig. 3.62). The additives used were not reported.

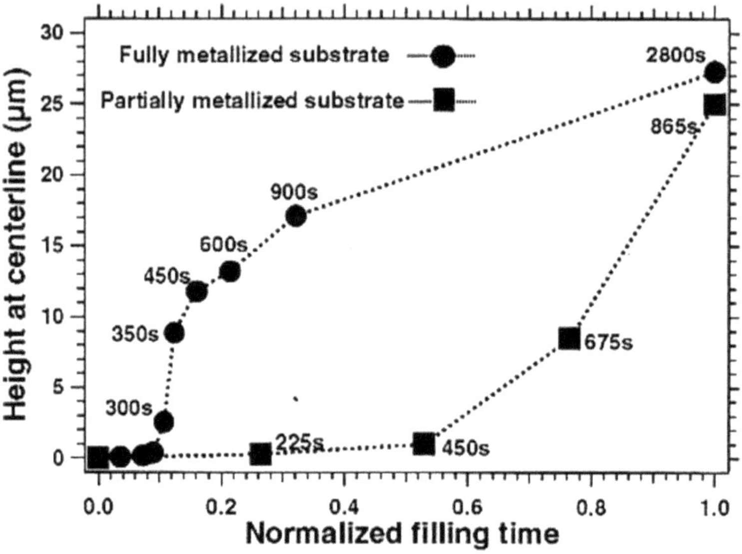


**Fig. 3.61** Filled copper center height and filling time. Fully covered and partially covered substrates center heights are shown

**Fig. 3.62** Cross section of 10-μm-diameter and 70-μm-deep via by Hitachi Kyowa

T. P. Moffat filled donut-shaped vias using $Cl^-$ and poloxamine additives at a potential of −0.650 V silver sulfate electrode (SSE) [60]. Seventeen minutes were required to fill the via. The via cross-section top was flat and had no V-shapes.

R. Beica, T. Ritzdorf et al. [61] filled 30-μm-diameter and 110-μm-deep and 12-μm-diameter and 100-μm-deep vias. Unfortunately, the electrodeposition conditions and additives are not described in the article. R. Baskaran et al. [62] filled the 8-μm-diameter and 100-μm-deep vias. The exchange current densities and electron transfer coefficients for their plating bath have been measured. From the current–voltage curve, the inhibition effect increases with increasing leveler concentration. A. Flugel et al. [63] measured an oscillatory deposition potential during galvanostatic deposition. Secondary ion mass spectrometry (SIMS) measurements showed oscillatory incorporation of contaminants. M. Arnolf et al. [64] showed that a type-I suppressor produces a temporary suppression effect. On the other hand, a type-III suppressor shows a sustained suppression effect.

J. D. Adolf and U. Landau have used numerical computation of current distribution to assess the effect of additive transport and adsorption on TSV filling [65]. They recognized that the strongly bound leveler displaces both PEG and SPS [66]. The role of leveler is analyzed [67]. In the model, the leveler concentration profile is determined as a function of TSV penetration depth, and it is shown that the leveler reaches a stagnation depth due to the balance of diffusion and incorporation. Figure 3.63a illustrates the effect of the leveler as it significantly reduces pinching

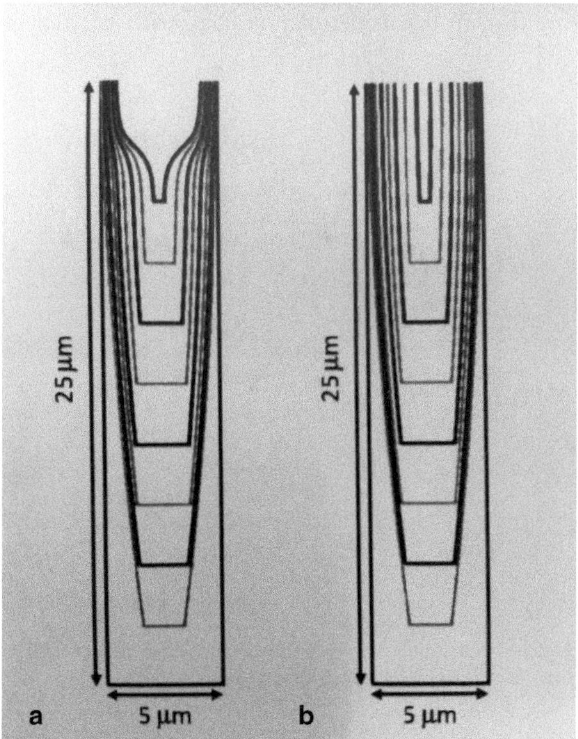


**Fig. 3.63** Illustration of the effect of the leveler. The leveler reduces pinching at the top of the TSV in **a**. Pinching occurs without leveler in **b**

at the top of the TSV [68]. This pinching simulation at the top of the TSV is very important from a practical viewpoint.

### *3.4.5 Reduction of Thermal Expansion Coefficient of Electrodeposited Copper for TSV by Additive*

3D packaging by using TSV is the next generation interconnection. There are two major processes. One of them is called as via-middle process. In between transistor formation and wiring, the vias are etched by reactive ion etching (RIE) and filled. The filled copper TSV expands because of the exposure to 400–600 °C heat of the wiring process. This is called TSV pumping. The other is called as backside, via-last process. The problem with via-last process is that the thin wafer after CMP is difficult for handling. In this report, we focused on via-middle process.

F. X. Che observed and numerically simulated the TSV pumping. The pumpings are caused by the mismatch of the thermal expansion coefficient of copper and silicon [69]. N. Kumar reported that a pre-annealing step is effective in reducing the TSV pumpings [70–72]. Seven times pre-annealing at 400 °C is suggested by P. Garrou [73]. TSV expands because of the exposure to 400–600 °C heat of the via-middle process. This expansion is called pumping and TSV destroys the upper wiring. Hence, we have to reduce the thermal expansion coefficient of copper TSV.

Figure 3.64 shows the change in thermal expansion coefficients with temperature. Brown line is the thermal expansion coefficient with pure copper. The red

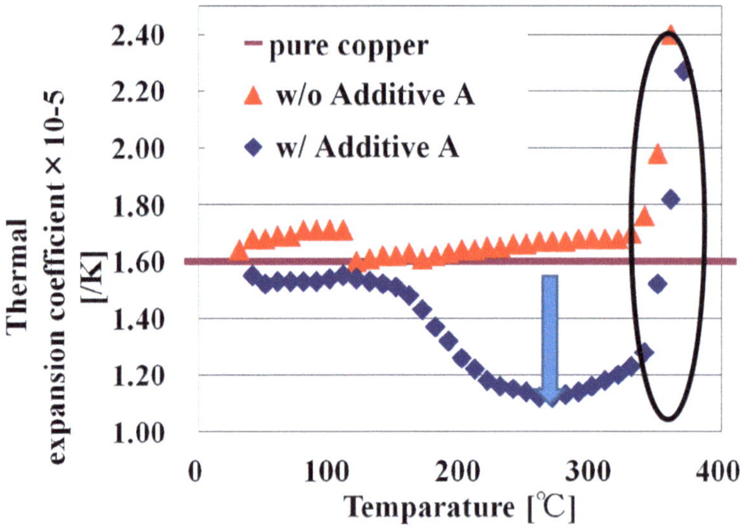


**Fig. 3.64** Relation between thermal expansion coefficient and temperature. Samples are pure copper, copper electrodeposits with and without additive A

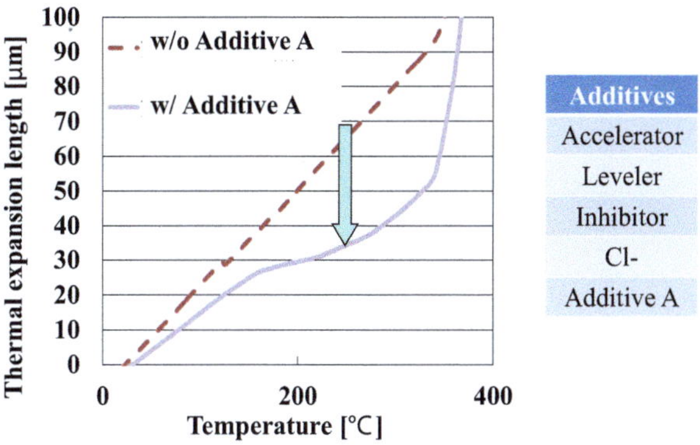


**Fig. 3.65** Relation between thermal expansion length and temperature. Samples are copper electrodeposits with and without additive A

triangle is without the additive A and blue square is with the additive A. A marked thermal expansion coefficient reduction is observed with the additive A between 150 and 350 °C (shown with blue arrow). The red triangle is without the additive A, the thermal expansion coefficient is same as pure copper. A sudden rise in thermal expansion coefficients over 350 °C is observed for both with and without additive A, as indicated with black circle.

Figure 3.65 shows the change in expansion length with temperature. Dotted curve is without additive A and the continuous curve is with additive A. Without additive A, the expansion length is proportional to the temperature. With additive A, a marked reduction in expansion length, again, is observed between 150 and 350 °C, as shown with blue arrow. About 50 % reduction of the expansion length is obtained at 250 °C. A sudden rise in expansion length at temperature more than 350 °C is observed with additive A.

Figure 3.66 shows the change in expansion length with temperature with different additive combinations. Continuous line A is pure copper and expansion length is proportional to temperature. Red dotted line is with inhibitor, chloride, and additive A and sudden rise in expansion length is observed at more than 350 °C. On the other hand, with blue dotted line with chloride and additive A, no sudden rise in expansion length is observed at more than 350 °C. This means that the inhibitor decomposes and produces a sudden rise in expansion length at more than 350 °C. About 22 % reduction of the expansion length was obtained at 400 °C and the reduction is indicated with red arrow.

Figure 3.67 shows the result of electron backscattering diffraction (EBSD). EBSD observes the backscattering diffraction by irradiating electron to the crystal sample.

The upper graph is without additive A and the lower graph is with additive A. Without the additive A at the room temperature, the mean grain size is as small as 113.6 nm. After annealing this sample up to 370 °C, the mean grain size grows up to as large as 158.6 nm. However, addition of the additive A miniaturized the mean grain size as 138.6 nm and this mean grain size is smaller than by without adding

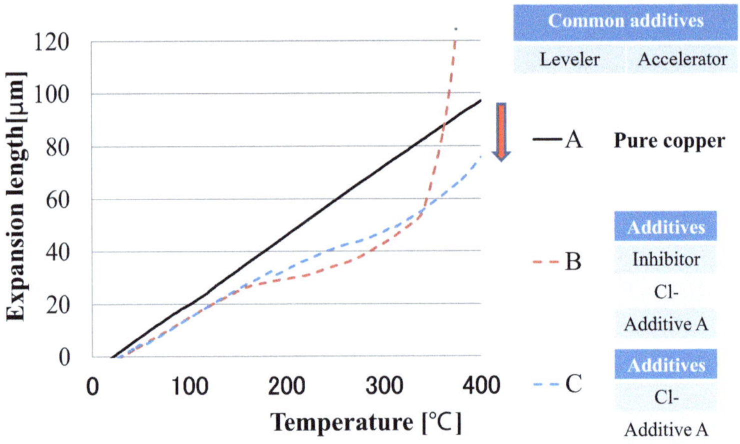


**Fig. 3.66** Relation between expansion length and temperature. Samples are copper electrodeposits with additive A, with and without inhibitor

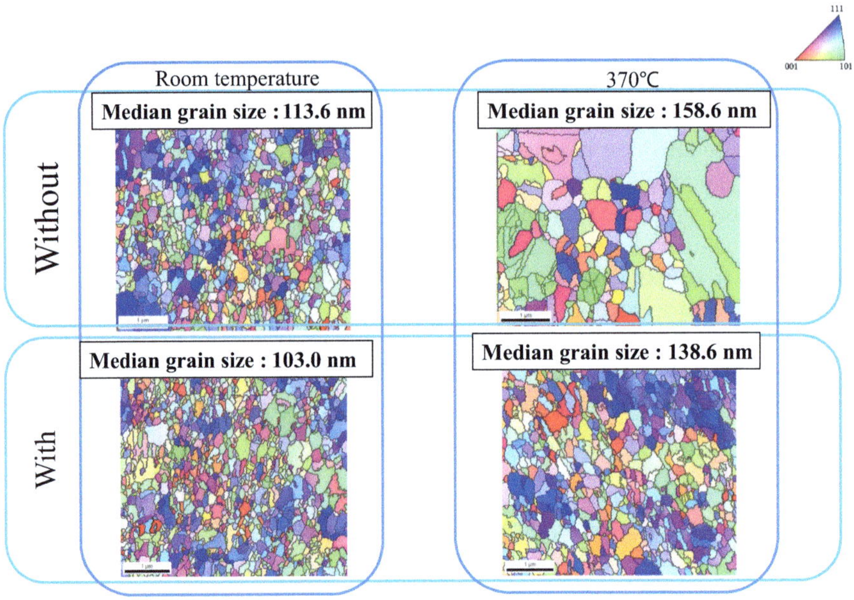


**Fig. 3.67** EBSD observation with and without additive A at room temperature and annealed temperature at 370 °C. Median grain sizes are compared

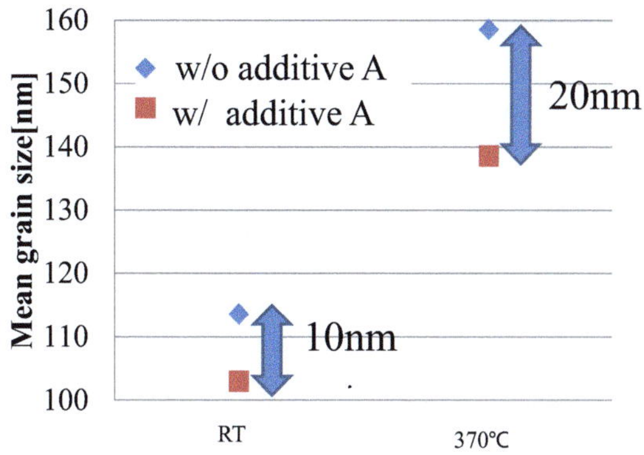


**Fig. 3.68** Relation between mean grain sizes at room temperature and annealed temperature at 370 °C

the additive A. Addition of the additive A miniaturized the mean grain size after annealing at 370 °C.

Figure 3.68 shows the effect of annealing temperature on the mean grain size of Cu. At room temperature, the difference in grain size with and without the additive A is 10 nm. However this difference enlarges to 20 nm at 370 °C. Additive A inhibits the grain growth at 370 °C.

Symbols (⊥) indicate the dislocation in Fig. 3.69. The impurity from the additive A segregates at the grain boundaries. This impurity prevents the slipping transformation of piled-up dislocations. However, without additive A, this impurity does not segregate at the grain boundaries and the slipping transformation of piled-up dislocation is large. So the grain growth occurs without the additive A.

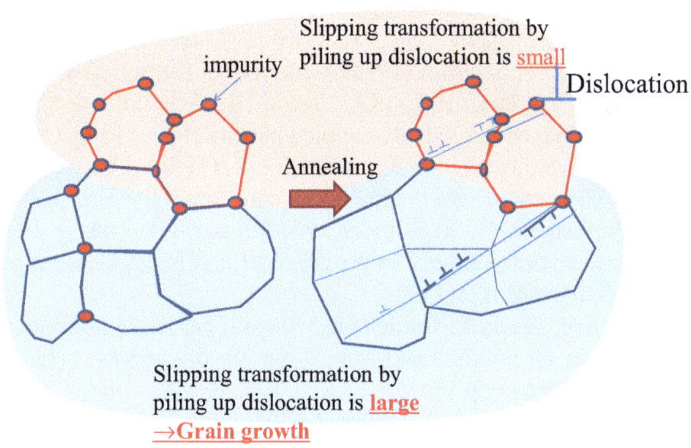


**Fig. 3.69** Schematic illustration of additive A impurity segregation and mechanism of grain growth inhibition

## References


1. Laemer FD, Schilp A (1992) A method for anisotropic etching of silicon. German Patent No. DE4241045
2. Zoschke K, Oppermann H, Manier CA, Ndip I, Puschmann R, Ehrmann O, Wolf J, Lang KD (2012) Wafer level 3D system integration based on silicon interposers with through silicon vias. In: Proceedings of 14th IEEE electronics packaging technology conference 5–7 Dec. 2012, pp 8–13
3. Mourier T, Ribiere C, Romero G, Gottardi M, Allouti N, Eleouet R, Roman A, Magis T, Minoret S, Ratin C, Scevola D, Dupuy E, Martin B, Gabette L, Marseilhan D, Enot T, Pellat M, Loup V, Segaud R, Feldis H, Charpentier A, Bally JP, Assous M, Charbonnier I, Laviron C, Coudrain P, Sillon N (2913) 3D integration challenges today from technological toolbox to industrial prototypes. Proceedings of IEEE interconnect technology conference, pp 1–3
4. Sheu SS, Lin ZH, Lin CS, Lau JH, Lee SH, Su KL, Ku TK, Wu SH, Hung JF, Chen PS, Lai SJ, Lo WC, Kao MJ (2012) Electrical characterization of through silicon vias (TSVs) with an on chip bus driver for 3D IC integration. Proceedings of IEEE 62nd electrical components and technology conference, pp 851–856
5. Peterson K (1982) Silicon as a mechanical material. Proc IEEE 70(5):420–457
6. Bhardwaj JK, Ashraf H, Hopkins J, Johnston I, McAuley S, Hall S, Nicholls G, Atabo L, Hynes A, Welch C, Barker A, Gunn B, Lea L, Guibarra E, Watcham S (1999) Advances in high rate silicon and oxide etching using ICP. MEMS/MST technology symposium at SEMICON West'99, San Francisco, CA, USA. July 12–16, 1999
7. Hynes AM, Ashraf H, Bhardwaj JK, Hopkins J, Johnston I, Shepherd JN (1999) Recent advances in silicon etching for MEMS using the ASETM process. Sens Actuators 74:13–17
8. Pang SW (2001) Dry processing of high aspect ratio Si microstructures for MEMS. Proceedings of international symposium on dry process, pp 49–55
9. Hashimoto K (1994) Charge damage caused by electron shading effect. Jpn J Appl Phys 33(10):6013–6018
10. Bhardwaj JK, Ashraf H, Khamsehpour B, Hopkins J, Hynes AM, Ryan ME, Haynes DM (2000) Method of surface treatment of semiconductor substrates. US Patent No. 6,051,503
11. Hartig MJ, Arnold JC (1997) Inductively coupled plasma reactor and process. US Patent No. US 5,683,548 A
12. Fukushima T, Bea J, Murugesan M, Lee KW, Koyanagi M (2013) Development of via-last 3D integration technologies using a new temporary adhesive system. 3D systems integration conference (3DIC), San Francisco, 2–4 Oct 2013, pp 1–4
13. Laermer F, Schilp A (1996) Method of anisotropically etching silicon. US Patent, 5501893
14. Ayón AA, Braff R, Lin CC, Sawin HH, Schmidt and MA (1999) Characterization of a time multiplexed inductively coupled plasma etcher. J Electrochem Soc 146(1):339–349
15. Tachi S, Tsujimoto K, Okudaira S (1988) Low-temperature reactive ion etching and microwave plasma etching of silicon. Appl Phys Lett 52(8):616
16. Pruessner MW, Rabinovich WS, Stivater TH, Park D, Baldwin JW (2007) Cryogenic etch process development for profile control of high aspect-ratio submicron silicon trenches. J Vac Sci Technol B25(1):21
17. Sakai I, Sasaki K, Tomioka K, Ohiwa T, Sekine M, Mimura T, Nagaseki K (2001) Proceedings of the 1st international symposium on dry process. The institute of electrical engineers of Japan, Tokyo, p 57
18. Sakai I, Sakurai N, Ohiwa T (2008) Proceedings of the international symposium on dry process, The Japan society of applied physics, Tokyo, p 125
19. Horiike Y, Okano H, Yamazaki T, Horie H (1981) High-rate reactive ion etching of $SiO_2$ using a magnetron discharge. Jpn J Appl Phys Part 2 20(11):L817
20. Hill ML, Hinson DC (1985) Advantages of Magnetron Etching. Solid State Technol 28:243
21. Kinoshita H, Ishida T, Ohno S (1986) Proceedings of the symposium on dry process. The institute of electrical engineers of Japan, Tokyo, p 36

22. Müller P, Heinrich F, Mader H (1989) Magnetically enhanced reactive ion etching (MERIE) with different field configurations. Microelectron Eng 10(1):55–67
23. Sekine M, Narita M, Horioka K, Yoshida Y, Okano H (1995) A new high-density plasma etching system using A dipole-ring magnet. Jpn J Appl Phys 34(11):6274
24. d'Agostino R, Flamm D (1981) Plasma etching of Si and $SiO_2$ in SF6-$O_2$ mixtures. J Appl Phys 52(1):162
25. Gomez S, Belen RJ, Kiehlbauch M, Aydil ES (2004) Etching of high aspect ratio structures in Si using SF6/$O_2$ plasma. J Vac Sci Technol A22(3):606
26. Shimizu H, Kimura D, Komiya H, Kawabata R (1984) Proceedings of the symposium on dry process. The institute of electrical engineers of Japan, Tokyo, p 121
27. Coburn JW, Chen M (1980) Optical emission spectroscopy of reactive plasmas: A method for correlating emission intensities to reactive particle density. J Appl Phys 51(6):3134
28. Amasaki S, Takeuchi T, Takeda K, Ishikawa K, Kondo H, Sekine M, Hori M, Sakurai N, Hayashi H, Sakai I, Ohiwa T (2010) Proceedings of the international symposium on dry process. The Japan society of applied physics, Tokyo, p 97
29. Amasaki S, Takeuchi T, Takeda K, Ishikawa K, Kondo H, Sekine M, Hori M, Sakurai N, Hayashi H, Sakai I, Ohiwa T Proceedings of the international symposium on dry process. The Japan society of applied physics, Tokyo, p 33
30. Nagai H, Hiramatsu M, Hori M, Goto T (2003) Measurement of oxygen atom density employing vacuum ultraviolet absorption spectroscopy with microdischarge hollow cathode lamp. Rev Sci Instrum 74(7):3453
31. Booth JP, Joubert O, Pelletier J, Sadeghi N (1991) Oxygen atom actinometry reinvestigated: comparison with absolute measurements by resonance absorption at 130 nm. J Appl Phys 69(2):618
32. Pereora J, Pichon L, Dussart R, Cardinaud C, Duluard CY, Oubensaid EH, Lefaucheux P, Boufnichel M, Ranson P (2009) In situ x-ray photoelectron spectroscopy analysis of $SiO_xF_y$ passivation layer obtained in a $SF_6/O_2$ cryoetching process. Appl Phys Lett 94:071501
33. Flamm DL, Donnelly VM, Mucha JA (1981) The reaction of fluorine atoms with silicon. J Appl Phys 52(5):3633
34. Lieberman MA, Lichtenberg AJ (2004) Principles of plasma discharges and materials processing, 2nd ed., p 587
35. Kusuda Y, Nonaka T, Motoyama S (2013) TSV process using DRIE and cathode coupled PECVD. ECS Trans 50(32):3–9
36. Kusuda Y, Minaguchi T, Miyashita T, Motoyama S (2009) Sidewall insulator film deposition for the TSV process using cathode coupled PECVD. The 9th international workshop on microelectronics assembling and packaging, p 41, Fukuoka, Japan
37. Hiramoto M, Minaguchi T, Motoyama S (2007) Deposition of $SiO_2$ film with excellent step coverage using PECVD. Mater Stage 7(5):16–20 (in Japanese)
38. SAMCO Inc. (2004) PECVD systems for optical devices—ST series. Opto devices technology outlook. Electr J 28:287–289 (in Japanese)
39. Laemer FD, Schilp A (1994) A method for anisotropic etching of silicon, German Patent No. DE4241045 C1, May 26
40. Nonaka T, Oda H, Noda Y, Kuratomi N, Nakano H (2012) A study of via hole etching for TSV process. The 11th APCPST abstract. Kyoto University, Kyoto, Japan, p 286
41. Andoricacos PC, Uzoh C et al (1998) IBM J Res Devel 42:567–572
42. Moffat TP, Wheeler D et al (2001) Electrochem Solid-State Lett 4:C26–C29
43. West AC, Mayer S et al (2001) Electrochem Solid-State Lett 4:C50–C53
44. Tantavishet N, Pritzker M et al (2003) J Electrochem Soc 150:C665–C669
45. Kondo K, Matsumoto T et al (2004) J Electrochem Soc 151:C250–C256
46. White JR (1987) J Appl Electrochem 17:977–1003
47. Nagy Z, Blaudeau JP (1995) J Electrochem Soc 142:L87–L92
48. Kondo K, Hamazaki K (2014) ECS Electrochem Lett 3(4):D3–D5
49. Kondo K, Nakamura T (2009) J Appl Electrochem 39:1789–1794
50. Kondo K, Yonezawa T et al (2005) J Electrochem Soc 152(11):H173–H177

51. Sun J-J, Kondo K et al (2003) J Electrochem Soc 150(6):G355–G358
52. Kondo K, Suzuki Y et al (2010) Electrochem S-S Lett. 13(5):D26–D28
53. Hayashi T, Kondo K et al (2011) J Electrochem Soc 158(12):D715–D718
54. Akolkar R (2013) ECS Electrochem Lett 2(2):D5–D9
55. Hayashi T, Kondo K (2013) J Electrochem Soc 160(6):D256–D259
56. Luhn O, Radisic A et al (2009) Electrochem & S-S Lett 12(5):D39–D41
57. Luhn O, Van Hoof C et al (2009) Electrochim Acta 54:2504–2508
58. Radisic A, Luhn O et al. (2011) Microelectronic Eng 88:701–704
59. Kadota H et al. (2010) JIEP 13(3):213–219 (in Japanese)
60. Moffat TP, Josell D (2012) J Electrochem Soc 159(4):D208–D216
61. Beica R, Sharbono C (2008) Through silicon via copper electrodeposition for 3D integration. Proceedings of ECTC conference
62. Baskaran R, McHugh P (2011) Characterization of the organic components in a commercial TSV filling chemistry. Paper presented at the 220th meeting of the Electrochemical Society, Oct 2011
63. Flugel A, Amold M (2011) Tailored design of suppressor ensembles for damascene and 3D-TSV copper plating. Paper presented at the 220th meeting of the electrochemical society, Oct 2011
64. Arnold M, Emnet C (2010) New concept for advanced 3D TSV copper plating additives. Paper presented at the 218th meeting of the electrochemical society, Oct 2011
65. Adolf JD, Landau U (2009) Scaling analysis of bottom up fill with application to through silicon via. Paper presented at the 216th meeting of the electrochemical society, Oct 2009
66. Landau U (2010) Electroplating of interconnects—scaling from nanoscale dual-damascene to micron-scale through silicon vias. Paper presented at the 218th meeting of the electrochemical society, Oct 2010
67. Adolf JD, Landau U (2010) Additive adsorption and transport effects on the void-free metallization of through silicon vias. Paper presented at the 218th meeting of the electrochemical society, Oct 2010
68. Adolf JD, Landau U (2011) Leveler effects on filling of through silicon vias. Paper presented at the 220th meeting of the electrochemical society, Oct 2011
69. Che FX, Putra W et al (2011) Numerical and experimental study on Cu protrusion of Cu-filled through-silicon vias (TSV). In: Proceedings of 3DIC 2011, 2011
70. Kumar N et al (2011) Advanced reliability study of TSV interposers and interconnects for the 28 nm technology FPGA. Proceedings of ECTC 2011
71. Huyghebaerta C, Coenena J et al (2011) Microelectr Eng 88:745–748 (5th May)
72. Croesa K, Varela O et al (2011) Microelectronics reliability 51 (9–11):1856–1859
73. Garrou P (2010) Cu protrusion, keep-out zones highlight 3D talks at IEDM. In: Solid state technology. Bruker corporation. http://www.electroiq.com/articles/ap/2010/12/cu-protrusion-keep-out.html. Accessed 22 Feb 2014

# Chapter 4
# Wafer Handling and Thinning Processes

**Takashi Haimoto, Eiichi Yamamoto, Takahiko Mitsui, Toshihiro Ito, Tsuyoshi Yoshida, Tsubasa Bandoh, Kazuta Saito and Masahiro Yamamoto**

## 4.1 Wafer Thinning Solution for TSV Devices

### *4.1.1 Introduction*

For 3D integrated chip (IC) using through-silicon via (TSV), wafer ultrathinning is an important technology due to technical and cost issues of the Cu via forming process. However, for TSV devices, less total thickness variation (TTV) and higher wafer cleanliness are required compared to that of conventional 2D devices in order to achieve cost savings by reducing the process time in post-processes. This chapter describes general thinning solutions and high-resolution thinning processes especially for TSV devices.


E. Yamamoto (✉) · T. Mitsui · T. Ito · T. Yoshida · T. Bandoh
Okamoto Machine Tool Works, Ltd., 2993 Gobara, Annaka, Gunma 379-0135, Japan
e-mail: e.yamamoto@okamoto.co.jp

M. Yamamoto (✉)
3DI Dept. ATS BU, Tokyo Electron Ltd., 3-1, Akasaka 5-chome,
Minato-ku, Tokyo 107-6325, Japan
e-mail: masahiro.yamamoto@tel.com

K. Saito (✉)
53M Japan Limited, 8-8, Minami Hashimoto 3-chome, Chuo-ku, Sagamihara-shi,
Kanagawa 252-5285, Japan
e-mail: ksaito@mmm.com

T. Haimoto
Dicer Grinder Marketing Team, Marketing Group, Sales Engineering Division,
DISCO Corporation, 13-11, Omori-Kita 2-chome, Ota-ku, Tokyo 143-8580, Japan
e-mail: thaimoto@disco.co.jp




K. Kondo et al. (eds.), *Three-Dimensional Integration of Semiconductors*,
DOI 10.1007/978-3-319-18675-7_4

### 4.1.2 General Thinning

The thinning process consists of grinding, also known as backgrinding (BG), and polishing. A typical fully automatic grinder has three spindles and four chuck tables

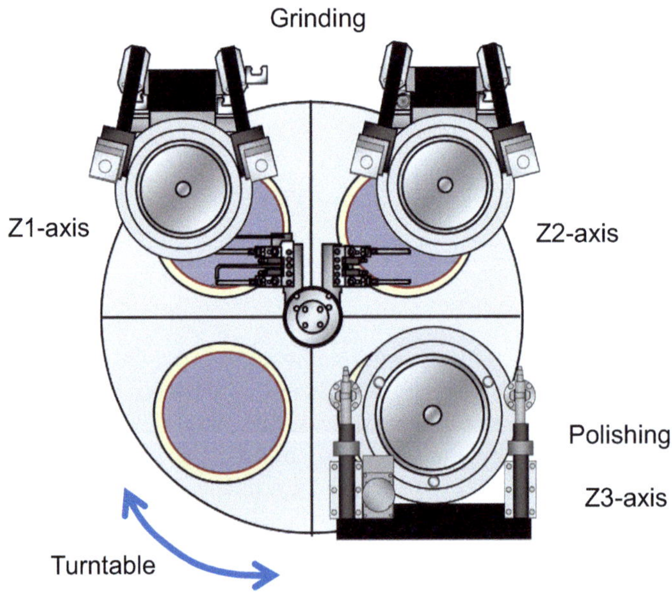


**Fig. 4.1** Typical fully automatic grinder

which are placed on a turntable, as shown in Figure 4.1. The Z1-axis is for rough grinding, the Z2-axis is for fine grinding, and the Z3-axis is for polishing. Normally, grinding is divided into two steps. Rough grinding removes most of the total grinding amount at higher process speed by using a wheel with rough diamond abrasive grains (e.g., #320) to improve productivity. And fine grinding removes the damage caused by the rough grinding at lower process speed by using a wheel with fine diamond abrasive grains (e.g., #2000) producing a smoother surface with less damage.

Figure 4.2 is a schematic diagram of wafer grinding. In order to protect the device layer from directly touching the chuck table, surface protection tape (called BG tape) is attached to the device side of the wafer prior to grinding. The wafer is held on the chuck table using vacuum. The chuck table and the grinding wheel rotate about their own axes, and the wafer is ground as the rotating wheel is lowered. By self-grinding the chuck table to a convex profile, the grinding wheel that comes into contact with only one half of the wafer is depicted in Fig. 4.2.

When the required finish thickness of a ground wafer is thinner than 100 μm, the stress relief, such as dry polishing (DP) or chemical mechanical polishing (CMP),

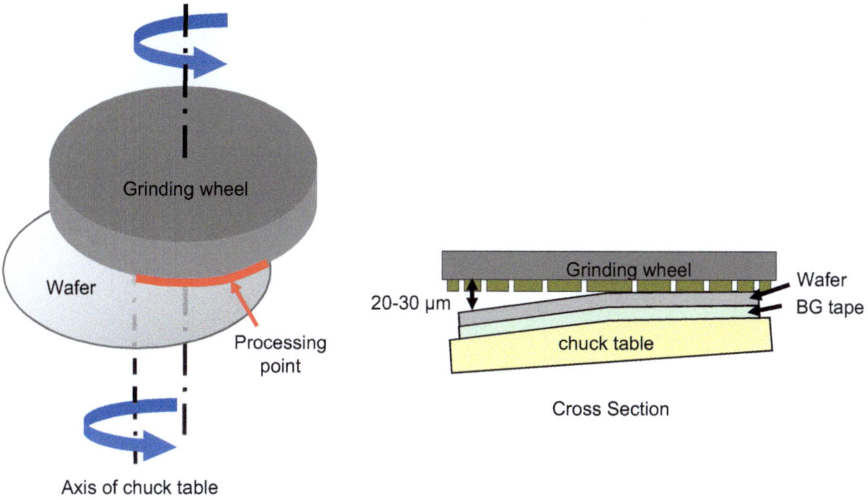


**Fig. 4.2** Schematic diagram of wafer grinding

becomes necessary. The polishing process improves die strength and reduces wafer warpage by removing subsurface damage caused by fine grinding.

### *4.1.3 Wafer Thinning for TSV Devices*

As described previously, wafer ultrathinning for TSV devices is required due to technical and cost issues of the Cu via forming process. To achieve a thinner wafer, TTV control becomes more important [1], because wafer thickness variations increase the process cost for post-processes, such as film forming, etching, and plating.

Furthermore, a ground wafer requires high-level cleaning instead of conventional cleaning with deionized (DI) water to be loaded into the front-end process equipment. By incorporating a CMP unit for the wet polishing process and a special cleaning unit with repetitive use of ozonated water and dilute hydrofluoric (HF) acid [2] into the grinder, contamination particles are removed from the ground wafer and high wafer cleanness is realized.

The next section focuses primarily on TTV control and describes "Auto TTV" adjustment.

### *4.1.4 TTV Control*

Generally, accurate TTV control is important for wafer ultrathinning, but conventional TTV adjustment [3] is done manually. The shape of a ground wafer is mea-

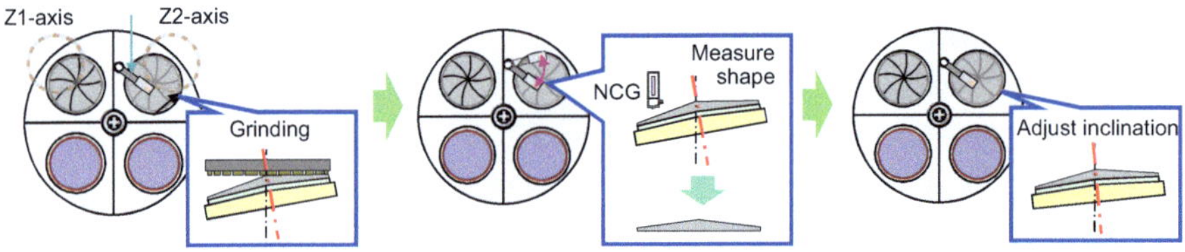


**Fig. 4.3** Auto-TTV sequence

**Fig. 4.4** Noncontact gauge (NCG)

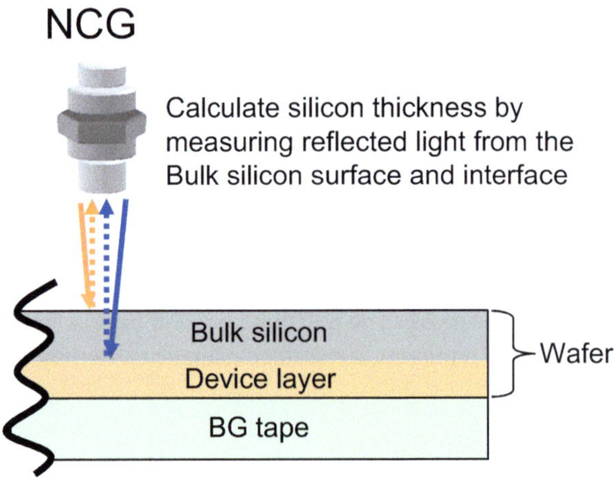


sured outside of the grinder, and the chuck table inclination is adjusted during maintenance. Manual operation causes equipment downtime, and adjustment may not be sufficient depending on the operator's ability. To achieve uniform wafer thickness and maintain a constant TTV during continuous processing, the "Auto TTV" function has been developed. This function measures the silicon thickness with a noncontact gauge (NCG) after grinding on the Z2-axis, and the chuck table inclination is adjusted during full-auto operation (Fig. 4.3).

NCG can isolate the measurement of bulk silicon thickness even if a wafer is attached to BG tape, as shown in Fig. 4.4, whereas a conventional contact gauge measures the total thickness of all materials. Therefore, the "Auto TTV" adjustment can be done based on only bulk silicon thickness without the influence of thickness variation of other materials and the risk of damaging ultrathin wafers.

Figure 4.5 shows the adjustment mechanism and supported wafer shapes. A chuck table has three axes—one is fixed and the other two can be adjusted. Based on silicon thickness measured by NCG, adjustments for the S-axis and D-axis are calculated, and the two axes are automatically controlled.

Figure 4.6 shows a comparison of TTV results of thinning wafers. Even silicon wafers, without BG tape, have more than 1.5 μm TTV before adjustment. How-

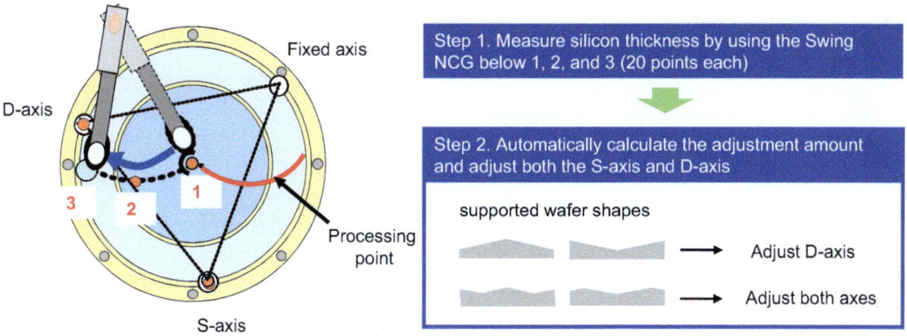


**Fig. 4.5** Adjustment mechanism and supported wafer shapes

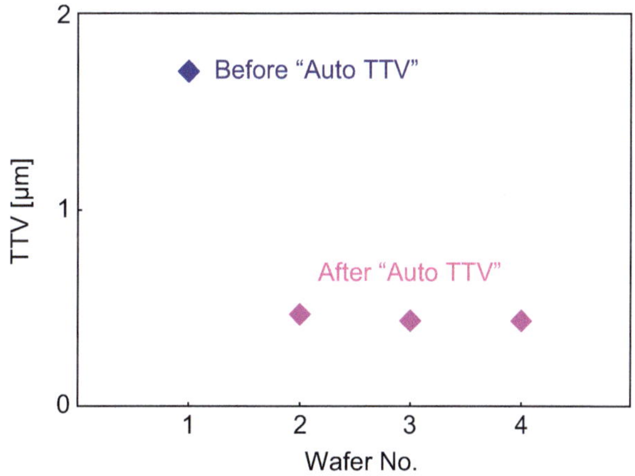


**Fig. 4.6** TTV results of thinning wafers

ever, by applying the "Auto TTV" adjustment, the TTV of thinning wafers can be reduced down to 0.5 μm.

Wafers for TSV devices are bonded temporarily to support carriers such as silicon or glass substrates with bonding adhesive for safe handling, as shown in Fig. 4.7. Due to the thickness variation of bonding adhesive, the TTV of a bonded wafer tends to be larger. To solve this issue, NCG becomes more effective, because it can isolate the measurement of bulk silicon thickness. Therefore, less TTV is realized without the influence of thickness variation of bonding adhesive.

If the bonding adhesive has a different thickness variation on each wafer, "R-Auto TTV" is more effective, where "R" represents regrinding. "Auto TTV" returns a TTV adjustment value for use with the next wafer. On the other hand, "R-Auto TTV" pauses grinding to measure the silicon thickness by NCG. After performing chuck table inclination adjustments on the Z2-axis, the function will resume fine grinding. Because of this, it is possible to apply the measured TTV adjustment value to the wafer being processed such that adjustments can be made even if the

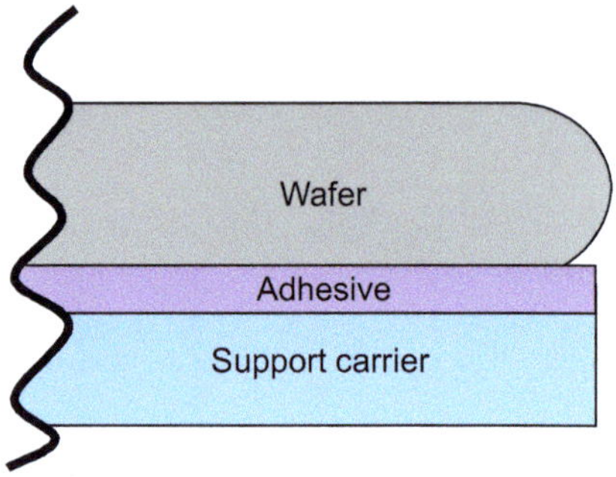


**Fig. 4.7** Bonded wafer

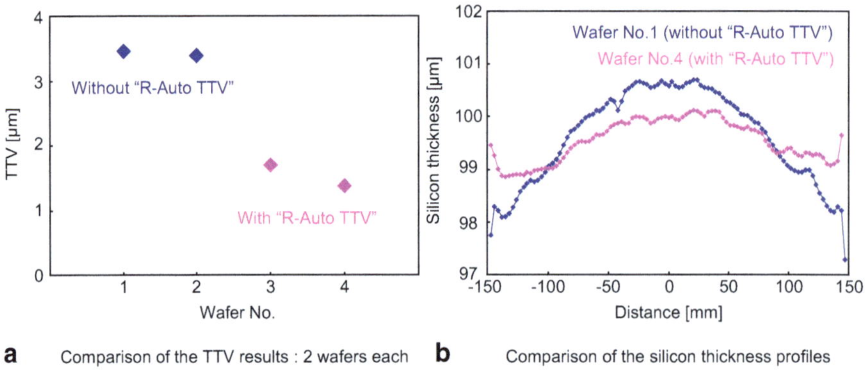


**Fig. 4.8** TTV results of bonded wafers

thickness varies between wafers. Figure 4.8 shows a comparison of TTV results of bonded wafers. By applying the "R-Auto TTV" adjustment, the TTV of a bonded wafer can be reduced, and uniform wafer thickness can be achieved.

Furthermore, wafer ultrathinning down to 4 μm for 3D multi-stack wafer-on-wafer (WOW) applications has been reported [4]. The TTV after thinning was approximately 1 μm within the class of 300-mm wafer thereby proving the real-world effectiveness of TTV control.

### *4.1.5 Summary*

Accurate TTV control and high-level cleaning are realized for wafer ultrathinning on TSV devices. However, TSV devices still have cost issues for mass production. To contribute to cost-reduction efforts, the wafer ultrathinning process will continue to be improved and optimized.

## 4.2 A Novel Via-Middle TSV Thinning Technology by Si/Cu Grinding and CMP

### 4.2.1 Introduction

The future demand for high-performance 3D packaging technology using TSV is strongly anticipated as the market continues its desire for a smaller form factor, higher speed, multifunction, and lower power consumption [5–10]. This technology is actually being used for backside-illuminated (BSI) complementary metal oxide semiconductor (CMOS) image sensor and high-performance field-programmable gate array (FPGA) [11, 12]. As new technologies become dependent on this future TSV process, it is imperative that we provide a higher yield in order to expand applications while lowering the baseline cost.

TSV fabrication process will be divided into four types as follows [13]:

1. Via-first process: TSV is fabricated prior to front end of line (FEOL).
2. Via-middle process: TSV is fabricated between FEOL and back end of line (BEOL).
3. Via-last process: TSV is fabricated after BEOL. Front via type prepared from the front side and back via type prepared from the backside.
4. Via after bonding process: TSV is fabricated after wafer (chip) stacked.

From the above four processes, via-middle process is the easiest process to repurpose. It uses existing semiconductor equipment and processes: TSV is fabricated between FEOL (such as transistors) and BEOL (such as multilayer interconnection).

Lastly, the via-middle TSV fabrication process before wafer thinning does not require a carrier substrate. This means high-temperature process such as spacer dielectric deposition is applicable.

This process advancement shown in Fig. 4.9 will be advantageous to the refinement of TSV that affects most of the device makers around the world as well as semiconductor assembly and testing contract manufacturer referred as out-sourced semiconductor assembly and test (OSAT) who are in research and development of the basic process.

Concerned about the contaminate problem which is caused by Si/Cu contact, in the via-middle process in which device wafer is processed, grinding and CMP of Si does not expose the TSV directly.

It is reported that Si grinding with resinoid bond grinding wheel to expose TSV caused clogging Cu onto grinding wheel surface, Cu smear (to which mops on Cu pulled on the grinding wheel), and Cu burned (oxidized Cu become brown) [14].

As a result, selective Si etching (dry or wet) and dielectric layer removal process are required to expose Cu. The etching process becomes very complex and increases cost.

Also, TSV formation length variation directly affects TSV-exposed height variation shown in Fig. 4.9 (5), so that, in the Fig. 4.9 (7) dielectric CMP process, TSV-exposed parts failure may occur or the vias may fail to be exposed [15].

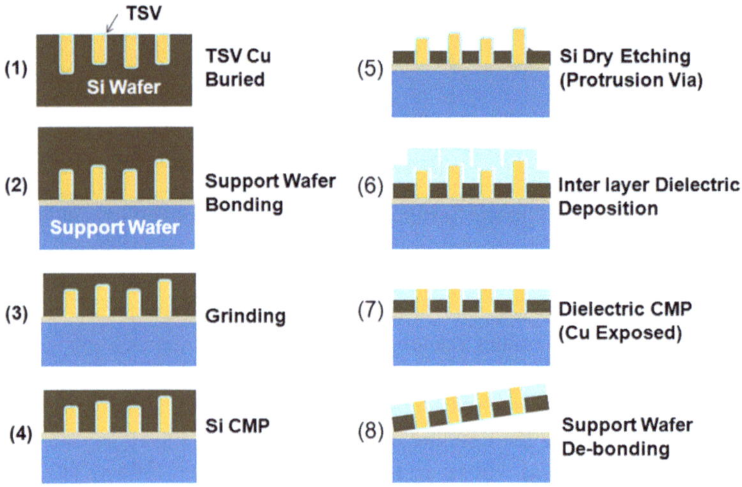


**Fig. 4.9** Conventional via-middle TSV exposure process

To solve these problems, the authors have developed Si/Cu simultaneous grinding by vitrified bond (glass-type bond) grinding wheels. This has proposed a very simple via-middle TSV formation process as shown in Fig. 4.10 [16]. At this stage, TSV is fabricated on wafer, and the TSV wafer is bonded to carrier substrate. Cu via is exposed by Si/Cu simultaneous grinding as shown in Fig. 4.10 (3). Next process is CMP as shown in Fig. 4.10 (4), (5). Therefore, our developed process consists of two CMP processes: (1) The first CMP flattens Si and Cu surfaces as well as eliminates any grind damage and (2) the second CMP exposes the bottom of the TSV.

We reported that in the case of low TSV density (1 %), we succeeded to process TSV without Cu smear or Cu burning by Si/Cu simultaneous grinding as shown in Fig. 4.12.

In this report, in the case of high TSV density (10–30 %), our proposal is to process TSV without Cu smear or Cu burning by Si/Cu simultaneous grinding. Also, we will include in the report the method of keeping the flat between Si surface and Cu via surfaces by first CMP and TSV protruding by second CMP.

We will include findings that Si/Cu simultaneous grinding makes Cu contamination depth 0.13 μm which is relatively shallow, and Cu contamination can be eliminated by Ni–B plating on Cu and Si wet etching process [17].

As a result, we are confident that our proposal will be recognized as the new standard of via-middle TSV wafer thinning technology.

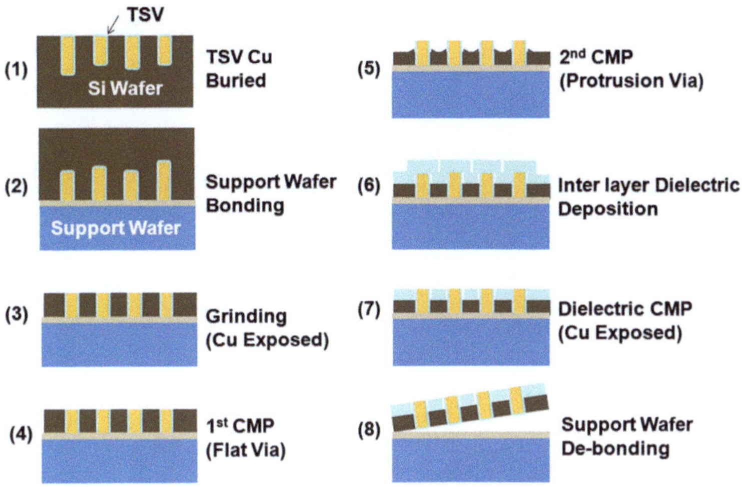


**Fig. 4.10** Via-middle TSV exposure process by Si/Cu grinding and CMP

## 4.2.2 *Experimental Method*

Figure 4.11 shows the cross-sectional structure of TSV wafers for process development:

- TSV diameter was from 10 to 100 μm
- TSV pitch was from 20 to 500 μm
- TSV depth was 20 μm

TSV liner oxide was plasma tetraethyl orthosilicate (p-TEOS) $SiO_2$ deposited at 350 °C, and its thickness was 0.5 μm. TiN barrier film was sputtered at room temperature (RT), and the thickness was 0.05 μm. Cu film was also sputtered on the

**Fig. 4.11** Structure of TSV wafer prepared for development

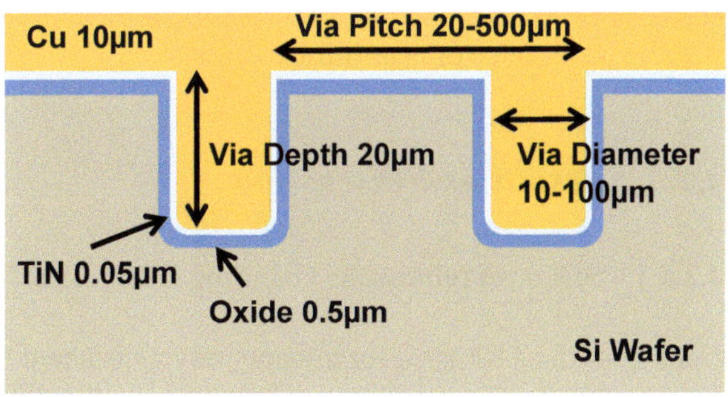

**Fig. 4.12** Optical micrographs of TSVs exposed Si/Cu grinding. **a** Resinoid bond grinding wheel (#2000 grit). **b** vitrified bond grinding wheel (#2000 grit)

barrier film. Electrodeposited Cu filled the vias. The areal density of TSV was 20 % in the TSV wafers.

By using the above condition, TSV wafer, Si/Cu simultaneous grinding, and first CMP and second CMP process were examined.

Laser microscope (Olympus OLM-3100) is used in the evaluation of the surface profile and roughness. Top of the TSV wafer (without substrate) thickness distribution, TTV, and polishing rate were measured by F50-XT (Filmetrics, Inc.) that uses spectral reflectance of near-infrared light.

## *4.2.3 Results and Discussion*

### 4.2.3.1 Si/Cu Simultaneous Grinding Using Low-Density TSV Wafers

Density of the TSV used for memory device is about 1 %. First, we describe low-density TSV Si/Cu simultaneous grinding results. Figure 4.12a shows the example that the resinoid bond grinding wheel with #2000 grit generates Cu smear and

**Table 4.1** Surface roughness after grinding (vitrified #2000 grit) and CMP

| Process | Measurement surface | Roughness Ra nm | | | | | | |
|---|---|---|---|---|---|---|---|---|
| | | Wafer #1 | | Wafer #2 | | Wafer #3 | | Average |
| | | Center | Edge | Center | Edge | Center | Edge | |
| Grinding | Cu | 34 | 27 | 32 | 31 | 29 | 33 | 31 |
| | Si | 36 | 20 | 16 | 17 | 28 | 22 | 23.2 |
| Polishing | Cu | 2 | 3 | 3 | 2 | 4 | 3 | 2.8 |
| | Si | 2 | 2 | 2 | 2 | 2 | 2 | 2 |

burning (Cu surface oxidation), which is not a good grinding result. In this case, residual Cu adheres and coats to the diamonds on the grinding wheel surface during grinding process which causes the burning and smearing. On the other hand, Fig. 4.12b shows the result of vitrified bond grinding wheel with #2000 grit. The vitrified bond grinding wheel showed good results by adjusting diamond density, porosity, and hardness bond properly to be grain balance properly.

Table 4.1 shows roughness (Ra) measurement of Si and Cu surface of these samples. The measurement results showed 31 nm on Cu and 23 nm on Si after grinding. The value is relatively large; thus, the surface is not good to consequent processes such as chemical vapor deposition (CVD) or reactive ion etching (RIE). After CMP, the roughness reduced improved from 2 to 4 nm, which is acceptable to CVD or RIE.

#### 4.2.3.2 Si/Cu Simultaneous Grinding Using High-Density TSV Wafers

Higher-density TSV is more difficult to grind Si/Cu simultaneously compared to lower-density TSV.

We examined the cause of Cu smearing, burning, and scratching; we tried to (adjust the bond type, porosity and hardness, diamond consistency of grinding wheel) Si/Cu simultaneous grinding, but we were unable to achieve good grind results.

Concurrently, we found out that the incidence of Cu smear and scratch depends on the grinding time of Si/Cu simultaneous grinding.

Therefore, we thought occurrence of Cu smear and scratch happened due to Cu clogging and coating grinding wheel surface.

In order to solve this problem, we investigated in situ cleaning method of the grinding wheel surface during Si/Cu simultaneous grinding.

As shown in Fig. 4.13 and 4.14, we evaluated the high-pressure micro-jet (HPMJ, Asahi Sunac Corp.) method (method of spraying high-pressure water to the grinding wheel surface), dry ice snow method (method of spraying micro-ice, which is made during $CO_2$ released into the atmosphere to the grinding wheel surface), dress board method (method for grinding wheel surface contact with another wheel and rubbed off mechanically). Full results shown in Table 4.2.

With dress board method, Cu adhering to the grinding wheel surface was enough to be removed, but grinding wheel wear is unusually fast, and not practi-

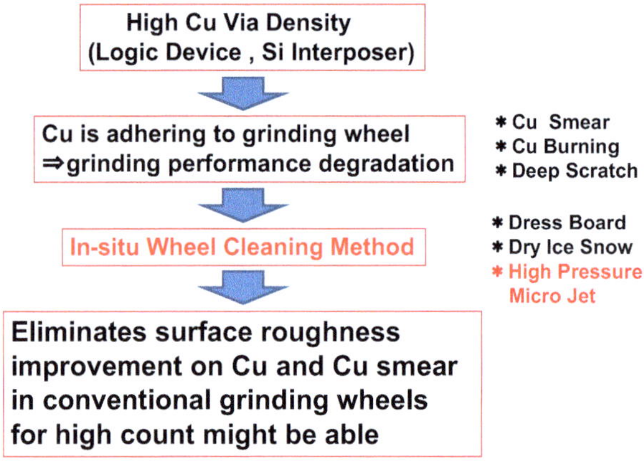


**Fig. 4.13** Si/Cu simultaneous grinding in high TSV via density problem and place wheel cleaning method

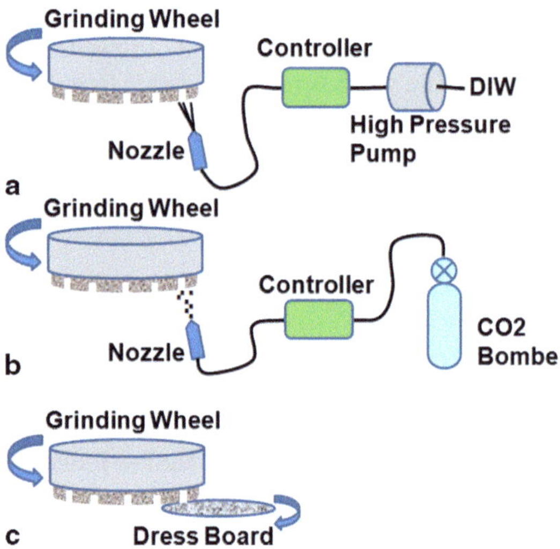


**Fig. 4.14** In situ cleaning method of grinding wheel. **a** HPMJ. **b** Dry ice snow. **c** Dress board

**Table 4.2** Feature of various cleaning method for grinding wheel

| Cleaning Methods | Characteristics and condition | Results |
|---|---|---|
| HPMJ | High pressure deionized water (DIW) 1-17 MPa deposits removal of DIW is 2 L/min in small amounts | Low wheel wear and tuneble Cu full cleaning(W/O remain) |
| Dry ice snow | $CO_2$ occurs upon release into the atmosphere 100 µm below the ice in 0.1–0.5 MPa pressure washing | Low wheel wear Cu spot remain |
| Dress board | Perfect dress boards facing the grinding wheel, rotate and mechanical removal deposits | High wheel wear rate (10 µm/wafer) |

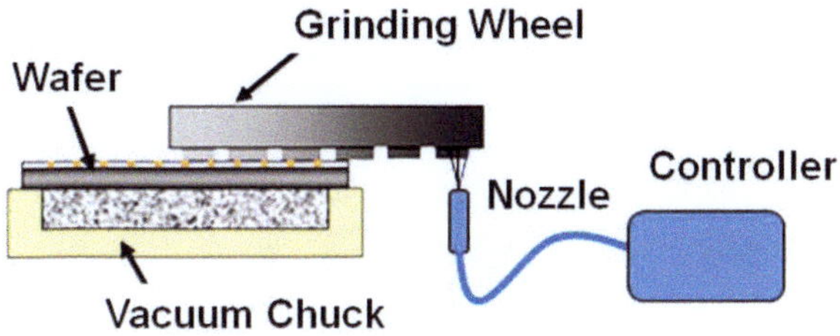


**Fig. 4.15** Schematic illustration of grinder with HPMJ cleaning

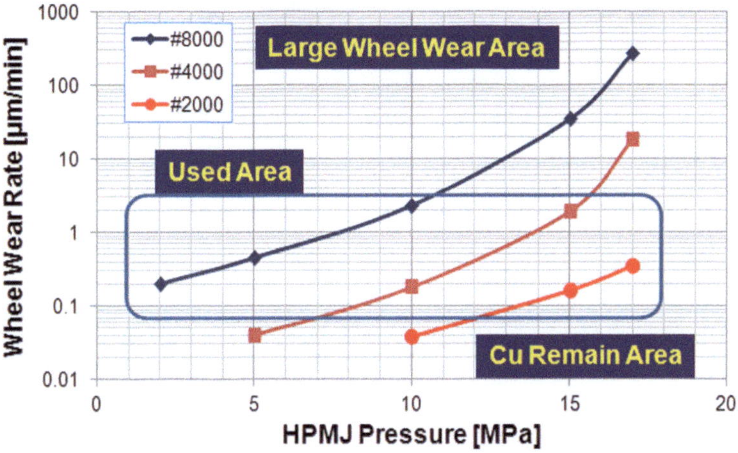


**Fig. 4.16** Relationship between HPMJ pressure and wheel wear rate

cal. With dry ice snow method, wear of grinding wheel was the lowest (less than 0.5 μm), but small amounts of Cu remained in the grinding wheel surface. On the other hand, HPMJ method, wear of the grinding wheel is small (less than 1 μm), and Cu on the grinding wheel surface was removed completely. According to these test results, we decided to use HPMJ method. We believe that this method enables the optimization of Si/Cu simultaneous grinding with adjusting pressure, nozzle opening width, and distance between nozzle and grinding wheel surface. Figure 4.15 shows a schematic illustration of the grinding equipment with HPMJ cleaning. First, TSV wafer is put on a vacuum chuck rotating at 300 rpm, and the grinding wheel rotating at 2000 rpm contacts the wafer at cutting down feed speed of 20 μm/min in the z direction with cooling water continually provided during this process.

The grinding wheels contact approximately half of the wafer diameter; thus, one sixth of the grinding wheel grinds the wafer. We set the spray nozzle out of the wafer to clean the surface of the grinding wheel.

Relationship between HPMJ pressure and wheel wear rate is shown in Fig. 4.16. We prepared vitrified #2000–#8000 grit for this evaluation. The results show that high water pressure accelerates the grinding wheel wear logarithmical-

**Fig. 4.17** Photographs of vitrified grinding wheel (#8000 grit) after Si/Cu grinding. **a** Without cleaning. **b** With HPMJ cleaning

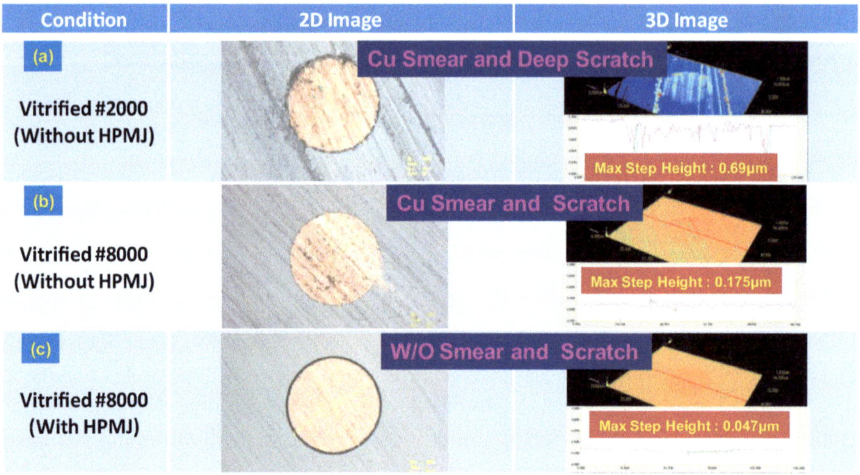


**Fig. 4.18** Optical micrographs of exposed TSV after Si/Cu grinding with vitrified grinding wheel (#8000 grit). **a** Without cleaning (#2000 grit). **b** Without cleaning (#8000 grit). **c** With HPMJ cleaning (#8000 grit)

ly. It was also found out that the wheel wear rate is drastically varied by the grit size of the grinding wheel. Higher mesh wheel tends to wear faster than lower mesh wheel. We found out that proper grinding wheel wear of high-pressure hydraulic power for each wheels as follows.

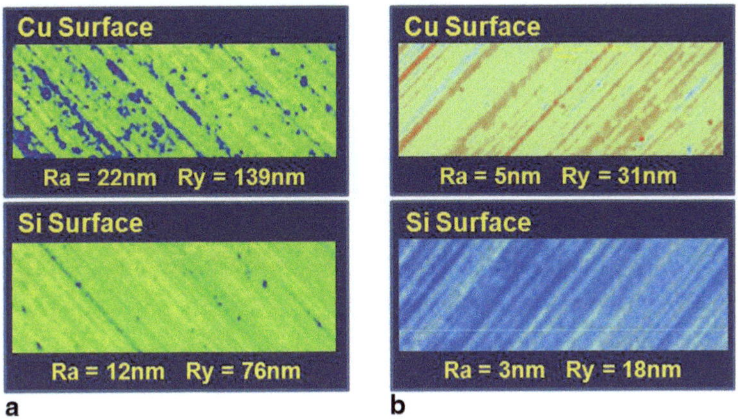


**Fig. 4.19** Roughness of Si and Cu surface after Cu/Si grinding with #8000 grit vitrified-bond grinding wheel. **a** Without cleaning

Vitrified #2000 grit is about 17 MPa, #4000 grit is close to 12 MPa, #8000 grit is around 5 MPa.

Figure 4.17 shows micrographs of vitrified grinding wheel (#8000 grit) after Si/Cu grinding without cleaning and with HPMJ cleaning. Without HPMJ cleaning, we found Cu clogging and oxidation.

By using in situ grinding wheel surface cleaning with HPMJ, the surface of the grinding wheel kept the initial grinding wheel surface condition.

Figure 4.18 shows optical micrographs of exposed TSV after Si/Cu simultaneous grinding with surface condition and scratch depth. The grinding wheel and cleaning conditions are (a) #2000 grit wheel without HPMJ cleaning, (b) #8000 grid wheel without HPMJ cleaning, and (c) #8000 grit wheel with HPMJ cleaning.

The #2000 grit wheel without HPMJ cleaning had burnt Cu surface, and 0.69-µm-deep scratch was observed.

The #8000 grit wheel without HPMJ; Cu surface condition was improved compared with #2000 grit wheel, but there was Cu smear on the Si surface. Scratch was also improved to 0.175-µm deep.

The #8000 grit wheel with HPMJ cleaning showed good surface condition. Cu smear was not observed, and scratch depth was as small as 0.047 µm.

Figure 4.19 shows the roughness of Si and Cu surface after Si/Cu simultaneous grinding with #8000 grit vitrified grinding wheel.

Without HPMJ cleaning, Cu surface roughness (Ra) was 22 nm and Si surface roughness was 12 nm. With HPMJ cleaning, Cu surface roughness was 5 nm and Si surface roughness was 3 nm.

The surface roughness is one fourth compared to that without HPMJ cleaning, and it is almost same surface roughness of ground surface of Si wafer without TSV.

Based on these results, we concluded that the reason of good surface condition of the ground TSV wafer with HPMJ cleaning and higher mesh wheel was that the wheel surface was always kept as the initial condition after dressing wheel.

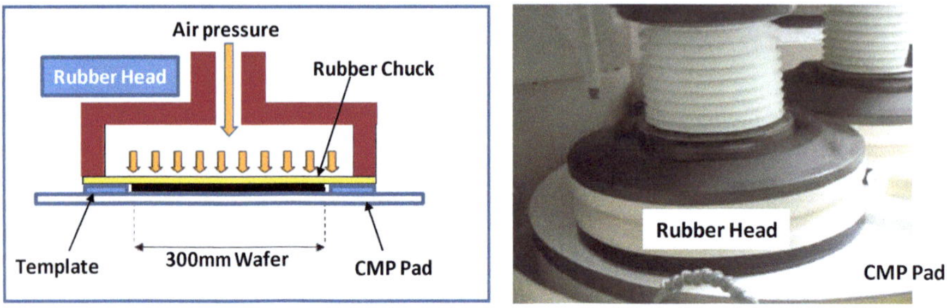


**Fig. 4.20** Structure and photograph of rubber head

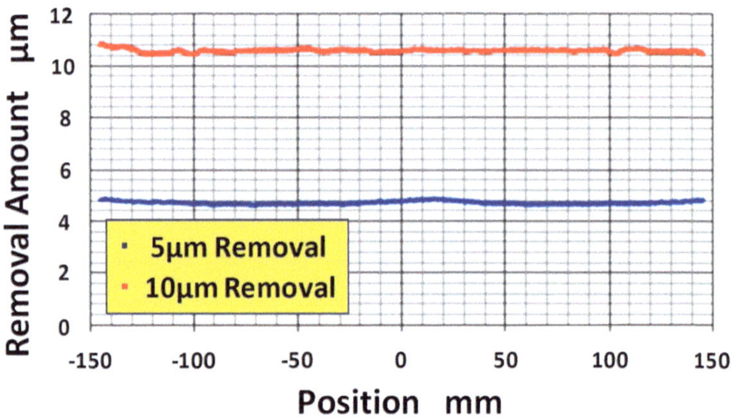


**Fig. 4.21** Profile of removal amount within a wafer

#### 4.2.3.3 Si/Cu Same Removal Rate CMP (First CMP): Non-Selective CMP Between Cu and Si

In this chapter, we describe first CMP for Cu via. The purpose of the first CMP is to eliminate damage of Si by grinding process. The removal amount of silicon is 2–3 μm.

The polishing head which was used for the test is shown in Fig. 4.20. Wafer thickness profile and its uniformity with removal amount are shown in Fig. 4.21. This rubber polishing head structure can give a uniform load in a larger area than wafer size that enables to keep the wafer thickness uniformity after CMP. The uniformity does not change even if the removal amount increases from 5 to 10 μm.

The TTV of the wafer after grinding is usually less than 1 μm. First CMP can keep this uniformity.

Another important purpose of the first CMP is to polish Cu, Si, and TEOS-$sio_2$ at a same polishing rate. It prevents step variation between Cu, Si, and TSV liner.

**Table 4.3** First CMP optimum condition for Cu/Si simultaneous CMP (same removal rate)

| Item | Condition |
|---|---|
| Slurry | RDS0902 (Oxidation $H_2O_2$ 5%)<br>Fujimi SFR 200 mL/min |
| Pad | IC 1400 (XY trench) nitta haas |
| Wafwe rotation | 41 $min^{-1}$ |
| Table rotation | 40 $min^{-1}$ |
| Load | 25 kPa |

CMP removal rate is controllable with the oxidizer ($H_2O_2$) concentration in the polishing slurry. Higher oxidant concentration gives faster Cu CMP removal rate, and lower concentration gives slower Cu CMP removal rate. Si CMP removal rate is not affected by concentration of the oxidizer. Thus, by adjusting the oxidizer concentration, Cu and Si can be polished at the same polishing rate.

Table 4.3 shows the optimized first CMP condition for Cu/Si simultaneous CMP with same removal rate.

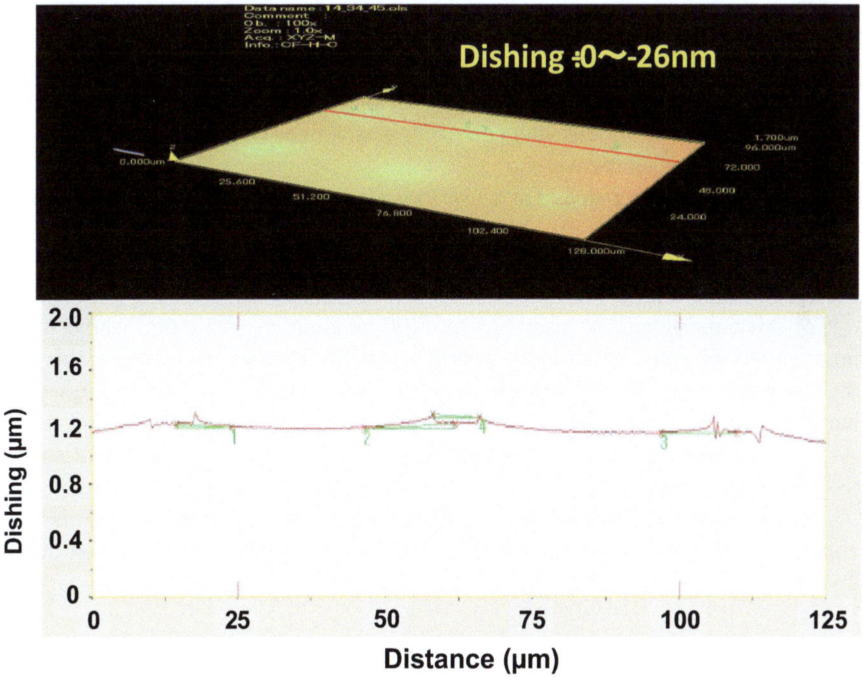


**Fig. 4.22** TSV shapes after first CMP

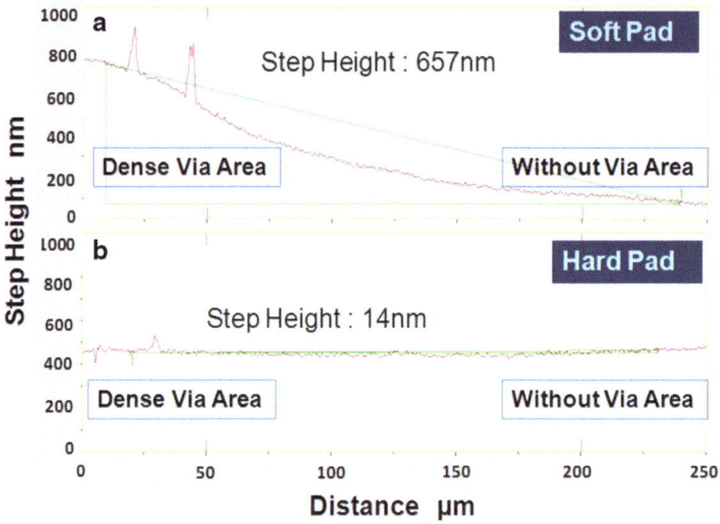


**Fig. 4.23** Surface profiles on the areas with dense TSVs and without TSVs. **a** Soft pad CMP. **b** Hard pad CMP

Figure 4.22 shows the 3D image of TSV polished under the optimum condition. The top of the Cu via from Si surface was well flat with a result of less than 30 nm of dishing.

As previously described, TSV density of memory devices is about 1 %. This means that TSV does not distribute on whole wafer area. There are dense TSV areas and non-TSV areas on the wafer.

We measured the step height at dense TSV areas and non-dense TSV areas which is 100-μm distance from TSV. The results are shown in Fig. 4.23.

Figure 4.23(a) shows the profile polished by a soft pad and (b) shows the result polished by a hard pad.

For this experiment, we used SUBA #400, made of woven cloth, whose hardness was 61 (Asker-C), compression ratio was 80 %, and it is 1.27 mm thick as the soft pad. The hard pad shown in (b) was IC1400 made of polyurethane, whose hardness was 57 (Shore-D), compression ratio was 2 %, and it is 1.27 mm thick. As shown in this figure, the step height was 657 nm by the soft pad and 14 nm by the hard pad.

We conclude surface deformation of the hard pad was small due to its mechanical properties and thus achieved good flatness in both dense and non-dense TSV area.

#### 4.2.3.4 TSV Protrusion CMP (Second CMP): Selective CMP Between Cu and Si

The second CMP is the processes for protruding TSV during CMP. In this process, it is needed to preserve the TSV shapes and polish only Si. To achieve this purpose,

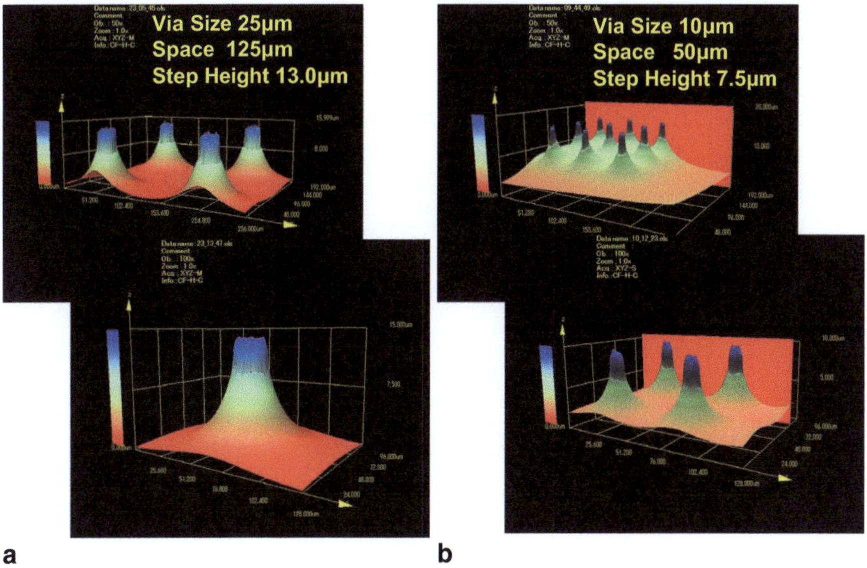


**Fig. 4.24** TSV protrusion shapes after second CMP. **a** Via size 25 μm. **b** Via size 10 μm

the polishing pad should be as soft as possible, and the slurry should have high selectivity to Cu and Si.

We chose slurry with high selectivity and polishing pads with soft-padded suede series.

Si polishing rate was 0.7 μm/min, and Cu polishing rate was 0.006 μm/min; thus, the selectivity was greater than 100.

Figure 4.24 and Figure 4.25 show the TSV protrusion after second CMP and protrusion height distribution, respectively. The second CMP condition is shown in Table 4.4. The shape of the protrusion and distribution is acceptable for TSV.

#### 4.2.3.5 Post Second CMP Cleaning

Post second CMP cleaning after protruding TSV is an essential technique to avoid electrical characteristics change of semiconductor devices due to Cu contamination. Typically, hydrochloric acid peroxide mixture (HPM) is used to remove metallic contaminant on silicon wafer. However, substantial Cu concentration on exposed TSV wafer is extremely high. Thus, chemicals to remove Cu from TSV wafer should be chosen carefully. After extensive investigation, we chose organic acid (TSV-1, Mitsubishi Chemical Corp.) and dilute HF acid as cleaning chemicals.

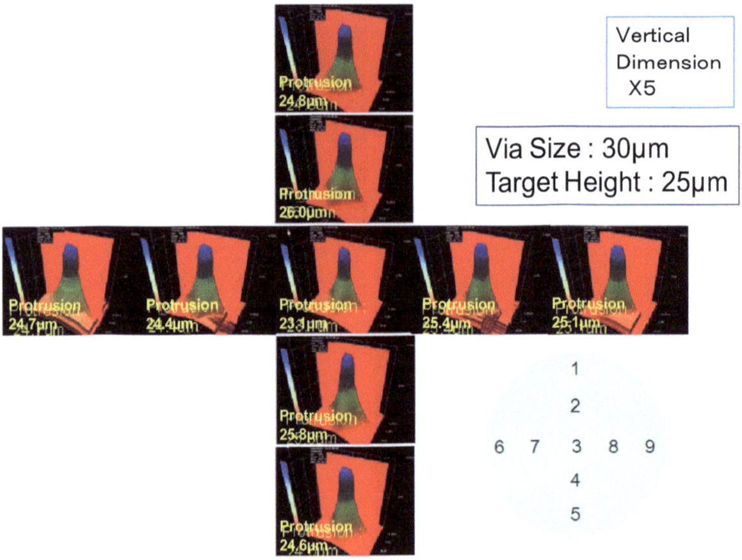


**Fig. 4.25** TSV protrusion height variations after second CMP

**Table 4.4** Second CMP condition for TSV protrusion

| Item | Condition |
|---|---|
| Slurry | RDS10906 Fujimi<br>SFR 200 mL/min |
| Pad | RN-H swede (XY trench) nitta haas |
| Wafer rotation | 21 $min^{-1}$ |
| Table rotation | 20 $min^{-1}$ |
| Load | 20 kPa |

The cleaning condition is shown in Table 4.5. Run #1 does not use surface scrubbing by brush, and Run #2 uses polyvinyl alcohol (PVA) brush to scrub the surface. Residual heavy metals on the silicon wafer were evaluated by time-of-flight secondary ion mass spectrometry (TOF-SIMS).

Measurement spots were wafer center and edge. Both spots were located between two TSVs whose size was 50 μm diameter and 150 μm pitch. The spot size was 20 μm diameter. Table 4.6 shows the results of measurement for residual heavy metal contamination. Contamination of Cu of run #1 was slightly greater than run #2 that indicates cross-contamination resulted via PVA brush scrubbing. Anyway, Cu contamination remaining on silicon was $3.3 \times 10^{11}$–$4.5 \times 10^{11}$ atoms/cm$^2$. It is out of user request value $5 \times 10^{10}$ atoms/cm$^2$. Further improvement is required for this subject. Contamination of other heavy metals was acceptable.

**Table 4.5 Cleaning condition after second CMP**

| Cleaning step | 条件 | |
|---|---|---|
| | Run #1 | Run #2 |
| ①Chemical 1 | TSV − 1*(× 40 Dilution)<br>100 $min^{-1}$, 2 min | TSV − 1*(× 40 Dilution)/PVA<br>Brush 100 $min^{-1}$, 2 min |
| ②Chemical2 | 0.5% DHF 100 $min^{-1}$, 1 min | 0.5% DHF 100 $min^{-1}$, 1 min |
| ③Chemical3 | Without | TSV−1*(× 40 Dilution)<br>100 $min^{-1}$, 2 min |
| ④DIW rinse | MS/DIW 100 $min^{-1}$, 1 min<br>DIW 100 $min^{-1}$, 1 min | MS/DIW 100 $min^{-1}$, 1 min<br>DIW 100 $min^{-1}$, 1 min |
| ⑤Spin dry | 2000 $min^{-1}$, 1 min | 2000 $min^{-1}$, 1 min |

TSV − 1* :Mitsubishi chemical

**Table 4.6** Metal contamination after cleaning using TOF-SIMS analysis

| Metal | Metal Contamination (× $10^{10}$ atoms/$cm^2$) | | | |
|---|---|---|---|---|
| | Run #1 | | Run #2 | |
| | Wafer center | Wafer edge | Wafer center | Wafer edge |
| Ti | <3.2 | <2.9 | <2.3 | <2.4 |
| V | <0.2 | <0.19 | <0.15 | <0.15 |
| Cr | <0.24 | <0.23 | 0.17 | <1.18 |
| Mn | <0.23 | <0.32 | <0.16 | <0.17 |
| Fe | <0.43 | <0.40 | <0.31 | <0.32 |
| Ni | <0.86 | <0.79 | <0.61 | <0.63 |
| Co | <0.54 | <0.50 | <0.38 | <0.40 |
| Cu | 36 | 33 | 42 | 45 |
| Zn | <3.5 | <3.2 | <2.5 | <2.6 |
| Sn | <1.5 | <1.4 | <1.1 | <1.1 |

### *4.2.4 Conclusion*

We developed a novel via-middle TSV thinning technology by Si/Cu simultaneous grinding, Si/Cu same removal rate CMP (first CMP), and Cu via protrusion CMP (second CMP).

First, in the Si/Cu simultaneous grinding, we developed a new grinding method with in situ grinding wheel surface cleaning by HPMJ, which enabled to grind high-density TSV wafer without Cu burnt and Cu smear

In first CMP after grinding (Si/Cu same removal rate CMP), we succeeded to remove wafer damage with keeping good uniformity by optimizing the ratio of oxidizer and hard pad.

In second CMP (TSV protrusion CMP), we got the good shape of via protrusion by using combination of ultrasoft suede pad and highly selective slurry.

Cleaning after second CMP, residual Cu contamination was over $10^{11}$ atoms/$cm^2$, but after this development process, we succeeded to have 100-nm protective membranes made from electroless Ni–5 %B on the TSV Cu surface, then lift off contamination by alkaline etching of Si [17]. We completed via-middle process without contamination.

Based on above results, we concluded that low-cost via-middle TSV with wet process was available.

To provide this process for production, we completed the multifunctional equipment which has grinding, CMP, post CMP cleaning in one equipment. We keep this equipment update to match the market requirement.

## 4.3 Temporary Bonding

### *4.3.1 Background*

Many kinds of temporary bonding adhesives and systems have been proposed by a number of companies in the industry. The words "temporary bonding adhesive" literary state that "debonding" is also required after roles are done. Those roles include supporting active IC device wafers beginning with BG process through-out backside processing for TSV formation. At BG process, mechanical force is applied for the adhesive layer, so adhesion needs to be appropriate. In addition, the adhesive layer has to have adequate hardness or strength. Requirements for processes of TSV formation are even more stringent. Thermal and chemical durability is one of such requirements. After the finish of TSV processes, the adhesive needs to be "debonded" by some means.

Conventionally, for a few decades of years, the so-called wax has been used as a temporary adhesive. The adhesive is categorized as an adhesive of hot-melt type and is common in abrasive industry for thinning application. For 3D TSV applications, such adhesive was tried to use at the early stage of development.

Semiconductor tapes such as BG tapes and dicing tapes are considered to be temporary bonding materials in a sense. Those tapes provide temporary support at each process step. Therefore, "double-sided sticky tapes" are being utilized for some temporary bonding applications such as glass thinning in liquid-crystal display (LCD) industry. Such sticky tapes have tacky adhesive layers formed on base plastic film. Tacky adhesive layer is, in general, tough to have high thermal resistivity. Conformity for surface topography is another issue. Thus, such tapes have not been selected for temporary bonding adhesive for 3D TSV application.

As of today, liquid-state adhesives are common for temporary bonding solution of 3D TSV application. There are, still, a few numbers of different systems which are proposed depending upon difference in curing mechanism of adhesive. In addition, various debonding methods proposed so far make the entire range of selection so wide.

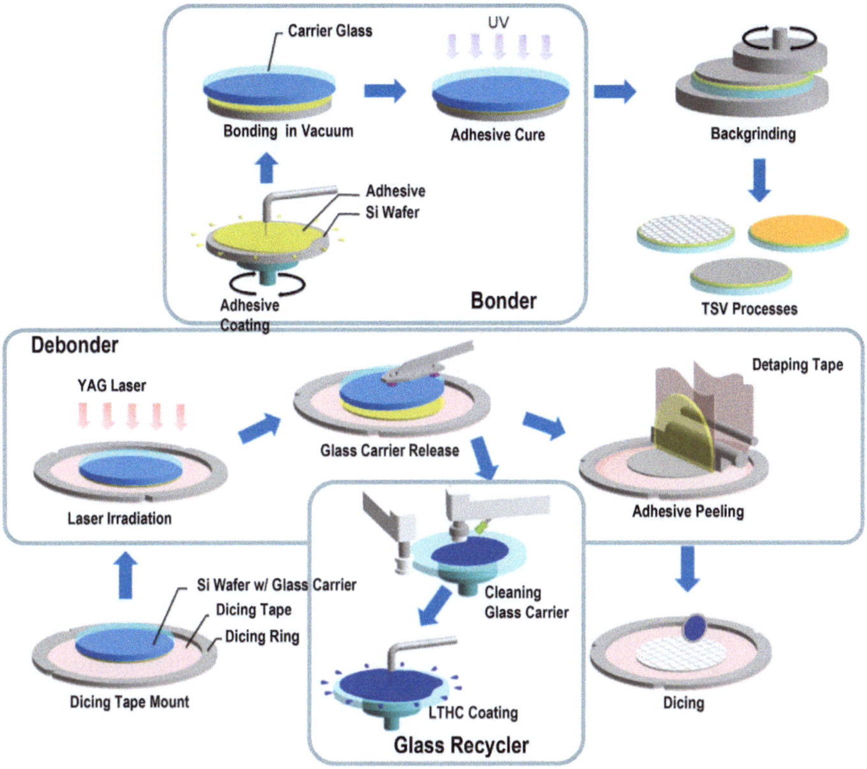


**Fig. 4.26** Process flow of the temporary bonding system. *UV* ultraviolet, *LTHC* light-to-heat conversion, *YAG* yttrium–aluminum–garnet, *TSV* through-silicon via

## 4.3.2 The 3M$^{TM}$ Temporary Bonding Materials

The 3M$^{TM}$ temporary bonding materials are basically combinations of ultraviolet (UV) curable temporary adhesive and unique laser-absorbing ink for carrier release. The UV curable adhesive is spin coated on a Si wafer surface, and the wafer is bonded to a glass carrier in vacuum. Then UV light is irradiated through glass carrier to cure the adhesive. The laser-absorbing ink is a coating liquid, and it can be spin coated onto glass carrier prior to bonding with Si wafer using the UV curable adhesive. The role of the ink is to form laser-absorbing layer and to release thinned and processed wafer from glass carrier by yttrium–aluminum–garnet (YAG) laser irradiation. The laser-absorbing layer is called as "light-to-heat conversion" or LTHC layer.

Figure 4.26 is a schematic view of the entire process flow of the temporary bonding system. Typical temporary bonding process is consist of following three processes.

**Bonding Process**

1. The adhesive is coated onto wafer by spin coating.
2. The wafer with adhesive coated is bonded to a glass carrier with laser-absorbing layer in vacuum chamber.
3. UV is irradiated through glass carrier in atmospheric pressure. This is the end of the bonding process

**Debonding Process**

1. YAG laser is scanned through glass carrier, such that LTHC layer is thermally decomposed.
2. Glass carrier is released from adhesive layer using vacuum suction cups.
3. Adhesive layer is peeled off from wafer surface using removing tape. This is the end of the debonding process.

**Glass Recycling Process**

1. LTHC residue is scrub cleaned by using 5 % ammonium hydroxide
2. LTHC ink is spin coated onto cleaned glass carriers. This is the end of glass recycling process.

Figure 4.27 shows the cross-section of bonded stack at each process step. Diameter of glass carrier is recommended to be larger by 1 mm than the side of the Si wafer. After BG, wafer edge is supported by adhesive. YAG laser irradiation is taken place with the thinned and TSV processed wafer being attached to dicing tape and set on a flat vacuum stage.

Figure 4.28 shows an example of large volume manufacturing tool. This is a picture of bonder for 12-in. wafers. A typical manufacturing tool is capable of bonding 25 wafers per hour or more.

### *4.3.3 The 3M™ Temporary Adhesive*

The adhesive was designed for temporary bonding where a thick enough coating layer can be spin coated at a time. The adhesive layer is required to be thick enough for conforming wafer surface topography. Bumps and pillars are sometimes formed on Si wafers. The adhesive needs to fill and cover such surface topography. Top of each bump or pillar is not allowed to physically contact glass carrier surface.

Since the adhesive does not contain any solvent or diluents, no drying process is required after coating. Immediately after spin coating, the wafer with adhesive coated is bonded to a glass carrier in vacuum. The so-called rotary pump level of vacuum is good enough to achieve voids-free bonding. Actual vacuum level for bonding is set to be 40–60 Pa or so. After bonding, the vacuum is broken followed by UV irradiation to cure adhesive.

Adhesive being UV curable delivers a number of significant advantages. It can be cured under RT, and it only takes a few tens of seconds to cure. In addition, there

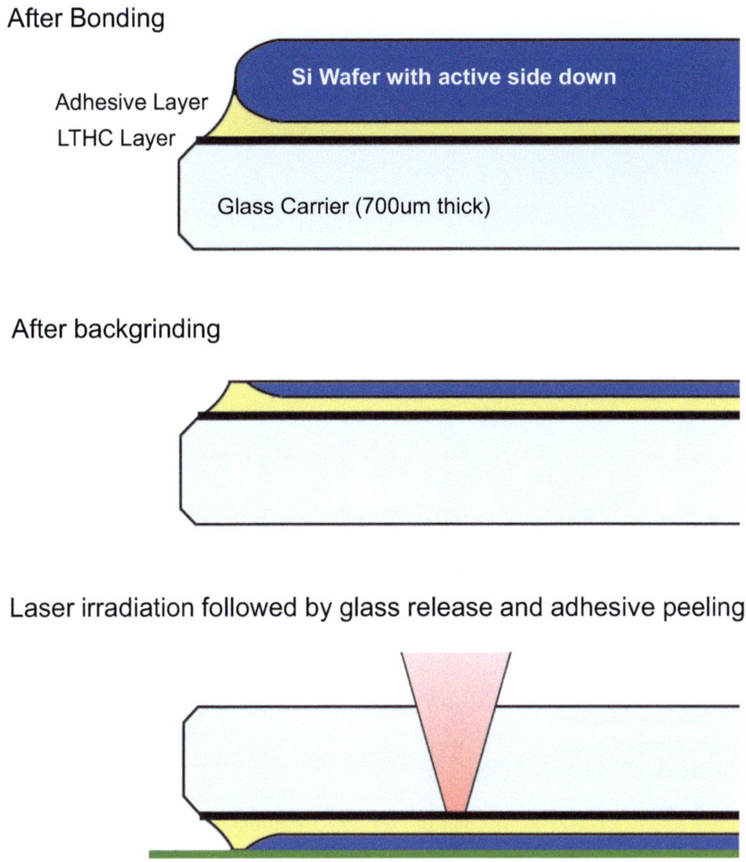


**Fig. 4.27** X-section of the bonded stack at each process step. *LTHC* light-to-heat conversion

is no need to press for bonding. Such features make dedicated manufacturing tools less complex.

Thermal resistance and chemical resistance are typical requirements for the temporary adhesive, once bonding is done and the adhesive is cured. Typical processes after bonding are BG or thinning followed by TSV processes for instance. TSV processes include several processes where temperature goes up as high as over 200 °C, and various chemicals are used for etching, cleaning, plating, and so on.

The company has a lineup of temporary adhesive products depending upon levels of such requirements. Figure 4.29 shows positioning of each adhesive with respect to heat and chemical resistance. LC-3200 is widely used for power IC manufacturing where thermal resistance requirement is not as high as that of TSV forming processes. LC-5300 is suitable for both thermal resistance and chemical resistance.

Table 4.7 shows some examples of test data for the temporary bonding adhesive. Those values of elongation at break point indicate how flexible cured adhesive film

**Fig. 4.28** An example of manufacturing tool of the temporary bonding system

**Fig. 4.29** Line-ups of the temporary bonding adhesive. *TSV* through-silicon via, *IC* integrated circuit

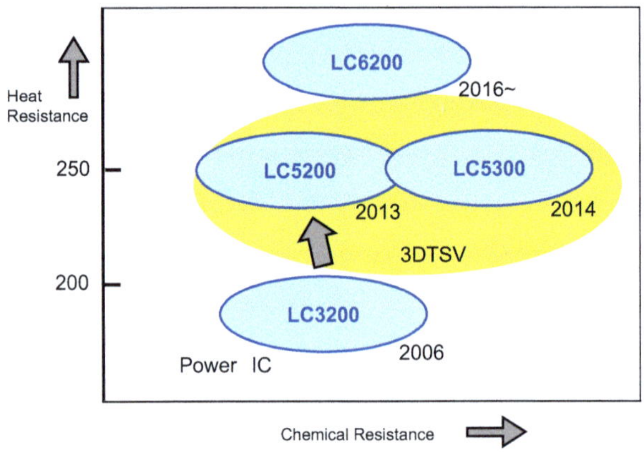


is. Cured adhesive film needs to be adequately soft and flexible for stretching when the film is peeled off from Si wafer surface at the end of processes. This characteristic is important especially for wafers with high topography such as bumps and pillars.

Those modulus numbers indicate that the adhesive layer is rigid enough for supporting Si wafer throughout BG process, until the wafer becomes thin as designed. The modulus needs to be within an appropriate range especially for BG where compression and sheer forces are applied. In general, BG speed can be increased with temporary bonding system compared to the use of conventional BG tape, especially

**Table 4.7** Test data of the temporary adhesive. *TGA*thermo-gravimetry analysis, *NMP* N-methyl-pyrrolidone, *WSSwafer support system, DMSO*dimethyl sulfoxide

| WSS adhesive | LC-3200 | LC-5200 | LC-5300 |
|---|---|---|---|
| Elongation at break (%) | 20 | 60 | 40 |
| Modulus (MPa) @ 25 °C | 297 | 1233 | 650 |
| TGA weight loss @ 250 °C,1 hr (%) | 5.1 | 1.4 | 1.2 |
| Tg (°C) | 50 | 45 | 50 |
| 5 %KOH/DMSO(wt gain %) | 12.5 | 37 | 1.6 |
| NMP @60C(wt gain %) | 32.3 | 59.8 | 6.6 |
| Viscosity (cps) | 3000 | 1970 | 2800 |

for thinner BG below 100 μm. This is attributed to rigidness of cured adhesive, and the support of wafer can be more solidly done.

Test method for thermal resistance is thermo-gravimetry analysis (TGA). It indicates degree of thermal decomposition. A 50-μm-thick cured adhesive film is prepared as the test specimen. The specimen is kept in nitrogen atmosphere, and the temperature ramps up to 250 °C. Due to thermal decomposition of organic components in the adhesive film at the temperature, the weight of specimen decreases. The lower percentage of the weight loss means the better thermal resistance. The LC5300 shows much lower weight loss than that of LC3200.

When the process temperature exceeds thermal resistance of the adhesive, what is expected to occur is delamination during processes where bonded wafers are exposed to heat. This is due to thermal decomposition of the adhesive layer which generates gases. Such gases cause delamination at either side of adhesive layer. Another phenomenon observed when temperature exceeds is that adhesion between adhesive layer and wafer surface becomes too high to peel, resulting in creating residue on wafer surface.

Table 4.7 also includes chemical resistance data. Weight gain by swelling in 5 %KOH/dimethyl sulfoxide (DMSO) solution and the same in 60 °C of *N*-methyl-pyrrolidone (NMP) solvent are measured for this test. Again, LC-5300 shows lower degree of swelling than LC-3200 does. When chemical resistance of adhesive is not good enough, the adhesive becomes swelling and causes delamination initiating at edge where adhesive layer is exposed to such chemicals and solvents.

The temporary bonding adhesive has relatively high viscosity, such that a thick enough coating can be made at a time by spin coating. Viscosity data are shown at the bottom of Table 4.7. Those relatively high viscosity numbers enable to deliver good TTV even if glass carriers and Si wafers have warp to some extent, because once the adhesive is coated thinner below 100 μm or so, it tends to resist to flow within the gap between glass carrier and Si wafer.

Another noticeable feature of the temporary bonding adhesive is that it can be peeled off from wafer surface at the end of processes. The adhesion strength is adjusted so such that peeling off as a film state, and unexpected delamination does not occur during other processes prior to the intended debonding. When adhesive is peeled off from wafer surface, the adhesive layer is still soft enough such that the layer can be bent with a small radius.

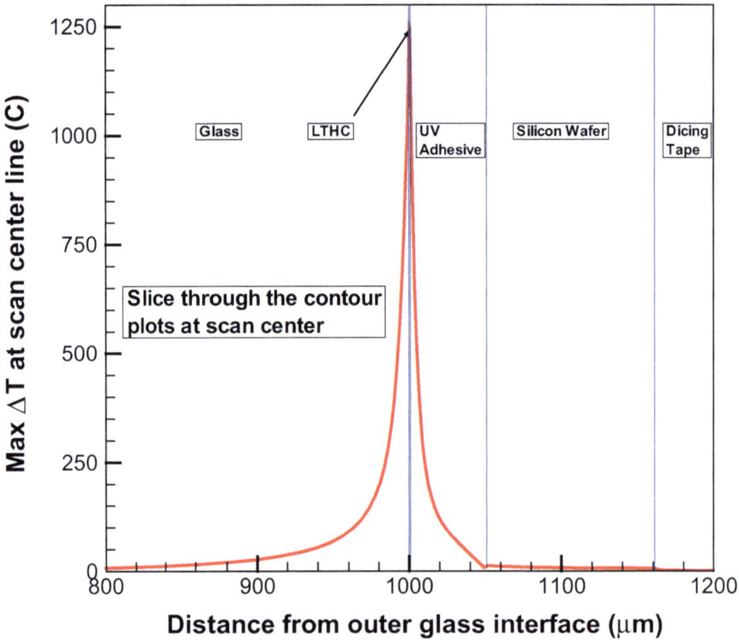


**Fig. 4.30** Thermal modeling of the temporary bonding system. *UV* ultraviolet

### 4.3.4 *Laser-Absorbing Layer*

The layer is formed on glass carrier by spin coating of laser-absorbing ink. The ink is a dispersion of carbon black and binder in some mixture of solvents. By spin coating the solution followed by drying, the laser-absorbing layer is formed. The layer typically has a thickness of 1 μm. When YAG laser is irradiated to the layer by scanning, it efficiently absorbs the laser energy and generates heat causing decomposition of organic binder in the layer. Figure 4.30 shows how high the temperature can theoretically rises at various points of the stack. This is a simple calculation based upon adsorbed energy versus thermal property of stack. Important point is that there is no thermal damage on wafer surface and glass carrier, which are already demonstrated to prove.

Lasers used for this process can be YAG laser which is familiar in the industry. The base wavelength of 1064 nm is just fine, and the second harmonic wavelength of 528 nm is also workable. Both continuous wave (CW) and pulse mode laser can be used. Scanning optics enable uniform laser irradiation on entire wafer surface to make a stress-free separation. Typical spot size of the laser at focus point is about 0.3 mm in diameter. Scan speed is typically 8.0 m/s. It takes about 90 s to scan a 12-in. wafer. Laser power is set to be the lowest required achieving a smooth glass carrier separation. The value is somewhere between 20 and 50 W depending upon

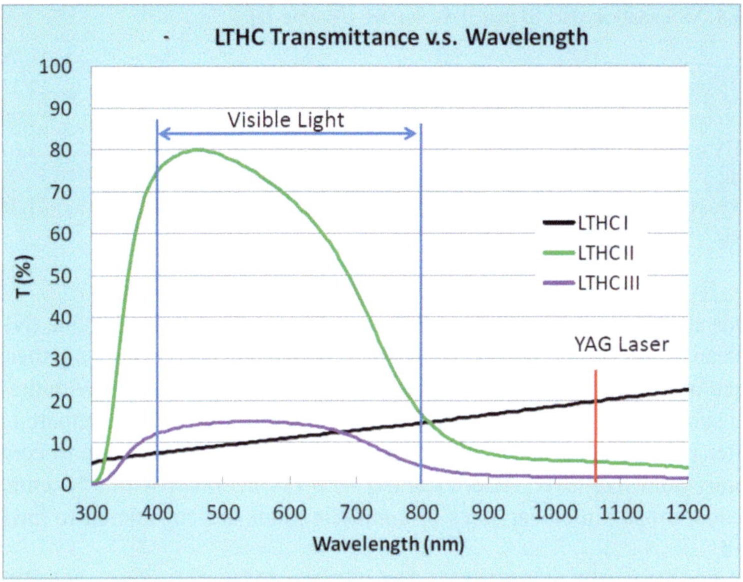


**Fig. 4.31** Light transmittance of LTHC layer. *LTHC* light-to-heat conversion

other parameter settings for scan speed and pitch. Excessive power results in generating debris of burned laser-absorbing layer, which is not preferable.

The laser-absorbing layer has typically thickness of 1 μm. Figure 4.31 shows light transmittance curve versus wavelength. The conventional laser-absorbing layer (LTHC I) has 20 % of transmittance at 1,064 nm of YAG laser wavelength, which means around 80 % of laser energy is absorbed by the layer. On the other hand, at visible light wavelength, the layer blocks about 90 % of light, therefore, wafer pattern can barely be seen through glass carrier with the layer. This is a disadvantage of the conventional laser-absorbing layer. With the newly developed laser-absorbing layer (LTHC II), the disadvantage has been resolved. LTHC III is for another type of purpose where YAG laser leak is further reduced.

Adhesive peeling after glass carrier release is processed by using a removing tape. The tape is another 3M™ product for the process. It is a kind of sticky tape and adheres well to temporary adhesive layer surface after laser release. The adhesive layer is peeled off together with the attached removing tape.

One of the most significant advantages of the 3M™ temporary adhesive is that the adhesive layer can be peeled off from wafer surface. No visible residue remains after adhesive peeling as long as adhesive is correctly selected and used. However, when surface analysis such as X-ray photoelectron spectroscopy (XPS) is performed on the wafer surface after adhesive peeling, molecular size levels of contaminant can be detected. This is a similar observation of such semiconductor tapes as BG tapes and dicing tapes which are directly attached to the active wafer surface. In case, such contaminants are not be accepted, and some simple cleaning

**Table 4.8** XPS analysis data of plated Au surface (Atomic %)

| | C | O | Si | Au | C/Au |
|---|---|---|---|---|---|
| Before processing | 38 | 26 | 3 | 33 | 1.15 |
| After processing | 53 | 23 | 2 | 23 | 2.30 |
| After UV ozone cleaning | 30 | 38 | 4 | 28 | 1.07 |
| After 5 % $NH_4OH$ cleaning | 47 | 15 | 1 | 38 | 1.24 |

methods are available. Surface rinsing using 5 % ammonium hydroxide ($NH_4OH$) well cleans such contaminants for instance. UV ozone cleaning is another effective method. Table 4.8. shows the effect of such cleaning methods. With the analysis, Au surface formed by sputtering was the background. When compared before and after processing, atomic concentration of carbon increased due to contact of temporary adhesive layer. After cleaning with UV ozone or with 5 % ammonium hydroxide, those molecular sizes of contaminations are considered to have been cleaned.

Material for carrier is limited to glass with the 3M system, because UV for curing adhesive and laser light for carrier release need to go through carrier. Typically, thickness of glass carriers is 0.7 mm. A larger diameter than that of wafer by about 1 mm is recommended. TTV of glass is an important index showing how good those glass carriers are. It is mainly attributed to manufacturing processes of glass carrier. A typical specification for TTV of glass carriers is 2.0 μm. However, majority of glass carriers have TTV values smaller than 1.0 μm.

Another important index for glass carriers is flatness. The flatness is commonly measured by means of a surface topography. However, when glass carriers are laid flat, gravity makes them flatter than they actually are. There are several means to cancel the effect of gravity. Another simple method is to perform surface topography as glass carrier being placed vertically.

Figure 4.32 shows examples of such measurement result of glass carrier flatness. This is a result without cancelation of gravity effect. Randomly selected 10 glass carriers were placed on flat measurement stage and tested. The result indicates that each glass carrier has around 100 μm of peak-to-valley value, which is the so-called flatness number. Those glass carriers were cut out from sheet glass which was made by means of floating method at a glass manufacturer. Such glass has front-side and backside. Warp toward upside can be seen for all of those 10 glass carriers. When a glass carrier is bonded to Si wafer, the glass carrier is placed on spin-coated adhesive layer, which is still liquid, in vacuum.

On the other hand, TTV value after bonding of a Si wafer and a carrier glass is almost always below 5 μm. The fact suggests that such flatness value as 100 μm or so is corrected or is flattened out by bonding with Si wafer. However, if such waviness exists within relatively small region on glass carrier surface, it would cause adverse effect on TTV after bonding.

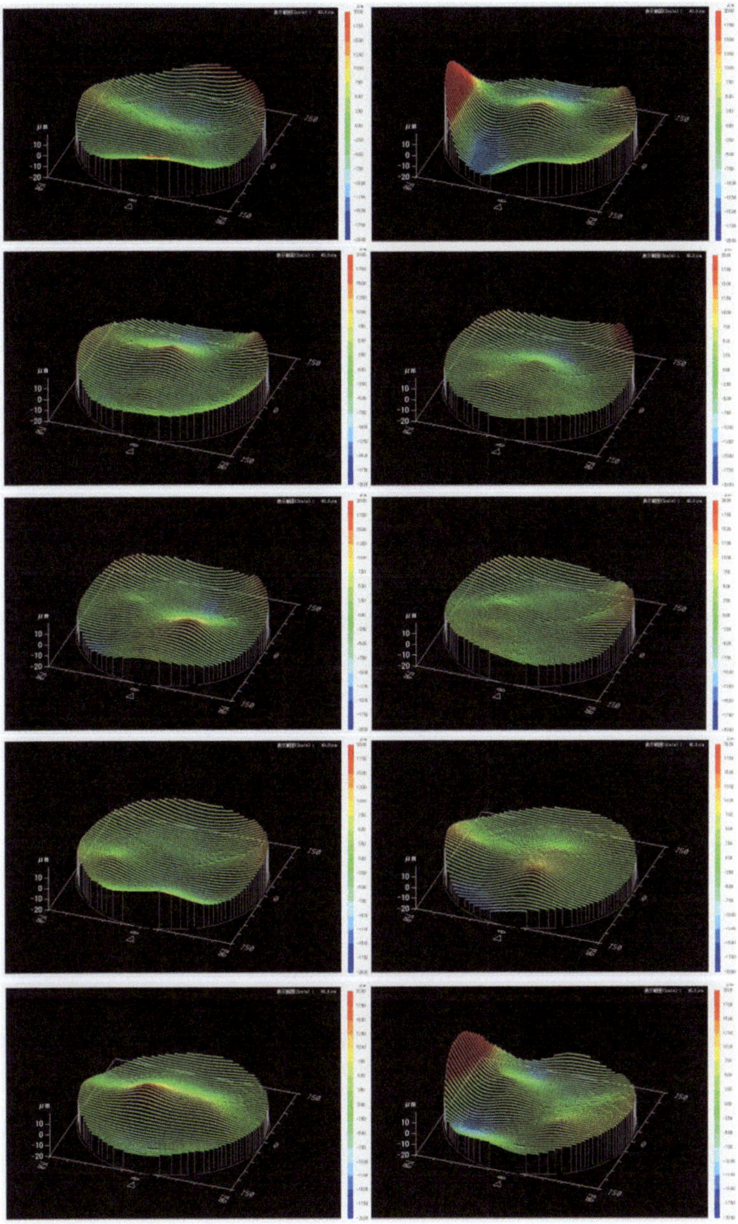

**Fig. 4.32** Surface topography measurement of carrier glass

TTV after bonding is a very important index to access the performance of temporary bonding system. The reason is it directly determines TTV of thinned wafer. TTV after bonding is determined as the result of stacking Si wafer, adhesive layer, and glass carrier. TTV of each needs to be minimized to make TTV after bonding better. Since adhesive layer is originally liquid and then cured to be a solid film, it is important that coating thickness uniformity of adhesive is good. The Newtonian nature of liquid adhesive delivers such result.

### 4.3.5 The Next Steps

The future advances of temporary bonding solution would be based on further developments on both material and process technologies. Heat and chemical resistivity of adhesive will further be improved. Since coat issue has been highlighted for TSV technology penetration, effort for cost down will also be sought for. A different debonding technology where the easy peeling feature of the temporary adhesive is effectively used will be introduced in the near future. The so-called mechanical debonding will be the one, with which cost of ownership (COO) can be expected to significantly reduce.

## 4.4 Temporary Bonding and Debonding for TSV Processing

### 4.4.1 Introduction

Improvements have been made in the semiconductor industry to performance, reliability, and productivity by means of miniaturization. Using TSV, equivalent scaling in 3D packaging for the lamination of device chips has been achieved as an extension of More Moore [18], and in recent years, 3D lamination products using TSV have been released starting with dynamic random access memory (DRAM) products, which are difficult to improve by geometrical scaling in terms of performance.

When manufacturing 3D lamination devices using TSV, processing must be carried out on thinned device wafers. When processing thin device wafers, the strength of the thin wafer is reduced, which may cause the wafer to warp, or to be chipped or broken, etc., during transport due to film formation stress. Therefore, in order to handle the wafer, the device wafer must be supported during processing. In general, this is referred to as thin wafer handling or the wafer support system (WSS), which is a technique where a support wafer is temporarily bonded to a device wafer, and, after completing the process, the support wafer is removed before being handed over for package processing.

In this chapter, we will explain the technical details of temporary bonding and debonding.

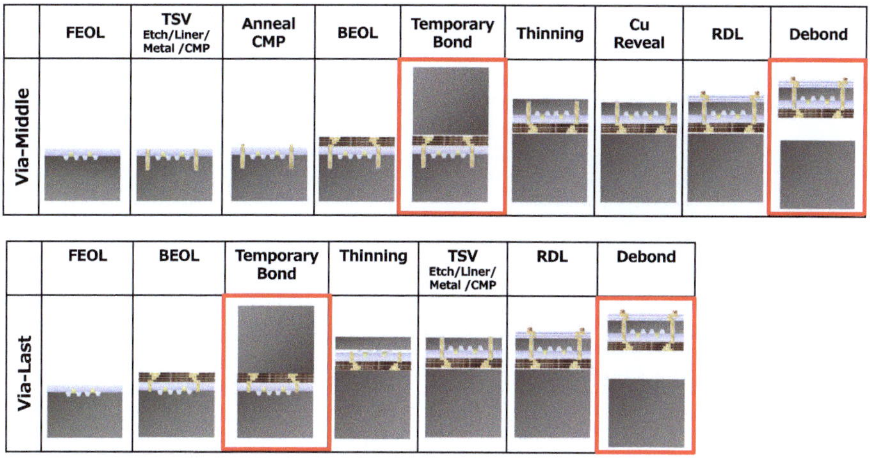


**Fig. 4.33** TSV process flow. *TSV* through-silicon via, *CMP* chemical–mechanical polishing, *BEOL* back end of line, *FEOL* front end of line, *RDL* redistribution layer

### *4.4.2 Temporary Bonding and Debonding Process*

Depending on the order of making the via, the TSV process generally uses a via formation process know as via-middle or via-last (Fig. 4.33). Each technique has its strong points and weak points, but in each process, wafer processing must be performed after temporary bonding. The main processes after temporary bonding are wafer thinning, CVD, lithography, etching, and the bumping process. After all of these processes, debonding of the support wafer is performed, and it is handed over for the dicing process. In the thin wafer handling process, the thin device wafer must be supported throughout the processes from bonding to debonding.

Table 4.9 shows the performance requirements for thin wafer handling. The performance requirements when carrying out wafer temporary bonding are void-free, TTV performance, bonding alignment accuracy, glue wafer edge coverage, and wafer warpage. If void occurs, holes will appear in the thinning process, swelling will occur in CVD, and shape defects will occur in the etching process. Large TTV will result in focus adjustments in the lithography process and diagonally cut tilting in the etching process.

Incorrect bonding accuracy will result in edge chipping in the thinning process and defects in notch searching during alignment in the lithography process. Incorrect wafer warpage will inhibit wafer handling. There are also a wide range of performance requirements for the adhesive agent, including wafer warpage, bond strength, heat resistance, chemical resistance, debonding performance, and cleaning performance. Most of this is covered in the bonding process, but the performance of the adhesive agent is a major factor in wafer warpage. Also, if the bond strength is insufficient, it may become unstuck during the thinning process or the CVD process due to stress on the device wafer. As chemical resistance will affect the lithography

**Table 4.9** Thin wafer handling key process and performance

| Key performance \ Key process | | Bonding | Thinning | CVD | Litho | Etching | Debondcleaning |
|---|---|---|---|---|---|---|---|
| | |  |  |  | 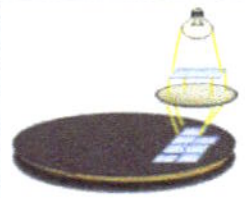 |  |  |
| Bond | Void | ✓ | Hole | Swelling | | Shape trouble | |
| | TTV | ✓ | | | Focus adjustment | Tilting | |
| | Bonding accuracy | ✓ | Chipping | | Notch search | | |
| | Edge coverage | ✓ | Chipping | | | | |
| | Warpage | ✓ | | | Handling | Handling | |
| Glue | Adhesive capability | | Delamination | Delamination | | | |
| | Thermal resistance | | | Delamination | | | |
| | Chemical resistance | | | | Chemical resistance | | |
| | Debond capability | | | | | | Debond capability |
| | Cleaning capability | | | | | | Cleaning capability |
| Debond | Debonding capability | | | | | | ✓ |
| | Cleaning capability | | | | | | ✓ |

process, debonding must be performed after completing the processes, and the adhesive agent must be cleaned without leaving a residue. In the debonding process, the support wafer must be removed without cracking or chipping the device wafer, and there must be no residual glue after removing the support wafer.

### 4.4.3 Debonding Method

In terms of temporary wafer bonding and debonding processes using an adhesive agent, there are various proposals for debonding methods. One method is to use a glass wafer as the support wafer, which is debonded by using a laser to eliminate the bond strength. Another method of debonding is to use a Si wafer as the support wafer and to slide the carrier wafer while applying heat, or to remove it mechanically.

Laser removal methods using glass wafers are mainly used in the fields of microelectro-mechanical systems (MEMS) and 2.5 D (silicon interposer). The key feature of this method is that the bond strength is removed using a laser, so that the support wafer is easily removed which means that throughput is fast. However, it is difficult to use this method on devices that are not heat resistant due to the great amount of heat generated by the laser. Also, as a glass wafer is used as the support wafer, certain issues have been raised, such as the allowance for conductivity that must be made due to adsorption to electrostatic chucks when processing in a vacuum chamber after temporary bonding, the effect of wafer warping due to the difference in coefficient of thermal expansion (CTE) between the glass wafer and Si wafer, the effect of device contamination due to metal contamination when foreign matters adhere to the glass wafer in order to deal with the above issues, and the steep increase in the cost of glass wafers with high TTV precision and other additional functions.

Using a Si wafer as the support wafer eliminates the issues raised with glass wafers, such as the adsorption to electrostatics chucks, the effect of wafer warping due to CTE mismatch, and metal contamination, and it can be easily expanded to existing devices. Also, TTV can be used at a comparatively low cost to equivalent device wafers. On the other hand, as a laser cannot be used as the removal method, a debonding method that does not use a laser is required.

Debond methods using Si carriers can largely be split into two categories: the thermal slide method and mechanical debonding method. The thermal slide method uses a thermoplastic adhesive agent that is softened in a heated chamber, so that the support wafer is removed by sliding. Although it is a relatively easy removal method, the downsides to this technique are that a thermoplastic adhesive agent that softens when heated is required due to the use of heat during debonding, and high-temperature processes cannot be used after temporary bonding.

Mechanical debonding is a mechanical debonding method where the adhesive agent used in the bonding process has an adhesive layer and a removal layer, which provides resistance in the post-temporary bonding processes. As detachment during debonding is carried out mechanically, it is easy to allow for a heat-resistant adhe-

**Table 4.10** Debonding method comparison

| Debond method | | Feature |
|---|---|---|
| Laser/UV debond | 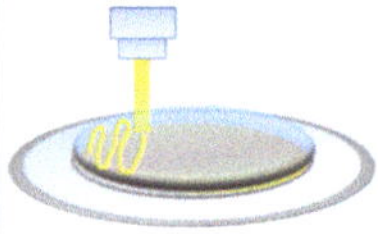 | Mainly 2.5D application |
| | | Easy to remove carrier wafer |
| | | Expensive good TTV glass wafer |
| | | Wafer warpage issue |
| Thermal slide |  | Mainly 3D application |
| | | Silicon carrier capable |
| | | Thermal resistance by thermoplastic glue |
| Mechanical debond |  | Mainly 3D application |
| | | Silicon carrier capable |
| | | Good wafer warpage < 100 um |

sive agent, and relatively, high temperatures can be used in the processes after bonding. However, in the processes after bonding, the support wafer must be removed in the debonding process while maintaining the adhesive strength so that the device wafer does not detach, which means that it is very important to control the bond strength of the adhesive agent.

A summary of the above methods is shown in Table 4.10. In the initial period of investigating the temporary bonding and debonding process, the laser release method and thermal slide method were used, but the RT debonding process has since been established due to improved device and adhesive materials. Currently, RT debonding is the most common method of TSV process.

### 4.4.4 Functions and Performance Requirements for Temporary Bonding Device

An example of the temporary bonding process is shown in Fig. 4.34. Although different modules are required depending on the different types of adhesive agent and the debonding method, basically, in the temporary bonding process, the main processes are the processes of applying the adhesive agent to the wafer, the baking process to evaporate the solvent, and the bonding process to bond the wafers.

In the adhesive agent application process, as the level differences on the device wafer (bump, Cu pillar) must be covered by the adhesive agent, the adhesive agent must be applied with a thickness of between 30 and 100 μm. Therefore, a 3000 mPa·s or higher viscosity adhesive agent is used, and there are requirements for a coater with a high-viscosity adhesive agent discharge capability, in-plane uniformity after adhesive agent application, the prevention and cleaning of adhesive agent wraparound produced on the side and rear of the wafer during spin application, and

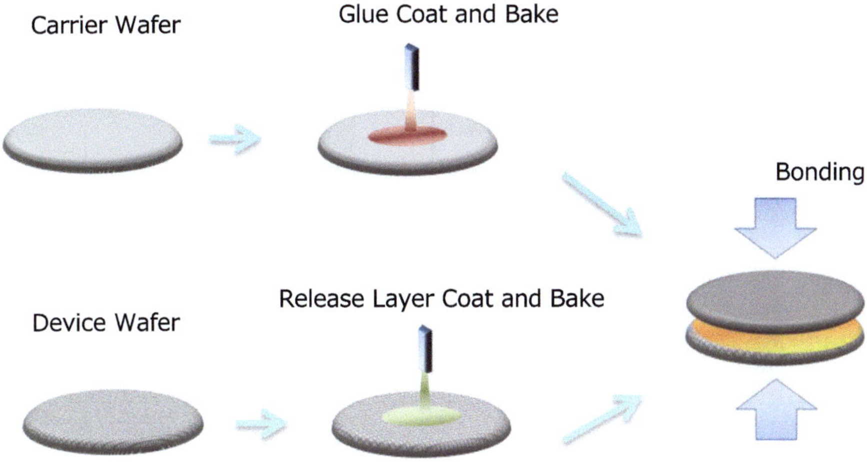


**Fig. 4.34** Temporary bonding process flow

edge-cut accuracy to control the amount of adhesive agent extrusion during bonding. Also, in terms of the required adhesive agent application performance, because it will be applied to a device wafer with level differences, and in order to facilitate film thickness uniformity after application, bubbling must be prevented from occurring.

The bake process is performed after coating the adhesive agent. In this process, the solvents contained in the adhesive agent are vaporized by means of heating on a hotplate. However, if heating momentarily surpasses a temperature beyond that of the solvent boiling point, bubbling caused by bumping will appear on the wafer surface, which will make it unsuitable for bonding. For that reason, in order that it does not boil suddenly, baking is generally performed in two stages. Also, as oxidation will occur if baked in an oxide atmosphere, depending on the adhesive agent, the materials must be baked in a nitrogen atmosphere. In this case, there is a requirement for the organization of the module to maintain the oxygen concentration within a few ppm.

For adhesive agents with many volatile chemicals, there is a requirement for a hotplate with a large ventilation capacity, or for a hotplate designed so that adhesive agents that have a strong effect on airflow are limited in terms of their effect on the airflow.

The adhesive agent application-baked wafer is handed over for the bonding process. In the bonding process, the device wafer with the adhesive agent applied is bonded to the support wafer. The device performance requirements for this are void-free, bonding alignment accuracy, adhesive agent edge coverage, and TTV. In order to bond the wafer without void, the in-plane uniformity of the wafer coated with the adhesive agent must be well maintained, and bonding must be performed in a bonder device with a vacuum environment with 10 Pa or less in order to prevent voids containing air.

Cut Width： 3000 μm ± 100 μm

| Degree | Micro Scorpe Picture (Connected) | Cut Width [um] | Degree | Micro Scorpe Picture (Connected) | Cut Width [um] |
|---|---|---|---|---|---|
| Notch |  | 2962.0 | 180 |  | 2989.8 |
| 45 |  | 2968.8 | 225 |  | 2975.1 |
| 90 |  | 3010.8 | 270 |  | 2973.3 |
| 135 | 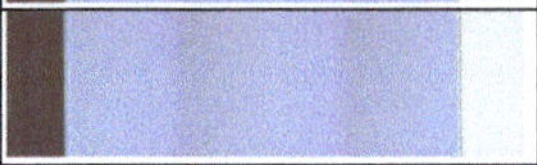 | 3005.5 | 315 |  | 2976.7 |
| | | | | Ave. | 2982.8 |
| | | | | Range | 48.8 |

**Fig. 4.35** Top-view picture after wafer edge treatment

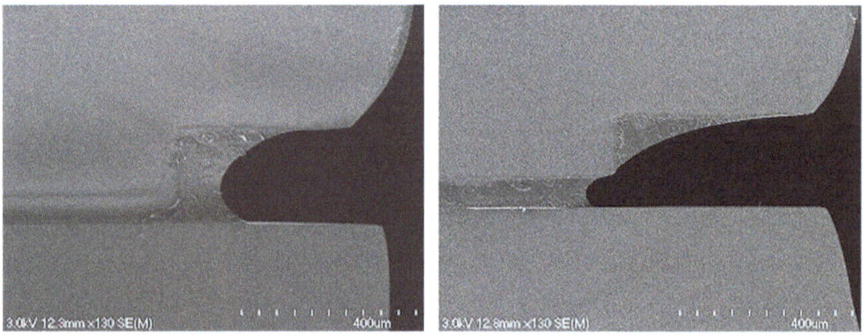

Optimized glue coverage Short of glue coverage

**Fig. 4.36** Cross-section view of glue edge coverage

Alignment accuracy when bonding the device wafer and the support wafer is required in order that problems do not arise in the bonding process. Although there are two methods for wafer alignment, which are mechanical alignment and optical alignment, contactless optical alignment is preferred from the perspective of mass production and maintenance due to the risk of adhesive agent adhering to the mechanical section in mechanical alignment.

In connection to adhesive agent edge coverage, after coating, the wafer edge glue is dissolved using a solvent and controlled. Figure 4.35 shows the result after cutting the wafer edge glue. Depending on the viscosity of the adhesive agent used during bonding, edge coverage can be controlled by adjusting the bonding force. The preferred state of wafer coverage is shown in Fig. 4.36. If the coverage is insufficient, the device wafer will come unstuck in the thinning process. Alternatively, if the coverage is too great, there is a risk that the bonding chamber will be contaminated with adhesive agent, so the amount must be properly regulated.

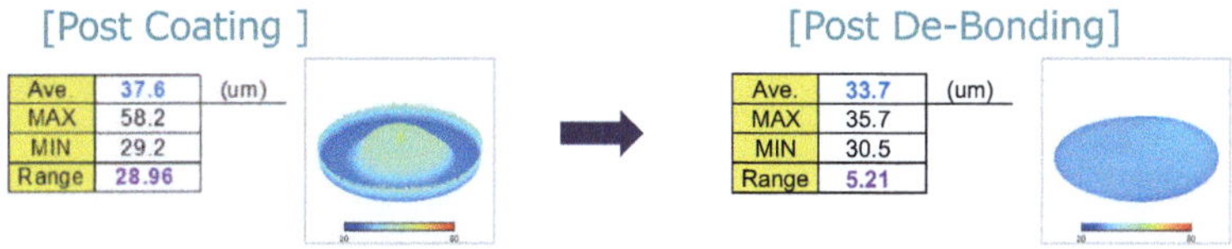


**Fig. 4.37** Within wafer glue uniformity between post-coating and post-debonding

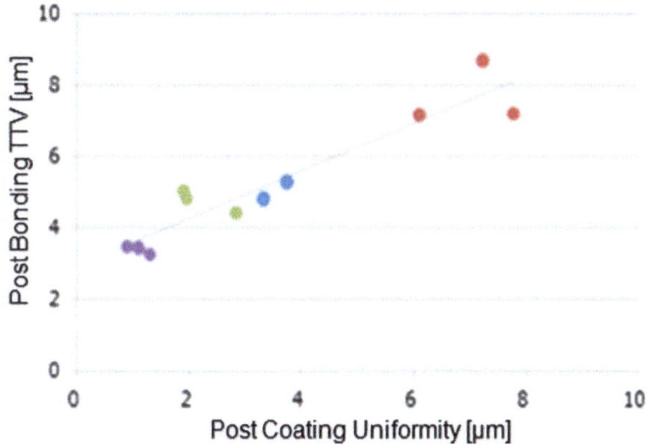


**Fig. 4.38** Correlation between post bonding TTV and post coating uniformity

Of the bonding devices that achieve the performances described above, the performance of TTV is getting the most attention. TTV has a direct influence over the device yield in later processes. In particular, the difficulty of the etching process is affected by the use of the TTV production process in via last. In order to perform bonding with proper TTV, there are requirements for maintaining in-plane uniformity when coating, and bonding must be performed at a temperature and pressure that matches the adhesive materials. Figure 4.37 shows the wafer in-plane uniformity after applying glue to a wafer with inferior coating conditions and the in-plane uniformity of the glue of the same wafer with optimized conditions after bonding and debonding. Optimizing the bonding conditions enables 1/5 in-plane uniformity. Figure 4.38 shows the correlation between the glue in-plane uniformity after coating and TTV after bonding. It is understood that glue in-plane uniformity must be correct after coating in order to have proper TTV due to the connection between glue in-plane uniformity and TTV.

### *4.4.5 Ability and Performance Requirements for Debonding Devices*

In this section, we will state the required abilities and performances in the debonding process in the RT debonding method. Figure 4.39 shows the RT debonding

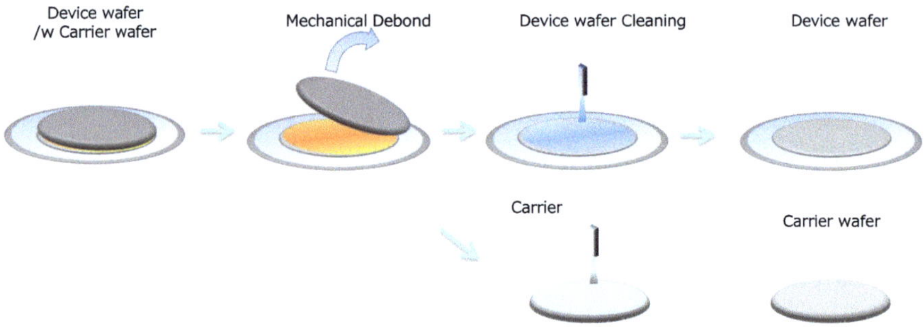


**Fig. 4.39** Debonding process flow

process flow. In the RT debonding process, in order to transport the device wafer after separation from the support wafer as a thin wafer, the bonded wafer is applied to the dicing tape before introduction to the debonder device. In this condition, the debonding process is performed after the wafer is introduced to the debonder device. After detachment from the support wafer, the device wafer on the dicing tape is cleaned, and the adhesive agent is removed before being extracted from the device.

In the debonder module, the dicing tape is introduced to the device wafer affixed to the support wafer, and the support wafer is mechanically removed in order to debond the support wafer. Device wafer cracks are a problem during debonding. If stress is placed on the device wafer when removing the support wafer, it can cause the device wafer to break. Therefore, there is a requirement in the debonder module for the ability to remove the support wafer without applying stress to the device wafer.

In addition to cracking counter measures in the debonder module, in the TSV process, the wafer can be easily damaged at low forces as the process is performed with a thinned device wafer, so it is easy for the device wafer to crack in the TSV process. If there are cracks in the device wafer, then they will become more prominent during debonding. Therefore, in the TSV process, special attention must be given to device wafer cracks.

In the cleaning module in the debonder device, cleaning using a solvent is carried out in order to remove the adhesive agent from the device wafer after debonding the support wafer affixed to the dicing tape and to remove the device wafer from the device in a condition where there is no adhesive agent. The required functions of the cleaning module are device wafer cleaning performance and damage-free dicing tape. Despite bumps and metal pads formed on the surface of the debonded device wafer, if there is any residue from the adhesive agent on the surface, then it will cause imperfections in later processes. Therefore, in the cleaning process, cleaning must be carried out, so that no adhesive agent residue is left on the surface. Also, as the device wafer is cleaned while affixed to the dicing tape, the dicing tape must have a chemical resistance to solvents.

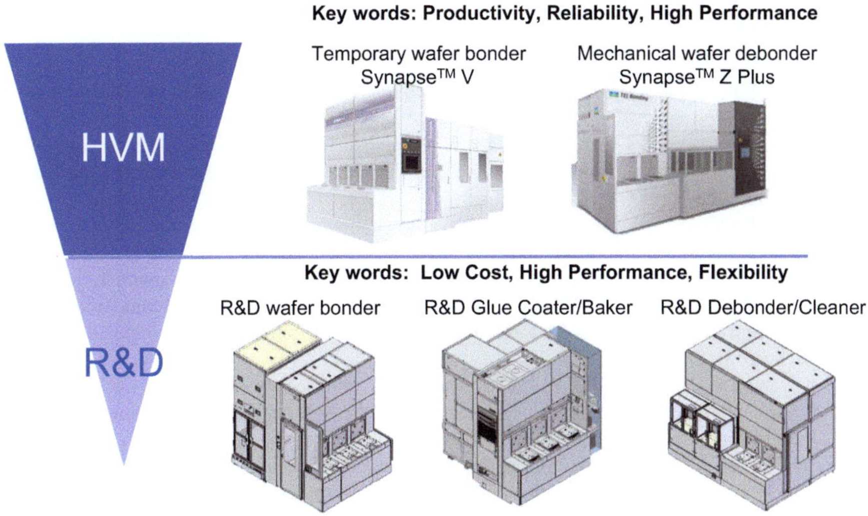


Fig.8 Tokyo Electron Ltd. Temporary bonder and debonder system lineup

**Fig. 4.40** Tokyo Electron Ltd. Temporary bonder and debonder system lineup. *HVM* hardware virtual machine, *R&D* research and development

### *4.4.6 Tokyo Electron's Temporary Bonder and Debonder Device Concept and Lineup*

Tokyo Electron is releasing the temporary bonder system Synapse™ V and the debonder system Synapse™ Z Plus as mass production TSV process devices. Synapse™ V is a system machine for the integrated performance of glue coating and baking and supports wafer bonding. Synapse™ Z Plus is a system machine that performs RT debonding, and device and carrier wafer cleaning. These devices use technology for which Tokyo Electron has mass production factories. They are being released through 300-mm TSV mass production factories, and the superior performance and mass productivity are widely supported by the market.

In addition to these devices, we are supporting TSV development with a lineup of glue bonders, glue coater/bakers, and debonder/cleaners that are useful for development and small-scale production (Fig. 4.40).

### *4.4.7 Future Outlook*

Having cleared problems with the process, the temporary bonding and debonding process have started to be used in DRAM 3D lamination products. In the future, we are expecting development to progress toward device thinning, low-consumption chemical solutions, and high throughput in order to reduce products costs.

## References

1. Kim YS (2013) Advanced wafer thinning technology and feasibility test for 3D integration. Microelectron Eng 107:65–71
2. Hattori T, Osaka T, Okamoto A, Saga K, Kuniyasu H (1998) Contamination removal by single wafer spin cleaning with repetitive use of ozonized water and dilute HF. J Electrochem Soc 145:3278–3284
3. Sun WP (2004) Fine grinding of silicon wafers: a mathematical model for the wafer shape. Int J Mach Tool Manuf 44:707–716
4. Kim YS, Kodama S, Mizushima Y, Maeda N, Kitada H, Fujimoto K, Nakamura T, Suzuki D, Kawai A, Arai K, Ohba T (2014) Ultra thinning down to 4-μm using 300-mm wafer proven by 40-nm node 2 Gb DRAM for 3D multiStack WOW applications. Symposia on VLSI Technology. Dig 26–27
5. Koyanagi M, Nakagawa Y, Lee KW, Nakamura T, Yamada Y, Inamura K, Park KT, Kurino H (2001) Neuromorphic vision chip fabricated using three-dimensional integration technology. ISSCC digital technical papers, pp 270–271
6. Kameyama K, Okayama Y, Umemoto M, Suzuki A, Terao H, Hoshino M, Takahashi K (2004) Application of high reliable silicon thru-via to image sensor CSP. Extended abstracts of international conference on solid state devices and materials, pp. 276–277
7. Morrow P, Black B, Kobrinsky MJ, Muthukumar S, Nelson D, Park CM, Webb C (2007) Design and fabrication of 3D microprocessors. Mater Res Symp Proc 970:0970-Y03-02
8. Patti RS (2006) Three-dimensional integrated circuits and the future of system-on-chip designs. Proc IEEE 94:1214–1224
9. Kim DH, Athikulwongse K, Healy M, Hossain M, Jung M, Khorosh I, Kumar G, Lee YJ, Lewis D, Lin TW, Liu C, Panth S, Pathak M, Ren M, Shen G, Song T, Woo DH, Zhao X, Kim J, Choi H, Loh G, Lee HH, Lim SK (2012) 3D-MAPS: 3D massively parallel processor with stacked memory. ISSCC digital technical papers, pp 188–189
10. SadakaM, Radu I, Cioccio LD (2010) 3D integration: advantages, enabling technologies & applications. In: proceedings of IEEE ICICDT, pp 106–109
11. Santarini M (2011) Stacked & loaded: Xilinx SSI, 28-Gbps I/O yield amazing FPGAs. Xcell J 74:8–13
12. Sukegawa S, Umebayashi T, Nakajima T, Kawanobe H, Koseki K, Hirota I, Haruta T, Kasai M, Fukumoto K, Wakano T, Inoue K, Takahashi H, Nagano T, Nitta Y, Hirayama T, Fukushima N (2013) A 1/4-inch 8Mpixel back-illuminated stacked CMOS image sensor. ISSCC Digital Technical papers, pp 484–485
13. Yole development (2010) 3D-IC & TSV Interconnects
14. Takahashi K, Terao H, Tomita Y, Yamaji Y, Hoshino M, Sato T, Morifuji T, Sunohara M, Bonkohara M (2001) Current status of research and development for three-dimensional chip stack technology. Jpn J Appl Phys 40:3032–3037
15. Olson S, Hummler K (2011) TSV reveal etch for 3D integration. In: proceedings of 3D systems integration conference (3DIC), pp 1–15
16. Yamamoto E (2011) TSV Wafer Thinning Technology. SEMATECH Symposium Japan
17. Watanabe N, Aoyagi M, Katagawa D, Bandoh T, Yamamoto E (2014) A novel TSV exposure process comprising Si/Cu grinding, electroless Ni-B plating, and wet etching of Si. Jpn J Appl Phys 53:05GE02
18. The International Technology Roadmap for Semiconductors (2011) Executive summary. http://www.itrs.net/Links/2011ITRS/2011Chapters/2011ExecSum.pdf. Accessed 25 Jun 2014

# Chapter 5
# Wafer and Die Bonding Processes

**Hiroaki Fusano, Yoshihito Inaba and Toshihisa Nonaka**

## 5.1 Permanent Wafer Bonding

### *5.1.1 Introduction*

Permanent wafer bonding is defined as the technology for the direct bonding of silicon and other wafer substrates without the use of, for example, bonding agents. This technology has been developed with the production of silicon-on-insulator (SOI) wafers as the primary target. SOI wafers are already in use as substrates in central processing units (CPUs) and other high-performance semiconductors. A heat-treatment process at a temperature of 1000 °C or higher was in use in the SOI wafer production process. If this high-temperature process was applied to microelectromechanical systems (MEMS), image sensors, semiconductor integrated circuits (ICs), etc., there would be a risk of damaging the device due to the heat. Therefore, there is a requirement for a low-temperature or roomtemperature bonding process. Wafer direct bonding using a low-temperature or moderate room temperature process is being applied to the production of high-performance substrates in various telecommunications devices that use non-silicon materials. It is also applied to the bonding of wafers formed of ultrastructures, and assembly packages for MEMS, for example. Recently, it has been applied to large-diameter 300-mm substrates and to the production process of backside illumination (BSI) complementary metal–oxide–semiconductor (CMOS) image sensor using BSI

Y. Inaba (✉)
Hitachi Chemical Co., Ltd., Ibaraki, Japan
e-mail: yoshihito-inaba@hitachi-chem.co.jp

T. Nonaka (✉)
Toray Research Center, Inc. (TRC), Shiga, Japan
e-mail: t-nonaka@hitachi-chem.co.jp

H. Fusano
3DI Dept. ATS BU, Tokyo Electron Ltd., Tokyo, Japan
e-mail: hiroaki.fusano@tel.com



K. Kondo et al. (eds.), *Three-Dimensional Integration of Semiconductors*,
DOI 10.1007/978-3-319-18675-7_5

technology. In the future, it is expected that this technology will be applied to semiconductor ICs through application to memory lamination and logic/memory lamination.

In this chapter, we explain the technology for wafer substrate direct bonding that uses a low-temperature or room temperature process.

## *5.1.2 Low-Temperature or Room Temperature Wafer Direct Bonding Method and Application*

In this section, we explain the method and application of low-temperature or room temperature wafer direct bonding. Low-temperature or room temperature wafer direct bonding methods, including fusion bonding, surface activated bonding, anodic bonding, and copper to copper (Cu2Cu)/oxide hybrid bonding, are currently either being applied to various device applications or being considered for application.

A summary of each bonding method is provided below.

### 5.1.2.1 Fusion Bonding

The first report of silicon wafer bonding at room temperature was made in1986 from IBM [1] and Toshiba [2], separately. With this technology, the bonding surfaces of two films to be bonded, such as wafers or silicon oxide films, are polished at an atomic level, and hydrophilic treatment is then carried out on the surface to form hydroxyl groups, which are then bonded together. The key feature of this method is that bonding can be carried out at room temperatures with low loads. The bonding strength of the bonded wafers can be increased using heat processing of 200 °C or higher in the post-processing after bonding [1]., as shown in Fig. 5.1.

### 5.1.2.2 Surface Activated Bonding

On the surface of two films to be bonded, such as wafers or silicon oxide films, sputter etching is conducted using an inert gas beam on, for example, the natural

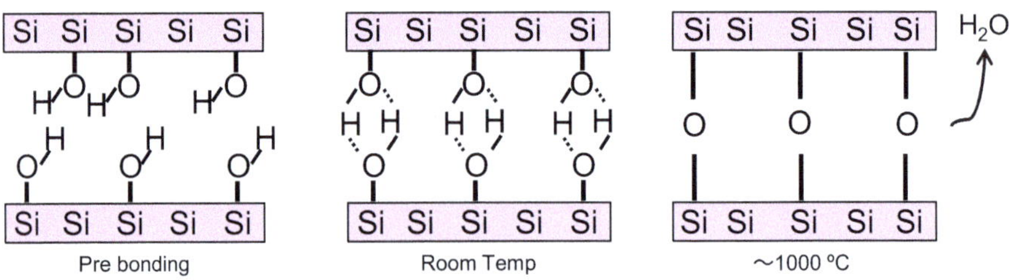


**Fig. 5.1** The wafer bonding method using hydrogen bonds

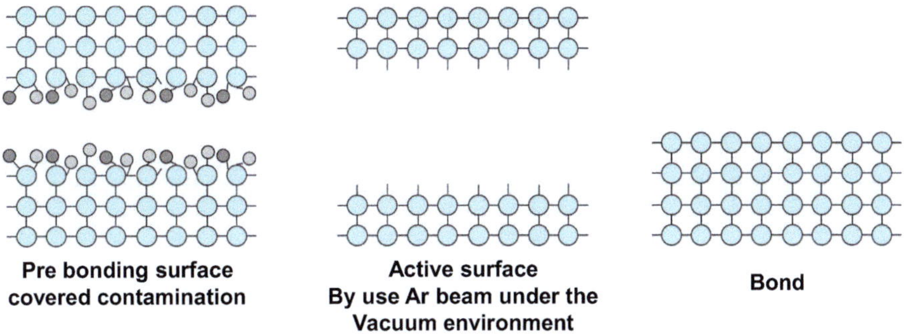


**Fig. 5.2** The wafer surface activated bonding at room temperature

oxide film and adsorbed gases covering the surface that are not required for bonding. Such unrequired natural oxide film and adsorbed gases are removed by means of this technology, so that the atoms on the bonding surface are exposed and are bonded together. The key feature of this method is that directly bonding the exposed atoms enables strong bonding at room temperature [1] as shown in Fig. 5.2.

#### 5.1.2.3 Anodic Bonding

While heating the polished surfaces of two wafers to be bonded, a glass substrate and a silicon substrate, a negative voltage is applied to the glass substrate so that an electrostatic attraction causes a covalent bond to arise between the surfaces of the glass substrate and the silicon substrate. This technology uses the described phenomenon for bonding (Fig. 5.3). A key feature of this technique is that the atomic-level substrate surface roughness required in the fusion bonding method is not needed. A detriment to this bonding method is the restriction in terms of selecting a substrate that is made of glass materials.

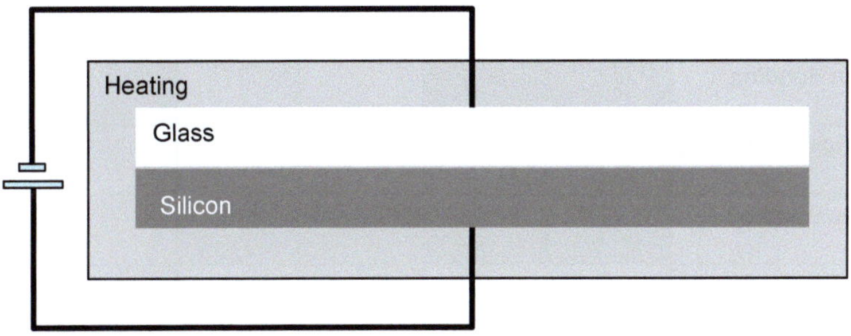


**Fig. 5.3** Anodic bonding of Si and glass

#### 5.1.2.4 $Cu_2Cu$/Oxide Hybrid Bonding

This technology is for bonding two wafers to be bonded, the bonding surfaces of which have an insulator film and a substrate with Cu electrodes. There are two major issues with this technology. The first is the alignment accuracy of the two wafers. The second is the bonding yield due to wiring resistance after Cu2Cu bonding.

#### 5.1.2.5 Conclusion of Low-Temperature or Room Temperature Wafer Direct Bonding Methods and Their Applications

A comparison of the application of the four low-temperature or room temperature wafer direct bonding methods above is shown in Table 5.1.

#### 5.1.2.6 Future Outlook for Bonding Application Using Low-Temperature or Room Temperature Wafer Direct Bonding Methods

We explain the future outlook for bonding application using low-temperature or room temperature wafer direct bonding methods by classifying wafer substrates into sizes of 200 mm or less and 300 mm or more.

First, we provide two examples of new applications using low-temperature or room temperature wafer direct bonding methods for wafer substrates with sizes of 200 mm or less.

The first application is to a solar cell. With low-voltage cost, high efficiency, and the realization of resource-saving solar cells as the aim, research is being promoted into hybrid silicon/nitride semiconductor multi-bonded tandem solar cells, which have a great affinity for concentrating solar power generation systems [2].

**Table 5.1** Bonding application versus low room temperature. Permanent bonding method

| Low/room temp. permanent bonding method | SOI | MEMS | LED | | Advanced Packaging | |
|---|---|---|---|---|---|---|
| | | | Carrier | HB-LED | CIS (BSI) | TSV Stacks |
| Fusion Bonding | Production | Production | | | Production | |
| Interatomic Bonding | | | | | | |
| Anodic Bonding | Production | Production | Production | Production | | |
| Cu2Cu/Oxide Hybrid Bonding | | | R&D | | R&D | R&D |

Production R&D

The second is an initiative that aims to reduce the cost of SiC substrate production. There are reports of initiatives for lower production costs without damaging the quality of SiC single crystals by adhering a high-quality single-crystal thin film to a low-cost supporting substrate [3].

Next, we provide two examples of applications of Cu2Cu/oxide hybrid bonding methods for wafer substrates with sizes of 300 mm or larger, which is a progressive area of research and development.

The first example is BSI application. The purpose of this application is to increase the speed of transmission of data received by photodetectors. The application method is to bond the Cu in the image sensor and the Cu in the logic circuit using the Cu2Cu/oxide hybrid bonding method, which increases the data transmission speed. This enables the separate production of process wafers in the analog image sensors and the digital logic circuit, which are completely different. By bonding the different process wafers, the chip occupancy area can be reduced.

The second example of application is to semiconductor ICs. Each company is still in the feasibility study stage for application to semiconductor ICs. Technical comparisons are still being performed for Cu2Cu/oxide hybrid bonding methods for the direct bonding of wafers and for lamination techniques for wafers and chips, etc. It is thought that comparisons including production costs as well as feasibility studies will continue for some time yet, but, as described above for BSI, device manufacturers are carrying out research and development all the time with the goal of improving data transmission speeds and reducing the chip occupancy area.

### *5.1.3 Requests Made to Equipment Makers and Initiatives Regarding Low-Temperature or Room Temperature Wafer Direct Bonding Methods*

In this section, we explain the requests made by device makers to equipment makers in connection to the bonding process for semiconductors ICs and BSI, etc., with a substrate size of 300 mm and the initiatives of equipment makers in this regard.

The substrate for bonding in this bonding process is the device substrate bonded in the preprocessing of the device production process. Therefore, this tends to be perceived as post-processing. However, to realize the bonding process, there are requirements for preprocessing technology and equivalent preprocess specifications. Specifically, in terms of process technology, there is a requirement for process development that matches the various kinds and structures of insulator films used by each company for the bonding surface. Therefore, as with process development in preprocessing, the best of knowledge method (BKM) of equipment makers cannot be applied easily. As for process specifications, this is also because, in the case of device substrate particle adhesion, contamination, or plasma processing, etc., the requirements for electrical damage or ramming damage to the device substrate are the same as the requirements for preprocessing.

In this section, we explain the process requirement items demanded from the equipment makers by the device makers to complete the bonding process. In the first part of bonding, the requirements include requirements pertaining to particles, contamination, and damage on the device substrate. In connection to the bonding process, post-bonding alignment accuracy (post BAA), scaling, distortion, bonding strength, and void, etc., are cited as requirements. In this section, we explain the requirements of the latter part of the bonding process.

#### 5.1.3.1 Post BAA

Post BAA refers to the alignment accuracy of two bonded wafer substrates.

The required specifications for post BAA vary according to the type of bonded device substrate. In the bonding process where pattern alignment is not required, it is tens of micrometers. In the bonding process where pattern alignment is required, depending on the pattern size, a highly precise degree of alignment accuracy is required. This requirement is based on concerns over displacement in connection to the via for bonding. For example, for an approximately 3-μm diameter via, the range is 1 μm. In the future, along with the miniaturization of pattern sizes, the required specifications will be for alignment accuracy within a range of 0.4 μm.

To fulfill the required specifications of device makers and to increase post BAA, equipment makers are trying to make improvements to the equipment in terms of, for example, the alignment sequence, the use of high-rigidity materials in the alignment unit, restricting external vibrations, and the addition of a temperature and humidity control unit.

#### 5.1.3.2 Scaling

Scaling refers to the phenomenon where the substrate to be bonded is stretched in the bonding process due to the bonding force. In the bonding process, one side of the wafer is maintained as the holding base, whereas the other side of the wafer is released from the holding base in the bonding process. For this reason, a phenomenon occurs where the released wafer is stretched from the wafer held by the holding base (Fig. 5.4). The required specification for scaling is 1 ppm or less. The reason for this requirement is that, if there is a large amount of scaling, in the following lithography process, an alignment will be required for the scaling amount. Also, if the scaling amount is reproduced, there are cases where the device maker can make adjustments in lithography.

To compensate for scaling, equipment makers are looking for solutions using physical phenomenon and improvements in the bonding preprocess.

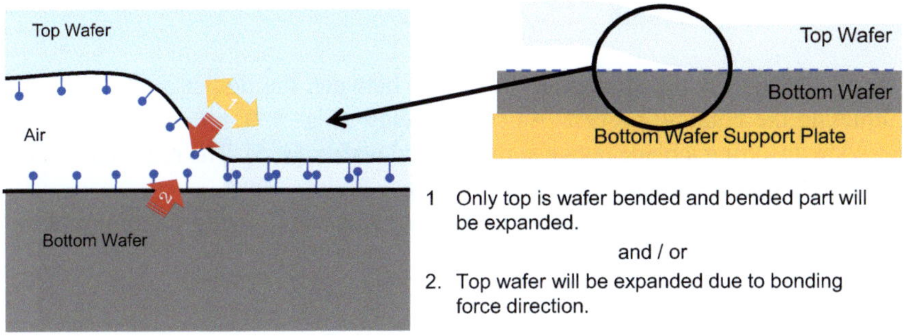


**Fig. 5.4** Scaling mechanism

### 5.1.3.3 Distortion

The term distortion as used here refers to a warp in the pre-bonding pattern due to a bend in the device substrate is caused by stress when bonding. With the difference between the ideal shot and distortion shot as shown in Fig. 5.5, the required specification is 10 nm or less. The reason for this requirement is that pattern alignment cannot be carried out in the following lithography process. As distortion is caused by the flatness of the wafer-holding base and the effect of particles on wafer substrates (Fig. 5.6), equipment makers are trying to improve wafer-holding methods, etc., to control distortion.

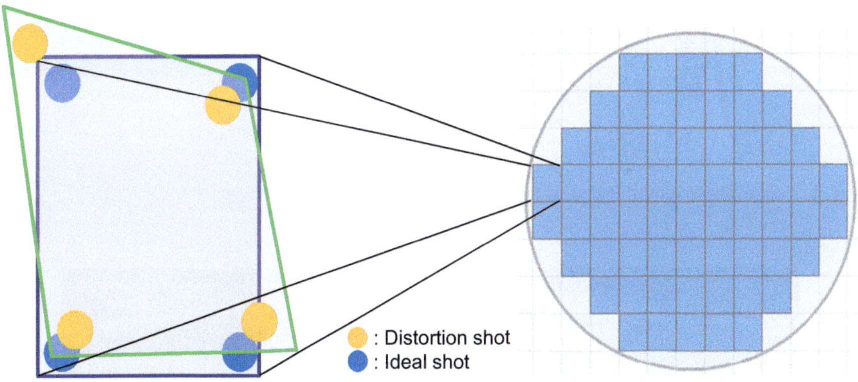


**Fig. 5.5** Distortion

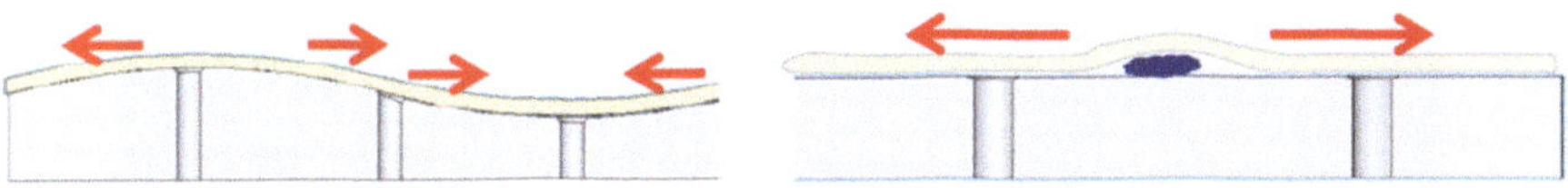

**Fig. 5.6** Cause of distortion

#### 5.1.3.4 Bonding Strength

Bonding strength is the strength of the bond between the device substrates after completing the bonding process. As shown below (Fig. 5.7), measurement is performed by inserting a blade between the bonded wafers [4, 5]. The required specification for bonding strength is 1.0 J/m$^2$ or more.

Equipment makers are making improvements and optimizing bonding preprocess conditions.

#### 5.1.3.5 Void

Void refers to air pockets formed on the bonding interface after completing the bonding process. Known examples of problems that occur due to void include damage to the wafer in post-chemical mechanical planarization (CMP) processing and defective wiring bonding. There are three causes of void. First is void that arises in the pre- and post-bonding process on the external bonding wafer (Fig. 5.8). The cause is wafer film surface roughness and flatness, as well as bonding preprocess conditions.

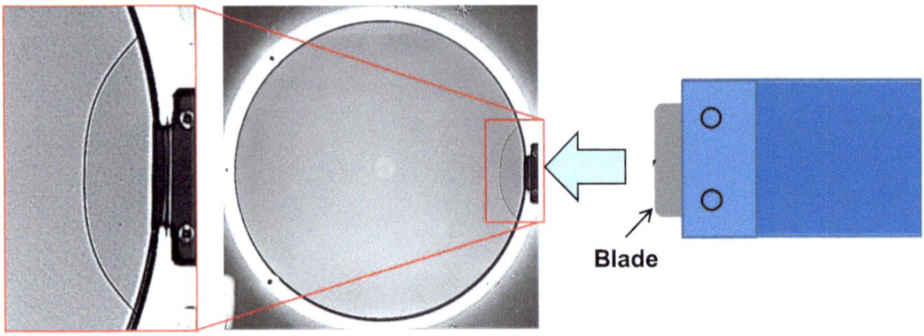


**Fig. 5.7** Bonding strength measurement method

**Fig. 5.8** Voids

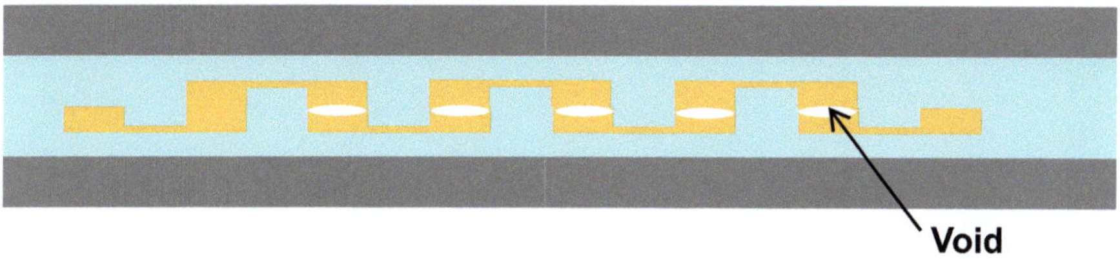


**Fig. 5.9** Void that depends on device wafer

The second is void that arises on the device substrate (Fig. 5.9). The cause is roughness due to CMP.

The third is void that arises due to particles.

The required specification for void is zero void. To control the occurrence of void, equipment makers are working on bonding pre- and post-process development, however, because it is thought that device substrates also cause void (film surface roughness/flatness), the process development on the device maker's side is also considered to be an important factor in achieving this specification.

### *5.1.4 Tokyo Electron Initiatives*

Tokyo Electron (TEL) is releasing the wafer-bonding device Synapse™ S[1] (Fig. 5.10).

The key features of this product are that the module unit structure is taken from other TEL's products that are in mass production for which there are mass-production

**Fig. 5.10** SynapseTM S

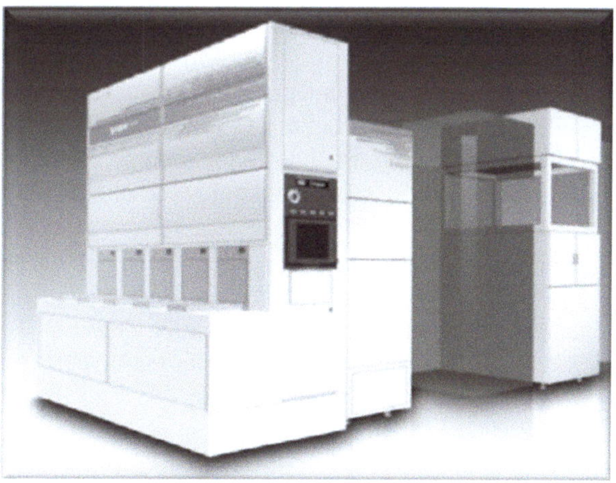

[1] Synapse is a registered trademark or a trademark of Tokyo Electron Limited in Japan and/or other countries

factories that achieve a 95 % or higher rate of operation and throughput of 15 wafer per hour or more.

### *5.1.5 Conclusion*

As noted in the previous section, it is expected that product devices and wafers using low-temperature or room temperature wafer direct bonding methods will have various industry uses, and they will bring about great changes in living and natural environments. We believe that various technical issues will have to be solved to achieve this. With device makers and equipment makers working together as one on these solutions, technology that contributes to society can be developed and realized. This ends the current section.

## 5.2 Underfill Materials

### *5.2.1 Technical Trend for Three-Dimensional Integration Packages and Underfill Materials*

In accordance with high-functionality trend of electronic devices, the semiconductor has enhanced its processing speed and capacity with miniaturization of wiring rule, bump pitch, and height in recent years. As shown in Table 5.2, flip chip (FC) connection design rule 130–150-μm pitches/30–40-μm bump heights were mainly used untill 2013, and the further narrowing pitch and gap will proceed to 60–80-μm pitches/20–30-μm bump heights in the near future. The bump materials are going to switch from lead-free (LF) or eutectic (Eu) to copper (Cu) bump to avoid solder bridge issue [8]. Package structure shifts from the conventional structure such as the bare die structure to the integration structures such as molded embedded package (MEP), 2.5D, and 3D structures (Fig. 5.11) [9].

There are two types of underfill materials, capillary underfill (CUF) and nonconductive paste (NCP). CUF is a liquid type underfill material, and it has been used for many kinds of packages that have various pitches and gaps [10]. The capillary phenomenon is well known as liquid flowing into narrow gap by its surface tension. Figure 5.12 shows the typical CUF encapsulation process. First is the dispensing of CUF under heated condition. After capillary flow, CUF is cured at high temperature.

There are two potential problems during this process. The first is the void problem. In case of narrow pitch and thin gap (less than 80 μm pitches/30 μm gaps), the void appears easily. The second is the warpage. The warpage of the package causes interconnect failure between chip-to-substrate (C2S) structure and stacked chip in 2.5D and/or 3D packages. Also, the trend of thinner chip makes the warpage problem worse. Both void and warpage problems have to be solved for advanced high-integration packages.

**Table 5.2** Trend of bump technology and design rule

| Item | Year | | | | |
|---|---|---|---|---|---|
| | 2013 | 2014 | 2015 | 2016 | 2017 |
| Bump-material | LF/Eu or Cu | Cu | | | |
| Bump pitch (μm) | 130–150 | 80–130 | | 60–80 | |
| Bump height (μm) | 30–40 | | 20–30 | | |

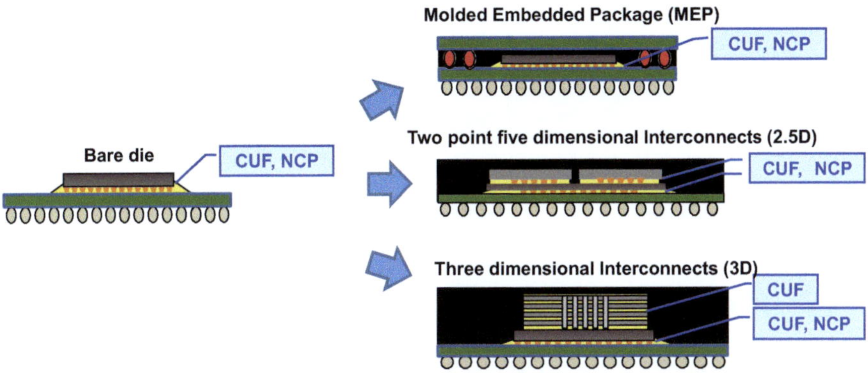


**Fig. 5.11** Trend of flip chip package. *CUF* capillary underfill, *NCP* nonconductive paste

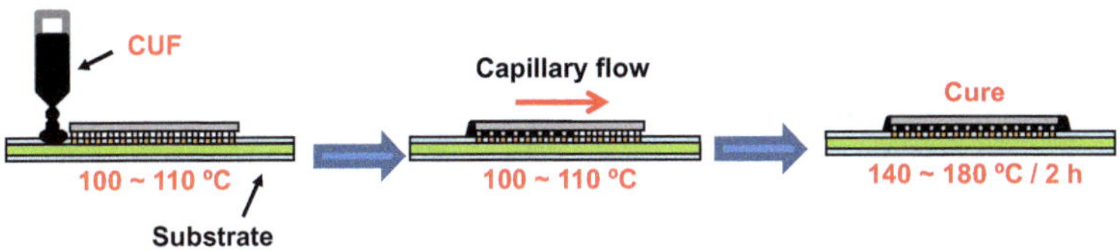


**Fig. 5.12** Flip chip process using CUF. *CUF* capillary underfill

NCP is suitable for packages having narrow pitch and gap such as 60–80 μm pitches/20–30 μm gaps, where CUF is difficult to apply. The key feature of NCP is that thermal compression bonding (TCB) process connects bump and NCP encapsulates between the gapes at the same time.

In addition, TCB process skips flux cleaning process. Therefore, NCP requires for Cu organic solderability preservative (OSP) cleaning capabilities in itself. Figure 5.13 shows typical NCP process. First is pre-dispensing NCP under heating condition. After TCB process, it is cured at high temperature.

NCP process also has two potential problems. They are void and solder connection reliability. The design trend such as narrow gap and pitch and high bump count causes difficulties of filling and tends to result in void inside NCP. TCB process is performed after pre-dispensing, so filler has a chance to be caught in between solder and Cu bump. This phenomenon eventually leads to solder bump connection failure. These potential problems in NCP process have to be solved for advanced high-integration packages.

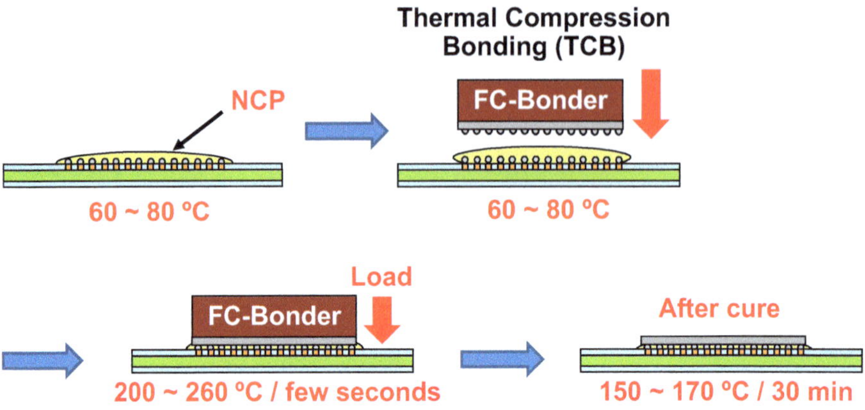


**Fig. 5.13** Flip chip process using NCP. *FC* flip chip

## 5.2.2 *Requirements for Underfill Materials*

### 5.2.2.1 Requirements for CUF and Material Technology Trend

2.5D and 3D packages using CUF require the void-less and low warpage. The trend of narrow pitch and gap of these packages can easily cause filling difficulties such as severe void entrapment. The mechanism of void entrapment is uneven flow of CUF. Figure 5.14 shows the image of the uneven flow and trapped voids.

To fix this problem, the use of smaller silica filler particles such as 1.0 μm average diameter (ϕ) or below was found to be effective as shown in Table 5.3. CUF-1 with silica filler larger than 1.0 μm ϕ has voids. On the other hand, CUF-2 with silica filler less than 1.0 μm ϕ showed no void at 25 μm gap.

The low warpage of the C2S structure is also required for 2.5D and 3D packages. Since stacked chips such as through silicone via (TSV) is connected on the C2S, the

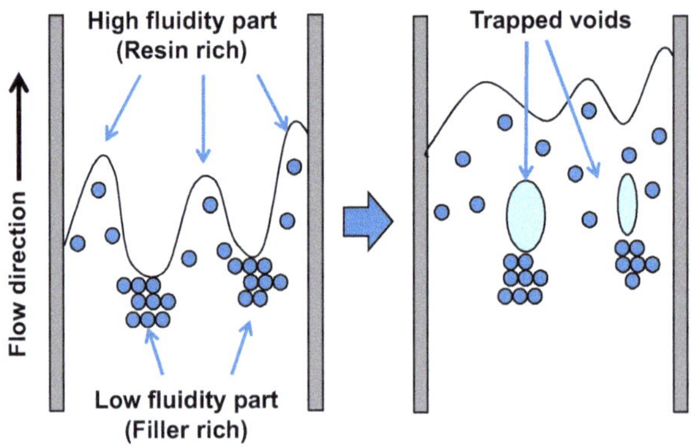


**Fig. 5.14** Image of the uneven flow and trapped voids

**Table 5.3** Void results of CUF-1 and CUF-2

| Item | Unit | CUF-1 | CUF-2 |
|---|---|---|---|
| Average filler size | μm | > 1.0 | < 1.0 |
| Filling gap: 25 μm | – |  |  |

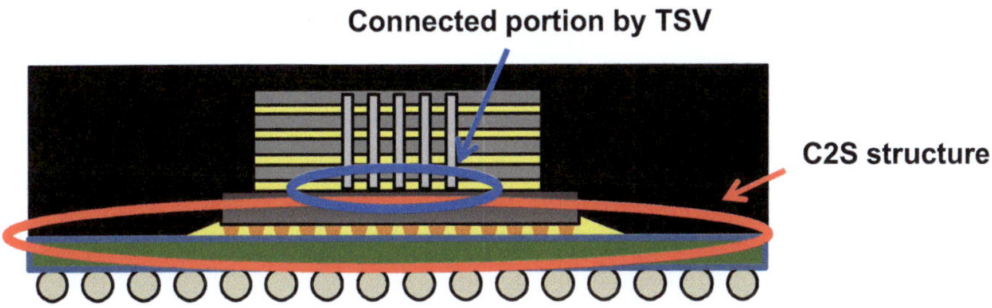


**Fig. 5.15** Connection with stacked chips on the C2S. *TSV* through silicone via, *C2S chip to substrate*

warpage of the C2S has the possibilities to cause connection failure with stacked chips (Fig. 5.15).

Various factors cause the warpage of the package. Finite element method (FEM) simulation is useful to estimate the target value for warpage control. We made the simulation model for developing suitable CUF for low warpage. Figures 5.16 and 5.17 show the simulation model and the direction of CUF properties, respectively.

From simulation, controlling the coefficient of thermal expansion (CTE) and glass transition temperature (Tg) are effective to reduce the warpage of the package at 25 °C. On the other hand, controlling the CTE and modulus are effective to reduce the warpage of the package at 260 °C. The direction for reducing the warpage of the package is different depending on the ambient temperature. Based on the simulation results, the materials that have different parameters are prepared and applied to the warpage reduction verification. The results are shown in Table 5.4. The results show the same tendency with the simulation.

Low Tg CUF-3 shows smaller warpage compared to control-1 at 25 °C. High CTE CUF-4 shows small warpage reduction at 25 and 260 °C. CUF-5 which has low modulus at high temperature shows low warpage at 260 °C. CUF-6 which has low Tg and low modulus at high temperature shows smaller warpage against control-1 at 25 and 260 °C. From these simulation and measurement results, various underfills can be designed for 2.5D and 3D application.

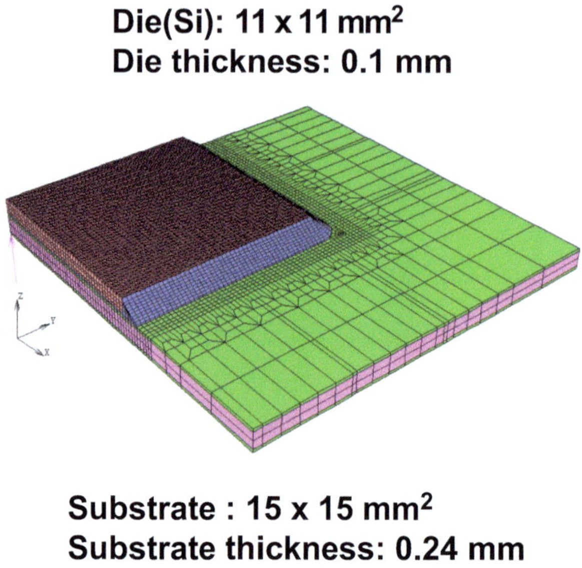


**Fig. 5.16** Simulation model

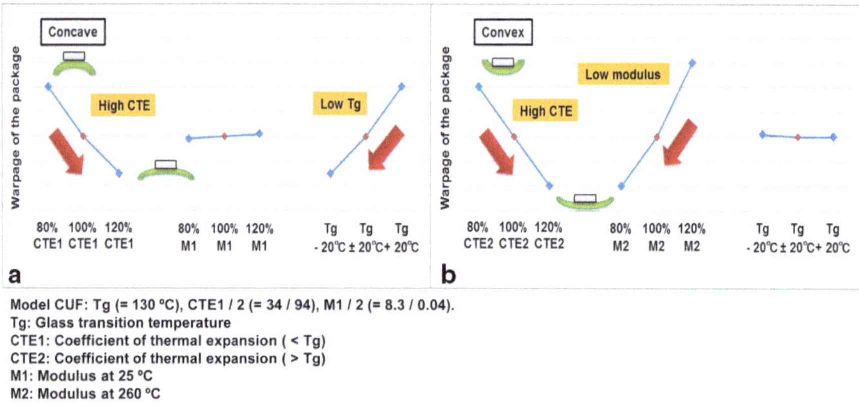


**Fig. 5.17** Direction of CUF properties (**a**) Simulation condition: 25 °C (**b**) Simulation condition: 260 °C. *CUF* capillary underfill

**Table 5.4** Properties and warpage results on CUF materials

| Item | | Unit | Control-1 | CUF-3 (low Tg) | CUF-4 (high CTE) | CUF-5 (low modulus) | CUF-6 |
|---|---|---|---|---|---|---|---|
| Property | Tg | °C | 111 | 89 | 110 | 110 | 101 |
| | CTE1/CTE2 | ppm/°C | 28/88 | 28/90 | 32/95 | 27/93 | 28/90 |
| | Modulus (25 °C/260 °C) | GPa | 8.7/0.10 | 8.5/0.10 | 8.5/0.12 | 8.7/0.06 | 8.9/0.05 |
| Warpage (25 °C/260 °C) | | μm | 167/−163 | 151/−160 | 160/−159 | 165/−155 | 157/−157 |

**Table 5.4** (continued)

| Item | Unit | Control-1 | CUF-3 (low Tg) | CUF-4 (high CTE) | CUF-5 (low modulus) | CUF-6 |
|---|---|---|---|---|---|---|

The measuring method for the warpage: Shadow moire
Die size: 9.2 × 9.2 mm$^2$
Die thickness: 0.1 mm
Substrate size: 15 × 15 mm$^2$
Substrate thickness: 0.3 mm
The warpage of the package at 25 °C: Concave warpage
The warpage of the package at 260 °C: Convex warpage

#### 5.2.2.2 Requirements for NCP and Material Technology Trend

2.5D and 3D packages require void-less and excellent solder bump connection reliability. Further narrowing of gap and pitch causes severe filling difficulties and results in void problem even when NCP process is used. Low-viscosity type NCP shows better wettability to substrate and bump, and it is effective in avoiding the void problem. Pre-dispensing of NCP and bonding process during TCB occasionally traps fillers between solder and Cu bump. The low-viscosity NCP is effective in reducing the problem. On the other hand, low-viscosity NCP has a potential to spread to the whole substrate after pre-dispensing. Therefore, the low-viscosity NCP is necessary to have thixotropic properties to restrain this issue. NCP with this property can hold the pre-dispensed shape after dispensing.

As shown in Table 5.5, high-viscosity material Control-2 showed voids and poor connectability. Low-viscosity and high thixotropic NCP-1 showed void-less and good bump formation.

**Table 5.5** Assembly results of NCP-1

| Item | | Unit | Control-2 | NCP-1 |
|---|---|---|---|---|
| Composition | Filler content | wt% | 55 | 55 |
| Property | Viscosity | Pa·s | 18 | 6 |
| | Thixotropic index | – | 4.0 | 4.2 |
| | Gel time (200 °C) | s | 2 | 2 |
| | Reliability result of initial (C-SAM) | – | 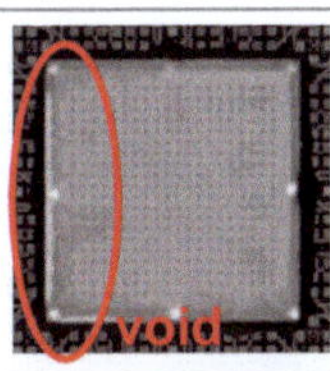  | 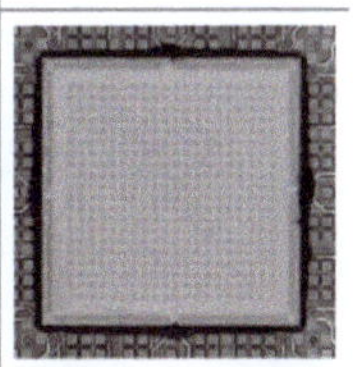 |
| | Solder bump connection | – |  | 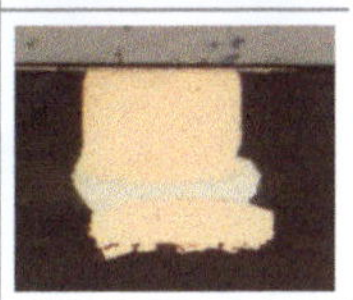 |

The measuring method for the solder bump connection: Optical microscope
Die size: 7.3 × 7.3 mm$^2$
Bump pitch: 80 μm (Peripheral) + 300 μm (Full array)
Bump gap: 40 μm
*C-SAM* Constant-depth mode scanning acoustic microscope

### *5.2.3 Application to CUF Between the Stacked Chips*

3D packaging draws attention as the next new high-integration technique, and its research and development has been active. 3D package is assembled by the C2S structure and the stacked chips structure. The desired properties of CUF on the C2S are discussed in Sect. 5.2.2. The application and property of CUF on the stacked chips structure are discussed in this section.

As shown in Fig. 5.18, the chip stacking process is categorized into three schemes such as chip-to-chip (C2C), chip-to-wafer (C2W), and wafer-to-wafer (W2W) process [9, 11, 12]. C2C and C2W are used by stacking a chip one by one or a chip on a wafer. W2W is performed by stacking a wafer onto another wafer. If the chip size in a wafer is the same as another wafer, the chip stacking in W2W becomes trivial, and it can maximize the chip stacking efficiency compared to C2C and C2W. W2W is the ideal process for production in terms of efficiency and cost.

After stacking the chips, 3D packaging is performed as shown in Fig. 5.19. Underfill is a key technology for improving the thermal and mechanical reliability of 3D packages. For the first step of 3D packaging, CUF is filled between stacked chips all at the same time [13]. Encapsulated stacked chips are assembled on the C2S.

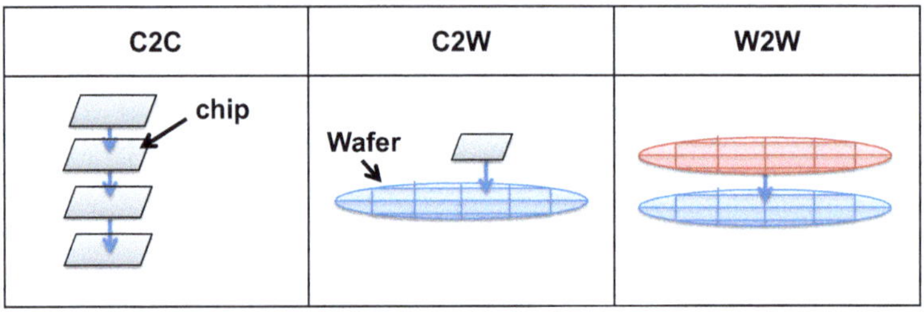


**Fig. 5.18** Assembly by C2C, C2W, and W2W. *C2C* chip to chip, *C2W* chip to wafer, *W2W* wafer to wafer

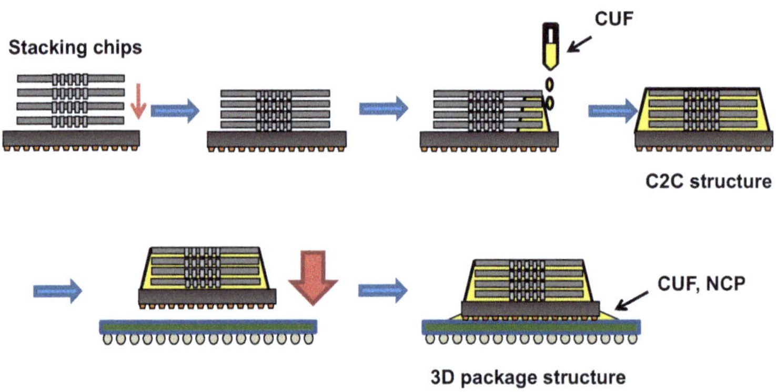


**Fig. 5.19** 3D packaging process using CUF. *CUF* capillary underfill, *NCP* nonconductive paste, *C2C* chip to chip

CUF process has an advantage for the C2C structure because CUF can be filled in several stacked chip gaps simultaneously. The most important requirement for CUF filled in between the stacked chips is void-less. To achieve this, the key technologies for 3D packaging is to control the properties of CUF, that is, viscosity, surface tension, contact angle between CUF, and chip surface for ~ uniform multilayer filling.

## 5.3 Nonconductive Films

### *5.3.1 Introduction*

In the semiconductor package field, adhesive resin has been used in between die and die or substrate to support the interconnection of them for more than 20 years. There are two types of such adhesive resins. One is a liquid type and the other is a film type. This section describes the latter one. Film type adhesive is only used as a pre-applied one for die assembly, although liquid type can be adopted both for pre- and post-apply, which needs high fluidity nature. Those three types underfill material' application of the processes are described in Fig. 5.20. This pre-applied adhesive film for FC assembly is usually called "nonconductive film (NCF)".

It is a kind of antonym of "anisotropic conductive film (ACF)," which contains conductive particle and shows electric conductivity only at the pressed point and the

| Process | Mass reflow | Thermal compression | |
|---|---|---|---|
| | CUF | NCP | NCF |
| Process flow of adhesive applying to substrate | 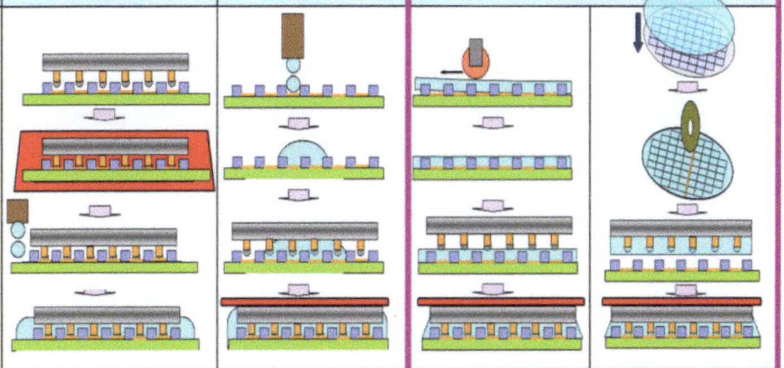 | | |
| Applying type | Post apply liquid | Pre apply liquid | Pre apply film |
| Resin heat resistance | No | Required | Required |
| Wafer level process compatibility | No | No | Yes |

**Fig. 5.20** Variation of flip chip assembly and underfill process. *CUF* capillary underfill, *NCP* nonconductive paste, *NCF* nonconductive film

direction. NCF is an insulation material as the name suggests. Pre-applied adhesive is usually used in the process of TCB. The process has an advantage of fine pitch bump die bonding. This film type is good for thin die assembly. Underfill material is usually expected to cover the die edge by forming a fillet around it. In the case of thin die, for example, 50 or less micrometer thick which 3D-IC die usually adopts, excess liquid underfill material often creeps up to the die backside. It is difficult to suppress it completely. When the film type is used, it is much easier to avoid it due to its higher controllability of material amount and flow. In the process, first, the interconnecting dies and/or substrate were aligned mechanically with each other within a few micrometer accuracy on the bonding stage. One die is usually cramped by an actively moving bonding heater head, and the other one is on a heating static or a passively level controlled stage. NCF is usually pre-applied on only the die of one side. Second, they were heated rapidly, and then the interconnection was formed. The NCF was cured and hardened simultaneously. During the cooling process, the thermal shrinkage difference among the materials used in the package causes the inner stress. The hardened NCF, which covers the full die area, works to avoid the stress concentration to the bump root.

### *5.3.2 Required Material Feature from Bonding Process*

NCF can be pre-applied in two different ways. One is applying NCF on whole wafer before dicing process [14], and the other is that NCF of singulated die size or similar is applied on the top surface of the die or the substrate which is placed on the thermal compression stage. Vacuum lamination process is preferably used for NCF applying to eliminate the air bubble at the interface of the NCF and the die or substrate. A die or a substrate usually has a certain topological structure on the surface, which sometimes catches the air bubble during NCF applying, especially in the case of large area lamination or ambient process. NCF should have a certain fluidity, which can fill up the bumpy structure of the die or the substrate at the lamination process.

One of the advantages of NCF as an underfill material is a compatibility of wafer level process [14], which can apply the material on hundreds of dies on a wafer in a single simple process. This process demands the material to have a compatibility of wafer dicing process. General blade dicing process demands that the material is hard enough to be cut by rotating blade with less deformation at the edge. As the process is a kind of grinding process, so a certain amount of powder dust from Si and NCF is generated. Such dust should be washed off from the NCF surface during the dicing or post-cleaning processes to avoid the contamination. It is preferable that NCF surface has a non-tackiness feature for the purpose. Fiducial mark detection on the wafer surface which is covered by NCF is needed for the blade alignment and precise die singulation. FC bonding also needs fiducial mark detection. It requires NCF transparency. Figure 5.21 illustrates a blade dicing process image of NCF on wafer, which is described above [15].

Figures 5.22 and 5.23 are an optical microscopic and a scanning electron microscopic (SEM) images of uncured NCF on wafer after blade [14].

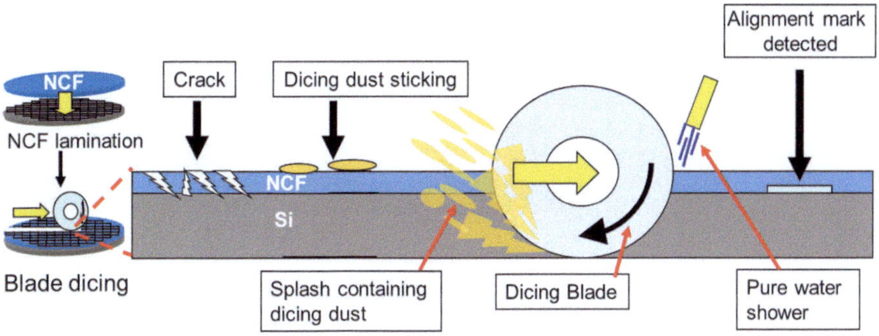


**Fig. 5.21** Blade dicing process of NCF on wafer. *NCF* nonconductive film

**Fig. 5.22** An optical microscopic image of NCF laminated wafer after blade dicing

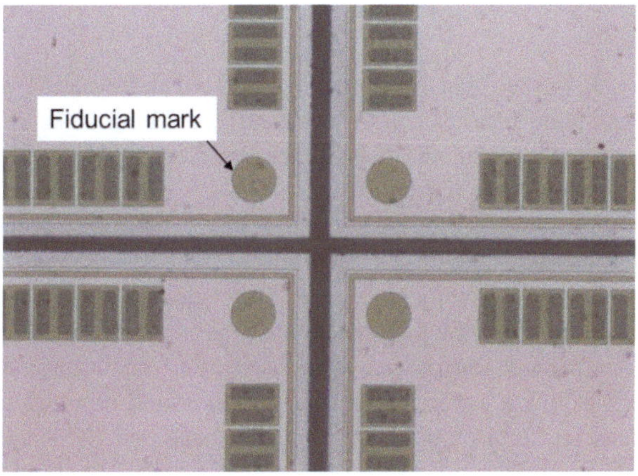


**Fig. 5.23** An SEM image of NCF laminated wafer after blade dicing. *NCF* nonconductive film

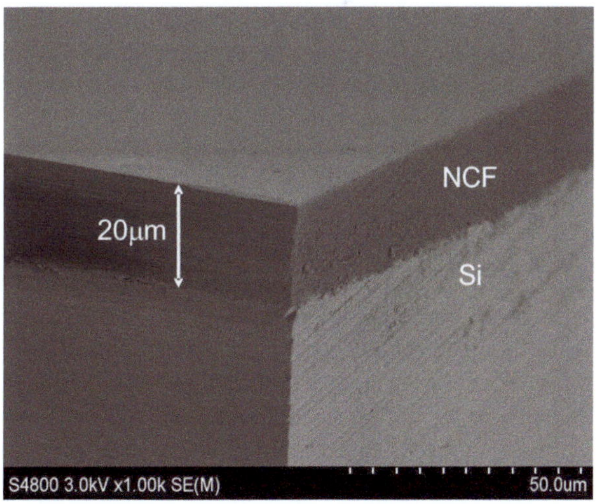

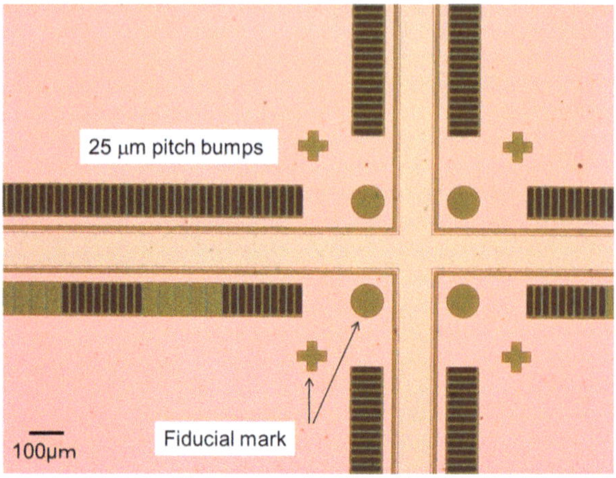


**Fig. 5.24** An optical microscopic image of a NCF laminated die

It is seen that NCF was cut without deformation or delamination at the edge, and there is no dust was on the surface. The images of fiducial marks on the wafer surface underneath the NCF are observed clearly in Fig. 5.24 [14]. Bumps are also clearly seen in the figure.

At the TCB process with NCF, the bonding die is picked up, aligned, and pressed on the other die or substrate on the stage of the equipment. NCF material is flown to fill the topological surface of the specimen. It is also pushed out from between the bump and the pad, which are interconnected by thermal compression process.

To meet the processes of lamination on bumpy surface without void, dicing of NCF on wafer without deformation and bonding to the other die or substrate making joint between the bump and the pad NCF material is usually designed to have a certain viscosity property. The viscosity should be very high around room temperature for the dicing process. On the other hand, high fluidity is needed to attain void-less lamination on topological surface of die or substrate. Die bonding process needs such high fluidity to attain void-less in NCF at the initial step and quick hardening by heating to show the stress relaxing function at the following step. Temperature dependence of NCF viscosity can be designed to meet the requirement. A typical example of it is shown in Fig. 5.25 [14]. The viscosity is high of $2 \times 10^6$ MPa at room temperature, dramatically coming down to $1 \times 10^3$ MPa by raising the temperature up to 100 °C and rapidly increases over 105 °C.

Lead free solder is one of the major interconnecting materials for 3D die stacking application. To form the joint, it should be melted and welded to the pad during the TCB process, which needs high-temperature heating of 230–260 °C. As the NCF between the die and the die or the substrate is also heated up to the same temperature, it should resist such a high temperature without material decomposition or excessive degradation. Such a heat-resistant performance is the special demand for pre-apply type underfill material. The post-apply process is carried out at much lower than solder melting temperature, for example, 60–120 °C. One of the

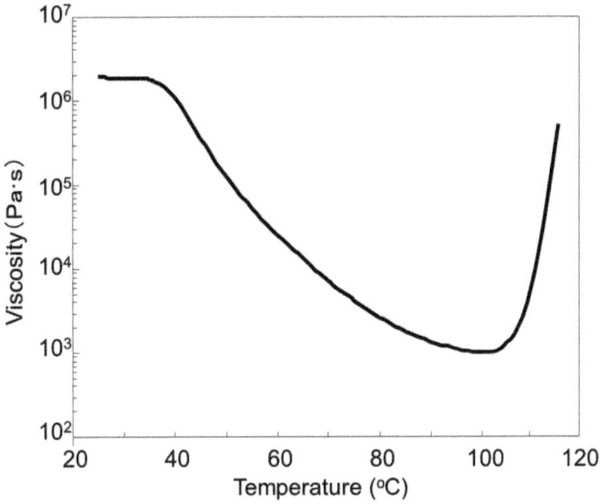


**Fig. 5.25** Typical temperature dependence of viscosity of NCF

advantages of the pre-apply underfill process is that, it has a stress relief function at interconnection bump root at the cooling step after the joint formation by solder welding of the bump and the pad. It is essential that the underfill material becomes large elastic modulus material at the step. Short process time is also demanded for TCB due to attainment of high productivity. So, the material is required to be hardened very rapidly and changed to the state of a high elastic modulus. Figure 5.25 reflects such a property. From the viewpoint of product, a longer floor life is preferable. It is desired that the curing reaction of NCF is very slow at room temperature and pretty high at bonding temperature.

Elastic modulus of epoxy resin, which is commonly used as an underfill material is a few GPa, which is much smaller than that of Si which is $\sim 1 \times 10^2$ GPa. Loading of an inorganic particle which has much larger elastic modulus than organic material to the resin is a general method to enhance the elastic modulus of underfill material. $SiO_2$ is generally selected as such a loading particle material which has the elastic modulus of $\sim 7 \times 10$ GPa. The refractive index of $SiO_2$ is 1.45. It is smaller than those of most of large elastic modulus resin components. One of the major origins of large modulus and high heat resistance of resin properties is the component atomic packing density, and it is also that of large refractive index in general. So, there is usually a refractive index difference between the underfill resin component and $SiO_2$ particle. The difference brings about a diffused reflection, and it tends to decrease the material transparency. Low transparency is not good for the fiducial mark detection, which is needed at both dicing and bonding processes. Using of smaller particle size $SiO_2$is one of the known ways to relax the transparency degradation, especially in the case of large amount of particle loading into the matrix resin. Loaded particle size influence on NCF transparency was investigated [15]. Submicron and nanosize particles were loaded 50 wt% to the same composition resin to make NCF. Figures 5.26a and b are optical micrographs of those 20 nm thick NCFs on die, respectively.

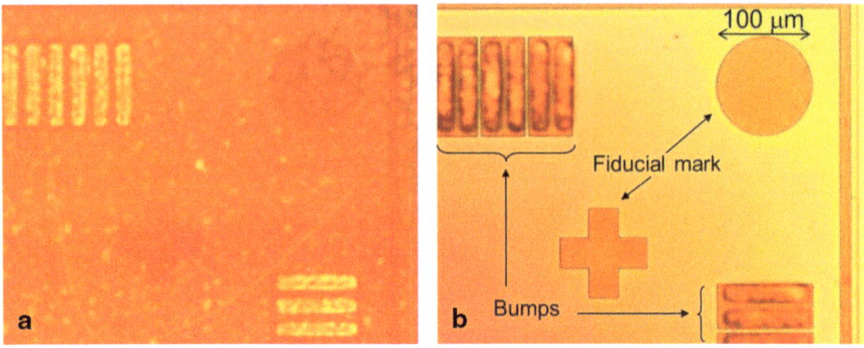


**Fig. 5.26** Optical micrographs of 20 nm thick NCF, where submicron and nanosize particles are loaded, respectively

## *5.3.3 Void Issue in NCF*

It is important to suppress the void of NCF without regard to applying method or material style. Such void may become a kind of moisture trap and induce electromigration between the wires, pads, and bumps, which are closely located and have electrically different potentials when it is positioned to bridge the gap of them. It may also be a cause of delamination of NCF. It is thought that there are plural origins of the voids. Gas may come out from inside of the NCF, which are made by decomposition or by-product of chemical reaction of the components, or by ejection of absorbed moisture. Other than NCF origin, void is also thought to be existed. Contamination on the die surface or organic substrate may exhaust the gas to the NCF. Trapped air at the interface of the NCF and the die or the substrate at the NCF lamination process may get into the NCF inside. Caught air at the TCB process, which is usually carried out in ambient circumstances, may be pushed into the NCF. In general, resin can absorb a certain amount of gas. If the adhesion strength of the interface between the NCF and the die or substrate is not strong enough, it can easily happens that the absorbed gas comes out, gathers there, and grows to void by a certain operation from outside like a stress or heat. Another property of the NCF influences the void emerging. As the cohesion strength of the NCF is weaker, the absorbed gas gathers and exists as void easier. NCF material properties can suppress the void.

Assembly process also significantly influences the void issue. It is preferable to suppress the void that every component to be assembled is kept dry before TCB is carried out. NCF lamination without any voids is also pretty important. TCB condition influences voiding a lot, which is the last chance of suppressing it. After the process, NCF is almost cured. It is very difficult to remove the void from the hardened material. Temperature and pressure are usually dynamically changed during TCB process under active control. Thermal and mechanical properties of the bonding specimen and the equipment strongly affect the TCB product. The TCB process usually has plural steps which have their own periods and conditions. Those are stage temperature, head temperature, applying force, controlled gap between the stage and the head, etc., which also influence the product.

TCB equipment usually has a performance bonding head, which can be very rapidly heated up to higher than 400 °C. This pretty dynamic temperature control can be one of the important parameters of NCF–TCB process. The void is suppressed by adjusting the TCB condition described above to meet each NCF material character and die and substrate specifications.

The die bonding process with pre-applied underfill is simpler than that with the post-applied one. At the underfill post-applied process, flux material is usually applied on the pad and washed off at the pre- and the post-bonding steps, respectively, when the solder is used for forming a joint. Pre-applied process does not have such steps. Flux function, which can chemically remove the oxide layer on the metal surface is preferable to make a well-soldered wetted joint. Certain specification of the bump and the pad require fluxing process. In such cases, NCF is needed to have the capability. An optical micrograph of a cross section of the joint of Cu pillar with solder cap bump and Cu pad with OSP, which was formed by TCB with NCF equipped with flux function, is shown in Fig. 5.27 [16]. It is seen that the bump solder wetted the Cu pad well. The Cu pillar was 38 × 38 μm$^2$ and 15 μm in height and the solder cap height was 15 μm.

**Fig. 5.27** A cross-sectional microscopic image of the joint Cu pillar with solder cap bump and pad formed by TCB with NCF equipped with flux function

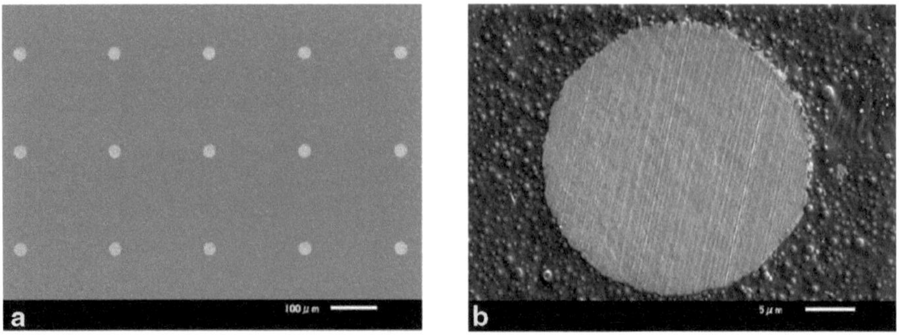


**Fig. 5.28** **a** and **b** are SEM images of the NCF laminated die surface after cut by cutting tool, respectively

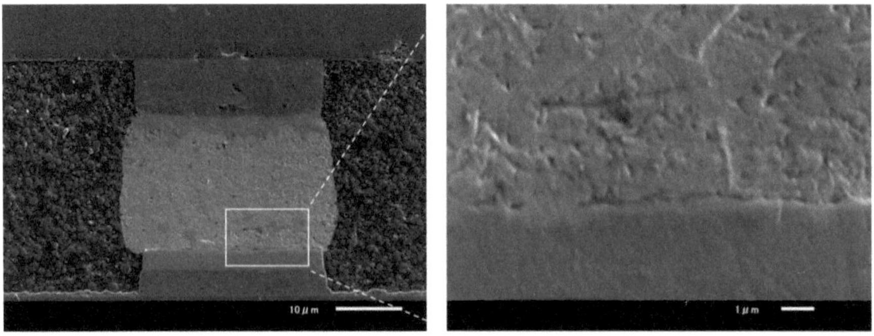


**Fig. 5.29** Cross-sectional SEM images of the joint made from a surface bit cut bump

Bonding process using pre-applied underfill should be designed to suppress the material trapping at the joint because the material is existed there beforehand. The optimization of the material property and the bonding condition can suppress it. There were more understandable researches, which aimed at avoiding it. Surface of the NCF-laminated die was cut or grinded, and the NCF material covering the bump top was completely removed before bonding. Figure 5.28a and b are SEM images of the NCF laminated die surface after cut by cutting tool, respectively [17]. Bump was composed of Cu pillar and SnAg solder cap. The top of the solder cap was cut accompanying with covering NCF, and then the flat SnAg surface was prepared. The die was bonded to another die with Au/Ni/Cu pad. Cross-sectional SEM images of the joint are shown in Figs. 5.29a and b in which material trap is not seen. Adding of a surface cut process before bonding is an understandable candidate to avoid the material trap at the joint [17].

### 5.3.4 *High Throughput NCF–TCB*

NCF is expected to mechanically support the reliability of the package after the die bonding is finished. Other than the stability of the mechanical property of the cured state, NCF insulation stability is very important for the purpose. The die for 3D stack often has a narrow bump gap, which is just a few to a few tens micrometer. Corrosion of the bump or the pad metal and the electrostatic attraction of the counter electrode may cause a forming of conductive path, which is a short circuit. Chemically active components and ion impurities are known to be possible enhancement agents of the corrosive reaction. NCF should be fabricated to contain least amount of such components.

Multi-memory die stacking, for example, of 4 or more dies, will be one of most expected applications in 3D-IC. The one-by-one die stacking process can be a sure approach. A more efficient stacking process will be needed to expand the market of 3D-IC by assembly cost reduction. One of the possible ways is dividing of pre- and main bonding and gang main bonding. This concept is illustrated in Fig. 5.30b compared with a conventional NCF–TCB process scheme, which is shown in Fig. 5.30a [18]. Most time-consuming step in TCB process is head cooling.

Die alignment and placement are performed at the pre-bonding process. Joints are formed and NCF is cured during main bonding. Constant temperature bonding head can be used for each process, respectively. Those two processes can be operated in parallel. Head cooling process is eliminated. Main bonding usually takes time longer than that of pre-bonding. It is a very simple process, which is just heat and press. If the main bonding time needs 15 times that of the pre-bonding, 15 dies

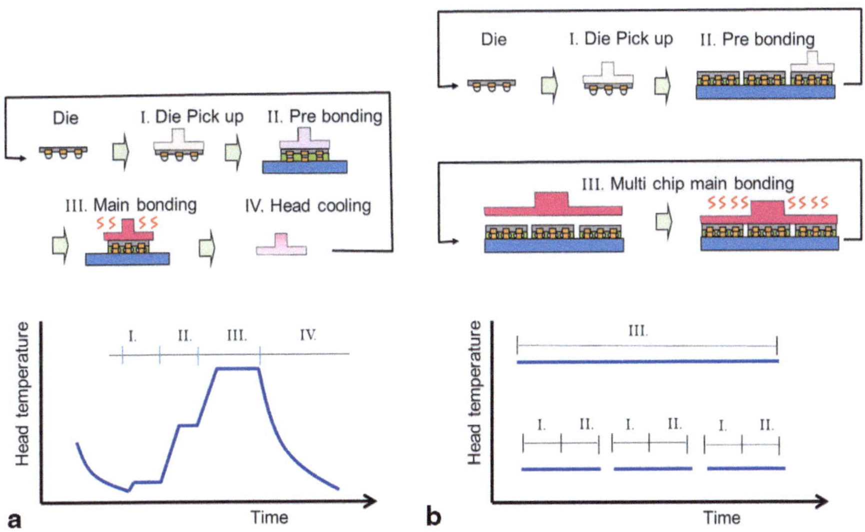


**Fig. 5.30** Conventional NCF–TCB process and dividing of pre- and main bonding and which is gang bonding are illustrated in (**a**) and (**b**), respectively

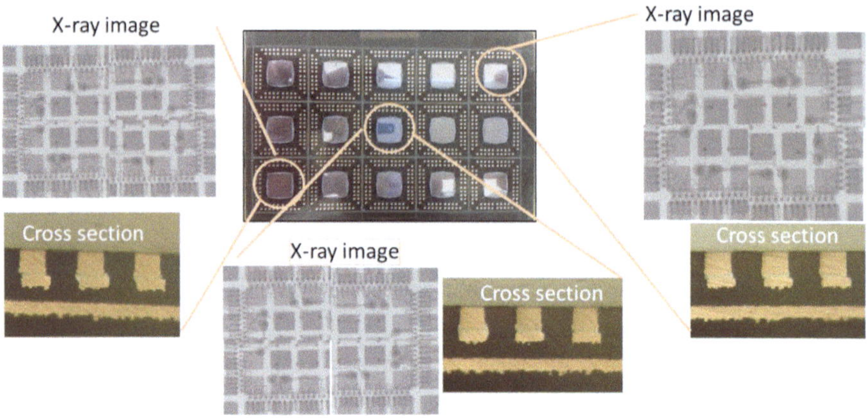


**Fig. 5.31** Pre- and main gang bonding demonstration results. Fifteen dies were bonded by main gang bonding simultaneously

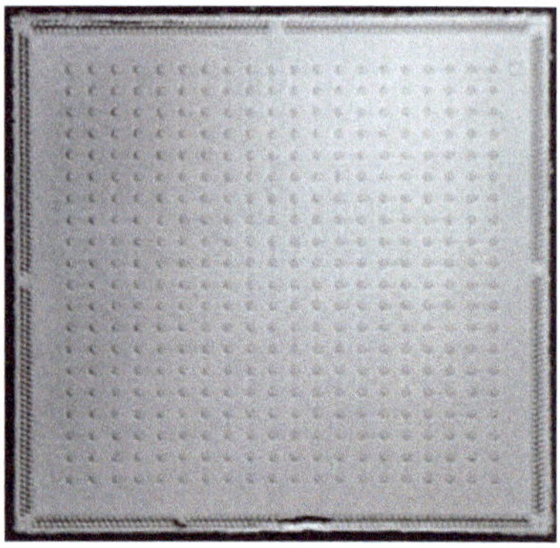

**Fig. 5.32** Typical C-SAM image of a die of 15 dies main gang bonded sample

main gang bonding can balance the total process. Figure 5.31 shows the result of the demonstration of this concept [18]. NCF was pre-applied on substrate. Pre-bonding was carried out with the head temperature constantly kept at 80 °C one by one for 15 dies. Then large main bonding head is kept at 240 °C pressed all the pre-bonded dies simultaneously. Each die size was $7.3 \times 7.3$ mm$^2$. Unusual die shift was not observed. The joint between Cu pillar with SnAg solder cap and Cu/OSP pad was formed well. The Cu pillar was $38 \times 38$ μm$^2$ and 15 μm height and the solder cap height was 15 μm. Void was not detected by constant-depth mode scanning acoustic microscope(C-SAM) observation, which is shown in Fig. 5.32 [18]. Gang bonding of plural dies in plane was demonstrated. There is another conceptual way to plural dies bonding simultaneously. As a pre-bonding, multi-dies can be aligned and stacked to the perpendicular direction, and then joints among each die can be formed by one heat and press. Wafer stacking can be a more efficient process. NCF material, which meets the new requirement, is going to be developed further and further to introduce more productive processes for 3D-IC to the market.

## References


1. Lasky JB (1986) Wafer bonding for silicon-on-insulator technologies. Appl Phys Lett 48:78
2. Shimbo M, Furukawa K, Fukuda K, Tanzawa K (1986) Silicon-to-silicon direct bonding method. J Appl Phys 60(8):2987
3. Takagi H (1999) Silicon wafer bonding by using surface activation method. Ph.D Thesis, The University of Tokyo. Aug. (hdl.handle.net/2261/54743)
4. Shigekawa N, Morimoto M, Nishida S, Liang J (2014) Surface-activated-bonding-based InGaP-on-Si double-junction cells. Jpn J Appl Phys. doi:10.7567/JJAP.53.04ER05. (53: Art. No. 04ER05)
5. Suda J, Okuda T, Uchida H, Minami A, Hatta N, Sakata T, Kawahara T, Yagi K, Kurashima Y, Takagi H (2013) Characterization of 4H-SiC homoepitaxial layers grown on 100-mm-diameter 4H-SiC/poly-SiC bonded substrates. The International Conference on Silicon Carbide and Related Materials. Sep
6. Maszara WP, Goetz G, Caviglia A, McKitterick JB (1988) Bonding of silicon wafers for silicon-on-insulator. J Appl Phys 64:4943–4950
7. Masteikaa V, Kowal J, Braithwaite NSJ, Rogers T (2014) A review of hydrophilic silicon wafer bonding. ECS J Solid State Sci Technol 3:Q42–Q54
8. Japan Electronics & Information Technology Industries Association (JEITA) (2013) Japan Jisso Technology Roadmap 2013 (in Japanese)
9. Knickerbocker JU et al (2008) Three-dimensional silicon integration. IBM J Res Dev 52(6):553–569, (November). http://citeseerx.ist.psu.edu/viewdoc/download?doi=10.1.1.419.1736&rep=rep1&type=pdf. Accessed 13 Jan 2015
10. Zhuqing Zhang Wong CP (2004) Recent advances in flip-chip underfill materials, process, and reliability. IEEE Transact Adv Packag 27(3):515–524. http://smartech.gatech.edu/bitstream/handle/1853/11437/CPWongIEEE30.pdf. Accessed 3 Jan 2015
11. Okuno A et al (2009) Material thechology that supports 3D mounting underfill. J Jpn Inst Electron Packag 12(2):114–119 (in Japanese). http://www.jstage.jst.go.jp/article/jiep/12/2/12_2_114/_pdf. Accessed 10 Nov 2014
12. Maeda N, Kitada H et al (2012) Development of Ultra-Thin Chip-on-Wafer Process using Bumpless Interconnects for Three-Dimensional Memory/Logic Applications. VLSI Technology (VLSIT), 2012 Symposium on, pp. 171–172. http://sogo.t.u-tokyo.ac.jp/ohba/news/VLSI1206.pdf. Accessed 3 Jan 2015
13. Denda S (2011) Stacking technologies of silicon chips with through silicon via electrodes. J Jpn Inst Electron Packag 14(7):571–577 (in Japanese). http://www.jstage.jst.go.jp/article/jiep/14/7/14_7_571/_pdf. Accessed 30 Jan 2015
14. Nonaka T et al (2008) Development of wafer level NCF (Non Conductive Film). Proceedings of 58th Electronic Components and Technology Conference, pp. 1550–1555, May
15. Nonaka T et al (2010) Wafer and/or chip bonding adhesives for 3D package. Proceedings of IEEE CPMtn3838T Symposium Japan, 10–14, Aug.
16. Kobayshi Y et al (2014) Flip chip assembly with wafer level NCF. Proceedings of International Conference on Electronics Packaging, Proceedings pp. 122–125, April
17. Nonaka T et al (2012) Low temperature touch down and suppressing filler trapping bonding process with wafer level pre applied underfilling film adhesive. Proceedings of 62th Electronic Components and Technology Conference, pp. 444–449, May
18. Nonaka T et al (2014) High throughput thermal compression NCF bonding. Proceedings of 64th Electronic Components and Technology Conference, pp. 913–918, May

# Chapter 6
# Metrology and Inspection

**Gilles Fresquet, Jean-Philippe Piel, Sylvain Perrot, Hideo Takizawa, Osamu Sato, Allen Gu, Michael Feser, Bruce Johnson, Raleigh Estrada and Yoshitaka Tatsumoto**

## 6.1 Principles of Spectroscopic Reflectometry

### *6.1.1 Introduction*

Spectroscopic reflectometry is a nondestructive technique widely used to analyze the properties of materials as thin layer thicknesses are used in advanced packaging manufacturing. The technique is based on the propagation of waves into media. If a discontinuity is encountered, a part of its energy is reflected back to the injection point according to the well-known law of reflection. The reflected signal gives useful information about the system and, in particular, thicknesses of thin layers.

Two types of reflectometry techniques can be distinguished: time-domain reflectometry (TDR) and frequency domain reflectometry (FDR). In the first case, the propagation of a pulse inside the medium is analyzed, and a possible echo could be observed. It is widely used for the characterization of defects in electrical cables. The second case analyzes the stationary waves in the medium. This second technology is the one used for semiconductor applications.

The usual configuration is to get the reflected light intensities on a broadband wavelength range. In most configurations, non-polarized light is used at normal

G. Fresquet (✉) · J.-P. Piel · S. Perrot
FOGALE nanotech, 125 Rue de l'Hostellerie, Ville Active, BâtA-Parc Acti+, 30900 NÎMES, France
e-mail: g.fresquet@fogale.com

H. Takizawa · O. Sato · Y. Tatsumoto
Lasertec Corporation, 10-1, Shin-yokohama 2-chome, Kohoku-ku, Yokohama 222-8552, Japan
e-mail: h_takizawa@Lasertec.co.jp

A. Gu · M. Feser · B. Johnson · R. Estrada
Carl Zeiss X-ray Microscopy Inc., 4385 Hopyard Road, Suite 100, 94588 Pleasanton, CA, USA
e-mail: allen.gu@zeiss.com



K. Kondo et al. (eds.), *Three-Dimensional Integration of Semiconductors*,
DOI 10.1007/978-3-319-18675-7_6

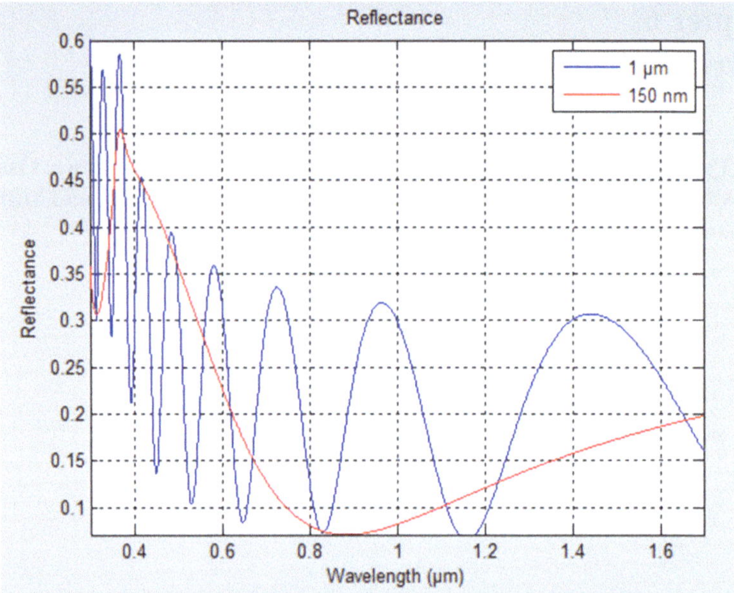


**Fig. 6.1** Comparison between 1 μm and 150 nm of $SiO_2$ layer on silicon substrate

incidence. Depending on the material, different spectral ranges of the electromagnetic field could be considered from the X-ray up to the far-infrared (IR) range.

For standard thin film characterization (10 nm–50 μm), the spectral range usually considered is from 300 nm to 1.7 μm (Fig. 6.1), and it could be extended easily up to 2 μm if the layers under investigation absorbed the light in the visible range like undoped silicon.

The biggest advantage of spectroscopic reflectometry is its simplicity and its low cost.

### *6.1.2 Measurement*

Practically, measurement is performed in two steps: In the first step, which corresponds to the calibration of the signal, a known sample, like bare silicon wafer, is employed for which the reflectance values are known: $R_{ref}(\lambda)$. The measure consists in acquiring the reflected signal coming from the calibration sample: $I_{ref}(\lambda)$.

In the second step, the reflected intensity from the sample under study is acquired: $I_{sample}(\lambda)$.

Finally, the absolute reflectance of the unknown sample versus wavelength is calculated using the following formula:

$$R_{sample}(\lambda) = \frac{R_{ref}(\lambda)\, I_{sample}(\lambda)}{I_{ref}(\lambda)}$$

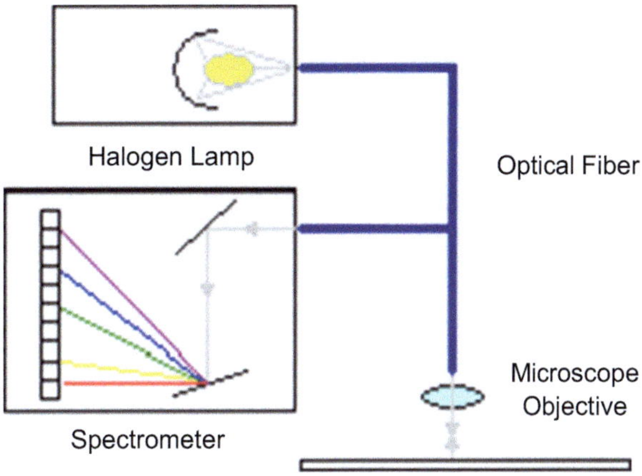


**Fig. 6.2** Reflectometer: typical setup

### 6.1.3 Setup (Fig. 6.2)

The source commonly used is a tungsten–halogen lamp (for visible and IR spectral range) or deuterium–halogen lamp (for deep ultraviolet (UV) and visible spectral range); the flux emitted is collected by a bundle of optical fibers and illuminates the sample under study. When a small spot is requested, for example, measures on a small area on patterned wafers, microscope objective is used to focus the beam on the surface on a few microns area. The reflected light is collected by the central fiber of the bundle and is sent to spectrometer equipped with a charge-coupled device (CCD) to cover 300–1100 nm spectral range or InGaAs array detector to cover 1000–2000 nm spectral range. The light spectrum acquisition is performed within few milliseconds depending on the signal to noise ratio (SNR) requirements.

### 6.1.4 Analysis

When the layers are thick, that means when the reflectance spectra presents multiple oscillations on the wavelength range, a Fourier transform of this signal is usually applied, and one gets rapid frequencies corresponding to the optical thicknesses of each layer. The knowledge of the refractive index allows us to get, finally, the physical thickness of each layer. This approach is very robust and is especially not sensitive to the reflectance amplitude which gives the possibility to work directly on the rough intensity directly.

But for the thin layer, typically for thicknesses less than 150 nm in the case of $SiO_2$ material, for example, as shown in Fig. 6.1, the Fourier transform does not work anymore, and it becomes necessary to use a multilayer stack approach.

In this last case, the measured reflectance curve is compared to a theoretical reflectance curve obtained for a multi-stack model. A fitting procedure is employed to minimize the distance between the two curves by varying automatically the physical parameters of the model like thicknesses of the layers.

More precisely, the model consists of the description of the sample using one or several layers if necessary on a substrate. For each layer, it is necessary to start with an estimation of the thickness, and one must introduce "$n$" value which is the refractive index and "$k$" the extinction coefficient. These values can come from tables if materials are well known. In other cases, it becomes necessary to estimate the index using mathematical equations to describe the variation of $n$ and $k$ versus the wavelength, and the literature is very rich for such mathematical laws. For example, dispersion laws like Cauchy law

$$n(\lambda) = A + \frac{B}{\lambda^2} + \frac{C}{\lambda^4}$$

or Sellmeier law

$$n^2(\lambda) = 1 + \sum_j \frac{B_j \lambda^2}{\lambda^2 - C_j}$$

are usually proposed for dielectric materials. If the layer is semi-absorbing, Drude–Lorentz laws are well adapted. When the layer could be considered as a physical mixing of two or more materials like porous silica (silica with inclusions of voids), effective medium approximation laws like Bruggeman's model is commonly used in this case.

From this information, it becomes possible to calculate the Fresnel coefficients for each interface of the stack. Between two interfaces, it must be taken into account the multiple reflections and transmissions of the waves and adding their amplitudes.

For example, for one layer deposited on a substrate, the global coefficient of reflection in amplitude is given by:

$$r = \frac{r_{0l} + r_{ls}\exp(-2i\delta)}{1 + r_{0l} r_{ls}\exp(-2i\delta)},$$

where, $r_{0l}$ and $r_{ls}$ are the Fresnel reflexion coefficient for the interfaces: Air/layer and layer/substrate, respectively, and $\delta$ is given by:

$$\delta = 2\pi \frac{t}{\lambda}(N^2 - N_0^2 \sin^2\varphi_0)^{\frac{1}{2}},$$

where, $t$ is the film thickness, $N$ the refractive index of the film, and $\varphi_0$ is the angle of incidence which is usually equal to 0 for reflectometry usual setup.

At the end, one can calculate the theoretical global reflectance versus the wavelength for the model proposed to describe the sample, which is:

$$R = |r|^2$$

The minimization procedure, usually the Levenberg–Marquard algorithm is used, provides optimized values for the fitted physical parameters under evaluation like thicknesses and coefficients of the index laws considered in the model. The quality of the minimization is expressed as a coefficient named, goodness of fit (GOF), for which the value is between 0 (poor adjustment) and 1 (perfect adjustment).

### *6.1.5 Conclusion*

Reflectometry, which is a nondestructive, noninvasive, and easy-to-use technique, requests, for thin layers, a model to estimate the physical parameters of each layer of the stack. The success of the technique depends on how well the numerical model can approximate the reality.

## 6.2 Low-Coherence Interferometry for Three-Dimensional Integrated Circuit Through-Silicon Via

### *6.2.1 Optical Measurement of Topographies and Thicknesses*

#### 6.2.1.1 3D IC TSV Needs Tomography

The use of through-silicon via (TSV) to conquer the third dimension (3D) enables to keep a very small form factor, and it opens up the way to stack far more than two dies. However, the more layers are stacked in a given volume as shown in Fig. 6.3, the more stringent is the requirement on thickness and topography management. Thus, a great issue in the field of metrology for 3D integration based on TSV is to be able to measure nondestructively, the thickness and topography of many superposed layers over a very large range. The total stack may reach a height of several millimeters while some thicknesses of interest are only of a few microns. Tomography, which consists in illuminating the sample with a penetrating waves and analyzing

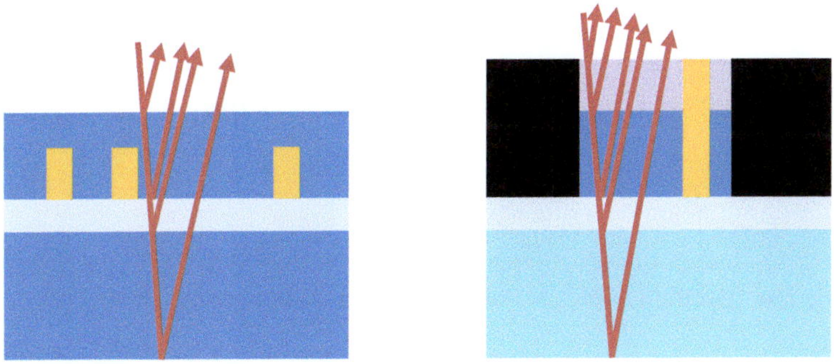

**Fig. 6.3** Illustration of the optical echoes in various multilayer structures

the echoes, is a very effective way to make this measurement. Many different waves can be used and are discussed in this article, such as X-rays, ultrasound, acoustic waves, and near-infrared (NIR) light. Optical measurement is of particular interest as it allows in reaching high resolution and high speed for a moderate cost. However, the detection principle is not straightforward and this chapter intends to give sufficient insight to whoever is interested in understanding the main parameters and the differences between various conceptual and practical implementations.

#### 6.2.1.2 Tomography with Low-Coherence Interferometry

Many materials used in microelectronic devices are transparent to NIR light, thus it is possible to illuminate the die from one side and have the light going through the layers. At each interface between two materials having different refractive indexes, a part of the light power will be reflected. If one uses a pulsed illumination and records the time of arrival of the reflected pulses on a photodetector, he/she can calculate the optical distances between the surfaces. However, the light travels 1 μm in as little as $3\times10^{-15}$ s. Such a method requires a femtosecond laser and a complex detection scheme, as chronometers do not reach that accuracy.

The alternative consists in using the coherence gating effect. When an optical beam goes through an interferometer, its power becomes modulated by the optical path difference that exists between the two arms. This modulation only occurs around the zero-difference optical paths along a distance called the coherence length. When using a low-coherence source, this distance can be as small as a few micrometers, and this modulation burst localizes precisely the position of an interface. To underline the link with the time of flight measurement idea, one can think of the pulse as being divided between a sample arm and a reference arm and of the detector as being able to detect precisely when two echoes arrive at the same time thanks to the fringe burst. Then, scanning the distance along the reference arm is equivalent to slowly reconstructing the time curve that a perfect chronometer would have given. This way, low-coherence interferometry enables to perform optical tomography with a great diversity of setups, making it possible to adjust the tradeoffs between cost, accuracy, and speed depending on the application. We now expose more precisely the underlying theory and focus on the main parameters governing the performances of this technique.

### *6.2.2 Theory of Optical Coherence Tomography*

#### 6.2.2.1 Basic Principle

Figure 6.4 shows a generic optical coherence tomography (OCT) setup that includes a low spectral coherence source, an amplitude division interferometer with

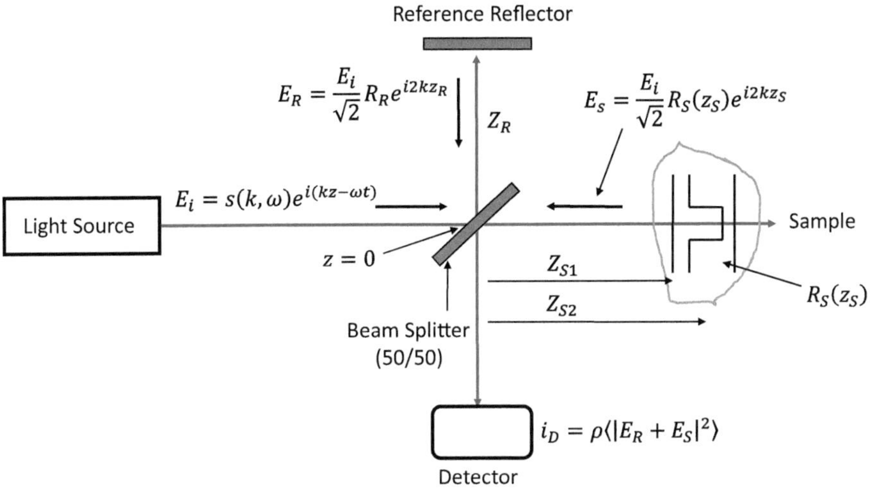


**Fig. 6.4** Michelson interferometer configuration commonly used in OCT

one arm lighting up the sample, and a detector. The incident light beam may be described by its electric field $E_i = s(k,\omega)e^{i(kz-\omega t)}$, where, $k = \frac{2\pi}{\lambda}$ is the wave-number and $\omega$ is the pulsation. For simplification, we consider a balanced beam splitter, thus the electric field coming back from the reference arm may be written as $E_R = \frac{E_i}{\sqrt{2}} R_R e^{i2kz_R}$, where $R_R$ is the complex reflection coefficient and $z_R$ is the optical distance from the beam splitter to the reference reflector. Similarly, for each step index in the sample, a reflected field propagates back to the detector that can be written as $E_S = \frac{E_i}{\sqrt{2}} r_s e^{i2kz_S}$. The strength of the reflections depends on the step index. For a normal incidence on the surface, we have the following expression: $R_{Sn} = \left( \frac{n_{Sn-1} - n_{Sn}}{n_{Sn-1} + n_{Sn}} \right)^2$. As a consequence, the reflection factor is quite high between air and silicon, around 30 % and very low between glass and glue.

The fields travel back through the beam splitter, then are summed on the detector and give rise to a current, which can be expressed as $I_D(k, z_R) = \frac{\rho(k)}{2} \left\langle | E_R + E_S |^2 \right\rangle$ with $\rho$ being the spectral response of the detector.

Eventually, the expression of the measured current intensity with regard to the wavelength and to the optical path difference is as follow:

$$I_D(k, z_R) = \frac{\rho(k)}{4}\left[S(k)[R_R + R_{S1} + R_{S2} + \ldots]\right]$$
$$+\frac{\rho(k)}{2}\left[S(k)\sum_{n=1}^{N}\sqrt{R_R R_{Sn}}\cos(2k(z_R - z_{Sn}))\right]$$
$$+\frac{\rho(k)}{4}\left[S(k)\sum_{n \neq m=1}^{N}\sqrt{R_{Sn} R_{Sm}}\cos(2k(z_{Sn} - z_{Sm}))\right]$$

For clarity, distances are expressed as a simple variable $z$, but one has to keep in mind that this represents the optical path, which integrates the product of the geometrical distance by the refractive index.

This expression is composed of three terms. The first one is independent from the optical path difference and corresponds to the reflection we would have with totally incoherent lighting. This is the continuous component supporting the fringes modulation. The second term, commonly called cross-correlation term, depends directly on the optical path difference. It conveys the information regarding the position of the sample surfaces with respect to the reference mirror. The third term, called the autocorrelation term, conveys the information of the different thicknesses in the sample. It also appears that the current is periodic with both $k$ and $z_R$. Consequently, there are two main ways to record an interferogram: by scanning the optical path difference or the wavelength. The first is referred as the time-domain OCT (TD-OCT) while the second is named Fourier-domain OCT (FD-OCT).

#### 6.2.2.2 Time-Domain OCT

The raw intensity signal obtained with a single wavelength is difficult to interpret, first because it mixes the information on the distance to the reference reflector with the thickness of the layers, and second because of the periodic nature of the cosine function. Using a broadband spectrum makes things much easier as it will blur the fringes when the optical path difference increases. This way, if the coherence length of the source is smaller than the layer thicknesses, the autocorrelation terms will disappear. To illustrate this phenomenon, we now express the current intensity in the case of a source with a Gaussian spectrum and a detector of constant sensitivity. After integrating the current over the wave number, we obtain:

$$I_D(Z_R) = \frac{\rho}{4}\left[S_0(R_R + R_{s1} + R_{S2} + \ldots)\right]$$
$$+\frac{\rho}{2}\left[S_0\sum_{n=1}^{N}\sqrt{R_R R_{Sn}}e^{-(z_R - Z_{Sn})^2\,\Delta k^2}\cos(2k_0(z_R - z_{Sn}))\right]$$

where, $S_0 = \int_{-\infty}^{+\infty} S(k)dk$ is the power emitted by the source over the whole spectrum, $k_0$ is the central wavenumber, and $\Delta k$ is the full width at half maximum of

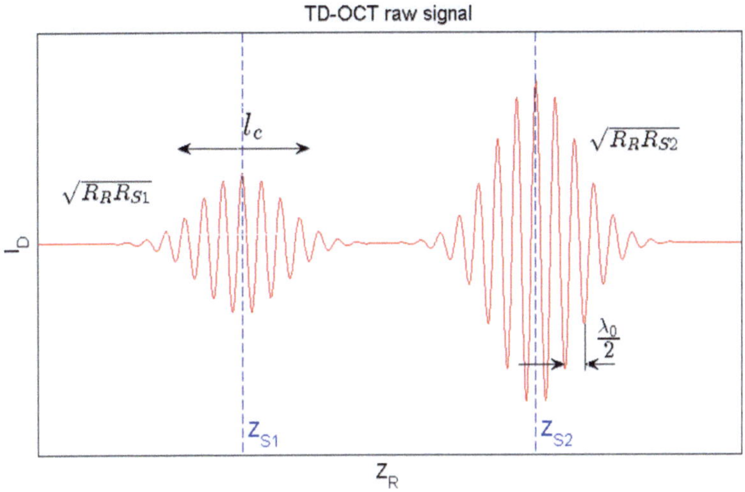


**Fig. 6.5** Illustration of the raw signal obtained in TD-OCT in the case of a sample with two surfaces. *TD-OCT* time-domain optical coherence tomography

the spectrum. The current intensity is now composed of a first term, constant, corresponding to the incoherent reflection case and of a second term, which gathers localized modulations.

From this expression, we can understand that the recorded interferogram is modulated with a period of $\lambda_0 / 2$, and that this modulation is altered by an envelope centered over the positions of the surfaces and whose width is equal to the coherence length as shown in Fig. 6.5. Signal processing consists mainly in demodulating the interferogram to get the envelope and then find the position of the peaks.

As a consequence, axial resolution is directly determined by the coherence length, which can be written as:

$$\delta z = l_c = \frac{2\ln(2)}{\pi}\frac{\lambda_0^2}{\Delta\lambda}$$

The full axial range that can be measured is limited either by the confocal depth of field (DOF) of the objective or by the scanning range of the reference mirror, whichever is the smallest. In most practical cases, the optical delay line is designed to cover the whole range authorized by the objective. Then, DOF and lateral resolution are the same as in confocal microscopy and only determined by the wavelength and the effective numerical aperture of the beam,

$$DOF = \frac{0.56\lambda}{\sin^2\left(\frac{\arcsin(NA)}{2}\right)}$$

$$\delta x = 0.37\frac{\lambda_0}{NA}$$

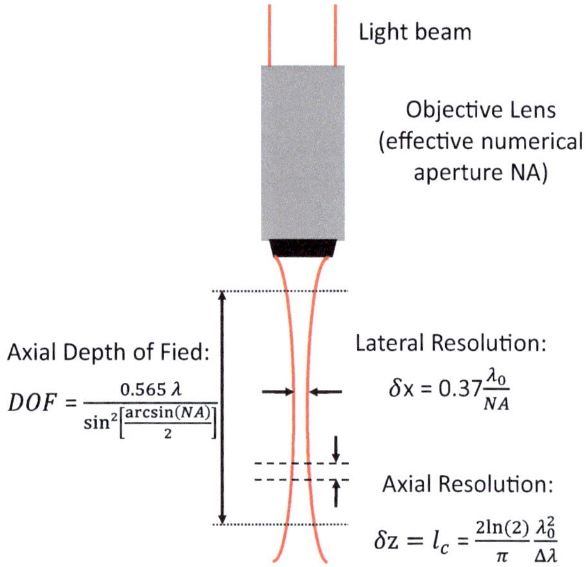


**Fig. 6.6** Summary of key features in optical coherence tomography. *DOF* depth of field

Figure 6.6 illustrates three features.

We can now express the SNR ratio. The useful part of $I_D$, which constitutes the signal, is the fringe amplitude and its maximum value is obtained when $Z_R = Z_S$.

$$Signal = I_D{}^2 = \frac{\rho^2 S_0^2 R_R R_S}{8}$$

Many noises can contribute to the measure, but the fundamental limit is driven by the shot noise.

$$noise = \sigma^2_{shotnoise} = \frac{e\rho S_0 \Delta f\left(R_R + R_S\right)}{2}$$

where $e$ is the electron charge and $\Delta f$ the electronic bandwidth.

Then, the SNR is:

$$SNR = \frac{\rho S_0}{4e\Delta f}\frac{R_R R_S}{R_R + R_S}$$

To compare systems on a practical basis, it can be convenient to express the SNR with respect to the measurement frequency $f_m$ and the number of resolution elements $M = \frac{\Delta Z}{\delta z} = \frac{\Delta Z}{l_c}$, considering $\Delta f \approx \frac{f_m M}{2}$

$$SNR = \frac{\rho S_0}{2ef_m M}\frac{R_R R_S}{R_R + R_S}$$

In more general cases, with non-Gaussian sources and non-spectrally uniform detectors, the behavior remains very close to that description. The main departure is caused by the wavelength dependence of the refractive indexes of the materials. As a consequence, it is not possible to obtain a zero optical path difference for every wavelength simultaneously, resulting in a larger envelope with decreased amplitude when the spectrum becomes too broad. Practically, this phenomenon puts a limit at the minimum coherence length that can be used when measuring large thicknesses.

#### 6.2.2.3 Fourier-Domain OCT

In FD-OCT, the reference mirror is not required to move, and the signal $I_D(k)$ is directly recorded. Again, two main ways can be followed to achieve that goal, by decomposing the light either after the interferometer or before. The first approach needs a spectrometer and is called spectral domain OCT (SD-OCT). The second approach had to wait for efficient coherent sources capable of rapidly and precisely changing their wavelength, and it is now called swept source OCT (SS-OCT). In that case, the wavelength is swept, and a unique detector records the intensity versus time. Both lead to an intensity spectrum and can be described with the same formalism.

Figure 6.7 illustrates the recorded spectrum in the case of one surface and two surfaces. In the right figure, we can identify the Gaussian shape of the emitted spectrum corresponding to the constant term of $I_D(k)$. This main line is modulated at a period of $\dfrac{\pi}{Z_R - Z_S}$ due to the interferences between the reflected beams from the reference arm and from the sample arm. In the figure on the left, a second surface has been added to show the increasing complexity. There are now at least three

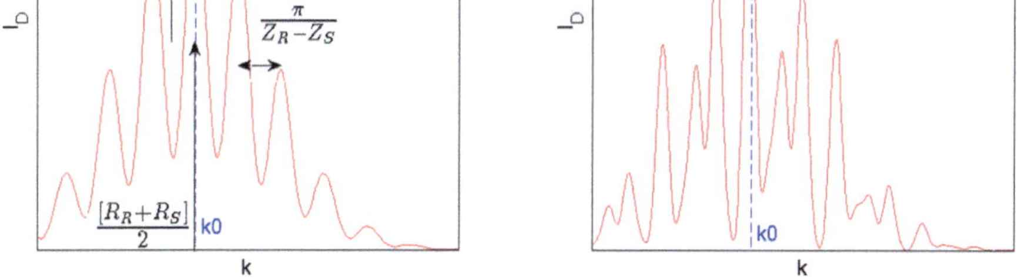


**Fig. 6.7** Illustration of the raw signal obtained in FD-OCT in the simple cases of one surface *(right)* and two surfaces *(left)* in the sample arm

modulations, two between the reference mirror and each surface, and one between the two surfaces. If the step index is high, as it is often the case in semiconductor applications, the multiple reflections may be strong enough to add more modulations at double and triple frequencies.

Signal processing is commonly conducted by remapping the spectrum from wavelength to wave number and calculating the Fourier transform. We show hereafter the analytical expression in the case of a Gaussian spectrum and a detector with constant sensitivity.

Considering that $\gamma(z)=e^{-z^2\Delta k^2}$ is the Fourier transform of the spectrum $S(k)=\frac{1}{\Delta k\sqrt{\pi}}e^{-\left[\frac{k-k_0}{\Delta k}\right]^2}$, $\delta$ is the Dirac function, and $\otimes$ stands for the convolution, we have

$$i_D(z)=\frac{\rho}{8}\left[\gamma(z)\left[R_R+R_{S1}+R_{S2}+\ldots\right]\right]+\frac{\rho}{4}\left[\gamma(z)\otimes\sum_{n=1}^{N}\sqrt{R_R R_{Sn}}\,\delta\left(z\pm 2\left(z_R-Z_{Sn}\right)\right)\right]$$
$$+\frac{\rho}{8}\left[\gamma(z)\otimes\sum_{n\neq m=1}^{N}\sqrt{R_{Sn}R_{Sm}}\,\delta\left(z\pm 2\left(z_{Sn}-Z_{Sm}\right)\right)\right]$$

The first term is a constant proportional to the reflected power. The second term shows that peaks appear at positions corresponding to the optical path difference between the reference mirror and the surfaces. The peaks width, $\gamma(z)$, is directly linked to the spectral coherence length of the source. It is noticeable that the peaks are located symmetrically around the zero optical path difference. Without additional information, it is not possible to determine the sign of the distance to the reference mirror. Practically, all surfaces have to be on the same side of the reference mirror and only one half of $I_D$ is used. Last term introduces peaks for the optical thicknesses between the sample surfaces. If one is only interested in measuring thicknesses, he can remove the reference mirror and only measure the autocorrelation peaks, which is equivalent to a frequency domain reflectometry setup. Schematic illustrations of signal processing in FD-OCT are given in Fig. 6.8.

It appears that axial resolution is limited by the coherence length, as for TD-OCT. The maximum range, in the limit of the axial depth of field, is determined by the spectral resolution $\delta\lambda$ as follow:

$Z_{max}=\frac{\lambda_0^2}{4\delta\lambda}\simeq\frac{N}{2}l_c$, where $N=\frac{\Delta\lambda}{\delta\lambda}$ is the number of resolved spectral channels.

Lateral resolution is still settled by the objective numerical aperture.

In order to evaluate the SNR ratio, we can now express the signal, which is obtained at a peak maximum:

$$signal=i_D\left(z=z_s\right)^2=\frac{\rho^2 S_0^2 R_R R_S}{16}$$

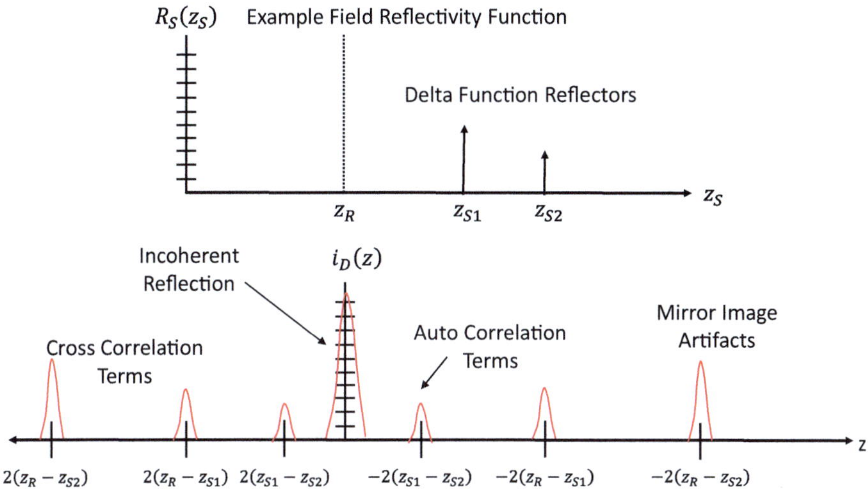


**Fig. 6.8** Illustration of the processed signal in FD-OCT in the case of two surfaces on the sample arm

The shot noise arises on each spectral channel. Its values are uncorrelated from one channel to another, and we can sum the variances to evaluate the noise after the Fourier transform.

$$noise = \sigma^2 = \frac{e\rho S_0 \Delta f \left(R_R + R_S\right)}{2}$$

Then, we deduce

$$SNR = \frac{\rho S_0}{8e\Delta f} \frac{R_R R_S}{R_R + R_S} \simeq \frac{\rho S_0}{8ef_{\mathrm{m}}} \frac{R_R R_S}{R_R + R_S}$$

Now that the fundamental principles have been described, we discuss their practical consequences.

#### 6.2.2.4 Practical Considerations

OCT deals with measuring distances and thicknesses of superposed layers. Thus, the main parameters of interest are the axial and lateral resolution, the axial range, the measurement repetition rate, and the ability to detect interfaces even with low reflectance or through rough and attenuating materials.

Because most of the OCT setups are using an optical fiber to carry the light from the main optical board to the sample, it shares some characteristics with confocal microscopy, such as the lateral resolution and the axial confocal gating. However, the use of the partial coherence enables a greater axial resolution, which does not depend on the numerical aperture but only on the spectrum width. It is now possible to use superluminescent diodes (SLD) emitting at 1, 1.3, or 1.5 µm, with a spectrum larger than 80 nm, resulting in a coherence length smaller than 10 µm in air. Thermal sources, such as domestic tungsten bulbs, are also used. Their broad spectrum, covering the whole spectral sensitivity of the sensors, allows to reach the smallest thicknesses, at the price of a lower SNR as the source is several orders of magnitude less brilliant than SLD.

Regarding those characteristics, TD-OCT and SD-OCT are the same. The very difference comes from the analysis of their SNR, for which we remind their expression for the same acquisition rate $f_m$.:

$$SNR_{TD-OCT} \simeq \frac{\rho S_0}{2ef_m M} \frac{R_R R_S}{R_R + R_S}$$

$$SNR_{SD-OCT} \simeq \frac{\rho S_0}{8ef_{\mathrm{m}}} \frac{R_R R_S}{R_R + R_S}$$

It appears that SD-OCT has an advantage of a factor $\frac{M}{4}$, where $M$ is the ratio between the axial scanned range and the coherence length. This factor may typically be as high as 100 and makes SD-OCT the best choice to measure through rough surfaces and interfaces with a very small step index, which generate very weak reflections, or to boost the acquisition rate while keeping the SNR constant. Acquisition rates greater than 1 kHz are commonly reached. SD-OCT also has the advantage of acquiring each wavelength independently, allowing to take into account the spectral variation of the refractive index and compensate for the dispersion during the signal processing. However, the signal turns out to be more difficult to use correctly. One has to separate the cross-correlation terms form the autocorrelation ones. This can be done by placing the mirror far from the surface so as to work in a very different range of thicknesses and to avoid any ambiguity. However, this reduces drastically the effective axial range. An alternative consists in recording several times the signal at one position while moving the reference mirror: This way, only the cross-correlation peaks are affected, and they can be safely identified.

Another aspect of SD-OCT is that some peaks may be superposed if two layers have similar thicknesses. Furthermore, when dealing with semiconductor materials with high refractive index contrast leading to high reflection factors, multiple reflections are also detected. There are peaks appearing at two times and three times the real thickness, with decreasing amplitude. Due to this multiple reflections and the presence of the autocorrelation terms, a very simple stack composed of two bare wafers bonded by a glue layer usually gives rise to a dozen of peaks, which makes the automatic measurement tricky.

On the contrary, TD-OCT guarantees to have all the peaks directly sorted, and no more numerous than the number of interfaces, preventing any ambiguity in the interpretation. Apart from the limitation of the objective, the scanning range in TD-OCT may virtually be as long as wished just by extending the delay line, whereas in SD-OCT the axial range is limited by the technology of NIR spectrometers or swept sources and evolves slowly. So far, TD-OCT remains a good solution to measure stacks, while SD-OCT is well adapted to the measure of small thicknesses, at higher speed, or of weak echoes due to a small refractive index contrast or a rough surface.

The main limitation of OCT is basically due to its optical nature. First, the beam can only go through transparent stacks, and it is not able to probe behind a metallic layer or an absorbing resist. Yet, it may be possible to measure from both sides of the stack to overcome this limitation in some cases. Second, the result is a measure of the optical path, and the group index has to be known to retrieve the absolute geometrical distance. Most of the materials used in three-dimensional integrated circuit (3D-IC) TSV have already been characterized, but the refractive index changes a little depending on the exact manufacturing process. Usually, this is not a practical limitation as most of the applications consist in monitoring processes, but it has to be considered when performing tool-to-tool matching with a different technology.

Among the major 3D integration process steps, some of them are indeed well known to create strong nonuniformity across the wafer, namely: TSV etching, wafer/carrier bonding, thinning, and TSV reveal etching. Each of these 3D processes actually generates a specific nonuniformity signature leading to a cumulative dispersion on the final wafer. It is very important to use accurate and repeatable metrology for both process implementation but also for process monitoring.

The interferometry technique can be used to characterize the TSV depth of the etching process. One method consists in illuminating the via (or group of vias) and analysing the reflected light coming back from the surface and the TSV bottom. The distance between interferogram gives directly the depth of the illuminated via(s) [1]. Another method consists on illuminating the wafer from the bottom and measuring silicon thickness of the wafer and the silicon thickness between the wafer backside and the TSV bottom. TSV CDs can be measured by microscopy in white light or NIR mode.

Wafer thinning is performed on a temporary carrier. The wafer is bonded on silicon or glass carrier and then thinned to a final thickness allowing the reveal of TSVs on the backside.

Such approach requires total stack thickness measurement but also individual thickness of each layer (wafer, glue, and carrier). IR interferometry is the technology of record for such process control. Another requirement concerns the remaining silicon thickness (RST) between the backside of the wafer and the TSV bottom. This measurement can also be performed by IR interferometry. Full field interferometry can measure the nails height uniformity and coplanarity after the reveal process. For the via-middle approach, the nails height uniformity across wafer surface depends on process uniformity of all previous process steps, from TSV etching, temporary bonding, to thinning [2].

All the process steps can be controlled with one metrology equipment (Fig. 6.9).

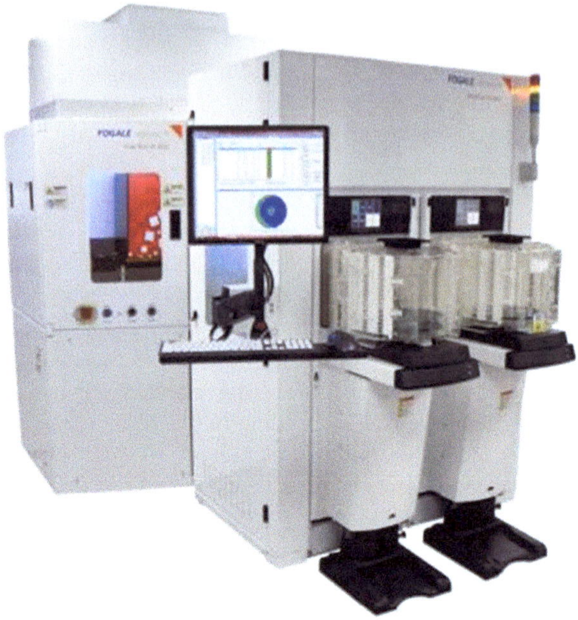

**Fig. 6.9** FOGALE Nanotech T-MAP DUAL 3D

### *6.2.3 Conclusion*

Optical coherence tomography techniques are very well adapted to the metrology needs of 3D-IC TSV as it allows to measure thicknesses from a few microns to a few millimeters, even for deeply buried layers, with microscopic resolution, good throughput, and nondestructively.

## 6.3 Silicon and Glue Thickness Measurement for Grinding

### *6.3.1 Introduction*

Using TSV for electrical connection between chips has a great advantage over the conventional wiring method because it can significantly reduce distance, offering such merits as higher speeds, lower power consumptions, smaller dimensions, and more signal connections. There are also expectations that highly functional semiconductor devices can be manufactured using TSV through layering of different types of chips including analog/digital ICs and memory/logic ICs. Because of these merits, many companies are making efforts to start commercial application of TSV. However, there are challenges that must be overcome before TSV can be widely used in the production of semiconductors.

One of such challenges is the need to know optimal conditions for back grinding of TSV wafers in order to achieve low production cost and high yield. Lasertec offers a solution for this challenge, a back grinding process measurement system called BGM300.

### 6.3.2 *TSV Wafer Manufacturing Method and Challenges of Grinding*

There are several ways to form TSV on semiconductor chips. Among them, the most common method adopted by semiconductor manufacturers is probably the one called "via middle".

In a typical via-middle method, wafers are manufactured in the following steps (see Fig. 6.10 for cross-sectional view of wafer at each of the steps).

1. Formation of transistors, deep etching for via holes, filling via holes with conductive material, and formation of a wiring layer.
2. A support wafer is bonded with glue to the surface of the TSV wafer.
3. Grinding is performed on the backside of the TSV wafer to make the wafer thinner. If the grinding goes too deep, it will chip off copper and cause copper contamination.

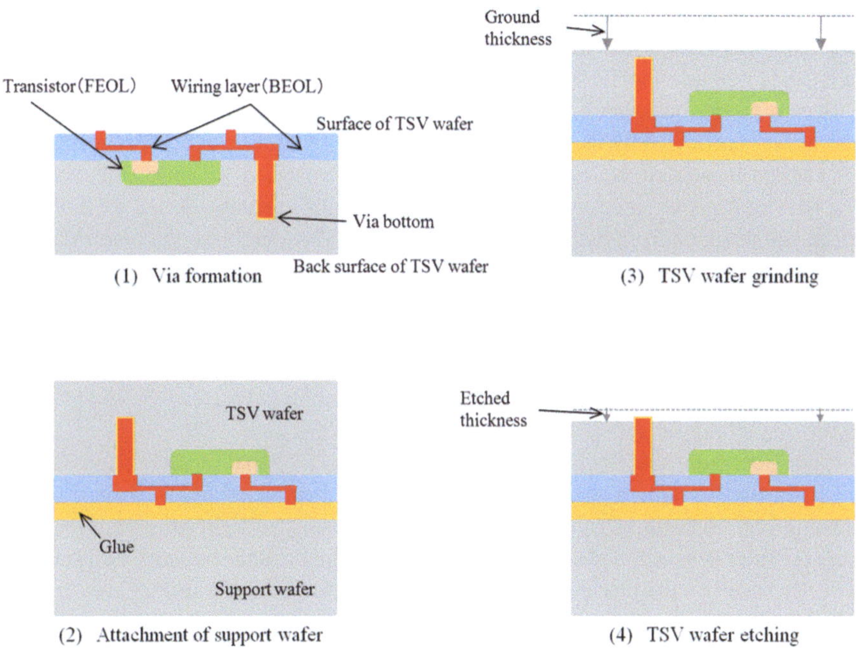


**Fig. 6.10** TSV wafer production process.

4. To prevent copper contamination, grinding needs to be stopped just before the via bottoms are exposed. The via bottoms are then exposed by etching. If the RST after grinding is too large, subsequent etching will be time-consuming and costly. Too much grinding, on the other hand, ends up with copper contamination.

To precisely perform grinding up to a preset end point to leave the right amount of RST, you need to know the distance from the back surface of the support wafer to the bottom of each via. The distance is equal to the sum of support wafer thickness, bond line thickness, and via depth. Bond line thickness and via depth tend to vary, depending on the location on the wafer. Nonuniformity of bond line thickness is especially large, spanning as much as several micrometers. It is therefore imperative to measure the distribution of distances from the back surface of support wafer to the end point of grinding throughout the wafer and detect abnormal bond line thicknesses or via depths. BGM300 developed by Lasertec offers accurate measurement capabilities to meet this requirement.

### 6.3.3 Features of BGM300

BGM300 employs two optical technologies: an IR interferometer optics used to measure the distance from the back surface of TSV wafer to a reflective object inside the wafer such as the bottom of a via and a sensor to measure the total thickness of the bonded wafer.

The IR interferometer optics of BGM300 consists of an interferometer equipped on our phase-shift measurement machine and a newly developed IR optics (Fig. 6.11). While the IR light illuminated through the objective lens is partially reflected by the back surface of the TSV wafer, the remainder of light penetrating the wafer is reflected by the bottom of the via. The two reflected light rays reach an IR camera located at the other end of the interferometer. The phase shift between the two light rays is measured from changes of light intensity. Based on this principle, the distance from the back surface of the TSV wafer to the via bottom can be measured accurately.

The distance from the back surface of the support wafer to a via bottom (distance "C" in Fig. 6.12) can be calculated highly accurately by subtracting the distance between the back surface of wafer and the via bottom ("A" in Fig. 6.12) captured by the IR interferometer from the total thickness of the bonded wafer ("B" in Fig. 6.12) measured by the sensor.

In areas where there is no printing pattern on surface that blocks the passage of IR light, the IR camera will capture interference of reflected light waves from the top and bottom surfaces of the glue. Using the intensity of light rays, you can measure the thickness of the glue. BGM300 measures the distance between two surfaces directly based on the phase shift of light rays reflected by those surfaces, ensuring a highly reliable measurement unaffected by via shapes or wafer layout.

Furthermore, BGM300 is furnished with a microscope to observe inside wafers with IR light as well as a function to automatically recognize patterns on a captured

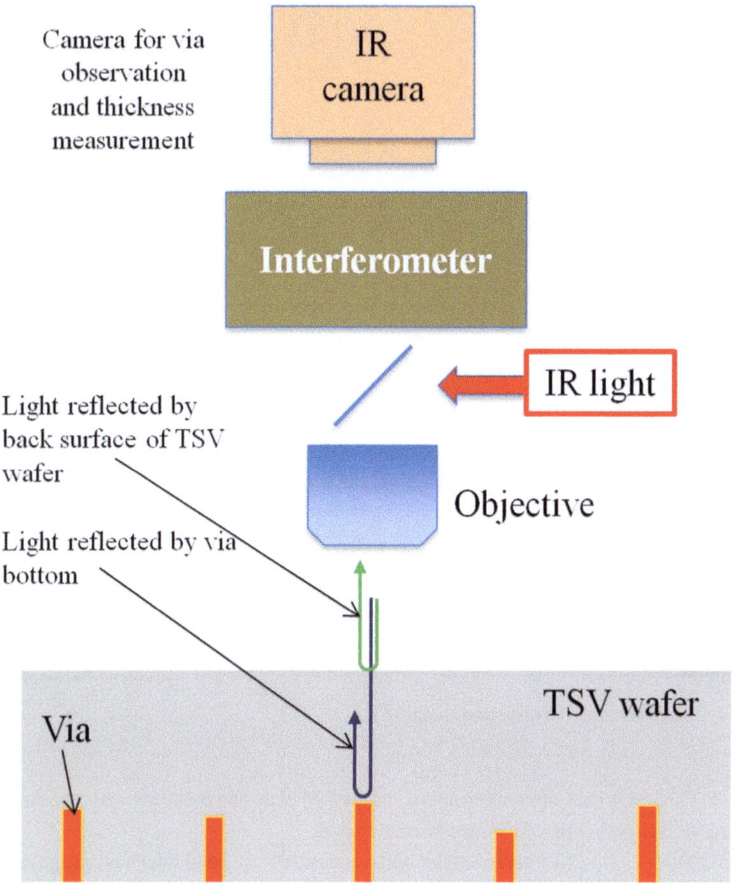


**Fig. 6.11** IR interferometer optics.

image. Using these functions, you can accurately measure the distance to a small-diameter via by pinpointing its position and automatically measure the distribution of distances throughout the wafer.

### *6.3.4 Verifying BGM300 Measurement Results*

To verify the accuracy of BGM300 measurement, we conducted a test using five different silicon plates with thickness data (uncertainty $\pm 0.05$ μm) purchased from a third party. We measured the thickness of silicon plates ten times, each using two different functions of BGM300, IR interferometer optics, and total thickness gauge. We then calculated the average of ten measurements and repeatability ($3\sigma$). The results are shown in Table 6.1. The results of the three independent measurements

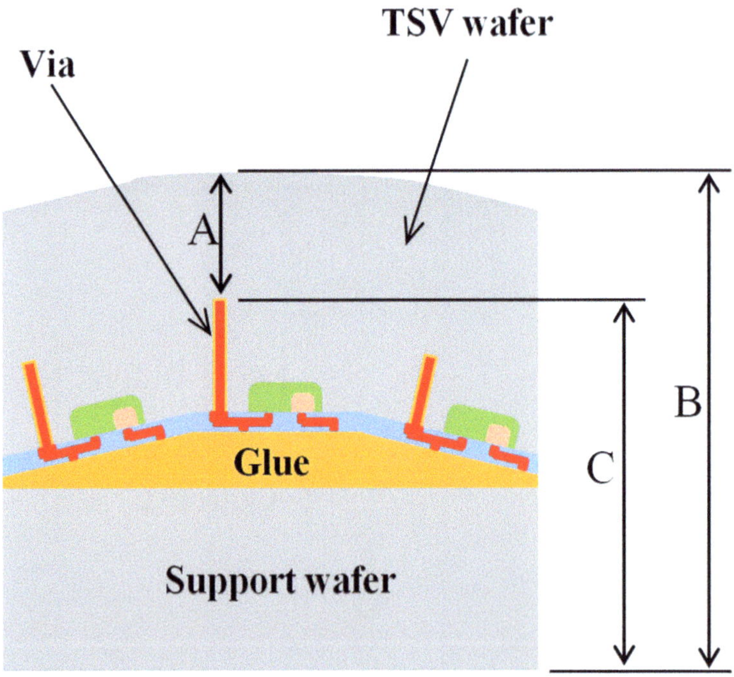


**Fig. 6.12** Measurement of via-bottom height "C."

**Table 6.1** Thickness measurement result of standard Si plate and measurement repeatability

| Standard Si thickness (±0.05) (μm) | BGM300 measurement result | | | |
|---|---|---|---|---|
| | IR interferometer optics measurement result | | Total thickness gauge measurement result | |
| | Average (μm) | 3σ | Average (μm) | 3σ |
| 25.28 | 25.25 | 0.02 | – | – |
| 50.39 | 50.36 | 0.10 | – | – |
| 99.46 | 99.35 | 0.06 | – | – |
| 674.57 | 674.63 | 0.05 | 674.42 | 0.02 |
| 774.96 | 774.94 | 0.02 | 774.86 | 0.01 |

match very closely. (Since the samples with a less than 100 μm thick could not be put on stage, there is no measurement result of the total thickness gauge for them.)

### 6.3.5 *Measurement After Grinding*

The IR interferometer optics of BGM300 directly measures a wide range of silicon thicknesses including bare wafers as much as 775 μm and thin layers as small as

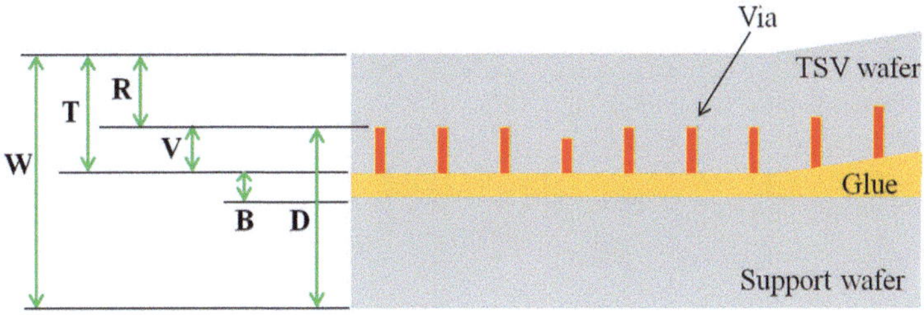


W: Total wafer thickness
T: TSV wafer thickness
R: Remaining Si Thickness [RST]
V: Via depth (T - R)
B: Boding layer thickness
D: Distance from the surface of support wafer to via bottom (W - R)

**Fig. 6.13** Measurement made by BGM300

a few micrometers. When a new process or condition is introduced, you may need to confirm if grinding is performed precisely to your target depth. In such a case, BGM300 can accurately measure the RST, which is about a few micrometers above the via bottom. You can measure various other thicknesses and distances accurately, as shown in Fig. 6.13, by choosing measurement points and comparing two values obtained from measurement in the proximity.

### *6.3.6 Optimized Wafer Grinding Based on Via Height Information from BGM300*

To prevent copper contamination while thinning a TSV wafer, you need to stop grinding before reaching the via bottom nearest to the back surface of the wafer. In the subsequent etching process, all via bottoms must be exposed (see Fig. 6.13). If there is a large variation in bond line thickness and via depth throughout the wafer, the distance from the back surface of wafer to via bottoms may vary greatly. This often results in a longer etching time and a higher cost. It may also lead to a large nonuniformity of via exposures after etching.

Some of the newest grinders for TSV wafers are capable of changing the grinding depth while thinning a wafer surface. Knowing the distribution of distances to via bottoms on wafers based on measurement of BGM300, you can utilize these grinders very effectively and mitigate the problem of etching.

We performed a test to see how effective a combined use of BGM300 and a new grinder is. Prior to grinding, we measured wafer thicknesses ("W" in Fig. 6.13),

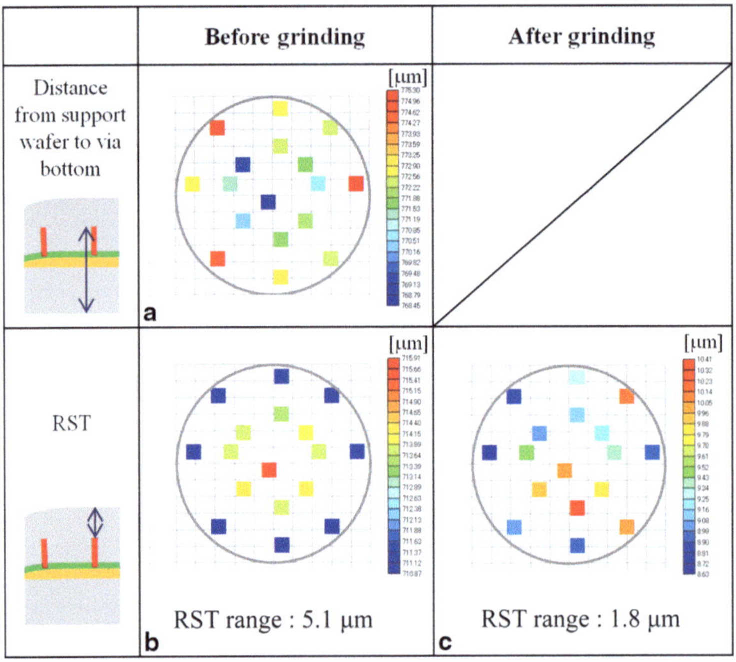


**Fig. 6.14** Examples of measurement with BGM300 before and after grinding. *RST* remaining silicon thickness

RSTs ("R" in Fig. 6.13), and distances from the back surface of the support wafer to via bottoms ("D" in Fig. 6.13) throughout the wafer. We then performed a TSV wafer thinning with variable grinding depths that are optimized based on the information from BGM300. Figure 6.14 shows the result of BGM300 measurements before and after the grinding. As shown here, with BGM300 and the new grinder used together, effective grinding without copper contamination can be performed on a TSV wafer whose bond line thickness and via depth are not uniformly distributed. This leads to a higher efficiency of etching, and, therefore, the TSV wafer can be processed to attain a higher uniformity of via exposures.

The wafer used in this test had a tendency that the distance from the support wafer to via bottom before grinding ("A" in Fig. 6.14) was thinner near the center than near the edge. Before the grinding, the distance from the surface of the support wafer to via bottoms ranged from 710.8 to 715.9 µm with a total thickness variation (TTV) of 5.1 µm, and the largest thickness was measured at the center (B). Based on this measurement, we made a setting to perform the deepest grinding near the center of wafer to make the RST distribution uniform. After the grinding, the TTV became 8.6–10.4 µm with a smaller TTV of 1.8 µm (C). We expect the deviation of via exposures to remain about 1.8 µm after etching, which removes silicon uniformly.

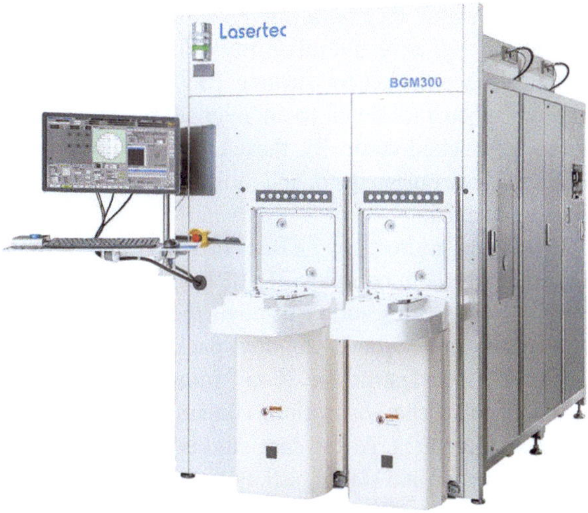


**Fig. 6.15** Appearance of BGM300

### *6.3.7 Conclusion*

Figure 6.15 shows the appearance of BGM300. BGM300 is an effective tool to enhance throughput and yield in the thinning process of TSV wafers and to reduce production cost. With a high precision XY stage and automatic pattern recognition, BGM300 can perform highly accurate measurement of a wide range of distances up to 775 μm on patterned wafers containing TSVs. The test results show that grinders can use via height information from BGM300 for copper contamination prevention and optimized grinding results. Additional inspection and measurement needs may arise from efforts of many companies engaged in commercial application of TSV. While facilitating wafer thinning process improvement with BGM300, Lasertec is ready to take on new challenges and contribute to an early realization of TSV devices.

## 6.4 3D X-ray Microscopy Technology for Nondestructive Analysis of TSV

### *6.4.1 Introduction*

Process development and failure analysis are becoming more challenging as the adoption of 3D-IC packaging and 2.5D interposer technology are increasing in the semiconductor industry. TSVs are more densely populated in compact circuits, and the structures are becoming more miniaturized. Since electrical behaviors of TSV such as impedance and capacitance strongly depend on TSV dimensions, shape, and defects, a viable technique for TSV structural characterization is of importance to circuit functionality and reliability. While IR and visible light interferometry offers a metrology solution for surface TSV and connected bumps, they do not provide enough subsurface detail to uncover subtle structural defects of TSV. Hence, as devices are stacked vertically, these techniques become less useful because structures of interest are buried deep, and most interconnecting materials are made of optically opaque metals.

3D X-ray microscopy (XRM) offers a nondestructive solution for structural and failure analysis in complex 3D-IC because it can visualize internal TSV structures with submicron resolution through the penetration power of X-rays. The analysis specimens range from individual packages to intact 300-mm wafers. This section introduces nondestructive X-ray imaging techniques with the emphasis on the mechanism of the absorption, contrast, and spatial resolution of different instrument architectures. A few case studies for TSV measurement and analysis in both 300-mm wafer and single packages are discussed.

## *6.4.2 Fundamentals of X-ray Microscopy*

### 6.4.2.1 Physics of X-ray Imaging

XRM has been established as a standard microscopic analysis technique along with optical and electron microscopies. The short wavelength and penetration power of X-rays allow visualization and quantification of internal structures of optically opaque or transparent specimens. In a simplest setup, X-rays are generated by a source, and pass through a sample, and the transmitted X-ray photons project a shadow image on an imaging detector. The collected images or radiographs are used to analyze specimen. Compared with the conventional physical cross-section technique, the X-ray imaging technique does not require samples to be trimmed, sectioned, or manipulated. Hence, it is a nondestructive technique.

The primary contrast mechanism for an X-ray imaging system is absorption contrast, which is generated by the difference in absorption of X-rays between different materials. Figure 6.16 shows the general behavior of the absorption ($1/e$ absorption length) of Si and Cu (two major elements in semiconductor TSV) as a function of X-ray energy. At low X-ray energies, the absorption length is dominated by the photoelectric absorption effect and increases with the third power of the X-ray energy away from the characteristic absorption edges that correspond to discrete electron energy levels of the material. At high energies, inelastic Compton scattering dominates. Although dual energy methods involving both photoelectric effect and Compton scattering have been explored to study chemical information, the photoelectric effect manifested by absorption contrast is mostly used in X-ray imaging techniques. As shown in Fig. 6.16, Cu fill is significantly more absorbing than the surrounding Si in TSV especially when the X-ray energies are between 10 and 100 keV used in XRM. As a result, Cu is shown with different intensity from Si in either 2D or 3D X-ray images.

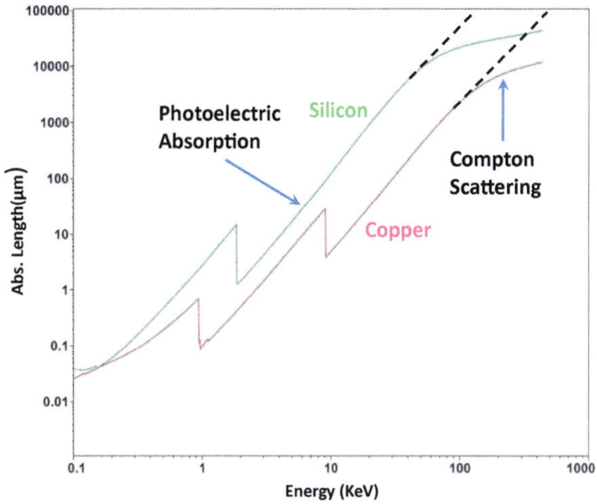


**Fig. 6.16** X-ray absorption of Cu and Si functions as X-ray energies. At lower energies, the behavior is dominated by photoelectric absorption. At high energies, Compton scattering dominates. The photoelectric absorption is mostly used for X-ray images

#### 6.4.2.2 3D X-ray Microscopy

The X-ray microscope, often referred to as a projection X-ray microscope, utilizes a simple shadow projection geometry. X-ray emits from the source, passes through a specimen, and the projection image is captured by a detector. By rotating a sample through 180° or 360°, one can acquire many projections from different angles of the object. These projections are reconstructed mathematically to obtain a 3D volume image of the object since each projection carries absorption information of the sample. Due to continuous downscaling of TSV in modern electronics, an XRM with submicron spatial resolution is a key application requirement. The resolution obtainable is a function of the X-ray source spot size ($S_{\mathrm{FWHM}}$), the detector resolution ($D_{\mathrm{FWHM}}$), and the projection geometry given by distances from source to sample $d_{ss}$ and sample to detector $d_{sd}$. Assuming simple Gaussian shapes for source spot and detector resolution with a given FWHM, the resolution can be written down as:

$$\mathrm{Img}_{\mathrm{FWHM}} = \frac{1}{M}\sqrt{\left(S_{FWHM}\frac{d_{sd}}{d_{ss}}\right)^2 + D_{FWHM}{}^2}$$

where, $M$ is the geometric magnification of an XRM system given by $1 + d_{sd} / d_{ss}$. The Gaussian blur of the sample image ($\mathrm{Img}_{\mathrm{FWHM}}$) is obtained by re-projecting the sample image on the detector, which convolves the Gaussian blur introduced by the source size and the Gaussian blur introduced by the detector resolution. In general, XRM spatial resolution is limited by either the $S_{\mathrm{FWHM}}$ or the $D_{\mathrm{FWHM}}$ through penumbral blurring.

There are two existing techniques to improve the spatial resolution of XRM: First, through the usage of a high-resolution detector; second, through the utilization of high geometric magnification, and an X-ray source with small spot sizes.

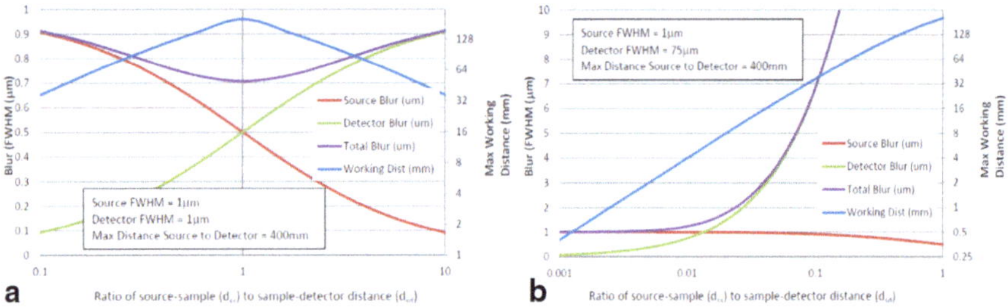


**Fig. 6.17** Total system resolution functions for varying source/detector geometry (**a**). *FWHM* full width at half maximum In XRM systems with both a high-resolution detector and small spot source that are matched in size, the best resolution (lowest blur) is achieved for a symmetric geometry (i.e., $d_{ss} = d_{sd}$). In this case, high resolution is possible even for very large samples or working distances. **b** In XRM systems with low-resolution detector, high resolution is only achieved when the sample is very small (low working distance). (Michael Feser. X-Ray Microscopy. In Encyclopedia of Optical and Photonic Engineering, 2nd Edition. Taylor and Francis: New York (In Press))

Figure 6.17a shows the ideal case for an XRM that utilizes both techniques: a small spot source and high-resolution detectors. It is possible to achieve a total system resolution exceeding the source or detector FWHM (full width at half maximum) by a factor of $\sqrt{2}$ when $d_{ss}$ equals to $d_{sd}$, assuming 1 μm for both source and detector FWHM. Working distance is not an issue in this type of systems and can be optimized for the TSV applications, in which the accommodation of very large sample manipulators, such as a 300-mm wafer holder, is required as described below.

Figure 6.17b shows the more traditional architecture only utilizing a small spot source combined with a low-resolution detector (flat panel of fiber coupled CCD), in which the working distance and hence maximum sample size for high-resolution imaging is severely restricted. The major disadvantage of this type of XRM is that the system resolution quickly deteriorates by the increase of working distance. Because the size of typical electronic or TSV packages is greater than 10 mm, the XRM that rely on high geometric magnifications for high resolution are not suitable for these applications. Hence, for full microstructure characterization of TSV in large packages or intact wafers, the architecture shown in Fig. 6.17a is well suited. This architecture is used in the case studies discussed below.

### *6.4.3 Applications for TSV Process Development*

In process development, TSV structures fabricated by a new or modified process often require validation prior to next step. 3D XRM imaging of specific critical structures gives direct physical evidence for the process engineer to analyze and unravel potential process issues. When a process adjustment is made, XRM may be used to check its effectiveness because structural analysis is possible at multiple process steps on the same wafer.

Since the XRM having a high-resolution detector maintains high system resolution as the source-to-sample distance increases (Fig. 6.17a), here, we employed this

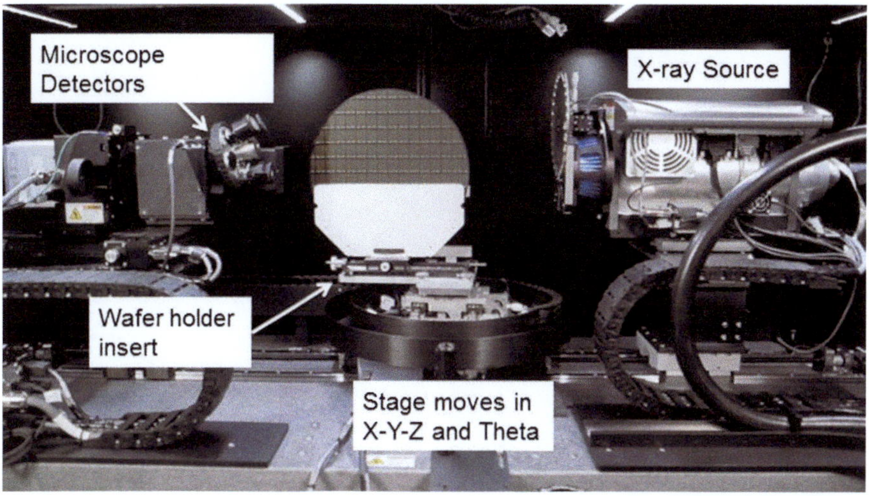


**Fig. 6.18** The XRM setup and the 300-mm wafer holder. The *arrows* indicate the translation knob and reference marks to position ROI. With the high-resolution detector, any location on the wafer may be imaged with 1 μm, or better, voxel resolution

type of XRM to image TSV in 300-mm wafer. Figure 6.18 shows that a 300-mm wafer holder was loaded into the XRM equipped with high-resolution detectors, ZEISS Xradia 520 Versa XRM. The wafer, gently held by a clamping system, is capable of rotating 360° during tomography acquisition. It is possible to image any region of interest (ROI) across the wafer with < 1 μm voxel resolution.

In this case study, 30 × 150 μm via arrays were patterned on chips across a 300-mm wafer and then filled with Cu. To test the TSV structural difference due to wafer warping, two tomograms were acquired to enable comparison of structures in the middle of the wafer to the structures toward the edge using 0.8 and 1.35 μm per voxel, respectively (Fig. 6.19a). On the test wafer, voids were found in both edge and center locations, which are visualized as dark areas in the middle of the virtually cross-sectioned TSV images (Fig. 6.19b). At the center location, the void was measured about 6.2 μm deep and 2.2 μm wide. For the ROI towards the wafer edge, voids were found at similar sizes. Since the slope of TSV plays a role for electrical activity, it is useful to visualize and measure it. As shown in Fig. 6.19c, the TSV at the wafer center region looked more sloped than the TSV at the wafer edge. Through its ability to image subtle structural differences, XRM may be helpful in evaluating electrical models derived by true 3D structures.

While XRM provide the best resultant volume data in all three geometric directions, the development of automated X-ray inspection (AXI) tools for high throughput, in-line inspection has been advancing sharply [3]. This type of technology equipped with automatic wafer handling robot offers a metrology solution for TSV. With 2D X-ray images taken from a vertical or an angular view of a wafer, structures of interest can be automatically measured and analyzed with high throughput. This approach gives good reconstruction results in the *z*-direction (into the plane of the sample) but less good results compared to full XRM in *x* and *y* directions,

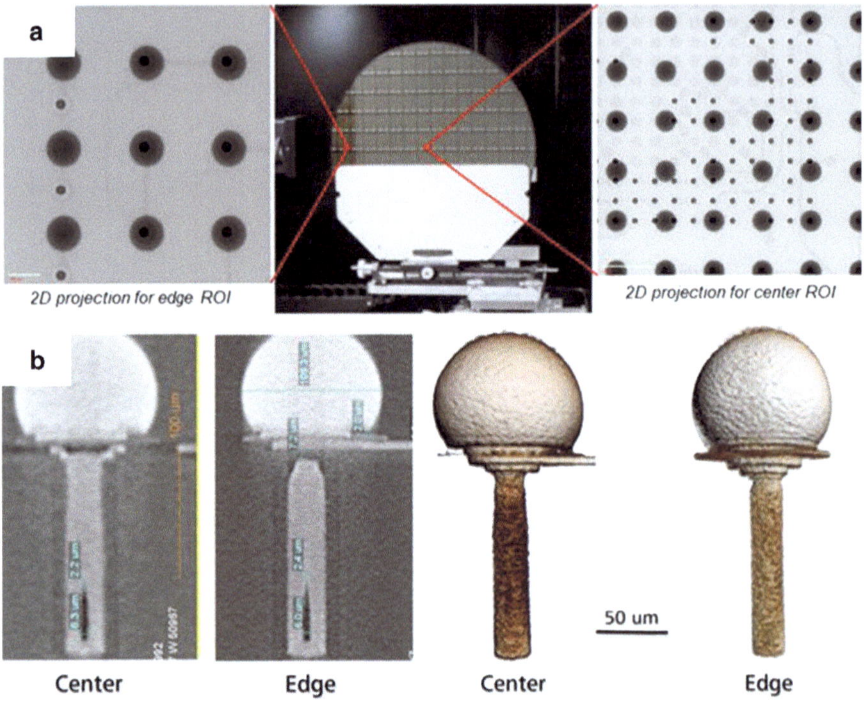


**Fig. 6.19** Structural analysis of TSV across a 300-mm wafer. **a** 2D projection images of an edge ROI (on the *left*) and a center ROI (on the *right*). The photo in the *middle* shows the relative ROI positions. **b** Two reconstructed slices at the center and edge showing the TSV shapes and voids. **c** 3D color-rendering images for comparison of TSV structures at the center versus edge. *ROI* region of interest. (Originally distributed at the International Wafer-Level Packaging Conference, San Jose, CA; November 12–13, 2014)

owing to the lack of available information in the original 2D images that this geometry provides. A cross-platform workflow may be developed to further benefit TSV process development, in which AXI is used in-line to identify defective regions and is subsequently followed by high-resolution imaging by XRM to uncover details of defective structures and improve the TSV process development.

### *6.4.4 Applications for TSV Failure Analysis*

It has been shown that the XRM technique has been effectively used for package failure analysis by providing 3D imaging of suspected failure locations without destroying the samples [4, 5]. With XRM imaging technique, the original microstructures of defects are preserved. This is a significant benefit for efficient failure typing and root cause analysis. The conventional, destructive physical cross section may result in "creating" a defect during the process. Further, additional analysis is not possible since the original part is destroyed.

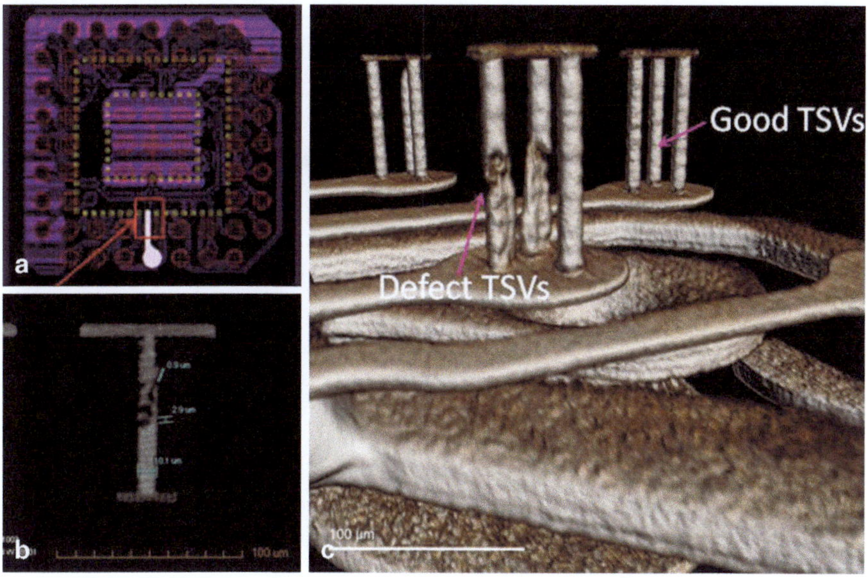


**Fig. 6.20** Nondestructive X-ray microscopic imaging for defect TSV. **a** Electrical test isolated the failed bump/TSV assembly; **b** an example of virtual cross-sectional slice of the TSV defects; and **c** 3D color-rendering image for good and defect TSV. *TSV* through-silicon via

In this final case study, nondestructive XRM is utilized to image a 10 × 10 mm TSV package. The initial electrical test found an open failure on the pin connector $V_{bat}$, shown in the highlighted area of Fig. 6.20a. The defective package was directly imaged by XRM with 0.57 μm voxel resolution and 1-h scan time. Figure 6.20b is one example of virtual cross-sectional slice, clearly showing the evidence of electrical breakdown in the target TSV. Submicron TSV voids were also visualized. 3D color-rendering image shown in Fig. 6.20c has revealed the 3D structures of the defective and the adjacent good TSV. This application has shown the capability of XRM to provide high-resolution images nondestructively, thus significantly enhancing the efficiency of TSV failure analysis.

### 6.4.5 Summary

Nondestructive imaging using XRM has been shown to be effective to study TSV process development and failure analysis. In addition, AXI was presented as a high-speed, in-line metrology complement. Contrast mechanism and several contributors to spatial resolution of an XRM system have been discussed in detail. To image large samples, the XRM using high-resolution detector has significant advantage over the systems using large geometric magnification. The core benefit of the XRM technique is to provide a nondestructive method to image TSV either in a 300-mm wafer or single packages. The subtle structural characteristics or defects of TSV may be visualized and measured by the XRM technique.

## 6.5 Wafer Warpage and Local Distortion Measurement

### *6.5.1 Introduction*

The manufacture of TSV devices involves various processes that put stress on wafers, including high-aspect hole etching for via formation, support-wafer bonding, and wafer grinding. The stress may cause wafers to distort locally or throughout the entire surface. If semiconductor chips are made from distorted wafers, they may end up with excessive warpage or unevenly distributed bump heights. Stacking such chips may result in unstable connectivity between layers, making end products unreliable or even defective. It is, therefore, important that the manufacturers of TSV devices understand the state of wafer distortions and control their processes to eliminate the effect of distortions as much as possible.

### *6.5.2 Basic Functions of WDM300*

Lasertec Corporation has launched WDM300, which accurately measures a wafer's warpage and local deformations based on high-density data. The tool exterior is shown in Fig. 6.21. WDM300 scans the surface of a wafer supported at periphery with three holding arms of the stage and captures height information at 1 x 1 μm intervals using an optical sensor. In the 1-μm-pitch measurement of a 300-mm wafer, the number of data points is about 70,000. The resolution in height component is 30 nm. The repeatability after ten measurements ($\sigma$) is within or less than 1 μm. WDM300 is applicable to the measurement of either patterned or non-patterned wafers.

WDM300 visualizes a wafer's global distortion or warpage in a height distribution map as well as in profile data for a cross section arbitrarily selected by the user (Fig. 6.22). For accurate profile data, the effect of gravity needs to be eliminated. In most cases, the user would operate WDM300 in its auto gravity cancellation mode, in which a wafer is measured twice—first on the front side and then on the backside after being flipped upside down. By comparing the data from both sides, WDM300 calculates the wafer's true distortions after cancelling out the effect of gravity contained in apparent distortions measured on both sides. Figure 6.23 shows the result of true distortion calculation.

WDM300 requires about 13 min in total for the 1-μm-pitch measurement of a wafer on both sides. A less dense 3-μm-pitch measurement is also available and takes about 6 min to complete the entire cycle. There is also an abbreviated mode of gravity effect cancellation with which you can halve the measurement time. For R&D users, on the other hand, a finer 0.3-μm-pitch measurement is also provided.

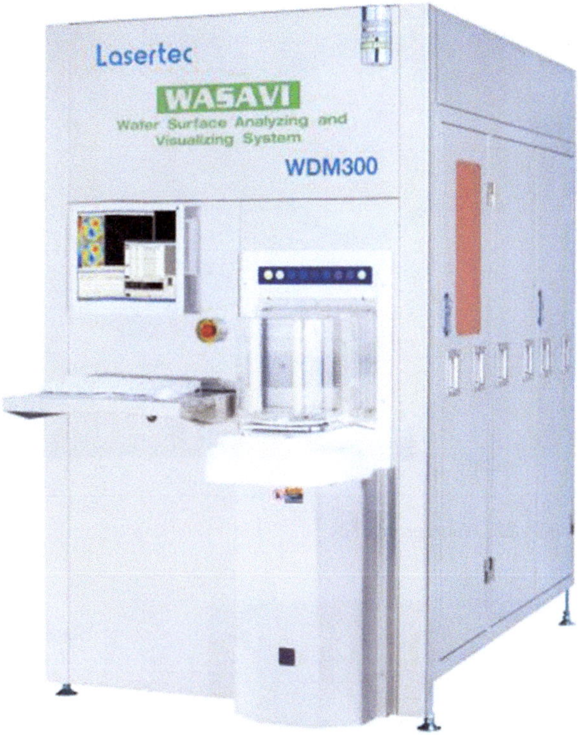


**Fig. 6.21** Exterior of WDM300

### *6.5.3 Measurement and Analysis of Local Deformations*

As mentioned above, WDM300 has about 70,000 measurement points in normal mode, providing not only large-area information such as wafer warpage but also smaller-scale information such as local deformations, gradients, and curvatures. It allows you to measure the difference in distortion level between areas of high via density and those of low via density.

One example is shown in Fig. 6.24, in which the curvature for each measurement point is depicted by an arrow. The direction of each arrow shows the direction of curvature while the length of an arrow indicates the degree of curvature. If the curvature is toward the backside (a concave wafer), it is shown as a red arrow. If the curvature is toward the front side (a convex wafer), it is shown as a blue arrow.

To calculate local deformations, WDM300 uses both a quadratic approximation curve for the whole wafer and a higher-dimensional approximation curve based on local measurement data. By subtracting the former from the latter, WDM300 can measure deviations from the global warpage trend, thereby capturing local deformations (Fig. 6.25).

It is expected that WDM300 helps find optimal settings quickly to minimize local deformations.

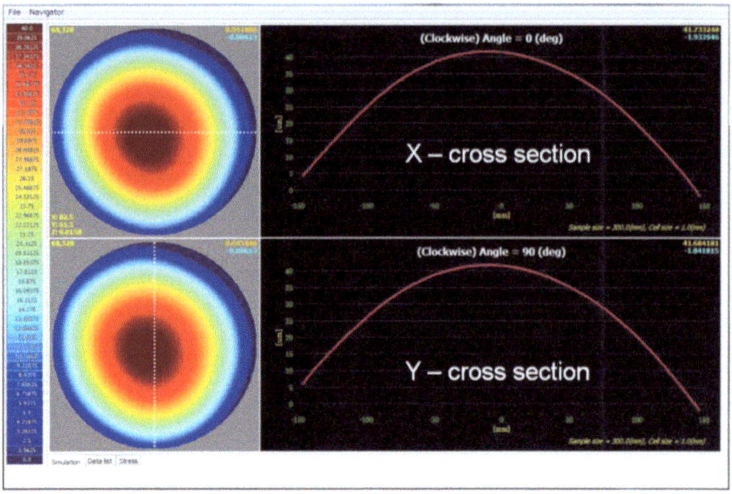


**Fig. 6.22** Warpage profiles

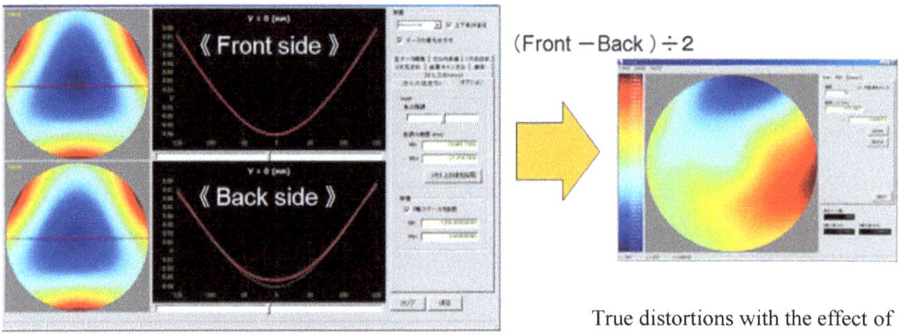


True distortions with the effect of gravity cancelled out

Apparent distortions including the effect of gravity (50 to 100 μm of distortion due to gravity)

**Fig. 6.23** Cancellation of gravity effect

## *6.5.4 Application*

Semiconductor devices are manufactured today by forming multiple layers of various device elements and wire connections on a 300-mm bare wafer. When a layer is placed on top of another during the manufacturing process, it is critically important to accurately align the upper layer with the lower one. As the design scale shrinks, the tolerance for overlay errors between layers is becoming extremely small. According to the latest ITRS roadmap, it is expected to be 5 nm ($3\sigma$) within a few years.

When new design nodes, new materials, and new methods are introduced, semiconductor device manufacturers require more stringent quality control for each layering step. For higher production yields, it is important to overcome the challenge of wafer distortions caused by stress from layers formed.

Fig. 6.24 Local curvature map

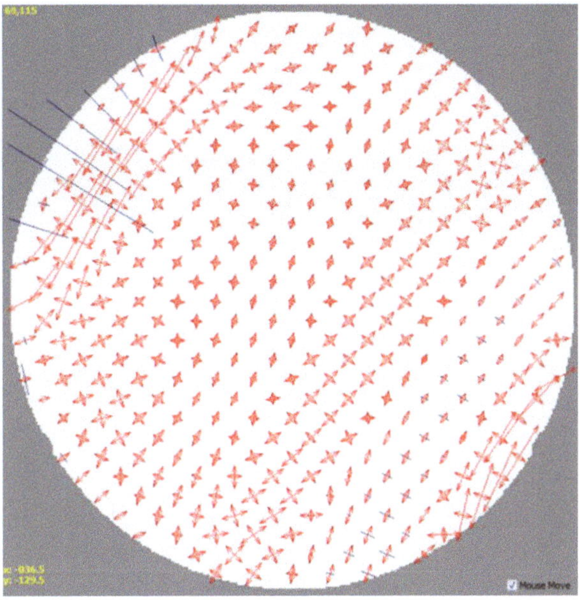

See the example in Fig. 6.26. Both wafer A and wafer B have a similar warpage trend for the whole wafer (left screen), but they have a totally different outlook in the distribution of local deformations (right screen). With the information provided by WDM300, the user can be made aware of any local deformations that may not be smoothed out even if the wafer is vacuum chucked by the scanner. Some believe that such residual local deformations on vacuum chucks have a strong correlation with overlay errors, which need to be avoided at any cost in the manufacturing process of highly miniaturized semiconductor devices.

WDM300 can measure site shape slope range (SSSR), which shows the degree of wafer warpage. SSSR indicates the slope of normal vector measured in each chip-sized range on wafer. You can set a SSSR (say, 200 nm/mm) as a threshold to determine whether a particular range on wafer, which is equivalent to a chip, is acceptable or not. In this way, you can use the measurement of SSSR for quality control purposes.

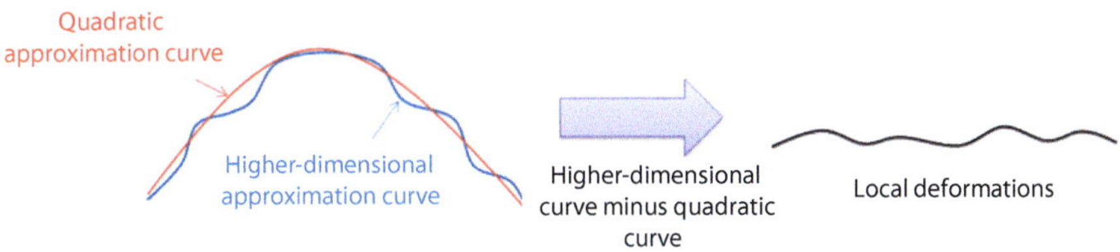


Fig. 6.25 Calculation of local deformations

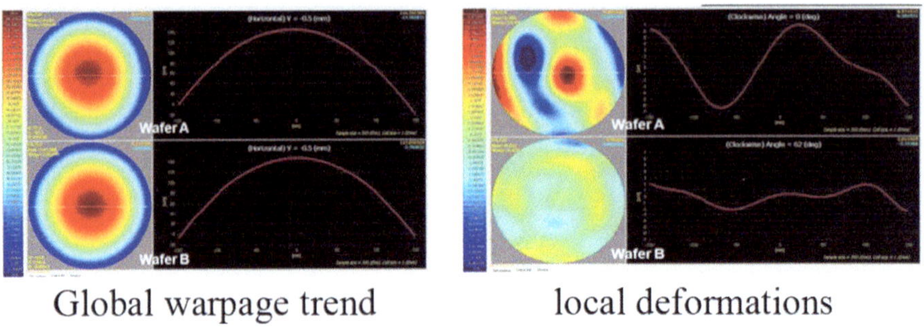


**Fig. 6.26** Global warpage trend and local deformations

## *6.5.5 Summary*

WDM300 calculates wafer distortions, provides key indicators about distortions, and visualizes them in an easy-to-see screen format. With the help of WDM300, the user can grasp the condition of wafer distortions in detail, identify problems, and take actions to improve process settings. WDM300 is an effective tool for the optimization of semiconductor-manufacturing process and contributes to better equipment usage and yield improvement.

# References

1. Courteville A (2005) Method and device for measuring heights of patterns. US Patent 7,782,468 B2
2. Le Cunff D, Tardif M, Piel JP, Fresquet G, Hotellier N, Le Chao K, Chapelon LL, Bar P, Eynard S (2014) Use of optical metrology techniques for uniformity control of 3D stacked IC's. In: Advanced semiconductor manufacturing conference, SEMI, Saratoga springs, New York, USA, May 2014
3. Bernard D, Golubovic D, Krastev E (2012) 3D board level X-ray inspection via limited angle computer tomography. In: Proceedings of SMTA International Conference, FL
4. Fahey K, Estrada R, Mirkarimi L, Katkar R, Buckminster D, Huynh M (2011) Applications of 3d X-ray microscopy for advanced package development. 44th International Symposium Microelectronics, Long Beach, CA
5. Sylvester Y, Johnson B, Estrada R, Hunter L, Fahey K, Chou T, Kuo YL (2013) 3D X-ray microscopy: a non destructive high resolution imaging technology that replaces physical cross-sectioning for 3DIC packaging. In: 2013 24th Annual SEMI advanced semiconductor manufacturing conference (ASMC), pp 249–255, Saratoga Springs, NY

# Chapter 7
# TSV Characteristics and Reliability: Impact of 3D Integration Processes on Device Reliability

**Kangwook Lee and Mitsuma Koyanagi**

## 7.1 Introduction

There are many challenges in 3D integration circuit (3D-IC) with through-Si via (TSV) and metal bumps to be solved before the volume manufacturing starts. The most serious concerns are implications of stress/strain and metal contamination on device reliabilities in 3D stacked chips. Influences of mechanical stress and strain are introduced in wafer thinning process for 3D integration. Cu TSVs and metal bumps introduce significant mechanical stress and strain into thinned Si wafer. Active region in the 3D-IC with a thinned Si wafer might be more easily affected by metal impurity contamination. Because an extrinsic gettering (EG) region for gettering metallic contaminants during IC process is eliminated by the wafer thinning process, Cu atoms diffuse from Cu TSV when the blocking property of the barrier layer in the TSV to Cu is not sufficient. These Cu atoms may diffuse into both dielectric and active region of Si substrate during the back-end process and cause the performance degradation and early breakdown of devices. In this chapter, the influences of mechanical stress/strain effects introduced by Si thinning and metal bump joining and Cu impurity contamination effect introduced from Cu TSV and grinded surface on device reliabilities in thinned IC chips are discussed. DRAM may be sensitively affected by various parameters introduced in a 3D integration process. The impacts of 3D integration processes on memory retention characteristics in thinned DRAM chip are introduced for reliable 3D DRAM.

K. Lee (✉) · M. Koyanagi
NICHe, Tohoku University, Miyagi, Japan
e-mail: kriss@bmi.niche.tohoku.ac.jp

M. Koyanagi
e-mail: koyanagi@bmi.niche.tohoku.ac.jp



K. Kondo et al. (eds.), *Three-Dimensional Integration of Semiconductors*,
DOI 10.1007/978-3-319-18675-7_7

## 7.2 Impact of Cu Contamination on Device Reliabilities in Thinned 3D-IC Chip

To fabricate 3D-IC, each functional IC wafer should be thinned to 10–50 μm thickness by mechanical grinding and stress-relief polishing methods. However, this may cause severe degradation in device reliability since an EG region for gettering metallic contaminants might be removed by the wafer thinning process, and consequently, active regions in the 3D-IC might be more easily contaminated by metallic impurities such as mainly Cu. Cu atoms stuck on the background surface are not completely eliminated even after the cleaning process. Cu atoms also diffuse from Cu TSVs when the blocking property of the barrier layer to Cu is not sufficient as shown in Fig. 7.1. These Cu atoms may diffuse into both dielectric and active region of Si substrate even at low-temperature post-processing and cause the performance degradation and early breakdown of devices [1, 2].

The impact of Cu contamination in the 3D integration has attracted attention in recent years. Hozawa et al. has reported that Cu diffusion effect is significantly enhanced in the thinned wafer thickness [3]. Secondary ion mass spectrometry (SIMS) is a popular method to evaluate metal diffusion behavior. Figure 7.2 shows Cu concentration profiles measured from the back surface and front surface of the thinned Si wafer of 50 μm thickness by SIMS measurement after the annealing at 300°C for various times [4]. Almost Cu atoms diffused into the region within 400 nm depth from the back surface after annealing at 300°C for 60 min. Cu concentrations measured from more than 400 nm depth from the back surface and the front surface show below $1 \times 10^{17}$ atoms/cm$^3$, which is the resolution limitation of SIMS measurement. The SIMS method may not accurately detect Cu impurities owing to the resolution limitation. Total reflection X-ray fluorescence (TRXF) analysis has attracted attention due to the high measuring sensitivity for Cu diffusion behavior [3, 4]. However, TRXF could not directly characterize Cu diffusion effect on device reliability in fabricated device wafers.

In order to measure sensitively Cu contamination effects on device reliability in 3D-IC, a transient capacitance measurement, which is called a capacitance–time *(C–t)* analysis, has been suggested as an electrical evaluation method [4]. This method can quantitatively define the generation lifetime of minority carrier in the

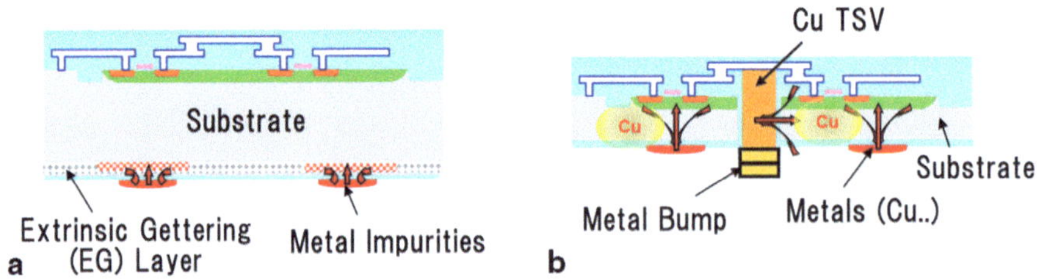


**Fig. 7.1** Conceptual structures of IC wafers before the wafer thinning (**a**) and after the wafer thinning for 3D-IC with Cu TSV and metal bumps (**b**). *TSV* through-Si via, *IC* integrated circuit

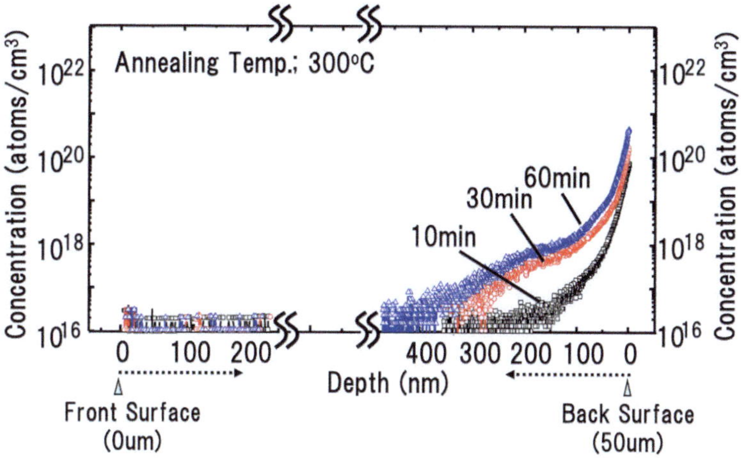


**Fig. 7.2** Cu concentration profiles measured from the back surface and front surface of the thinned Si wafer of 50 μm thickness by SIMS measurement after the annealing at 300°C for various times

depletion region, where Cu impurities generate deep levels at mid-gap in energy band and increase the generation–recombination probability between electrons and holes [5, 6]. In the *C–t* analysis, the capacitance change of the metal–oxide–semiconductor (MOS) capacitor is measured after applying a step voltage to the gate electrode as shown in Fig. 7.3. Just after applying the step voltage, the MOS capacitor is in the deep depletion condition; hence, it represents a smaller capacitance. The capacitance increases with time to the final value of $C_f$, as minority carriers are generated in the depletion region. When $C$ reaches $C_f$, the transient time $t_f$, which is the time required for the capacitance to reach the inversion state from the initial deep depletion state, is important in the *C–t* analysis.

The transient time $t_f$ decreases as the generation lifetime of minority carrier becomes shorter. Hence, a shorter $t_f$ implies more metallic contamination. Therefore, it can sensitively and electrically characterize the lifetime degradation of minority carriers caused by Cu contamination in the fabricated device wafer. In this session,

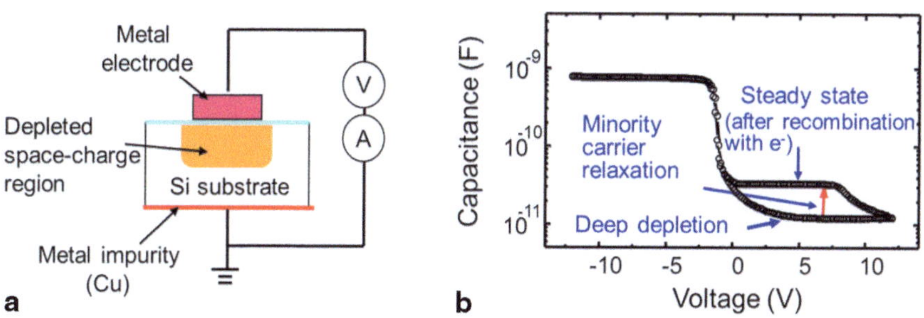


**Fig. 7.3** Configuration of *C–t* analysis using a MOS capacitor (**a**) and *C–V* plot obtained for the minority carrier relaxation.

we review the Cu diffusion characteristics and the impacts of Cu contamination on device reliabilities in 3D-IC by the *C–t* analysis.

### 7.2.1 Impact of Cu Diffusion at Backside Surface in Thinned 3D-IC Chip

Figure 7.4 shows the measured *C–t* curves of the capacitor formed in the thinned wafer of 50 μm thickness comprised with a defect-free denuded-zone (DZ) layer (a) and the generation lifetime ($\tau_g$) of minority carrier measured *C–t* analysis versus surface concentration of Cu atom after intentionally diffuse Cu atoms from the backside surface at 300°C (b). The *C–t* curves show a severe degradation with a shorter $t_f$ even after the initial annealing for 5 min. It means that Cu atoms easily diffuse into the active region from the back surface of the polished wafer, and consequently, the generation lifetime of minority carrier is significantly reduced. The quantitative relationship between the generation lifetime of minority carrier and surface concentration of Cu atom is shown in Fig. 7.4b. The generation lifetime is significantly reduced from the as-deposition condition with surface concentration of Cu atom after the annealing [4].

#### 7.2.1.1 Effect of Intrinsic Gettering Layer

Intrinsic gettering (IG) layer can effectively prevent Cu diffusion, because it is a defected zone with sufficiently high density of oxygen precipitates formed by high-temperature annealing treatment in Ar ambient. In order to electrically characterize the blocking property of IG layer to Cu diffusion by the *C–t* analysis, the MOS capacitor is fabricated on an annealed wafer of 50 μm thickness comprising DZ layer and IG layer, where IG layer region is 20 μm depth from the back surface as shown in Fig. 7.5.

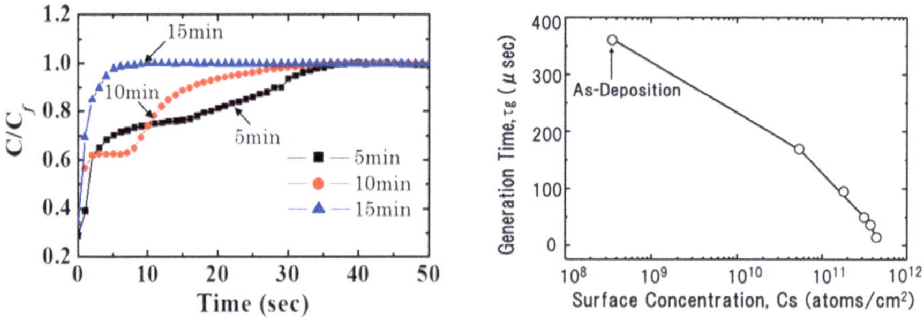


**Fig. 7.4** Measured *C–t* curves (**a**) and the generation lifetime ($\tau_g$) of minority carrier versus surface concentration of Cu atoms (**b**) after the intentional Cu diffusion from the backside surface

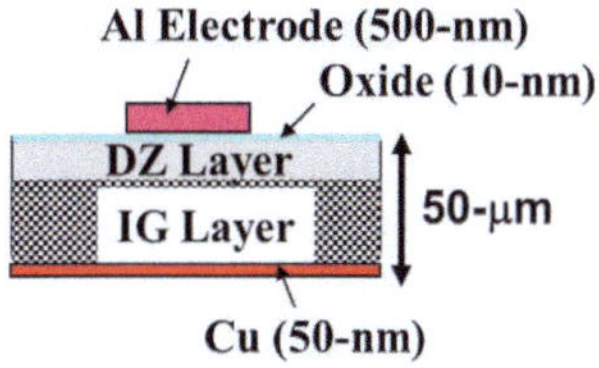


**Fig. 7.5** Cross-sectional structures of the MOS capacitor fabricated on the Si substrate of 50 μm thickness with 20-μm-thick IG layer. *DZ* denuded zone, *IG* intrinsic gettering

The *C–t* curves of the MOS capacitor exhibit only a little change from the as-deposition condition even after annealing up to 350 min, as shown in Fig. 7.6. It indicates that Cu atoms hardly diffuse into the active region owing to Cu retardation by IG region, and the generation lifetime of minor carrier is only minutely reduced even after the long annealing time. IG layer can effectively prevent Cu diffusion, since it is a defected zone with sufficiently high density of oxygen precipitates [4, 7].

The generation lifetime, $\tau_g$, of minority carrier is plotted versus annealing time as shown in Fig. 7.7. In the case of the MOS capacitor formed on the wafer composed of DZ layer, the generation lifetime of the minority carrier is rapidly decreased with the annealing time. Meanwhile, in the case of the MOS capacitor formed on the wafer composed of DZ and IG layers, the generation lifetime of the minority carrier exhibits only a little decrease even after 40-min annealing time [8].

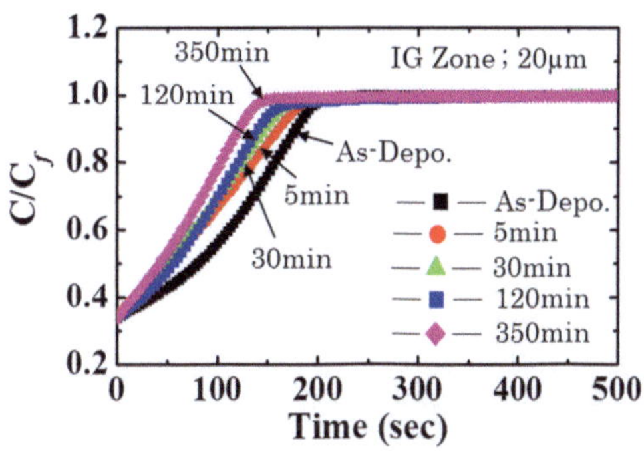


**Fig. 7.6** Measured *C–t* curves of the MOS capacitor formed on the Si substrate of 50 μm thickness with IG layer after the intentional Cu diffusion from the backside surface at 300 C. *IG* intrinsic gettering

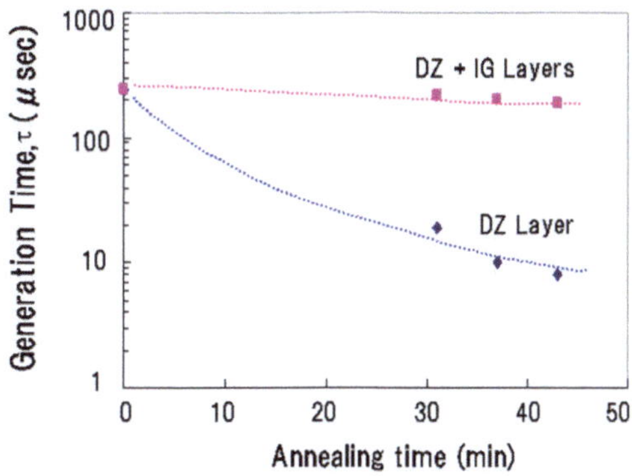


**Fig. 7.7** Generation lifetime, $\tau_g$, of minority carrier obtained from the *C–t* analysis versus annealing times at 300°C. *DZ* denuded zone, *IG* intrinsic gettering

#### 7.2.1.2 Effect of EG Layer

Generally, the stress-relief polishing process is required to release a residual stress in the thinned wafer after a mechanical grinding process. Depending on the polishing condition, the defect band zone with the range of between 0.1 and 1 μm thickness is remained near the grinded surface of the thinned wafer. This defect zone could be used as a gettering layer to Cu diffusion, because it has high-density point defects and dislocations. The blocking properties to Cu diffusion at the backside surface of the thinned wafers with various EG layers are electrically evaluated by *C–t* analysis. To compare the blocking property of EG layer, several types of EG layers are prepared at the back surface of the thinned wafer by mechanical grinding and following chemical–mechanical polishing (CMP), dry polish (DP), ultra-poligrind (UPG), poligrind (PG), and #2000 methods, respectively. The range of the defect band zone is strongly affected by the grinding condition. The surface morphologies on the back surfaces of thinned wafers treated with mechanical grinding and following CMP, DP, UPG, PG, and #2000 methods are measured by atomic force microscopy (AFM) analysis. Figure 7.8 shows AFM images on the back surfaces at the thinned wafers where the surface roughness (RMS) is averagely 8 nm (UPG), 2 nm (DP), 0.30 nm (CMP), 15 nm (PG), and 22 nm (#2000).

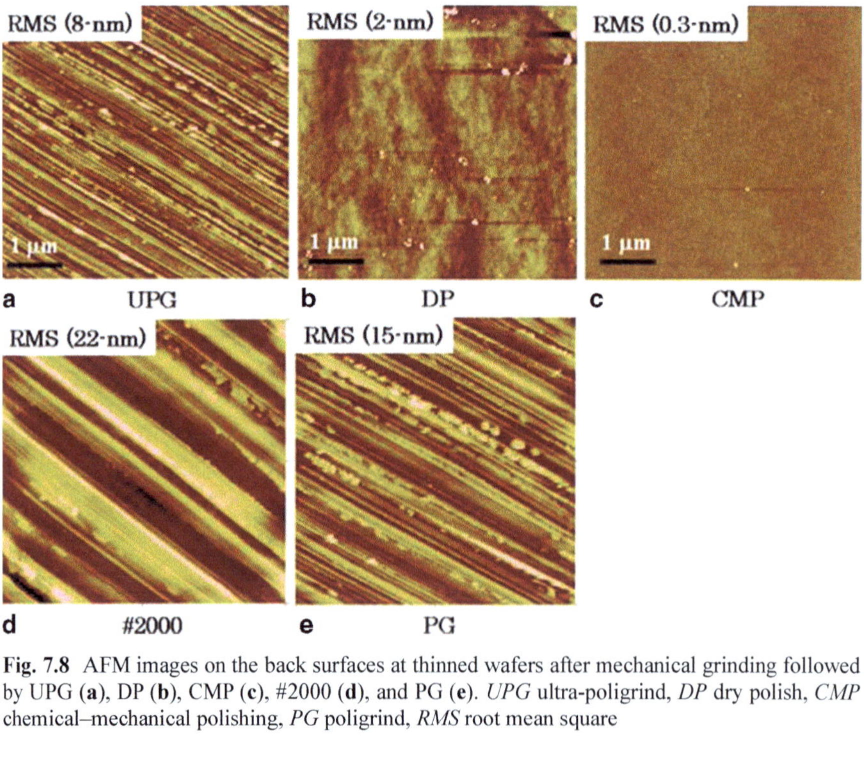


**Fig. 7.8** AFM images on the back surfaces at thinned wafers after mechanical grinding followed by UPG (**a**), DP (**b**), CMP (**c**), #2000 (**d**), and PG (**e**). *UPG* ultra-poligrind, *DP* dry polish, *CMP* chemical–mechanical polishing, *PG* poligrind, *RMS* root mean square

The crystal defects on the back surfaces of the thinned wafers are evaluated by transmission electron microscopy (TEM). Figure 7.9 shows TEM cross-sectional images of the back surfaces at the thinned wafers. After mechanical grinding, the defect band zone has around 1 μm depth from the surface and severe deep micro-cracks, point defects, and dislocations as shown in Fig. (7.9a). After the stress-relief polishing process, the thickness of the defect band zone is decreased to 400 nm (UPG), 100 nm (DP), 50 nm (CMP), 300 nm (PG), and 450 nm (#2000), as shown in figures, respectively. The CMP-treated wafer has smaller surface roughness and thin crystal defect damaged zone. The DP-treated wafer has 100 nm thickness crystal defect damaged zone. But point defects and dislocations are relatively small in the damaged zone. Meanwhile, the UPG, PG, and #2000 treated wafers have much coarser surface roughness and thicker defect damaged zone including severe detects and micro-cracks compared to the CMP- and the UP-treated wafers. These

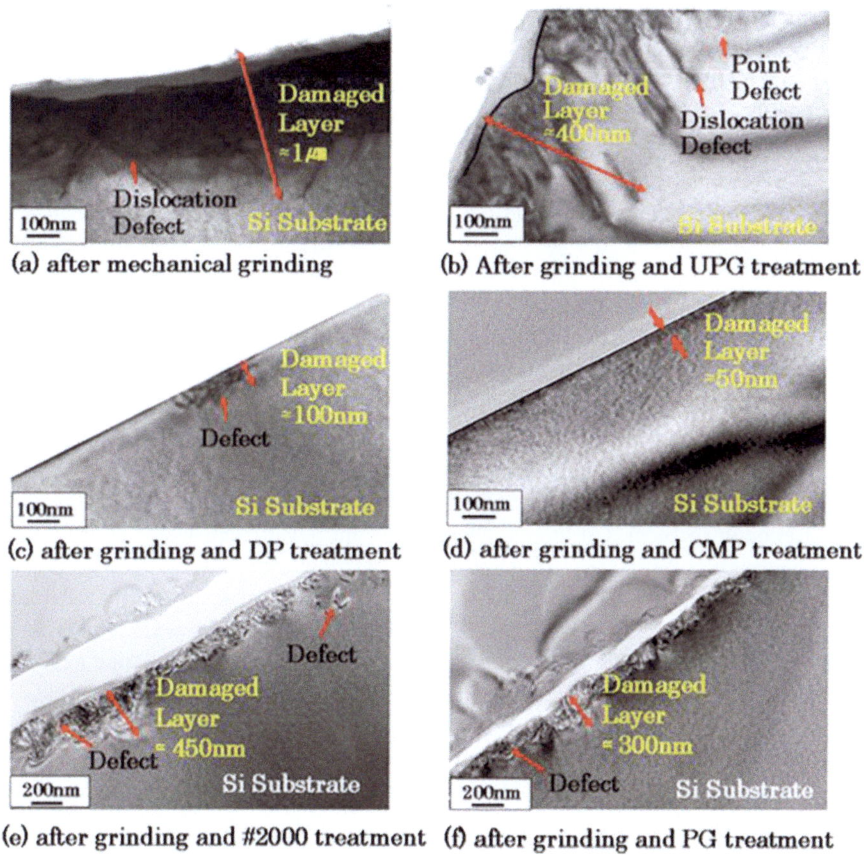


**Fig. 7.9** TEM cross-sectional images on the back surfaces in thinned wafers after mechanical grinding (**a**) followed by UPG (**b**), DP (**c**), CMP (**d**), #2000 (**e**), and PG (**f**), respectively. *UPG* ultra-poligrind, *DP* dry polish, *CMP* chemical–mechanical polishing, *PG* poligrind

processes, especially #2000, may create subsurface damages such as disordering, point defects, and dislocations in Si crystal and vertical micro-cracks.

The MOS capacitors are fabricated on the thinned wafers which are thinned down to 100 μm thickness by mechanical grinding and following various stress-relief polishing processes. For accelerated Cu diffusion test, 50-nm-thick Cu layer is deposited on the back surface as a contamination source. To intentionally diffuse Cu atoms into the active region, the wafers are annealed at 300°C in $N_2$ ambient for various annealing time.

Figure 7.10 shows the measured *C–t* curves of the MOS capacitors formed on the thinned wafers treated by various stress-relief processes. In case of the MOS capacitors formed on the CMP, UPG, PG, and #2000 treated wafers, the *C–t* curves show the degradation with shorter generation lifetime of the minority carrier from the as-deposition condition after annealing for 20 min as shown in figures. It means that Cu atoms diffuse into the depletion region from the back surface, and consequently, the generation lifetime is reduced. By increasing the annealing time, the *C–t* curves show more severe degradation resulting in more serious decrease, because more Cu atoms rapidly diffuse into the depletion region. In case of the MOS capacitor formed on the DP-treated wafer, however, the *C–t* curves exhibit relatively a little bit change from the as-deposition condition after annealing up to 60 min. It indicates that Cu atoms relatively hard to diffuse into the depletion region owing to Cu retardation by the EG layer formed by DP treatment, and the generation lifetime is minutely reduced even after relatively longer annealing time.

The normalized generation lifetime of the minority carrier is plotted versus annealing time as shown in Fig. 7.11 [9]. In the CMP, DP, PG, and #2000 treated wafers, the lifetimes are rapidly decreased with the annealing time regardless of the polishing conditions. Meanwhile, the DP-treated wafer shows relatively a little bit change because of the strong gettering efficiency to Cu diffusion of the DP-treated wafer. The CMP-treated wafer has the shallow defect damaged zone of 50 nm thickness with smooth surface roughness and no defects. It indicates that the CMP is useful method to enhance the strength of the thinned wafer, because it has advantages in terms of stress release and damage removal. However, the CMP-treated wafer shows the poor gettering ability to Cu diffusion, because the shallow damaged zone is not enough to block Cu atoms. Meanwhile, the DP-treated wafer shows most good gettering ability to Cu diffusion. It indicates that the crystal defects damaged zone of 100 nm thickness which was formed near the grinded surface by DP treatment acted as a good EG layer. The UPG, PG, and #2000 treated wafers have deep micro-cracks, severe point defects, and dislocations in the damaged zone near the grinded surface. These severe damaged zones show relatively unstable retardation performance to Cu diffusion; therefore, Cu atoms may diffuse easily into the depletion region passed through these severe defects during the annealing treatment. However, even the DP-treated wafer, the damaged zone is not completely uniform as shown in Fig. 7.12, and it induces relatively unstable retardation to Cu diffusion for longer annealing time and higher annealing temperature [8].

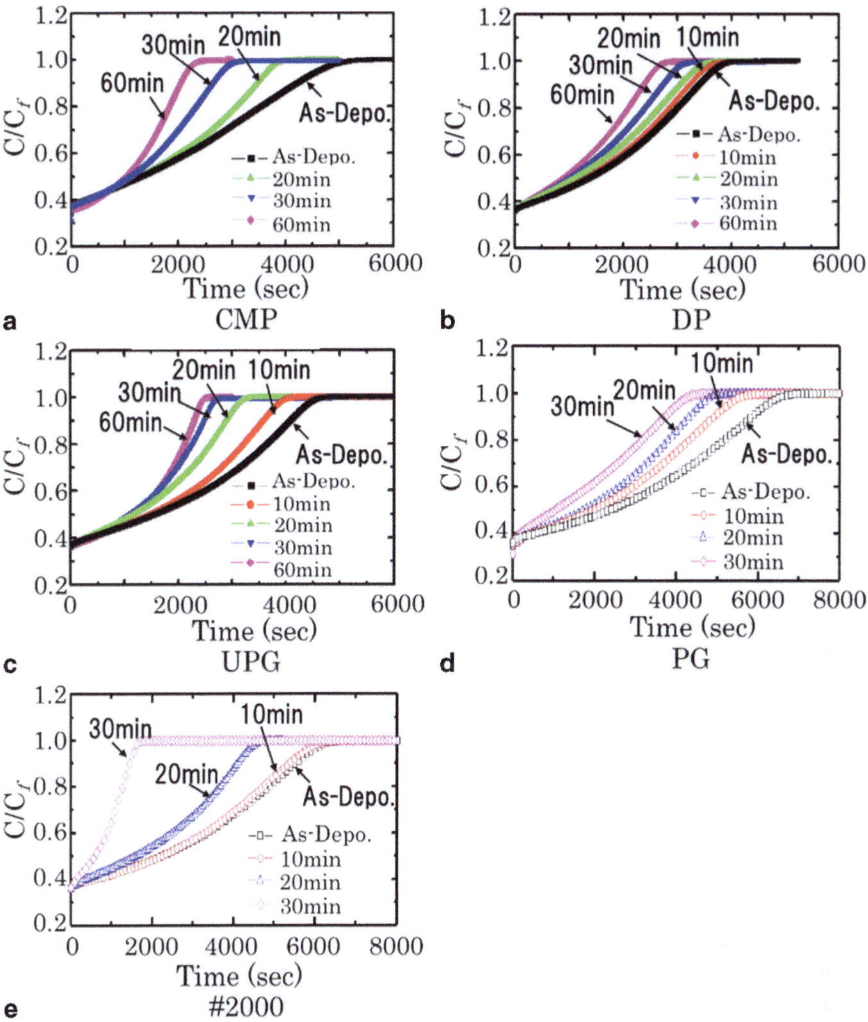


**Fig. 7.10** Measured *C–t* curves of the MOS capacitors formed on the thinned wafers treated with CMP (**a**), DP (**b**), UPG (**c**), PG (**d**), and #2000 (**e**). *UPG* ultra-poligrind, *DP* dry polish, *CMP* chemical–mechanical polishing, *PG* poligrind

In the case of line A with 50-nm-thick damaged zone, Cu atoms diffused into Si substrate passed through relatively shallow damaged zone. In the case of line B with 100-nm-thick damaged zone, meanwhile, Cu atoms are blocked within the damaged zone and not diffused into Si substrate. It indicates that the crystal defects damaged zone of around 100 nm thickness is required to act as a good EG layer to Cu diffusion. However, it is another challenge to form 100-nm damaged zone uniformly by the conventional polishing method.

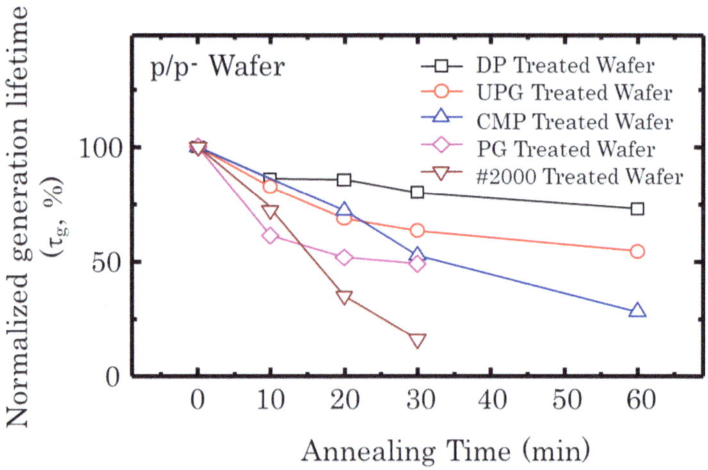


**Fig. 7.11** Normalized generation lifetime, $\tau_g$, of the MOS capacitors formed on the thinned wafers treated with DP, UPG, CMP, PG, and #2000, after the intentional Cu diffusion at 300°C. *UPG* ultra-poligrind, *DP* dry polish, *CMP* chemical–mechanical polishing, *PG* poligrind

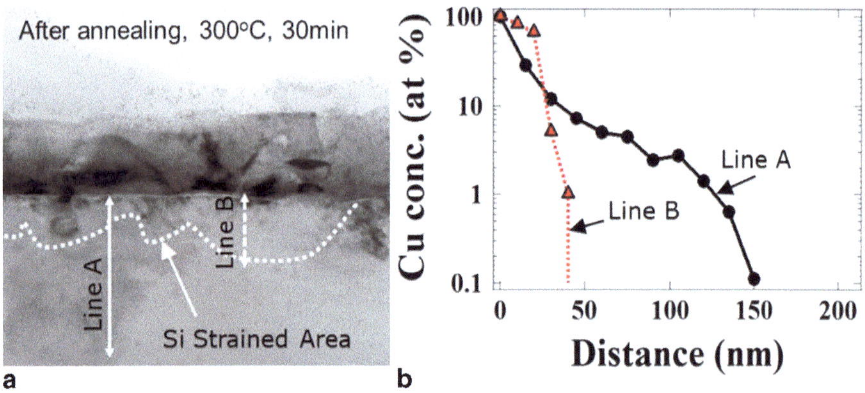


**Fig. 7.12** TEM cross-sectional image of the back surface in the PG-treated wafer after Cu diffusion at 300°C for 30 min (**a**) and Cu concentration profiles measured by SIMS (**b**)

### *7.2.2 Impact of Cu Diffusion from Cu Via*

TSV formation is a key technology for the 3D-IC fabrication, because TSV is an important factor to determine 3D-IC performance. Lately, Cu TSV has been attracting attention because Cu has a low resistance that significantly reduces the resistive-capacitive (RC) delay and high filling speed that increases the production throughput. Cu TSV is most commonly fabricated by deep Si etching, lining with dielectric layer and sputtering with barrier and seed layers followed by Cu electroplating. However, Cu atoms from TSV can easily diffuse into the active

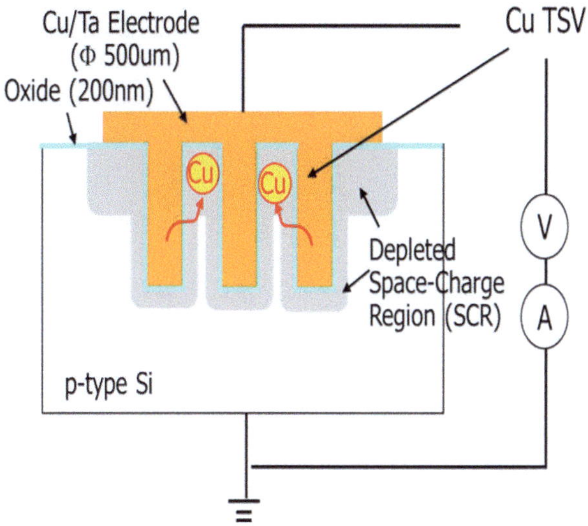


**Fig. 7.13** Conceptual structure of a trench MOS capacitor composed of Cu/Ta gate electrode and Cu TSVs. *TSV* through-Si via

area and contaminate the nearby devices when the blocking property of the barrier layer to Cu diffusion is not sufficient, and it causes the degradation of the device characteristics. The Bosch process commonly used to form high-aspect-ratio vias. However, the BOSCH via etching results in a sidewall surface roughness called scalloping, caused by the cyclic etching and passivation process. If the scalloping roughness is high, then it would be challenging for the conformal deposition of dielectric liner and barrier layer. Especially, poor coverage of barrier layer may induce Cu diffusion from Cu TSV during a post-annealing process. The influence of Cu diffusion from the Cu TSV is electrically characterized by the *C*–*t* analysis using a trench MOS capacitor composed of Cu/Ta gate electrode and Cu TSVs as shown in Fig. 7.13 [10].

#### 7.2.2.1 Effect of the Barrier Thickness and the Scallop Roughness

To compare the influences of the Si scallop roughness, two types of the sidewall scalloping with average roughness of 30 and 200 nm are prepared by changing the cycling step of $SF_6$ etching and $C_4F_8$ passivation in the BOSCH process, as shown in Fig. 7.14.

After the formation of 100-nm-thick oxide liner into via holes, two types of Ta barrier layers with thickness 10 and 100 nm at the surface are deposited into via holes to compare the influence of the step coverage, where the minimal Ta barrier layer thickness at the trench sidewall is approximately 3 and 20 nm. Figure 7.15 shows scanning electron microscope (SEM) cross-sectional images of scallop portion with the minimal thick Ta layer of approximately 30 nm in trench via with 200-nm scallop roughness. After the deposition of seed layer, Cu layer which acts as both gate electrode and Cu via conductor, is formed by electroplating. The metal

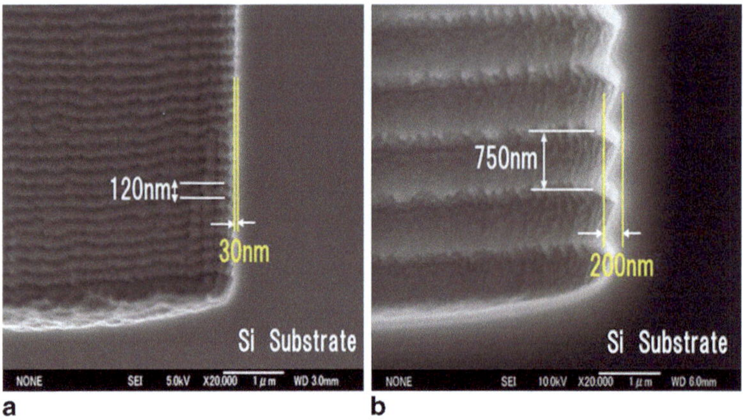


**Fig. 7.14** SEM cross-sectional views of Si trenches with different sidewall scalloping roughness. *SEM* scanning electron microscope

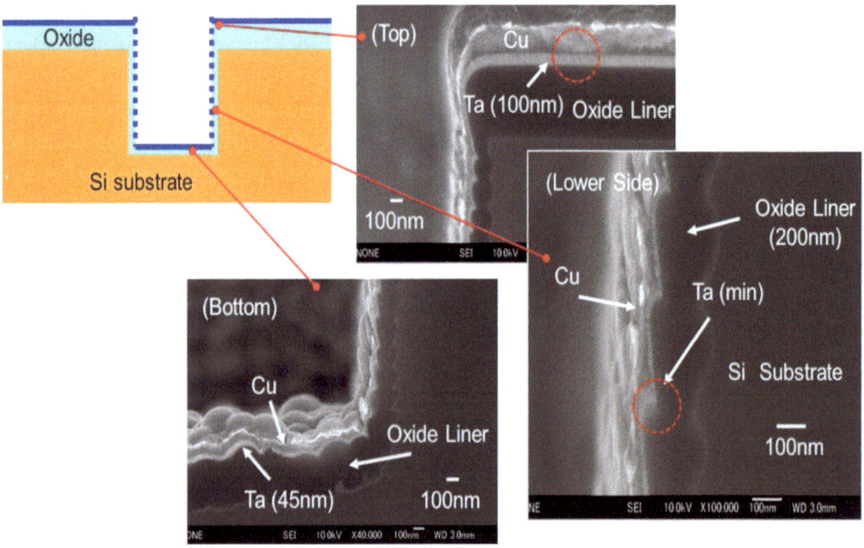


**Fig. 7.15** SEM cross-sectional views of Si trenches with different sidewall scalloping roughness

gate electrodes composed of Cu and Ta layers are patterned into 1000 μm square size to form the trench MOS capacitor including 50 × 50 via array.

To intentionally diffuse Cu atoms from Cu electrodes into the substrate, the wafers are annealed at 300°C for various annealing time. Figure 7.16a and b shows the measured *C–t* curves of the trench capacitors with 10-nm-thick Ta layer (at the surface) after annealing. The *C–t* curves of the trench capacitor with 30-nm scalloping roughness show a severe degradation with a shorter generation lifetime after the initial annealing for 5 min as shown in Fig. 7.16a. It indicates that Cu atoms diffuse

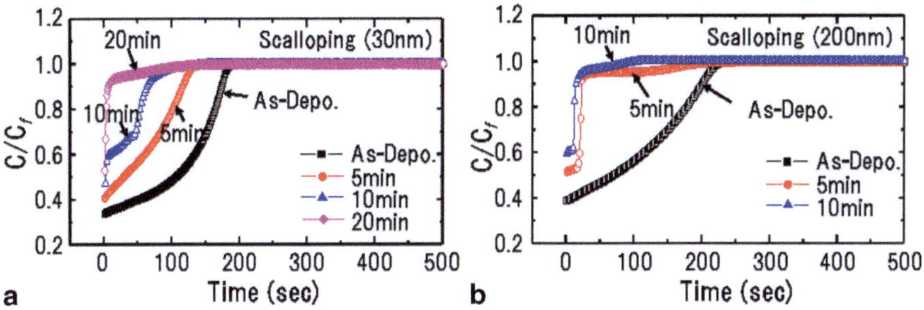


**Fig. 7.16.** Measured *C–t* curves of the Cu/Ta gate trench MOS capacitors having thinner Ta layer, with 30-nm scalloping (**a**) and with 200-nm scalloping (**b**) after annealing at 300 °C and various times

into the active area from the Cu electrode through scallop portions with extremely thin Ta layer in vias. Increasing the annealing time, the *C–t* curves show more severe degradation with decreasing the generation lifetime as a result of more serious Cu diffusion into the active area. The *C–t* curves of the trench capacitor with 200-nm scalloping roughness show far more severe degradation with a much shorter generation lifetime as shown in Fig. 7.16b, compared to those of 30-nm scalloping roughness. It means that Cu atoms diffuse more rapidly into the active region from the Cu electrode, owing to worse coverage or a poor quality of Ta layer resulting from large scalloping roughness.

Figure 7.17a and b shows the measured *C–t* curves of the trench capacitors with 100-nm-thick Ta layer (at the surface) after annealing. The *C–t* curves exhibit no change from the as-deposition condition even after annealing up to 60 min, regardless of the scalloping roughness. It means that it is hard to diffuse Cu atoms from Cu electrode into the active region through scallop portions with Ta layer of approximately 30 nm thickness.

The generation lifetime of the minority carrier is plotted versus the annealing time as shown in Fig. 7.18. In the trench capacitors with 10-nm-thick Ta (at the

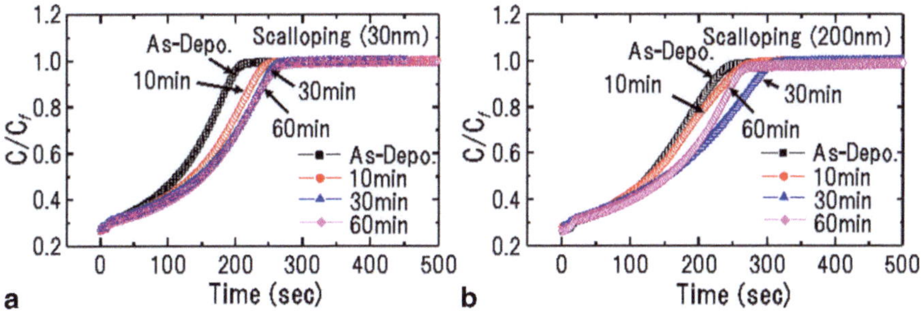


**Fig. 7.17** Measured *C–t* curves of the Cu/Ta gate trench MOS capacitors having thicker Ta layer, with 30-nm scalloping (**a**) and with 200-nm scalloping (**b**) after annealing at 300 °C and various times

surface), the generation lifetime is significantly reduced from the as-deposition condition after annealing for 5 min even in the small scalloping roughness of 30 nm. However, in the trench capacitors with 100-nm-thick Ta (at the surface), the generation lifetime is not reduced after annealing up to 60 min at 300°C even in the large scalloping roughness of 200 nm.

#### 7.2.2.2 Effect of the Annealing Temperature

Recently, via middle Cu TSV has attracted attention, because it has strong potential for the mass production in IC foundry and DRAM vendors [11]. Via-middle Cu TSV is most commonly fabricated by deep silicon etching, lining with dielectric layer and sputtering with barrier and seed layers followed by Cu electroplating after front-end-of-line (FEOL) process. Recently, to avoid the stress effect induced by Cu TSV, the annealing step at higher than 400°C temperature is applied before back-end-of-line (BEOL) process. Moreover, in conventional IC fabrication process, it requires post-BEOL $H_2$ sintering anneal process at 400°C. Therefore, the blocking property of the barrier layer to Cu diffusion is evaluated at 400°C annealing temperature. Figure 7.19 shows the *C–t* curves of the trench capacitors with 100-nm-thick

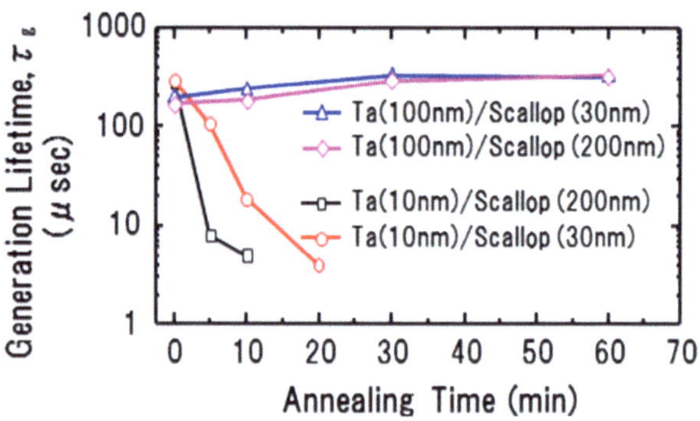


**Fig. 7.18** Generation lifetime of minority carrier obtained from the *C–t* analysis versus annealing times at 300°C

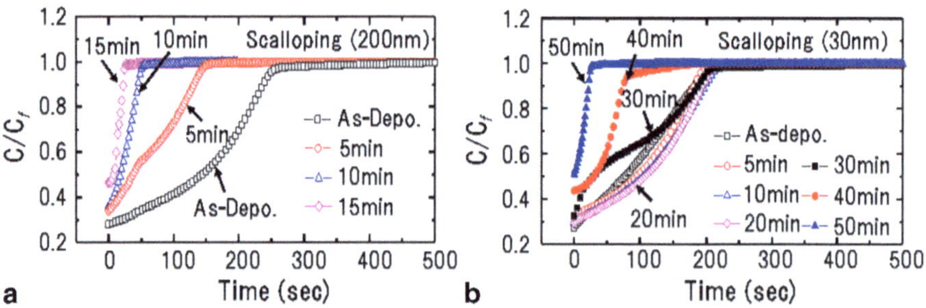


**Fig. 7.19** Measured *C–t* curves of the Cu/Ta gate trench MOS capacitors having thicker Ta layer, with 200 nm scalloping (**a**) and with 30 nm scalloping (**b**) after annealing at 400°C and various times

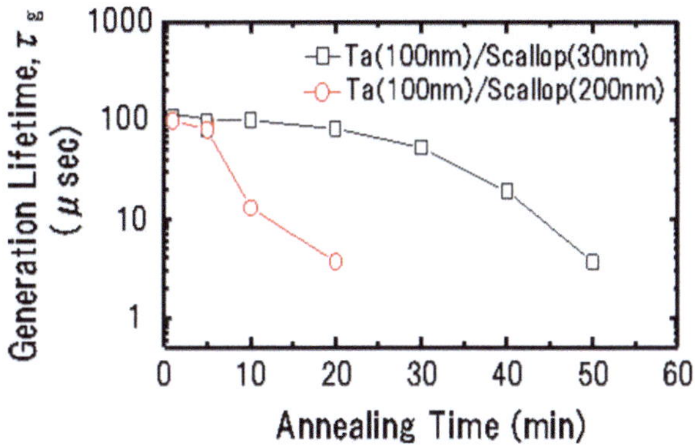


**Fig. 7.20** Generation lifetime of minority carrier obtained from the *C–t* analysis versus annealing times at 400°C

Ta layer at different scallop roughness after annealing at 400°C. The trench capacitor with 200-nm scallop roughness shows severe degradation with a shorter transient time even after the initial annealing for 5 min as shown in Fig. 7.19a. Meanwhile, the trench capacitor with 30-nm scallop roughness shows no degradation until 20-min annealing time, but it started to degrade after 30-min annealing time as shown in Fig. 7.19b.

Figure 7.20 shows the generation lifetime of the minority carrier, obtained from the *C–t* analysis versus annealing times at 400°C. The generation lifetime of minority carrier is reduced to 50 % level from the as-deposition condition after annealing for 30 min at 400°C in the MOS capacitor even with a shallow scallop roughness of 30 nm and thicker Ta barrier layer. It means that Cu atoms diffuse easily into the active region from via-middle Cu TSV under several post-annealing process at higher 400°C and cause the severe degradation of the device performances

#### 7.2.2.3 Keep-Out-Zone Characterization by Cu Diffusion from Cu Via

To characterize keep out zone (KOZ) induced by Cu diffusion, the planar MOS capacitor which is placed near Cu TSVs with various distance from 10 to 100 μm is used as shown in Fig. 7.21. To intentionally diffuse Cu atoms into the depletion region of the planar MOS capacitor, TSV has rough sidewall scalloping of 200 nm roughness and thin Ta barrier layer of 10 nm thickness (at the surface).

Figure 7.22 shows the minority carrier lifetime versus the distance between the MOS capacitor and Cu TSVs after annealing at 300°C for 5 and 10 min, respectively. The minority carrier lifetime of the MOS capacitor which is departed 50 μm from Cu TSV decreases to 50 % level from the as-deposed condition after annealing for 5 min. Moreover, after annealing for 10 min, the lifetime dramatically decreases even the distance between the planar capacitor and Cu TSV is departed 100 μm distance [8]. It means that Cu atoms easily diffuse into the active device area from Cu TSV with poor coverage of barrier layer and degrade the device characteristics even after annealing at 300°C. Therefore, the Cu diffusion issue from the Cu TSV is more serious concern in via-middle Cu TSV technology, because via-middle

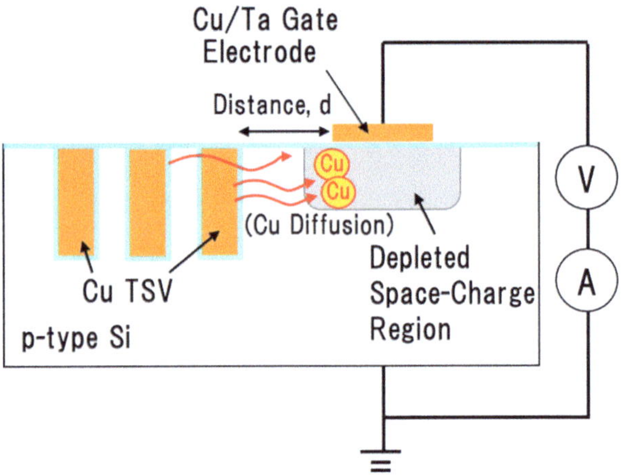


**Fig. 7.21** Configuration cross-structure of a planar MOS capacitor placed near several Cu TSVs. *TSV* through-Si via

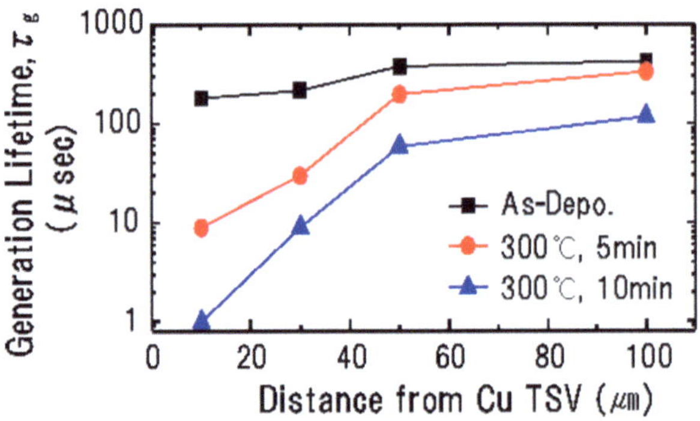


**Fig. 7.22** Generation lifetime of minority carrier versus the distance from Cu TSVs after annealing at 300°C. *TSV* through-Si via

process requires several post-annealing steps at higher 400°C. It is a key issue to form high-reliable TSV by optimizing thickness, crystalline structure, and material of barrier layer in real 3D-IC.

## 7.3 Impact of Mechanical Stress/Strain on Device Reliability in Stacked IC

3D integration involves a vertical stacking of thin dies or wafers with Cu TSVs and metal micro-bumps. To realize a compact-sized 3D-IC, each functional wafer should be thinned to 10–50 μm thickness. However, the ultra-thin nature of Si substrate leads to several problems such as weak mechanical strength, warping, and local deformation in the stacked die [12, 13]. Moreover, the weak mechanical strength of the extremely thin die/wafer itself has a potential concern lead to die breaking for 3D integration, because thin IC chip with high-density TSVs is highly fragile and more easily damaged. Metal micro-bump joining becomes severe concern decreasing die thickness, because of large coefficient of thermal expansion (CTE) mismatch of Si, Cu, and organic underfill material. In this chapter, the influences of mechanical stress/strain inducing by Si thinning and metal micro-bump joining on device reliabilities in thinned 3D-IC are discussed.

### *7.3.1 Micro-Bump-Induced Local Stress in Stacked IC*

Metal micro-bumps in the 3D-IC may cause two different stresses, namely, thermo-mechanical stress (TMS) and locally induced mechanical stress (LMS). TMS is due to the difference in the CTE values between Si and the bump metal as shown in Fig. 7.23a [12, 14]. Meanwhile, LMS is the result of local deformation in the thinned die around the micro-bump region owing to the CTE mismatch between the bump metal, Si, and organic underfill as shown in Fig. 7.23b. The magnitude of the TMS induced by TSVs and micro-bumps as well as the LMS produced in the stacked die around the micro-bump region can be calculated via finite element analysis [15]. Therefore, one can predict the KOZ for the devices in the 3D-IC well in advance. The quantitative measurement of TMS and LMS in the stacked IC is generally carried out nondestructively either by using piezoresistive stress sensor [16] or by employing micro-Raman spectroscopic technique [17].

In general, in the typical Raman spectrum of crystalline Si, mainly one single degenerated longitudinal optical (LO) peak is observed, whose frequency is located at around 521 $cm^{-1}$. The Raman spectrum results are obtained using the excitation laser with the wavelength of 488 or 785 nm. It is well known that a tensile strain will shift the Si Raman peak toward lower frequency ($\Delta\omega > 0$), while a compressive stress will result in the high-frequency shift ($\Delta\omega > 0$) for Si Raman peak. Since the frequency of the lattice vibrations of a material will change when the material is subjected to compressive stress and/or tensile strain, the shift in the Si Raman peak frequency can be directly related to the kind of stress present in the material. Also, from the magnitude of the peak frequency shift, one can quantify the amount of stress. It should be noted that the relation between stress/strain in Si and the Raman frequency is quite complex, since all the nonzero strain tensor components

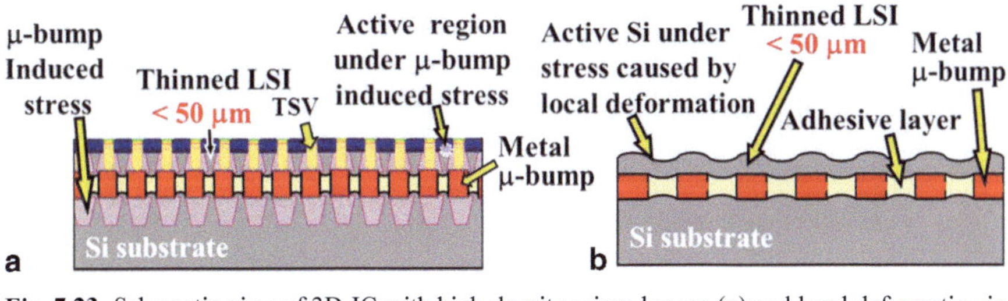


**Fig. 7.23** Schematic view of 3D-IC with high-density micro-bumps (**a**) and local deformation in 3D stacked IC (**b**)

influence the position of the Raman peak. However, in most of the cases, it is assumed as linear for stress determination. Here, we have used $\sigma(\text{MPa}) = -434 \times \Delta\omega$ ($\text{cm}^{-1}$) and $\sigma xx + \sigma yy$ (MPa) $= -434 \times \Delta\omega$ ($\text{cm}^{-1}$), respectively, for uniaxial and biaxial stress in the (100) plane of Si. In the following section, the reliability issues arising due to the micro-bumps in 3D stacked ICs are reviewed. Cu/Sn micro-bump is inalienable interconnect material for low-temperature BEOL integration process. Although interconnect metals induced TMS in 3D-ICs are known for decades, only very few reports have discussed about TMS caused by metal micro-bumps; nevertheless, such micro-bumps are also copiously used in the die/wafer stacking for face-to-face bonding.

In what follows, 2D-stress distribution is discussed for cross-sectional 3D-IC samples containing an array ($100 \times 10$ micro-bumps) of Cu/Sn micro-bumps with different sizes such as $5 \times 5$, $10 \times 10$, and $20 \times 20$ $\mu\text{m}^2$. The top and bottom chip sizes are, respectively, $5 \times 5$ and $7 \times 7$ $\text{mm}^2$, and the die thickness is around 280 μm. The cross-sectional SEM image of bonded CuSn micro-bump, where the under bump metal Cu is formed by electroplating followed by the evaporation of Sn, clearly reveals the formation of $Cu_6Sn_5$ and $Cu_3Sn$ inter-metallic compound (IMC) at the interface as shown in Fig. 7.24a [20]. The cross-sectional 2D-TMS distribution image for 3D-IC with $20 \times 20$ μm sized Cu/Sn bumps is shown in Fig. 7.24. Similar to Cu TSVs, there exists severe compressive stress in the Si resides in the immediate vicinity of micro-bump. At the bump-space region, it induces either less compressive stress or tensile stress. Though the magnitude of the induced stress increased after annealing, the stress distribution pattern remained the same even after the post-heat treatment [19]. The maximum stress values of 125, ~250, and > 350 MPa compressive stress for before bonding, after bonding, and after post-heat treatment have been observed, respectively. Both the magnitude and the in-depth distribution of the compressive stress induced by the micro-bump increased with the increase of the bump size. In the case of finer size ($5 \times 5$ μm) and high-density Cu/Sn bumps, the compressive stress produced by the two adjacent micro-bumps along the plane parallel to Cu–Si interface overlapped to each other at the bump space region.

As compared to the TMS induced by metal micro-bumps and TSVs in the active Si, the magnitude of LMS arising due to the local deformation of thinned IC after underfill injection and curing is tremendously large. This causes serious reliability

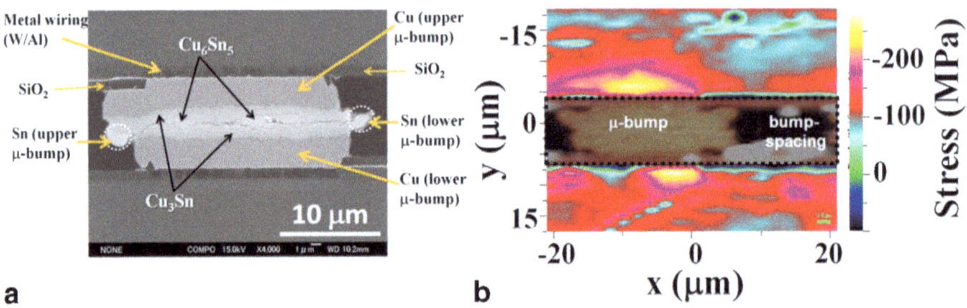


**Fig. 7.24** Cross-sectional view of EEB formed Cu/Sn micro-bump **a** SEM image and **b** 2D-stress distribution. *EEB* electroplating–evaporation bumping

Fig. 7.25 Line profile revealing the local deformation formed in 3D-IC chips with variously stress relieved

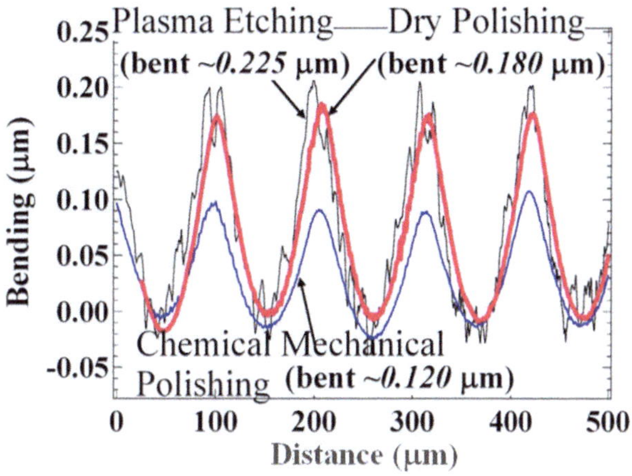


problems before realizing the high-density 3D-ICs. The degree of local deformation depends on various factors such as CTE and modulus of underfill as well as the micro-bump density, pitch and height, and to some extent, the surface morphology of the stress-relieved thinned die.

The magnitude of local bending induced by micro-bump varies with the stress-relief method. The 10-μm-thick die experienced a maximum for ~225 nm bending for the plasma etched relief process, while it was nearly half for CMP process shown in Fig. 7.25. It is proved that the grinding grooves left behind on die surface after the stress relief are the main cause for this kind of maximum bending [15]. Such local bending induces a large amount of LMS at the active Si of the stacked IC as shown in Fig. 7.26, where + 1.8 GPa of tensile stress around the micro-bump region and < −0.5 GPa of compressive stress at the bump space are noticed.

Fig. 7.26 2D stress distribution image obtained for the top die integrated over bottom die having an array of micro-bumps

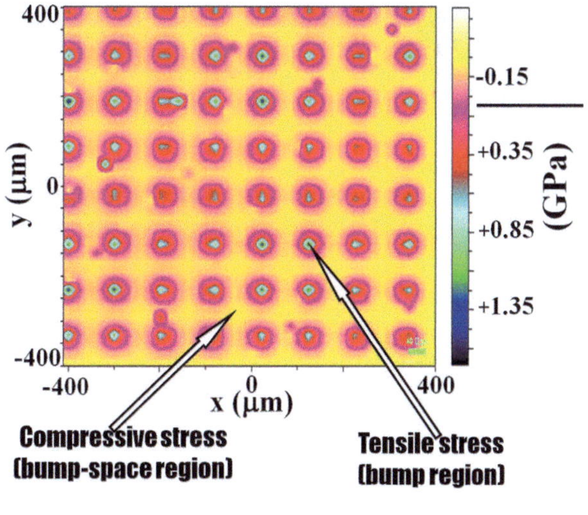

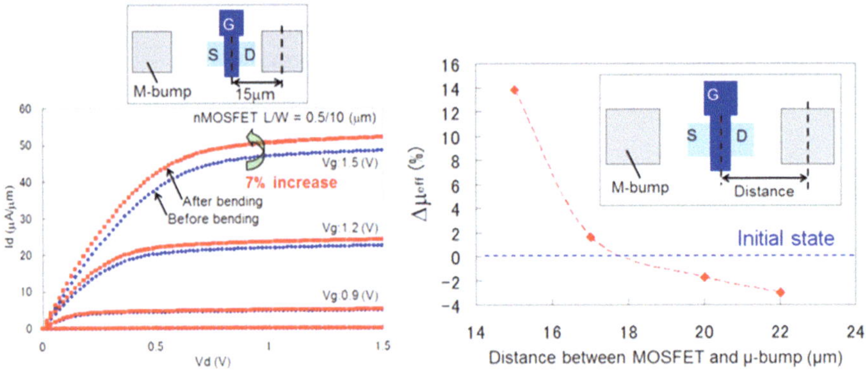


**Fig. 7.27** Effects of local bending stress introduced by micro-bumps in 30-μm-thick die on the $I_d$–$V_d$ characteristics **a** and electron mobility **b** of NMOSFET. *MOSFET* metal–oxide–semiconductor field-effect transistor, *nMOSFET* n-channel metal–oxide–semiconductor field-effect transistor

The metal–oxide–semiconductor field-effect transistor (MOSFET) characteristics are affected by the mechanical stress introduced by micro-bumps. The electron mobility and hence drain current increase near micro-bump due to the tensile stress, whereas they decrease as approaching the center of bump space region due to the compressive stress as shown in Fig. 7.25. It is more challenging to minimize the effects of the mechanical stress in thinned 3D-IC with high-density micro-bumps [18].

### 7.3.2 Si Mechanical Strength Reduction by Thinning

To basically understand the mechanical property of thin Si substrate, the Young's modulus (E) of Si substrate is evaluated by nano-indenter method. The relationship among the crack length, the maximum load, and material parameters is described by Lawn et al. [19] that $K\chi = \alpha(E/H)^{1/2}\, \pi/\chi^{3/2}$ , where *E*, *H*, and $K_c$ are Young's modulus, hardness, and fracture toughness of the indented material, respectively, *P* is the maximum load, *c* is the length of the radial crack trace on the surface, after the indenter is totally unloaded, α is a constant depending on the type of indenter. The nano-indenter tip of pyramid geometry is projected into the surface of thin Si substrate. The load is continuously increased to a designated maximum value. Next holding segment is introduced to allow the material to relax before unloading. The process is repeated four times at relatively low applied load of 24 mN, and the position of the indenter tip on the surface is monitored with a capacitance meter.

We evaluated the Young's modulus of Si substrates with different thickness varying from 100, 50, and 30 μm, where the backside surfaces of each substrate were CMP processed after mechanical grinding. The Young's modulus of Si substrate is decreased depending on the reduction of the Si substrate thickness as shown in Fig. 7.28. The Young's modulus values of Si substrates of 100 and 50 μm thicknesses are 180 and 176 GPa, respectively. These values are similar to the Young's modulus

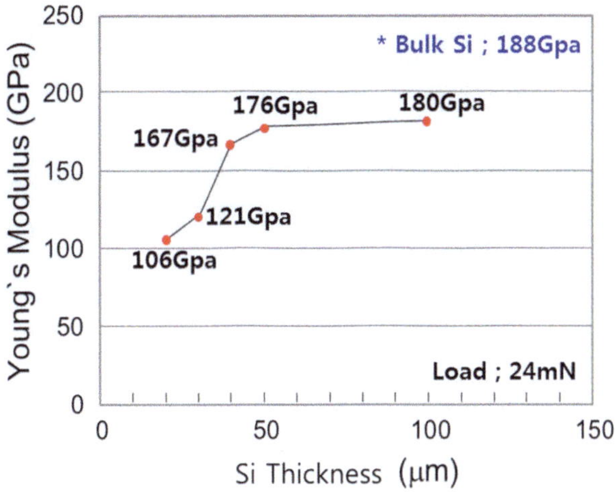


**Fig. 7.28** The dependency of Young's modulus with Si substrate thickness

of bulk Si (188 GPa) [20]. However, the Young's modulus is started to decrease at 40 μm substrate thickness (167 GPa) and noticeably decreased by approximately 30% (121 GPa) and 40% (106 GPa) at 30 μm and at 20 μm Si thicknesses compared to the modulus of 50 μm thick Si substrate [21], respectively.

It indicates that the Si substrate with 50 μm thickness has enough mechanical strength to achieve the high-reliable 3D-IC, because the mechanical property did not degrade compared to a bulk Si substrate. However, below 30 μm thickness, it has potential reliability challenge to stack many layers, because the mechanical property is noticeably deteriorated.

Wafer/die of less than 50 μm thickness leads to a many fold decrease in E and H values. Figure 7.29 shows the variation in Young's modulus (E) values of 30-μm-thick Si substrates with the stress-relief methods [22]. The CMP-processed substrate maintains higher E values among several types of stress-relieved substrates. Meanwhile, the mechanical properties are to be poor for the dry polishing (DP) and plasma etching (PE) processed substrates. CMP stress-relief method could

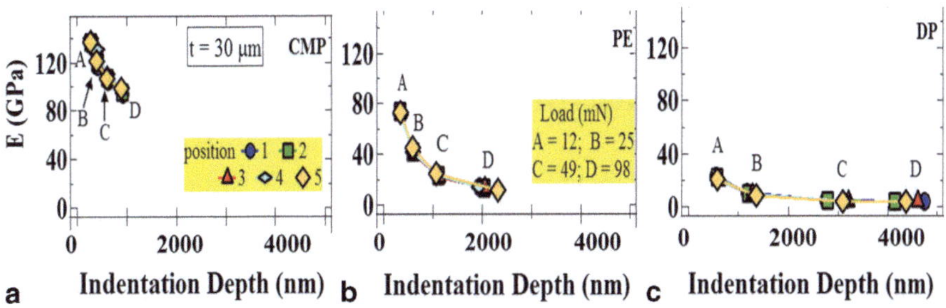


**Fig. 7.29** Modulus variation of 30-μm-thick wafers with stress-relief methods **a** CMP, **b** PE, and **c** DP, respectively. *DP* dry polish, *CMP* chemical–mechanical polishing

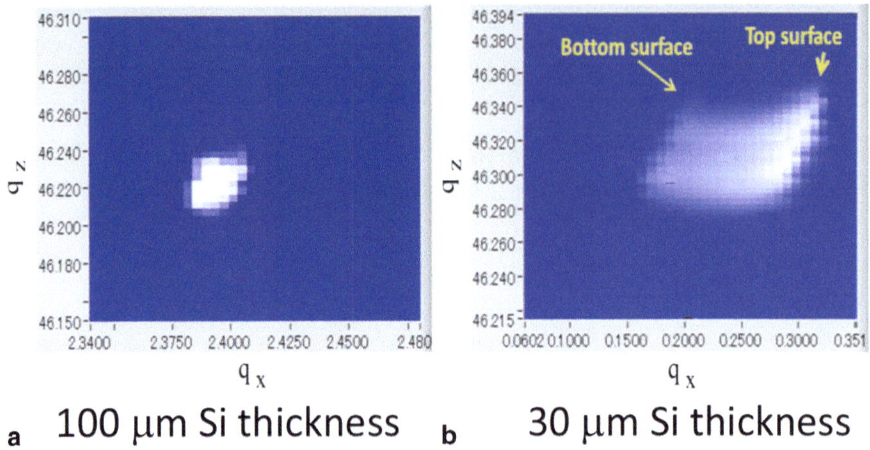


**Fig. 7.30** The reciprocal lattice images obtained by micro-diffraction method with Si substrate thickness

minimize the impact of the stress-induced damage on the front surface and consequently enhance the reliability for the high-performance 3D-IC.

To understand the impact of Si thinning on the deterioration of mechanical property in the thin Si substrate, the lattice structure of Si atoms is evaluated [21]. Figure 7.30 shows reciprocal lattice images of 100 and 30 µm thicknesses Si substrates obtained by micro-diffraction method. In the case of 100-µm-thick Si substrate, the spread in $q_x$ (which is closely related to the lattice tilt in Si crystal) and the spread in $q_z$ (which is correlated to the d-spacing in Si lattice) are approximately 0.025 rad and 0.02 rad. In the case of 30-µm-thick Si substrate, $q_x$ and $q_z$ values are approximately 0.15 rad and 0.08 rad. The larger spread in $q_x$ and $q_z$ values means that Si atomic lattice is more distorted. The lattice structure of Si substrate is highly distorted in the 30-µm-thick substrate compared to the 100-µm-thick Si substrate. We assume that the large distortion of the lattice structure in thin Si substrate induces the reduction of Young's modulus, consequently weakens the mechanical strength of Si substrate with 30 µm thickness.

## 7.4 Impact of 3D Integration Process on DRAM Retention Characteristics

Dynamic random-access memory (DRAM) stores electronic charge in capacitor as information data. The control of the retention time (refresh) of the stored charge is important in DRAM. However, the retention characteristics of DRAM may be sensitively affected by various parameters introduced in a 3D integration process. Therefore, the control of the retention time (refresh) for the stored charge is a key issue for realizing reliable 3D DRAM. This requirement derives from needs to keep

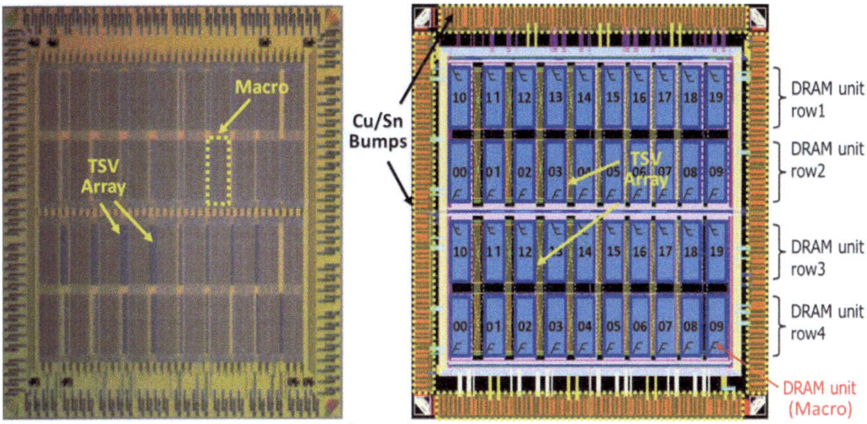


a b

**Fig. 7.31** The photograph (a) and the configuration (b) of the fabricated DRAM chip. *DRAM* dynamic random-access memory, *TSV* through-Si via

the refresh interval constant even if thin DRAM chips stack vertically to achieve 3D DRAM. One of the most critical reliability issues for high-reliable 3D DRAM is the data retention characteristics attributed to the electron leakage in the storage capacitor. The electron leakage is induced by several mechanisms introduced during 3D integration processes such as wafer thinning, thin chip bonding, and Cu TSV formation, etc. In this chapter, the impacts of 3D integration processes on memory retention characteristics in thin DRAM chip are discussed. To evaluate the influence of 3D integration processes on memory retention characteristics, a DRAM test element group (TEG) chip comprising n-channel metal–oxide–semiconductor (nMOS) cell arrays is fabricated by using 90-nm CMOS technology as shown in Fig. 7.31 where a DRAM chip is organized by 40 macros and 38 TSV arrays.

Figure 7.32 shows the circuit layout and configuration of each macro, which is composed of a memory cell array, a decoder, a sense amplifier, and buffer circuits. A planar-type cell is employed as a DRAM memory cell structure, because it may be sensitively affected by various parameters introduced in a 3D integration process.

To compare the effect of cell structure, the memory chip is fabricated using a p/p-Si substrate with twin well and triple well (introduced by deep n-well) structures as shown in Fig. 7.33.

### *7.4.1 Impact of Mechanical Strength on Retention Characteristics in Thin DRAM Chip*

The impacts of Si thinning on memory retention characteristics in thin DRAM chip are electrically characterized as shown in Fig. 7.34.

At first, known good die (KGD) memory chip of 760 μm thickness is thinned down to 200 μm thickness by mechanical grinding. Cu/Sn metal bumps are formed

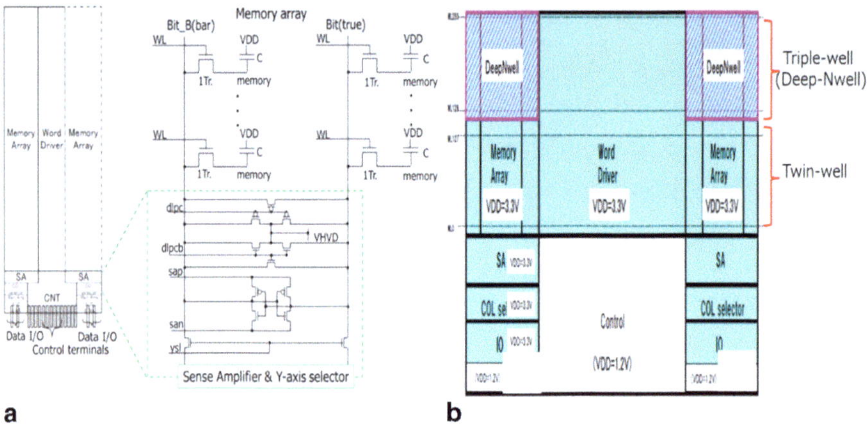


**Fig. 7.32** The circuit layout (**a**) and configuration (**b**) of memory macro in the fabricated DRAM chip

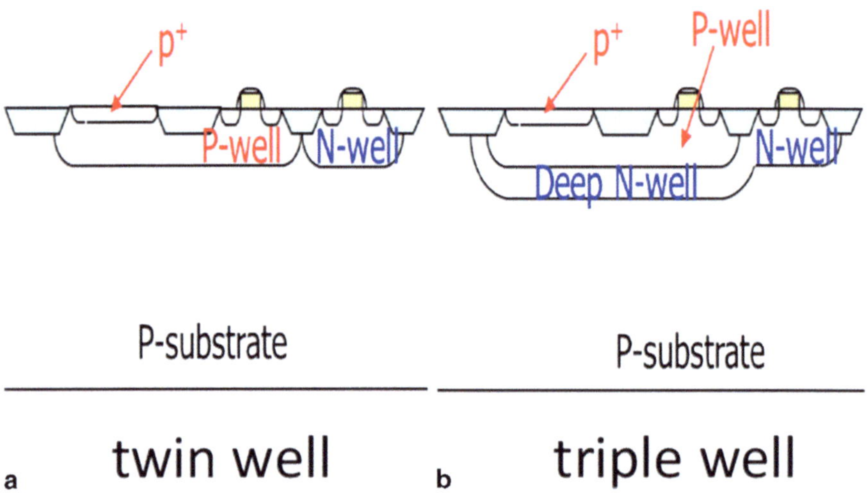


**Fig. 7.33** Cross-sectional views of well structures in the fabricated DRAM chip

on metal pads at the peripheral area in a die level. The memory chip is face down bonded to a Si interposer chip through Cu/Sn bumps by thermo-compression bonding method. To compare the effectiveness of underfill material, one sample is underfilled by an epoxy material and cured for 30 min at 200˚C as shown in Fig. 7.35. The memory retention characteristics of DRAM chip of 200 μm thickness are evaluated as a reference using the evaluation pads on the Si interposer.

To electrically characterize the impacts of Si substrate thickness on device reliabilities, the DRAM chip is thinned down to 50 μm thickness by mechanical grinding and following CMP process. After evaluation of the retention time of DRAM cell, the DRAM chip is further thinned down to 40, 30, and 20 μm thicknesses

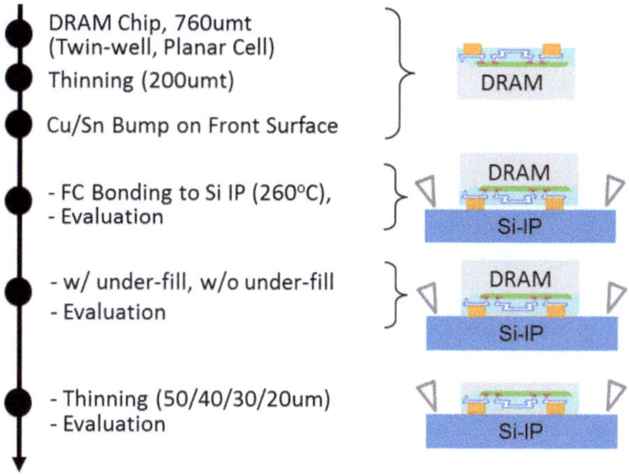


**Fig. 7.34** Process flow for the evaluation of impacts of Si thinning on memory retention characteristics. *DRAM* dynamic random-access memory

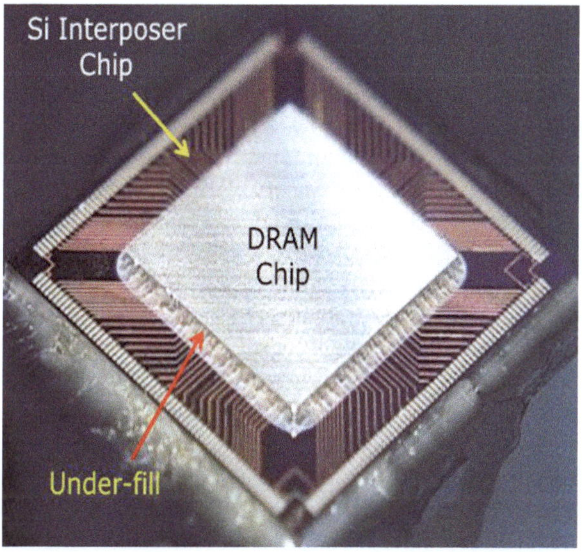


**Fig. 7.35** The photograph of DRAM TEG chip (200 μm thickness) after the face down bonding and the underfill

by plasma Si etching to avoid abundant mechanical-induced stress on the thinned chip, respectively. The retention characteristics of DRAM cells are evaluated at each chip thickness to characterize the dependence of the substrate thickness. The DRAM cell in the center area, which was departed 2000 μm distance from Cu/Sn bumps on the peripheral area, is measured to avoid local mechanical stress/strain effects induced by metal bump joining [14]. To avoid another path for storage capacitor charge from leaking out, the retention time is measured at *24°C* and 0.8 $V_{DD}$ without substrate bias. Figure 7.36 shows the failure rates of the DRAM cell array ($W/L$=3.50/0.30 μm) fabricated in the thinned DRAM chips with underfill as a function of static retention time. The retention characteristics of DRAM cell array ($W/L$=3.50/0.30 μm) in the DRAM chip with underfill are degraded depending on

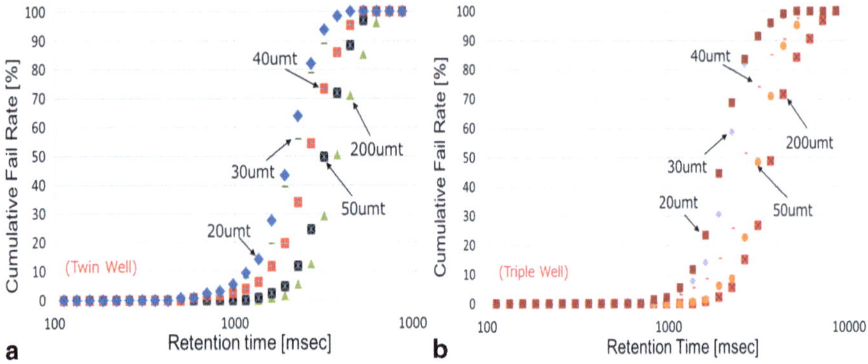


**Fig. 7.36** The failure rate of DRAM cell array fabricated in twin well (**a**) and triple well (**b**) as a function of retention time measured after chip bonding (200 μm thickness) with underfill and chip thinning to 50/40/30/20 μm thicknesses, respectively

the decreasing of the chip thickness. In the case of twin well (a), the DRAM cell array shows 50 % failure at *3601*, *3253, 2370,* 2027, and 1842 msec at *200, 50, 40, 30, and 20 μm thickness* conditions, respectively. In the case of triple well (b), the DRAM cell array shows 50 % failure at 3688, 3219*, 2580, 1946, and 1918 msec at* 200, 50, *40, 30*, and 20 μm thickness conditions, respectively. As seen in the figure, the retention characteristics of DRAM cell are degraded depending on the reduction of the chip thickness. Especially, below 40-μm chip thickness, the retention characteristics of DRAM cell are noticeably degraded regardless of the well structure (triple well, twin well) in the Si substrate. The retention time of DRAM cell in the 20-μm-thick chip is dramatically shortened by approximately 40 % compared to the 50-μm-thick chip.

Fig. 7.37 shows the failure rates of the DRAM cell array fabricated in the thinned DRAM chips without underfill. The retention characteristics of DRAM cell relatively not so degraded until 30 μm thickness, but abruptly degraded below 20 μm thickness regardless of the well structure (triple well, twin well) in the Si substrate. In the case of twin well (a), the DRAM cell array shows 50 % failure at 3756, 3562, 3485, 3228, and 1773 msec at 200, 50, 40, 30, and 20 μm thickness conditions, respectively. In the case of triple well (b), the DRAM cell array shows 50 % failure at 3807, 3633, 3516, 3368, and 1796 msec at 200, 50, 40, 30, and 20 μm thickness conditions, respectively. The retention time of DRAM cell in the 20-μm-thick chip is dramatically shortened by approximately 50 % compared to the 50-μm-thick chip.

Figure 7.38 shows the correlation of the median retention time of DRAM cell array at 50 % failure versus the Young's modulus of Si substrate depending on the chip thickness. Young's modulus of Si substrate begin to abruptly decrease below 30 μm thickness. In the case of the DRAM chip with underfill, the retention times of DRAM cell at each thickness are shorted by approximately 12 % (50 μm), 33 % (40 μm), 46 % (30 μm), and 50 % (20 μm) compared to the 200 μm thickness chip, regardless of the well structure. In the case of the DRAM chip without underfill, meanwhile, the retention times of DRAM cell at each thickness are shorted by approximately

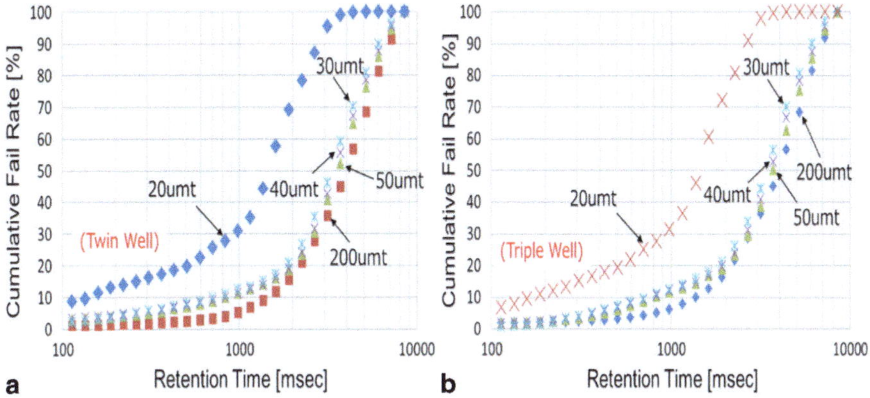


**Fig. 7.37** The failure rate of DRAM cell array fabricated in twin well (**a**) and triple well (**b**) as a function of retention time measured after chip bonding (200 μm thickness) without underfill and chip thinning to 50/40/30/20 μm thicknesses, respectively

**Fig. 7.38** The median retention time of DRAM cell array at 50 % failure versus the Young's modulus depending on the chip thickness

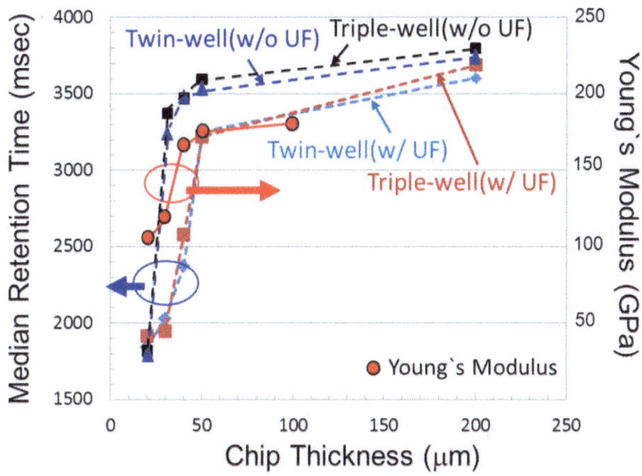


5 % (50 μm), 7 % (40 μm), 13 % (30 μm), and 53 % (20 μm) compared to the 200 μm thickness chip, regardless of the well structure. Below 20 μm chip thickness, the retention characteristics are dramatically degraded, regardless of the underfill process. It is hard to perfectly characterize the reason of the data retention time shortening, because the degradation of the retention characteristics has many complicated factors. Based on the evaluation results in this study, but we assume that the distortion of the lattice structure in the thin Si substrate induced by wafer thinning is one of primary factors to degrade retention characteristics of DRAM cell. Below 20 μm thickness, the lattice structure of thin Si substrate is severely distorted; hence, it may cause an effect on fundamental properties such as Young's modulus and band gap energy. The reduction of Young's modulus induces the deterioration of the mechanical strength. The change of band gap energy may effect a minority carrier lifetime, consequently shortening the retention time of DRAM cell [23].

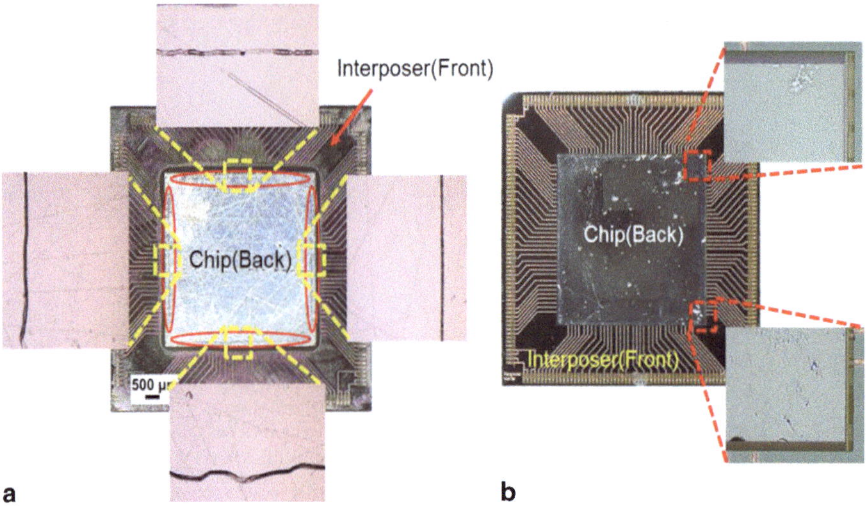


a b

**Fig. 7.39** Photographs of the stacked DRAM chip which was thinned down to 20 μm thickness with underfill (**a**) and without underfill (**b**)

Figure 7.39 shows the photographs of the stacked DRAM chips with underfill (a) and without underfill (b), which was thinned down to 20 μm thickness, respectively. In the case of with underfill, the cracks are not happened until 30 μm thickness, but generated in the 20 μm thickness at around Cu/Sn bump joining positions on the peripheral region and propagated between them. In the case of without underfill, meanwhile, the cracks are not happened even after thinned down to 20 μm thickness. We assume that the reduced mechanical strength of the thin chip with 20 μm thickness did not endure a local mechanical stress/strain [13, 14] induced by CTE mismatch between Cu/Sn bump, Si substrate, and epoxy underfill; hence, it causes cracking at bumping joining position. Meanwhile, the mechanical strength of 20-μm-thick chip still endure the local stress/strain induced by CTE mismatch between Cu/Sn bump and Si substrate; hence, it do not cause cracking in the case of without underfill.

It may require further to minimize the local mechanical stress/strain effects by optimizing underfill material and bump joining structure (material, array pattern, etc.) for the reliable thin chip stacking under 30 μm thickness.

### *7.4.2 Impact of Cu Contamination on Memory Retention Characteristics in DRAM Chip*

One of the critical origins of the electron leakage is metallic impurities such as Cu introduced in the 3D integration process. In particular, Cu is extremely mobile in silicon and silicon dioxide [23, 24]. It exhibits the highest diffusivity in silicon among all metal elements. When present in active device regions, Cu impurities could cause functional failures through a variety of mechanisms increasing leakage currents or carrier recombination rates and loss of functionality [25]. In extremely sensitive devices such as DRAM cell, even small contamination levels can shorten retention time. The influence of Cu contamination at the backside surface of the thinned DRAM chip on retention characteristics is electrically characterized after intentional Cu diffusion at various annealing condition.

To compare the EG characteristics of the grinded surface (Fig. 7.11), the DRAM chips are thinned down to 50 μm thickness by mechanical grinding and following CMP and DP treatments, respectively. Each thinned DRAM chips are face down bonded to the Si interposer through Cu/Sn bumps by thermo-compression bonding. Thin Cu layer of 50 nm thickness is deposited on the grinded surface as a contamination source. Cu atoms are intentionally diffused into the active area from the backside grinded surface for 30 min at 200, 250, and 300°C in $N_2$ ambience, respectively, as shown in Fig. 7.40. The retention characteristics of DRAM cell in the thinned DRAM chip (CMP-treated) dramatically degraded after annealing at 300 °C, regardless of the well structure as shown in Fig. 7.41. Meanwhile, the retention characteristics of DRAM cell in the thinned DRAM chip (DP treated) not degraded even after annealing up to 300°C temperature, regardless of the well structure as shown in Fig. 7.42.

Figure 7.43 shows the correlation of the median retention time of DRAM cell array at 50% failure depending on the annealing temperature. Until 250°C annealing condition, the retention characteristics of DRAM cell in the thinned DRAM chip not so degraded, regardless of the well structure and the post-treatment methods (CMP, DP). However, in the case of CMP treated chip, the retention time dramatically degraded after annealing at 300°C, regardless of the well structure. The DP-treated chip has better gettering property to Cu diffusion compared to the CMP-treated chip. It indicates that the triple well structure introduced by deep n-well is not enough to protect Cu diffusion. Meanwhile, an EG layer formed by DP treatment shows a good blocking property to Cu diffusion [26].

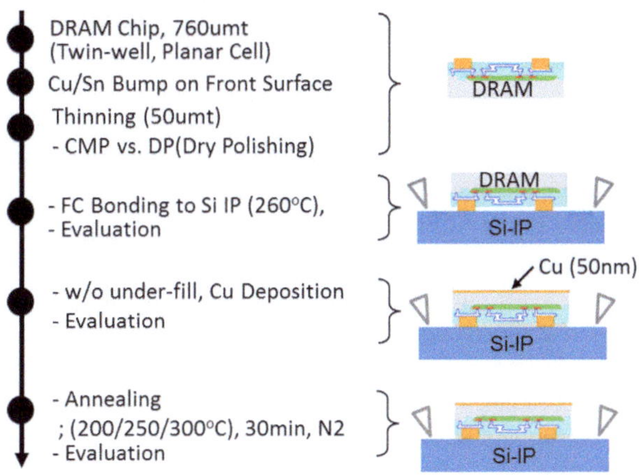


**Fig. 7.40** Process flow for the evaluation of impact of Cu diffusion on memory retention characteristics. *DP* dry polish, *CMP* chemical–mechanical polishing, *DRAM* dynamic random-access memory

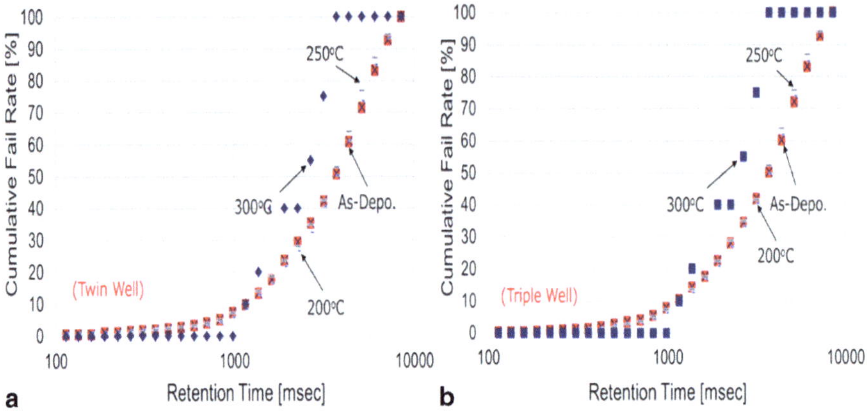


**Fig. 7.41** The failure rate of DRAM cell array fabricated in twin well (**a**) and triple well(**b**) as a function of retention time measured after Cu as-deposition and annealing at various temperatures for 30 min (CMP-treated chip)

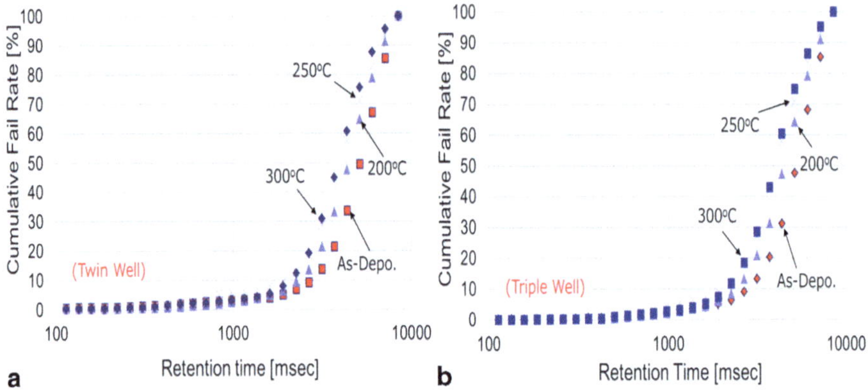


**Fig. 7.42** The failure rate of DRAM cell array fabricated in twin well (**a**) and triple well (**b**) as a function of retention time measured after Cu as-deposition and annealing at various temperatures for 30 min (DP-treated chip)

The influence of Cu contamination from Cu TSV on memory retention characteristics is characterized. Figure 7.44 shows the process flow to evaluate the impact of Cu contamination from Cu TSV. The DRAM chip of 50 μm thickness is face down bonded to the Si interposer through Cu/Sn bumps by thermo compression bonding. After the formation of P-tetraethyl orthosilicate (TEOS) layer and PR patterning on the backside surface of the thinned DRAM chip, via holes of 10 μm diameter with 30-nm scallop roughness are fabricated from the backside using BOSCH process. The P-TEOS liner is formed into via holes, and the bottom area of liner layer in via holes is contact etched using plasma etching process. After formation of sputtered 50-nm-thick Ta barrier and 500-nm-thick Cu seed layers (at the surface), via holes are filled by Cu electroplating. After PR patterning of the wiring, Cu and Ta layers are etched to form

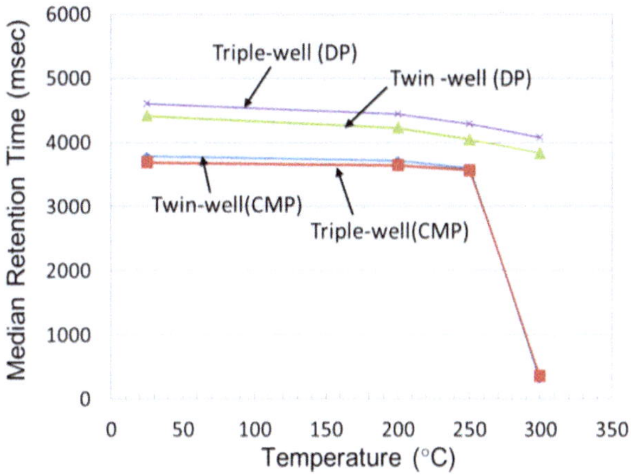


**Fig. 7.43** The median retention time of DRAM cell array at 50 % failure versus various annealing temperatures for 30 min. *DP* dry polish, *CMP* chemical–mechanical polishing

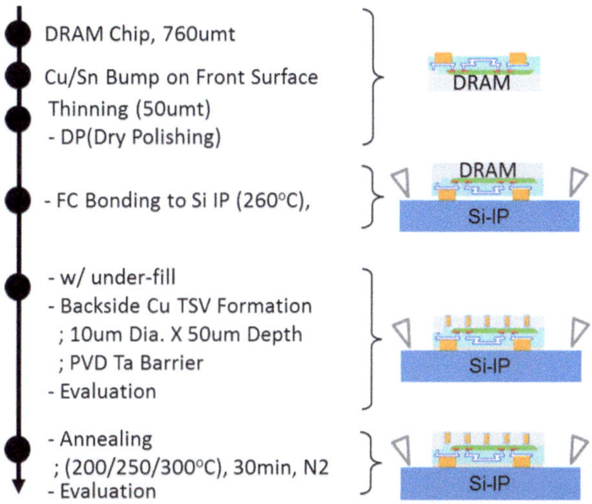


**Fig. 7.44** Process flow for the evaluation of impact of Cu contamination from Cu TSV on memory retention characteristics. *DRAM* dynamic random-access memory, *TSV* through-Si via

the backside TSV array. In this experiment, relatively thin Ta barrier layer is intentionally deposited into via holes in the DRAM chip to accelerate the Cu diffusion effect.

Figure 7.45 shows the mapping data of the retention characteristics of memory cell array at each memory macro after 30 min annealing at various temperature, before (a), 200˚C (b), 250˚C (c), and 300˚C (d), respectively. As seen in the figure, until 250˚C annealing condition, the retention characteristics of DRAM cells at both memory macros not degraded. However, the retention characteristics of some memory cells at the edge area in each memory macro, which were departed 20–50 μm distance from Cu TSV array, are began to degrade after annealing at 300˚C, 30 min [27]. As the annealing temperature increases and the annealing time longer, the damaged memory cell area will be spread into all macros in the DRAM chip. We assume that Cu atoms easily diffuse from some Cu TSVs with Ta barrier layer of poor coverage into the active device area and then the retention characteristics of

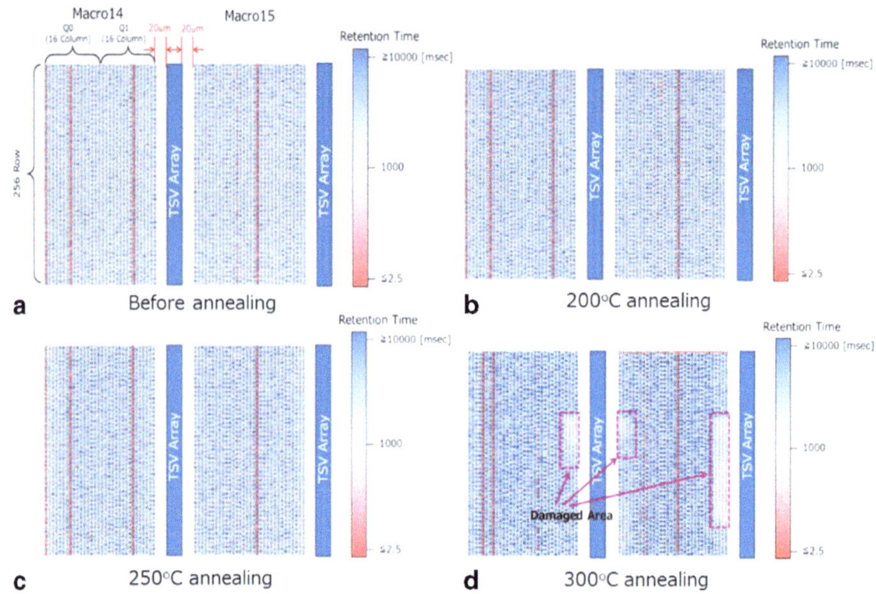


**Fig. 7.45** The mapping data of the retention characteristics of memory cell array in macro 14 and macro 15, which are separated 20 μm distance from Cu TSV array, after 30 min annealing at various temperature; before (**a**), 200˚C (**b**), 250˚C (**c**), and 300˚C (**d**), respectively

some memory cell are degraded even after relatively low temperature annealing at 300˚C. As the annealing time and the annealing temperature increase, more memory cells will be damaged because Cu atoms more spread out into larger area. Therefore, it becomes more challenging to form via-middle Cu TSV because it needs several post-annealing steps at higher than 400 ˚C temperature to control stress-induced effects introduced by Cu TSV. It has future concern to form the reliable Cu TSVs with small diameter, high aspect ratio, and high density by using current physical vapor deposition (PVD) method for the formation of barrier layer.

## References

1. Hozawa K et al (2002) True influence of wafer backside copper contamination during the back-end process on device characteristics. IEEE IEDM, pp 737–740, Dec
2. Istratova AA, Weberb ER (2002) Physics of copper in silicon. J Elecrochem Soc 149(1): G21–G30
3. Hozawa K et al (2009) Impact of backside Cu contamination in the 3D integration process. IEEE VLSI Symp, pp 172, June
4. Bea J et al (2011) Evaluation of Cu contamination at backside surface of thinned wafer in 3-D integration by transient-capacitance measurement. IEEE Electron Device Lett 32(1):66–68
5. Heiman FP (1967) On the determination of minor carrier lifetime from the transient response of an MOS capacitor. IEEE Trans Electron Devices 14:781–784

6. Lee SY et al (1999) Measurement time reduction for generation lifetime. IEEE Trans Electron Devices 46:1016
7. Lee KW et al (2011) Evaluation of Cu diffusion characteristics at backside surface of thinned wafer for reliable three-dimensional circuits. Semicond Sci Tech 26:025007
8. Lee KW et al (2014) Impact of Cu contamination on device reliabilities in 3-D IC integration. IEEE Trans Dev Mater Reliab 14(1):451–462
9. Lee KW et al (2011) Cu retardation performance of extrinsic gettering layers in thinned wafers evaluated by transient capacitance measurement. J Elecrochem Soc 158(8):H795–H799
10. Lee KW et al (2012) Impact of Cu diffusion from Cu through-silicon via (TSV) on device reliability in 3-D LSIs evaluated by transient capacitance measurement. IEEE IRPS, pp 2B.4.1
11. Yu CL et al (2011) TSV process optimization for reduced device impact on 28 nm CMOS. IEEE VLSI Symp, pp 138
12. Murugesan M et al (2010) Wafer thinning, bonding, and interconnects induced local strain/stress in 3D-LSIs with fine-pitch high-density microbumps and through-Si vias. IEEE IEDM, pp 2.3.1
13. Murugesan M et al (2012) minimizing the local deformation induced around Cu-TSVs and CuSn/InAu micro-bumps in high-density 3D-LSIs. IEEE IEDM, pp 28.6.1
14. Murugesan M et al (2010) Impact of micro-bumps induced stress in thinned 3D-LSIs. IEEE 3D–IC. doi:10.1109/3DIC.2010.5751432
15. Aditya P et al (2011) Microbump impact on reliability and performance in TSV stacks. MRS proceedings, pp 1335
16. Kumar A et al (2011) Residual stress analysis in thin device wafer using piezoresistive stress sensor. IEEE Trans Comp Packag Manuf Technol 1(6):841
17. De WI (1996) Micro-Raman spectroscopy to study local mechanical stress in silicon integrated circuits. Semicond Sci Technol 11:139
18. Kino H et al (2013) Impacts of static and dynamic local bending of thinned Si chip on MOSFET performance in 3-D stacked LSI. IEEE ECTC, pp 360
19. Chantkul P et al (1981) A Critical evaluation of indentation techniques for measuring fracture toughness: 11, strength method. J American Ceramic Society 94(9):539–543
20. Matthew A et al (2010) What is the Young's modulus of silicon. J Microelectromechanical Systems 19(2):229–238
21. Lee KW et al (2013) Degradation of memory retention characteristics in DRAM chip by Si thinning for 3-D integration. IEEE Electron Device Lett 34(8):1038–1040
22. Murugesan M et al (2013) Mechanical characteristics of thin dies/wafers in three-dimensional large-scale integrated systems. 24th SEMI-Advanced Semiconductor Manufacturing Conference, pp. 66–69
23. Weber ER (1983) Transition metals in silicon. Appl Phys A 30:1–22
24. Helneder H et al (2001) Comparison of copper damascene and aluminum RIE metallization in BICMOS technology. Microelectron Eng 55:257–268
25. Ramappa DA, Henley WB (1999) Effects of copper contamination in silicon on thin oxide breakdown. J Electrochem Soc 146(6):2258–2260
26. Lee KW et al (2014) Impacts of 3-D integration processes on memory retention characteristics in thinned DRAM chip for high reliable 3-D DRAM. IEEE Trans Electron Devices 61:379–385
27. Lee KW et al (2014) Impacts of Cu contamination in 3D integration process on memory retention characteristics in thinned DRAM Chip. IEEE IRPS, pp 3E.4.1

# Chapter 8
# Trends in 3D Integrated Circuit (3D-IC) Testing Technology

**Hiroshi Takahashi, Senling Wang, Shuichi Kameyama, Yoshinobu Higami, Hiroyuki Yotsuyanagi, Masaki Hashizume, Shyue-Kung Lu and Zvi Roth**

It has been proved that the manufacturing cost of 3D integrated circuit (3D-IC) is about 2.4 times to that of conventional stacked packaging. In 3D-IC, it is reported that the chip loss, cost referred to yield, accounts for about 48 % of the total production cost. Therefore, development of testing technology for 3D-IC becomes essential for reducing the manufacturing cost of 3D-IC. In this chapter, we describe the testing technologies for 3D-IC. In Sect. 8.1, we marshal the issues that must be handled in the 3D-IC testing. From Sects. 8.2 to 8.4, we introduce the

H. Takahashi (✉)
Computer Science, Graduate School of Science and Engineering, Ehime University, Ehime, Japan
e-mail: takahashi@cs.ehime-u.ac.jp

S. Wang · Y. Higami
Ehime University, Ehime, Japan
e-mail: wang@cs.ehime-u.ac.jp

Y. Higami
e-mail: higami@cs.ehime-u.ac.jp

S. Kameyama
Fujitsu Limited, Kanagawa, Japan
e-mail: kameyama@jp.fujitsu.com

H. Yotsuyanagi · M. Hashizume
The University of Tokushima, Tokushima, Japan
e-mail: yanagi4@ee.tokushima-u.ac.jp

M. Hashizume
e-mail: tume@ee.tokushima-u.ac.jp

S.-K. Lu
National Taiwan University of Science and Technology, Taipei, Taiwan (R.O.C.)
e-mail: sklu@mail.ntust.edu.tw

Z. Roth
Florida Atlantic University, Boca Raton, FL, USA
e-mail: rothz@fau.edu



K. Kondo et al. (eds.), *Three-Dimensional Integration of Semiconductors*,
DOI 10.1007/978-3-319-18675-7_8

outlining of the proposed 3D-IC testing technologies in so far. From Sects. 8.5 to 8.7, we provide the 3D-IC testing technologies that are proposed by our research group in Japan.

## 8.1 Crucial Issues and Key Technologies for 3D-IC Testing

In general, 2D integrated circuit (2D-IC) testing is applying full scan design to the circuit under test (CUT) so as to increase the controllability and observability from the external. However, in 3D-IC, external probing is the only way to pre-bond test. In addition, only the stacks on either the top die or bottom die have external accesses in post-bond test.

We marshal the issues of 3D-IC testing in the following.

**Issue 1** In 3D-IC testing, numerous test moments should be configured. Specifically, we have to test the 3D-IC as the pre-bond test before stacking, the mid-bond test during partial stacking, the post-bond test for the full stacked ICs, and the final test after packaging.

**Issue 2** In 3D-IC testing, we have to test the internal die and through-silicon via (TSV) which construct the 3D-IC individually. For internal die, the defects caused by cracks and separation due to stacking, and the abnormal device parameters have to be tested. For TSV, a test should be performed on the premise that direct probing each TSV on the die is impossible. The defects referred to the break, the micro void, and the pinhole in TSVs should be treated as the open faults, the bridging faults, and the delay faults, respectively. Moreover, it should be noted that defects at TSVs not only occur in the manufacturing process but also may occur in the normal operation after shipping due to the aging.

**Issue 3** Since the controllability and observability of the 3D-IC are very low, a test access mechanism (TAM) used to deliver test data between the stacked dies is required.

Next, we consider the required technologies in 3D-IC testing. A TSV testing in 3D-IC can be separated into two distinct categories: the pre-bond test and the post-bond test. In the 3D-IC testing, the pre- and post-bond test requires effective and optimized approaches. In the pre-bond test, the techniques for built-in self-test (BIST) and probing are required. In the post-bond test, the techniques for BIST and scan-based TAM are necessary. Also, methods for optimizing the test time, TAM, and test constraints in terms of power dissipation and thermal are also required. Moreover, methods for generating test patterns for various fault models at TSV are expected.

## 8.2 Research Trends in Pre-Bond Test for 3D-IC

When we construct a 3D-stacked IC with known good die (KGD) , we have to test TSV during the pre-bond wafer test. In the pre-bond test, it is crucial to detect the defective TSVs which contain defects as the break, the micro void, or the pinhole. However, the current probe technology cannot have contact with the individual TSV due to the small pitch and high density of TSVs. Furthermore, the TSVs are single ended which means that one end of a TSV of a die before stacking should be either an open or a ground connection.

In order to apply the pre-bond TSV test in such a testing environment, many sophisticated methods with/without probing have been proposed. Test mechanisms for pre-bond testing of TSV using probe technology have been proposed in [1, 2]. In [1], the authors present a new technique for pre-bond TSV testing that is compatible with the current probe technology. It utilizes many single probe needle tips to make contact with multiple TSVs. The authors propose the design-for-test (DFT) architecture and the technique for the probe card that can measure resistance and capacitance, as well as to perform stuck-at and leakage tests. The proposed method is applied to a network of TSVs. The authors introduce a gated scan flop (GSF) to each individual TSV so as to enable the pre-bond test for TSVs through probing. Moreover, in [2], the authors propose the scan test method to contact TSVs without the need for probe pads during the pre-bond testing. The authors focus on the scan test of the die logic utilizing scan chains that can be reconfigured for the pre-bond test to allow scan in and scan out through TSVs. This method utilized a die wrapper that consisted of the boundary-scan flops with GSFs. The GSFs are the boundary-scan registers at the TSV interface.

On the other hand, numbers of BIST architectures for the pre-bond test are presented to avoid probing the TSV directly. In [3], the authors present two schemes for testing TSVs by performing the pre-bond test. The first method uses a "charge-sharing" technique commonly seen in DRAM for testing the "blind TSVs" which have one end floating. The second method uses a "voltage-dividing" technique commonly seen in ROM for testing the "open-sleeve" TSVs which has one end shorted to the substrate. In [4, 5], the authors propose the design method and test structure to characterize and to repair TSV defect. The proposed test structure consists of the voltage division circuit for detecting the faulty TSV and the comparator circuit for repairing from the faulty TSV to the repairable TSV. In [6, 7, 8], the authors propose a method for noninvasive pre-bond TSV test using ring oscillators and multiple voltage levels. The proposed methods can detect resistive opens and leakage faults by measuring variations in the delay of wires connected to TSVs. In the study, the open defects at TSVs are modeled as capacitive loads of their driving gates. The propagation delay through the TSVs is measured by means of ring oscillators. The authors provide a method to create a regression model to predict the defect size for a given measured period of the ring oscillator. In [9], the authors propose the test structures based on an input sensitivity analysis technique in which they estimate the TSV capacitance through measuring the change in clock period of ring oscillator.

The main concern of the BIST architectures for the pre-bond testing is to require careful calibration and tuning. The BIST architectures for the pre-bond testing also occupy a relatively large die area, when we consider the tens of thousands of TSVs between dies.

Recently, the authors of [10] develop a spring-type probe using microelectromechanical systems (MEMS) technology to test the TSVs. In this method, the MEMS probe can directly touch down the narrow TSV and reduce the scrub marks as far as possible.

## 8.3 Research Trends in Post-Bond Test for 3D-IC

As you may know, IEEE 1500 and IEEE 1149.1 (boundary scan) based test mechanisms for 2D-IC have been standardized as die test wrapper. In 3D-IC, the IEEE P1838 workgroup is going to develop a standardization of the test mechanisms for post-bond test after stacking (3D-SiC is used as the abbreviation in IEEE P1838 workgroup).

In [11, 12], the authors propose the DFT architecture and the test approaches for 3D-IC test. The DFT architecture supports a modular test approach, in which, the cores, dies, stacks, printed circuit board (PCB), the inter-die TSV-based interconnects, and the external input/outputs (I/Os) can be tested individually as separate units. The DFT architecture reuses the existing DFT hardware at the core, die, and product level. The proposed die-level wrapper is based on IEEE 1500. The specific mechanisms for modular testing include: (1) configuring dedicated probe pads for probing the non-bottom dies, (2) configuring a test elevator on the lower dies for transporting test control and data signals up and down to the other dies after stacking, and (3) hierarchical wrapper instruction register (WIR) chains which can access only the cores where test is needed.

In addition, methods to optimize the TAM and the number of TSVs used for a given test architecture of 3D-IC have been proposed. Moreover, methods for optimizing a test schedule while considering the constraints of power consumption and thermal during the test have been proposed. In [13, 14], the authors address the problem of the test architecture optimization for 3D-ICs to minimize overall test time. The authors propose optimization techniques that consider die-external and die-internal tests during test time minimization. An optimal number of TAM between tiers of the 3D-IC can be obtained through the mathematical method. The authors verified the impact of the number of test pins in the bottom die and the number of TSVs used for testing on the test length referred to the number of test clocks. In [15], the authors propose a method that is used to minimize the test time for the post-bond testing of 3D-IC under the constraints of TSV count and TAM bandwidth. In the method, integer linear programming (ILP) is utilized.

Previous work on the scan chain design methodology for 3D-IC has been proposed for wire length optimization only. In [16], the authors propose a method adopting a genetic algorithm (GA) to optimize the trade-off between delay and power for scan chain reordering by introducing a weighted cost function in terms of delay and power consumption metrics. In [17], the authors discussed the test flow of wafer level and package level in detail and studied the test contents, wafer level probe access, and on-chip DFT infrastructure. In [18], the authors propose an approach that automates the insertion of die wrappers. The proposed test architecture enables the pre-bond test as well as the post-bond and the interconnect test to a die. In the method, the users can perform the automated insertion of an IEEE 1500-based 3D wrapper by using existing electronic design automation (EDA) tools. The bottom dies can be extended with IEEE 1149.1 to reduce the pin count and to enable board-level interconnect test. In [19], the authors propose the method for optimizing the test architecture for 3D-IC using TSVs. The proposed test architecture can significantly reduce the test length compared to the conventional method of sequentially testing all dies. However, the number of dies that can be tested in parallel is limited by the given TAM width. Therefore, the authors utilize an ILP model to formulate the architecture optimization problem to derive the optimum solutions. The minimization of test length can be achieved by the derived optimal test architecture with an optimal test schedule. Experimental results indicate that increasing the number of test pins typically dominates the test length reduction compared to increasing the number of test TSVs. Also, larger dies at the lowest tiers in the stack lead to shorter test lengths. Moreover, in the chapter, the authors indicate three different cases of test architecture optimization problems involving hard dies, soft dies, and firm dies. In the case of the hard dies, an optimal solution is derived under a constraint that 2D TAM design already exists and has been fixed. In the case of the soft dies, the optimal solution is derived while to consider the 2D TAM design is flexible that means the TAM of each die can be codesigned during the test architecture design for the stack. For the firm dies in which the test architecture already exists, adding extra hardware is allowed, an optimal solution is derived as serial/parallel conversion hardware is added to the die so that fewer test elevators are used than the fixed 2D TAM width.

Prior studies on the test architecture optimization for 3D-IC reduce the test cost without considering the uncertainties in the input parameters. Prior methods consider only a single point in the input parameter space. The assumed values of the test power and pattern count of logic cores used for the optimization of the architecture may differ from the realistic values, which can only be determined after the design stage. In [20], the authors propose an optimization method for the test architecture of 3D-IC taking into account the uncertainties in the input parameters in terms of test power and available bit width. The authors develop an ILP model to formulate the robust test architecture optimization problem. In [21], the authors propose a cost model used for evaluating the test cost for 3D-IC. In order to minimize the overall test cost, the authors also propose a heuristic solution by explicitly enumerating the test flows. The authors formulate the problem of selecting a cost-effective test flow in terms of matrix partitioning based on the heuristic procedure. In [22], the authors propose a test scheduling for 3D-IC while considering the thermal aspect. In the study, the authors assume that the 3D-IC is stacked by many floors (layers) which consist of multiple cores. The test scheduling is implemented to ensure that the temperature rise during the test does not exceed the limits. In scheduling the test, a single core for testing is selected, and the temperature rise of its neighboring cores is computed and compared with the limit of $P_{max}$. The core with lower temperature rise will be tested first.

In [23], the authors extend the existing 3D DFT architecture for the purpose of supporting the wrapped embedded intellectual property (IP) cores and multi-tower stacks by using the industrial EDA tools. In [24], the authors propose a low-cost testing scheme for TSVs in 3D-IC. The proposed scheme reuses the pre-bond test scheme for the post-bond testing of TSV because each die in a 3D-IC is pre-bond testable. Since most of the test patterns used in the pre-bond test are also applied in the post-bond test, the test flow becomes simple, and the pattern counts can be reduced as well. In [25], the authors address the problem of scan chain ordering under a limited number of TSVs. The authors also propose a fast two-stage algorithm to compute a final order of scan flip-flops (FFs) by formulating the scan chain ordering as traveling salesman problem (TSP). The proposed algorithm consists of two stages. In the first stage, the proposed method constructs an initial simple path through all scan FFs using a modified greedy algorithm, multiple fragment heuristic, via a dynamic closest-pair data structure. In the second stage, the authors propose two new techniques, 3D planarization and 3D relaxation, to minimize the wire (and/or power) cost and to reduce TSV usage, respectively. In the method, a greedy algorithm, multiple fragment heuristic is modified and combined with a dynamic closest-pair data structure to derive the initial solutions, in turn, reduce the wire/power cost and relax the number of TSVs.

## 8.4 Research Trends in Automatic Test Pattern Generator and Test Scheduling for TSVs in 3D-IC

In this section, we describe the current researches on DFT technologies for test generation, fault diagnosis, and parameter variation detection of TSVs. Test cost is a major concern in IC testing. In 3D-IC, it may be impossible to test the ICs of each layer in parallel due to the constraints of the features of the TSV structure. Therefore, in the test generation for 3D-IC, a test-scheduling problem has been investigated in many studies for reducing the test time as well as test cost.

In [26], the authors propose a test scheme for detecting interconnect faults by using circuit characteristics, inherent test resources in design, and test patterns of interconnects between embedded cores. Since chips are tested before interconnecting them, the effects of interconnect faults can be propagated through fault-free dies. Test patterns for interconnect faults are selected among the test set for cores which are generated in advance. As a result, the test generation problem for interconnect faults is simplified. In [27], the authors propose the test infrastructure for transition faults that supports both the pre-bond and the post-bond testing for the 3D-IC. Furthermore, the authors propose a method to test TSVs in the post-bond test, without regenerating test patterns. The authors also discuss the infrared (IR) drop issue of 3D-IC during transition test. In [28], the authors indicate that the thermo-mechanical stress caused by TSV fabrication processes may change the timing behavior and may affect the delay fault detection in turn. They designate that the quality of the test patterns which are generated without taking TSV stress into account is significantly degraded. In order to evaluate the impact of TSV stress on delay test in consideration of the layout, the authors propose a test pattern generation flow using the TSV stress aware cell libraries. From the experimental results, the authors confirmed that the test quality was improved by using the TSV stress aware test generation. In [29], the authors propose a method for generating functional test patterns that can specify the maximum temperature for the target locations and layers in the 3D-IC. Information of maximum hotspot temperature is necessary to select the appropriate package and to optimize production costs. The authors develop a thermal modeling for predicting temperature profile at the end of a program phase. An ILP formulation is applied to derive the program phases, which can create maximum temperature at a target location. From the experimental results, the authors show that a much higher temperature at a hotspot can be achieved than that obtained by a functional tracing approach.

In [30], the authors propose a scan-based test interface for TSVs and a procedure for generating fault diagnosis test sequences for post-bond testing. The proposed method can identify single and multiple defective TSVs in 3D-IC. In this chapter, the authors consider the stuck-at, bridge, open, delay, and crosstalk faults. Walking 1/0 test sequences are used as diagnosis test sequences. The walking 1/0 test sequence denotes a sequence consisting of $n$ bit long binary vectors constructed by a single 1s(0s) followed by $n-1$ 0s(1s), where $n$ is the number of TSVs. For identifying the crosstalk-related delay faults, checkerboard test patterns such as "01010101…" are used.

It is shown in [31] that defective power TSV can cause excessive IR drop, which results in path delay faults. The authors have proposed a test generation method for detecting such path delay faults. They designate that open detects in power TSV do not lead to excessive IR drop, but that the leakage defects in power TSV should be tested. In the test generation method, IR drop simulation is performed with injecting a defect so that IR drop is converted to extra gate delay in the surrounding logic gates. Timing violation paths are identified by static timing analysis, and the timing-aware automatic test pattern generator (ATPG) is applied for the extracted path delay faults. This method has no extra requirements in hardware overhead of the DFT.

In [32], the authors propose the DFT technique using variable output thresholding (VOT). The VOT scheme is implemented with a ring oscillator architecture consisting of two TSVs and a number of logic gates. The authors demonstrated the availability of VOT in detecting parametric delay faults on the TSVs by observing the oscillation period of the ring oscillator by dynamically switching the output of a TSV from a normal inverter to a "Schmitt trigger" inverter. In [33], the authors propose a method and architecture for characterizing the propagation delays of TSVs in a 3D-IC. In the proposed architecture, every two TSVs are paired up to form an oscillation ring. Moreover, a technique called sensitivity analysis is used to further derive the propagation delay of each individual TSV involved in an oscillation ring.

In [34], the authors propose a method for generating the digital test sequences for short and open faults at TSVs. The defects in the TSVs are modeled at the electrical level by considering neighboring TSVs, drivers (inverters), and buffers. The authors derive that the numbers of faults cannot be detected only through voltage comparing. Therefore, quiescent power supply current (IDDQ) test and the test applying various levels of voltage are required to detect the faults at the TSVs.

In [35, 36], the authors propose a test architecture adopting an additional bypass mode to the GSF for avoiding the extra clock stages. Also, they conducted retiming at both die and stack level to recover the extra delay added to TSV paths by a DFT insertion technique. The experimental results using a stuck-at ATPG tool indicate that the wrapper insertion and the retiming do not affect the resultant pattern count.

In [37], the authors propose a method for testing and characterizing the delay of interposer wires. In their method, a data analysis flow based on oscillation test is developed to adapt the variation of wire length of the interposer wires. In [38], the authors propose a reconfigurable in-field repair solution that can tolerate the impacts induced by TSV defects on reliability.

In [39], the authors propose techniques for screening open defects while considering the impact of open defects on TSVs. They estimate that the additional delay introduced is due to a resistive-open defect as well as due to rerouting based on spare TSVs. The authors also propose an optimization method based on ILP formulation that allocates spares to functional TSVs so that the additional delay due to the rerouting does not exceed an upper limit.

Test scheduling for 3D-IC also has been investigated deeply to handle the test cost (referred to test time or test length). Many optimization solutions for scheduling tests have been proposed.

In [40, 41, 42], the authors study the constraints of maximum power consumption and thermal in 3D-IC testing. The authors also propose solutions for optimizing the test scheduling for minimizing the test time. In [42], the authors propose two test-scheduling methods denoted by partial overlapping and rescheduling. They aim to minimize the total test application time (TAT) under power constraints, which involves the pre-bond test time and the post-bond test time. By the partial overlapping approach, different dies are tested concurrently in the post-bond testing. The rescheduling approach partitions the test sessions during pre-bond test, and it recreates a post-bond test schedule so as to increase parallel test sessions for reducing the total TAT. In [41], the authors addressed the issue of test time minimization with test scheduling while taking temperature constraints into consideration and formulated the test scheduling under thermal constraints using the superposition principle. In [40], the authors introduce a test partitioning method for reducing the test time of the 3D-IC under thermal constraint. The proposed partitioning method allows partitioned tests to be applied in parallel, though original tests cannot be applied in parallel due to the limitation of temperature. Each test is partitioned when the temperature of the test exceeds the specified temperature.

## 8.5 An Accurate Resistance Measuring Method for TSVs in 3D-IC

In this section, we introduce the research progress provided by our research group in Japan. We describe a new method to measure the resistance of high-density post-bond TSVs including serial micro-bumps and bond resistance. The key idea of the new technology is to use Electrical Probe embedded in stacked silicon dies. It is a measuring circuit based on analog boundary scan (IEEE 1149.4). We modify the standard analog boundary-scan structure to realize high measuring accuracy for TSVs in 3D-IC. The main contribution of the method is to measure the resistance of high pin-count (e.g., >10,000) post-bond TSVs accurately. Electrical Probe correspond to the high density of TSV (e.g., <40 μm pitch) and work like a Kelvin probe. The measurement accuracy is less than 10 mΩ.

We have already published papers, [43] ©2014 IEICE and [44, 45] ©2014 JIEP, about this method. We quoted and summarized the contents of the papers with permission from these societies; refer the original papers for the details.

### 8.5.1 Background of Our Study

Manufacturing defects related to TSVs (e.g., voids or pinholes of TSV contamination or height variation of micro-bumps, misalignment during bonding, etc.) cause interconnect failures between dies in 3D-IC [17]. Boundary-scan test (IEEE 1149.1) is normally used to detect interconnect failures of post-bond dies in 3D-IC (see Fig. 8.1) [23]. The boundary-scan test is useful for detecting complete open/short

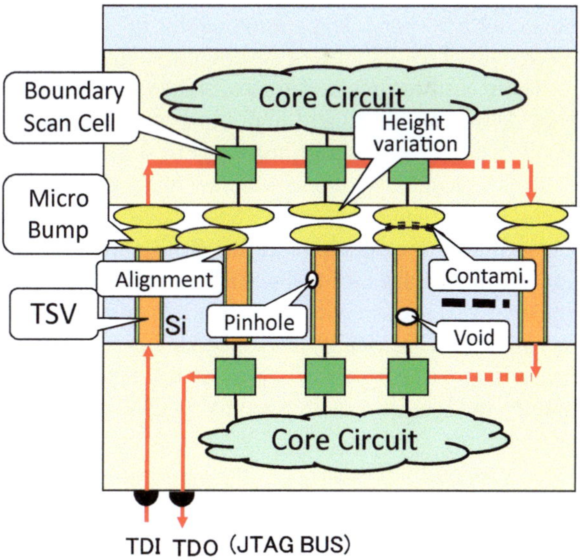


**Fig. 8.1** Manufacturing defects and boundary-scan test in 3D-IC. (With permission from IEICE)

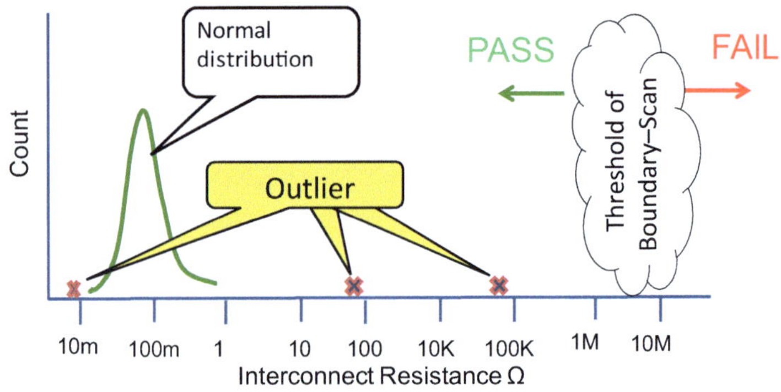


**Fig. 8.2** Outlier of TSV-based interconnects. (With permission from JIEP)

faults, but it is not able to detect resistive-open/short faults. Furthermore, it is not able to measure the value of the resistance. However, it is very important to measure the TSV-based interconnect resistance (TSV, micro-bumps/Cu-pillars, and bonding resistance) for the evaluation of manufacturing processes. If it is possible to measure the value of each TSV-based interconnect resistance individually, we can find out the outliers of a standard deviation (see Fig. 8.2). By eliminating the cause of the outliers, we can improve the manufacturing process, and then we can get a high yield of 3D-IC.

The conventional typical measuring methods for TSV resistance are:

1. Measuring, all together, many TSVs connected as a daisy chain [46]
2. Kelvin measurement using four test pads for one TSV with Mechanical Probe [47]

Such methods are not able to be applied in measurements of high-density and high pin-count post-bond TSVs individually in 3D-IC, so they are used for pre-bond mechanical TSV samples or several pin-count measuring in 3D-IC.

The measuring method for resistance of TSV-based interconnect resistance in high density (<40μm) and high pin-count (>10,000) TSVs at post-bond stack in 3D-IC has not been established yet (Fig. 8.3).

In this chapter, we describe a new method that is able to measure the resistance of the high-density post-bond TSV-based interconnects individually. The scope of the resistance measurement in this section includes TSV, micro-bumps/Cu-pillars, and bonding. Back-end-of-line (BEOL) and re-distribution line (RDL) are not included in this section because they are mature technologies. The proposed technology is to use Electrical Probe embedded in the stacked silicon dies. The Electrical Probe is a measuring circuit based on the analog boundary scan (IEEE 1149.4). We modify the standard analog boundary-scan structure to realize high-measurement accuracy for TSVs in 3D-IC. We believe that the proposed method is useful for yield learning of 3D-ICs.

The remainder of this section is organized as follows: Section 8.5.2 describes the resistance measuring method using conventional analog boundary-scan technology.

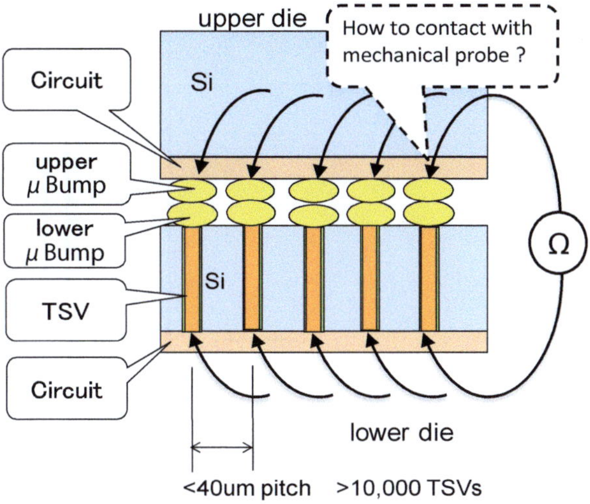


**Fig. 8.3** Internal probing of a 3D-IC. (With permission from JIEP)

This section also describes the problems of the conventional analog boundary scan at measuring the small resistance value of TSVs. Section 8.5.3 proposes methodologies to solve the problems. Finally, Sect. 8.5.4 summarizes the proposed method.

## *8.5.2 Problems of Conventional Analog Boundary Scan for TSV Resistance Measurement*

This section describes the resistance measuring method using conventional analog boundary-scan technology and then describes problems in measuring TSV-based interconnect resistance.

### 8.5.2.1 Analog Boundary Scan

Analog boundary-scan IEEE 1149.4 [48] is a standard that is extended to analog circuits on the basis of digital boundary-scan IEEE 1149.1 [49]. Analog boundary scan has the capability of not only interconnection test of analog nets but also measuring the L, C, or R value of components connected between ICs [50]. Figure 8.4 shows the overall architecture of analog boundary scan. An analog boundary module (ABM) is implemented between an analog I/O pin and the analog core. ABMs and a test bus interface circuit (TBIC) are connected by an internal analog BUS (AB1, AB2). Analog signals on AB1 and AB2 pass through the TBIC to an external analog BUS (AT1, AT2). AT1 and AT2 are connected to external analog measurement equipment. Figure 8.5 shows the standard ABM circuit, which consists of six analog switches (A-SW) and one comparator.

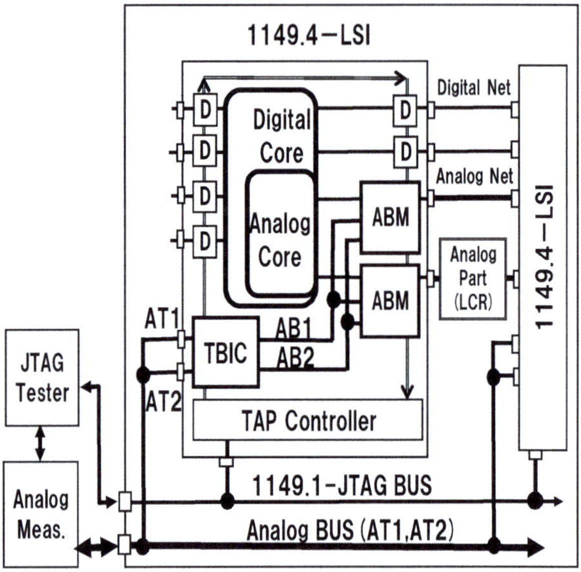


**Fig. 8.4** Analog boundary-scan circuit. (With permission from IEICE)

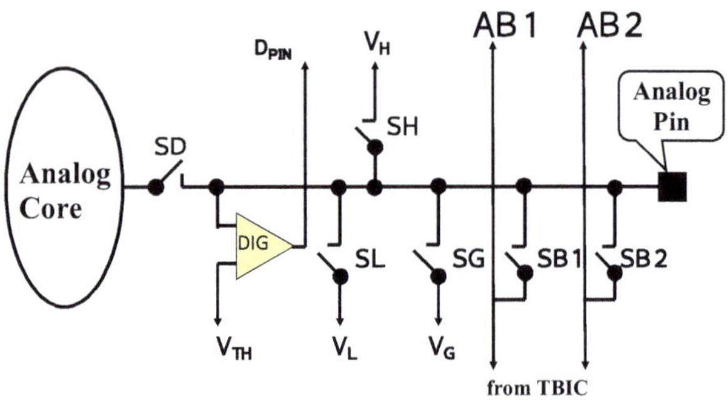


**Fig. 8.5** ABM circuit. (With permission from JIEP)

### 8.5.2.2 Standard Resistance Measuring Method by 1149.4

Figure 8.6 shows the measurement circuit configuration and principles of the IEEE 1149.4 standard. Measuring procedure is as follows:

*Step 1* All A-SWs of the ABMs and TBICs are set as shown in Fig. 8.6a, and the ground-referenced voltage $V1$ at the left side of resistance $Z$ is measured by voltmeter $V$ while constant current $I_S$ is applied to $Z$ by a constant current source.

*Step 2* All A-SWs of the ABMs and TBICs are set as shown in Fig. 8.6b, and the ground-referenced voltage $V2$ at the right side of resistance $Z$ is measured by voltmeter $V$ while constant current $I_S$ is applied to $Z$ by a constant current source.

*Step 3* A control computer calculates by Eq. (8.1) and provides the resistance value of $Z$.

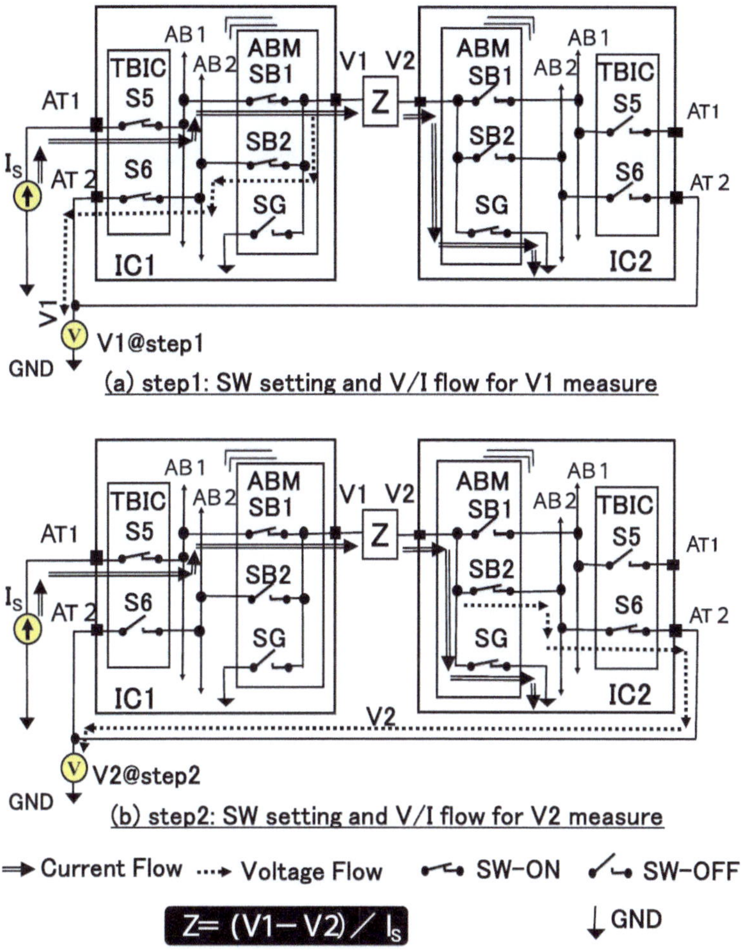


**Fig. 8.6** Standard resistance measuring method. (With permission from JIEP)

$$Z = (V1 - V2)/I_S \tag{8.1}$$

*It is important to note that V1 and V2 measurements are referenced to ground level.*

#### 8.5.2.3 Problems of Conventional Analog Boundary Scan for TSV Resistance Measuring

IEEE 1149.4 Standard document [48] at Chap. 9 describes as "An objective of this standard is to facilitate measurement of complex impedances with an accuracy better than ±1 % when measuring impedances 'between 10 Ω and 100 kΩ'." However, the TSV-based interconnect resistance value is *several tens or hundreds of milliohms*, which is far smaller than the expectation of the standard. Therefore, if we try to measure the small TSV-based interconnect resistance using the conventional 1149.4 standard as is, the measurement accuracy will be unacceptable or the resistance will be immeasurable in the worst cases due to the following problems:

a. Problem of the ground-referenced voltage measuring method

   The TSV-based interconnect resistance value (several tens or hundreds of milliohms) is less than one thousandth of the resistance value of A-SWs in the ABMs and TBICs (several hundreds or thousands of ohms) . Therefore, a voltage drop across the resistance *Z* by applying constant current $I_S$ is less than one thousandth of the voltage drop across the A-SWs. If the accuracy of measurement for *V*1 and *V*2 in Fig. 8.6 is 0.1 % (one thousandth), the voltage measurement error and the voltage drop across the resistance *Z* will be almost the same. As shown in Fig. 8.7, if the measurement error ($\varepsilon 1$, $\varepsilon 2$) of the ground-referenced voltages *V*1 and *V*2 are equal or larger than the difference between *V1* and *V*2, the calculation results by Eq. (8.1) for the resistance *Z* will have a large error or does not make sense (e.g., negative resistance). Hence, the ground-referenced voltage measuring method is not practical for measuring the TSV-based interconnect resistance whose resistance value is quite small.
b. Problem of wiring resistance in the die
   Wiring (BEOL and RDL) from an ABM circuit to a micro-bump (or TSV) in the die has a resistance whose value might be several tens or hundreds of milliohms. This causes large errors in measuring the TSV-based interconnect resistance value (several tens or hundreds of milliohms), because the resistance of the wiring is added to the value of the measured resistance *Z*. If an electro-static-discharge-protection (ESD-protection) resistor exists in the wiring, the measurement error is further increased.
c. Problem of leak current of A-SWs in the case of large number of TSVs
   In the case of measuring resistances of a large number of TSVs (e.g., tens of thousands), the ABMs same as the number of TSVs are connected to the internal analog BUS (AB1, AB2), and consequently the effect of the leak current of A-SWs increases, therefore the measurement error increases.

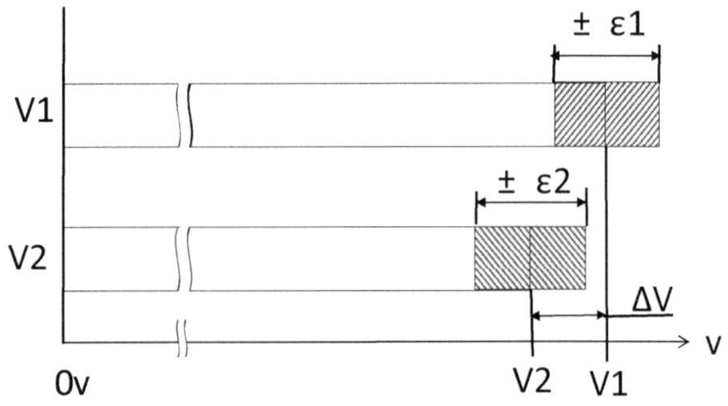


**Fig. 8.7** Measurement error in ground-referenced voltage measuring method. (With permission from IEICE)

### 8.5.3 Proposed Measuring Method

As shown above, the relatively high value of A-SW resistance in ABMs makes it impossible to accurately measure extremely small TSV-based interconnect resistances using conventional analog boundary-scan methods. Therefore, we propose a new method to enable measuring the TSV-based interconnect resistance accurately by modifying the conventional analog boundary-scan method.

a. Floating Measurement Method

The conventional measuring method causes significant measurement error by measuring ground-referenced voltage. We propose a new measuring method utilizing floating measurements to minimize the error. The circuit configuration and setting of A-SWs to realize this method is shown in Fig. 8.8. In this way, both constant current application and voltage measurement are enabled to float from ground. This enables direct measurement of the voltage drop $\Delta V$ across the resistance $Z$. The small voltage drop across the resistance $Z$ can, thus, be measured with high resolution. It also enables highly accurate measurement since it is not affected by ground noise. Furthermore, it needs to measure only one time, and then the test time is shortened.

b. Complete Isolation of the Current Path and the Voltage Path

In order to measure the voltage across the resistance $Z$ at the closest position, the method isolates the current application path and the voltage measurement path completely. One of the specific routing methods is to separate one measuring line in the ABM circuit to two lines (current and voltage) and connect these lines to the terminal of resistance $Z$ through BEOL and RDL in the die. The other method is to use two ABMs for one terminal of resistance $Z$ and to wire the line from each ABM to the $Z$ terminal through the BEOL and RDL in the die. In the case of the latter, redesign of the ABM macro is not required, and the ESD protection resistance is not a factor. The proposed method allows measuring the $Z$ accurately without measuring error caused by the resistances of BEOL, RDL, or the ESD-protector.

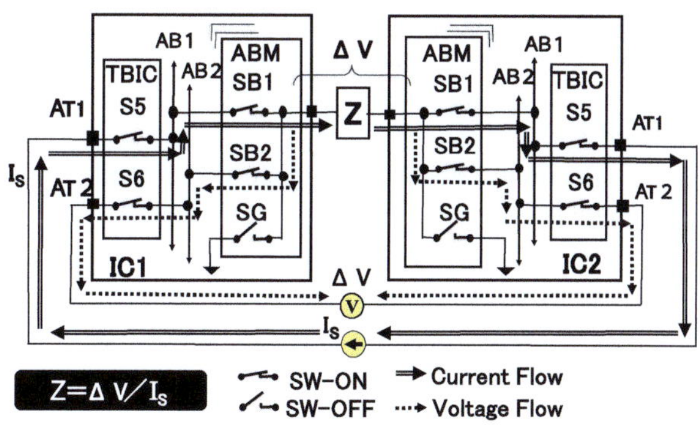


**Fig. 8.8** Floating measuring method for TSVs. (With permission from JIEP)

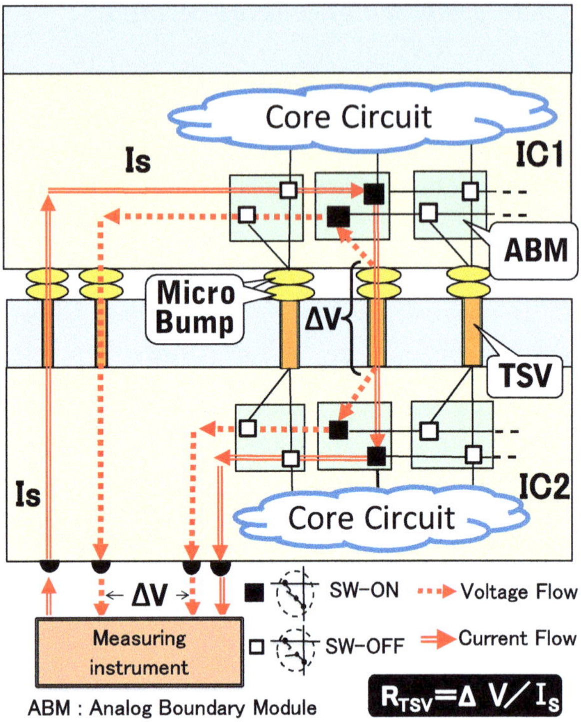


**Fig. 8.9** Floating measurement and *V*/*I* path isolation. (With permission from JIEP)

An example circuit of 3D-IC implemented in the proposed method (A) and (B) is shown in Fig. 8.9. Note that TBIC and control logic circuits are omitted.

c. Segmenting the Internal Analog BUS (AB1, AB2)

In order to avoid the influence of A-SW leak currents, a number of the ABMs (thousands or tens of thousands) connected to the internal analog BUS (AB1, AB2) are grouped, and the internal analog BUS is segmented. Then a TBIC circuit is used in each segmented internal analog BUS. An example of an internal analog BUS divided into **m** segments is shown in Fig. 8.10. By the segmentation, the penalty of the leakage current to measurement error decreases to 1/m. The desired number of segments is determined by the leak current characteristic of the A-SW in the die.

As mentioned above, the proposed method can decrease measurement error caused by A-SW leak current.

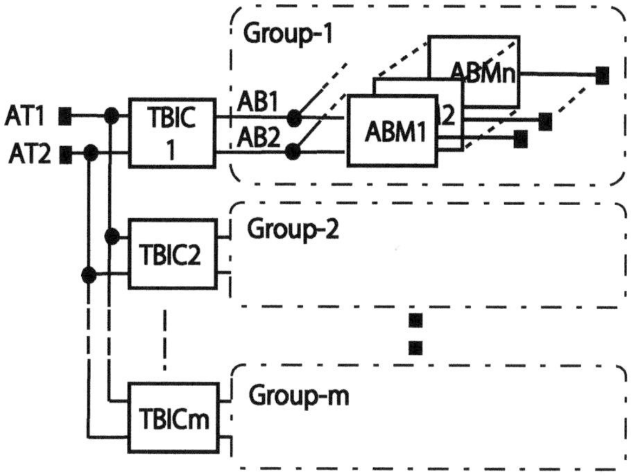


**Fig. 8.10** Grouping the internal analog buses. (With permission from IEICE)

### *8.5.4 Summary*

In this section, we have provided a highly accurate measuring method for the small resistance value of the TSV-based individual interconnect between dies in post-bond 3D-IC as the outcome of our research group. The proposed method uses an Electrical Probe, which is composed of modified analog boundary-scan circuit.

Tom DeMarco said, "You can't control what you can't measure." In order to improve the yield of 3D-IC, it is necessary to measure the TSV-based interconnect resistance accurately.

## 8.6 Delay Measurement Circuits for Detecting TSV Delay Faults

In this section, we describe a test method that utilizes a time-to-digital converter (TDC) embedded in boundary-scan design to estimate the excessive delay occurs at interconnects in 3D-IC presented by us. The boundary-scan cells are used as an input generation block in the lower die and as a delay detection block in the upper die. The other testing method is also proposed to utilize the effect of adjacent TSVs to enhance the fault effects.

### *8.6.1 Application of Time-to-Digital Converter Embedded in Boundary Scan for 3D-IC Testing*

We have proposed a post-bond testing method to detect delay caused by an open TSV using a TDC embedded in boundary-scan design [51]. To detect small delay

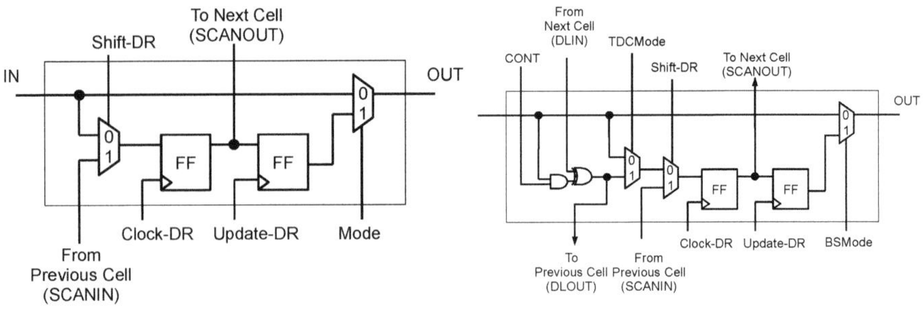


**a** Basic Boundary-Scan cell(BC_1) **b** TDCBS cell(BC_TDC)

**Fig. 8.11** A boundary-scan cell with TDC. *FF* flip flop

faults, we proposed a boundary-scan cell, called TDCBS cell, which can form a time-to-digital converter to measure the delay of incoming paths [52]. In 3D-IC testing, it can be used for detecting an excessive delay by measuring the delay of a path in a lower die and a TSV.

A TDCBS cell is designed based on a basic boundary-scan cell BC_1 in Fig. 8.11a [53]. In our DFT method, the capture FF in a boundary-scan cell is also utilized to form a TDC. The delay line used in a TDC is composed of embedded XOR gates. Figure 8.11b shows the new boundary-scan cell called BC_TDC cell. By setting TDCMode = 0, the BC_TDC cell can be used as a basic boundary-scan cell. By setting TDCMode = 1, BC_TDC cells can form a delay line loop as shown in Fig. 8.12. The transition signal of the *i*th output of core A is selected to be observed by CONT[*i*] = 1. The delayed responses are captured at FFs. Difference of timing slack between fault-free and faulty circuits can be detected. Figure 8.13 shows an example of detecting the delay of a path. In this example, the rising transition at output OUT[1] is selected to be observed. The control circuit selects the output by providing the control signals CONT[1] = 1, CONT[*j*] = 0 ($j \neq 1$), and LOOPCNT = 1. The transition signal goes through the delay line loop. The number of FFs, Nslack, at which H is captured (i.e., the FFs captured the signal after the rising transition finished), is observed. The timing slack, Tslack, can be estimated as follows:

$$\text{Tslack} = \text{Nslack} * \text{td}, \tag{8.2}$$

where td is the delay of an XOR gate in the delay line.

Figure 8.14 shows the overall architecture of a boundary-scan circuit with TDCBS cells. It includes the standard JTAG test access ports (TAP), which are Test Mode Select (TMS), Test Clock (TCK), Test Reset (TRST), Test Data In (TDI) and Test Data Out (TDO). The signal LOOPCNT is used to connect the terminals of the delay line to form a loop. Using a delay line loop, the delay of the transition signal can be observed even if the input is connected near the end of the delay line.

In 3D-IC testing, TDCBS cells are placed in both the lower and upper dies as shown in Fig. 8.15. The TDCBS cells at the input of the middle/top die are used for post-bond test to measure the delay of the core in the bottom die and the TSVs

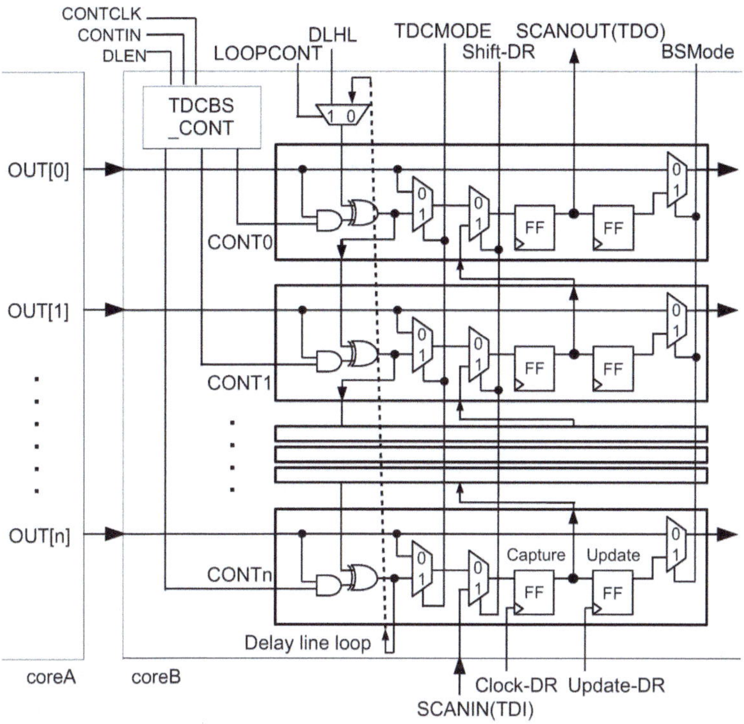


**Fig. 8.12** A delay line loop configured by TDCBS cells. *FFs* flip flops, *TDCBS cell* TDC boundary-scan cell

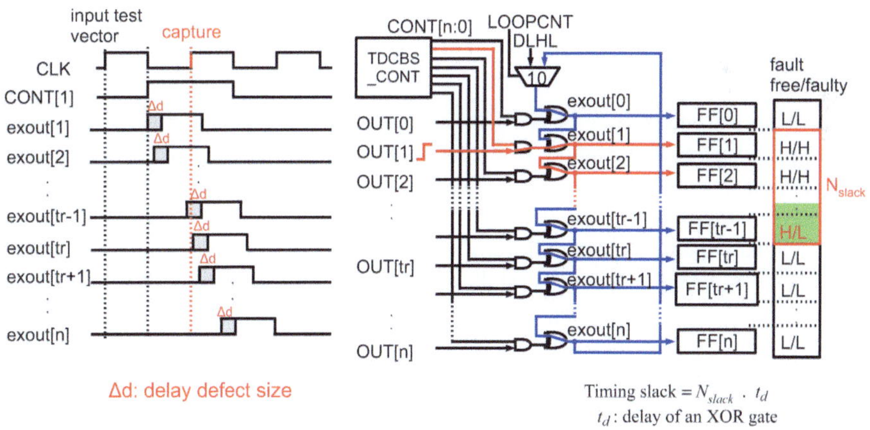


**Fig. 8.13** Detection of delay fault using a TDC configured by TDCBS cells. *FFs* flip flops, *TDCBS cell* TDC boundary-scan cell

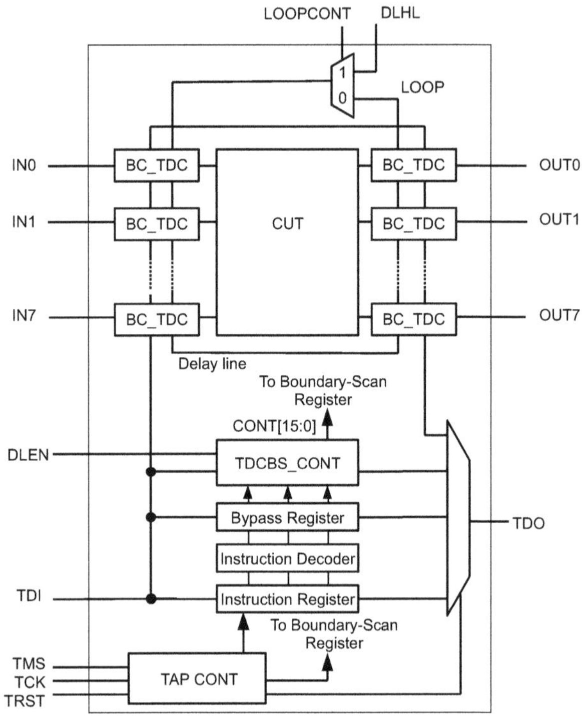


**Fig. 8.14** The overall architecture of a boundary-scan circuit with TDCBS cells. *BC_TDC* boundary-scan cell, *CUT* circuit under test

between dies by bypassing the transition at the paths inside a CUT at the lower die and providing it to a TSV.

### 8.6.2 Delay Measurement Circuit Using the Vernier Delay Line

Since delay caused by a defective TSV is usually very small, it is difficult to detect it. We propose another method for detecting an open defect occurred at a TSV considering the effects from adjacent TSVs using Vernier delay line (VDL) which can measure the delay with higher timing resolution than TDC [54]. Figure 8.16 shows a model of an open TSV with adjacent TSVs used in our work. The signal at an open TSV is strongly affected by crosstalk compared with a fault-free TSV, especially when it is in a TSV array [55].

Figure 8.17 shows the proposed measurement circuit. It can provide test signal from the bottom die through TSVs to middle/top die in the test mode. In test mode, a transition signal and the inverted transitions are provided to the target TSV and its adjacent TSVs, respectively, in order to increase the delay occurs at a defective TSV. The difference of the delay is measured using VDL that can measure the delay in more detail than TDC [56].

In test mode, all TSVs are separated from the cores and connected to VDL by the transmission gates when applying TESTEN=H.

The rising transition is provided from TESTIN to the target TSV. The signal is inverted by an XOR gate for the other TSVs to provide the falling transition at

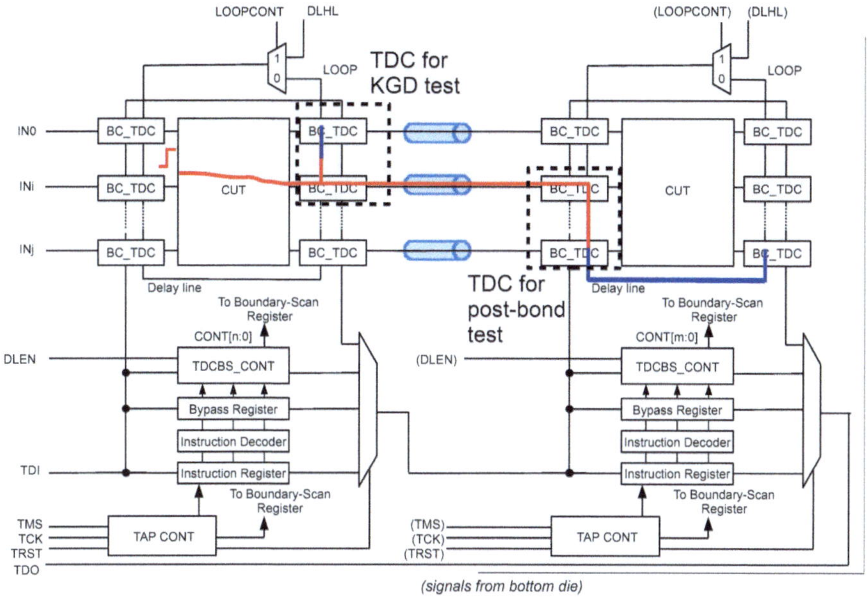


**Fig. 8.15** An example of testing a TSV using TDCBS circuit. *TDC* time-to-digital converter, *KGD* known good die, *BC_TDC* boundary-scan cell, *CUT* circuit under test

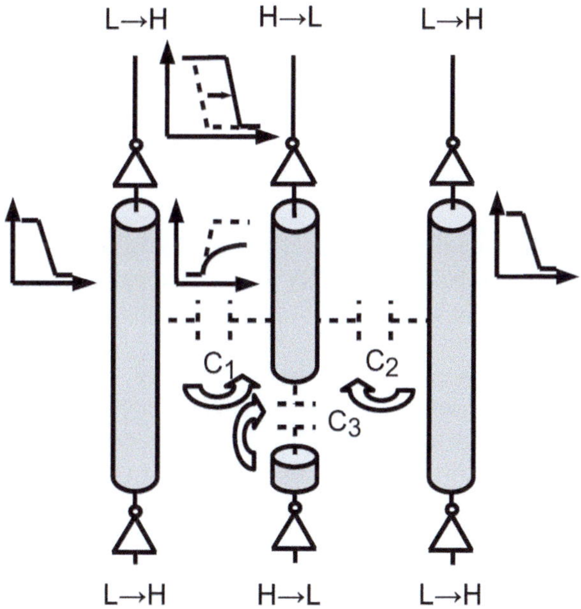


**Fig. 8.16** A model of an open TSV with adjacent TSVs

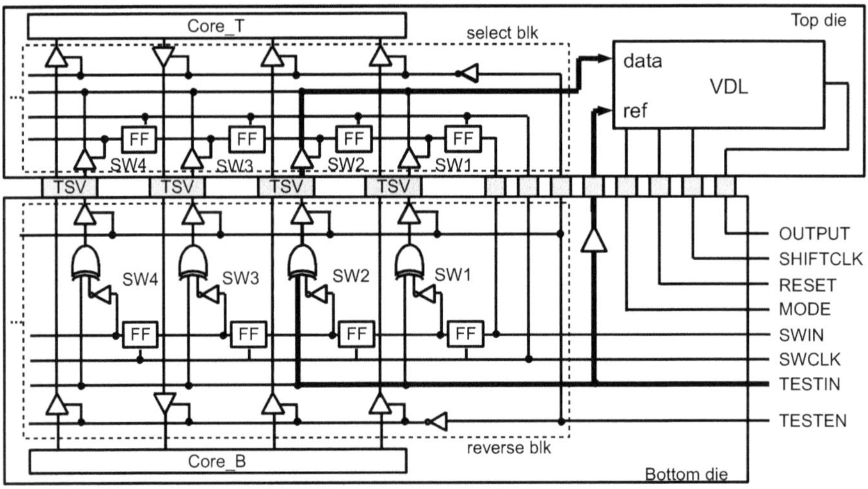


**Fig. 8.17** Fault detection circuit using VDL for detecting delay at an open TSV. *FFs* flip flops, *TSV* through-silicon via, *VDL* Vernier delay line, *SW* switch

the TSVs adjacent to the target TSV. A target TSV is selected by a shift register controlled by SWIN and SWCLK. The signal from TESTIN is propagated through the target TSV to the delay measurement circuit including a VDL. The signals SHIFTCLK, RESET, and MODE are used for controlling the VDL. The delayed transitions captured at the FFs in the VDL are observed through OUTPUT terminal. By comparing the delay obtained for the other TSVs, the target TSV is identified as faulty if the excessive delay is observed.

### 8.6.3 Estimation of Defect Size Detectable by the Test Method

To estimate the delay size detectable by the delay measurement circuit, we apply simulation based on the TSV array layout shown in Fig. 8.18. The faulty TSV with a soft-open defect is placed in the center of the 3 × 3 TSV array. The resistive-open defect is inserted in a TSV as shown in Fig. 8.19. The size of a TSV is based on ITRS 2011 roadmap. The resistive open inserted in a TSV is modeled as an open that is partly connected by 0.5 μm square area and height dh (nm) as shown in Fig. 8.19. S-parameter of the layout in Fig. 8.18 is extracted using the 3D electromagnetic simulator EMPro for several different dh values. We designed a circuit layout of select_blk and reverse_blk blocks in the proposed test circuit in Fig. 8.17 and extracted a SPICE net-list with RC components. Monte Carlo simulation is applied to estimate the effects of process variations for detecting an open TSV. The rising transition is provided to the defective TSV and is observed by VDL. Table 8.1 shows the results of the delay measurement using VDL for the fault-free circuit and the faulty circuits with a defective TSV. Column dh shows the size of soft opens inserted in the target TSV. Column $T_{FF}$ shows the number of FFs in the VDL that can capture H value. Column $N_{Sim}$ shows the number of simulations obtained in the results among

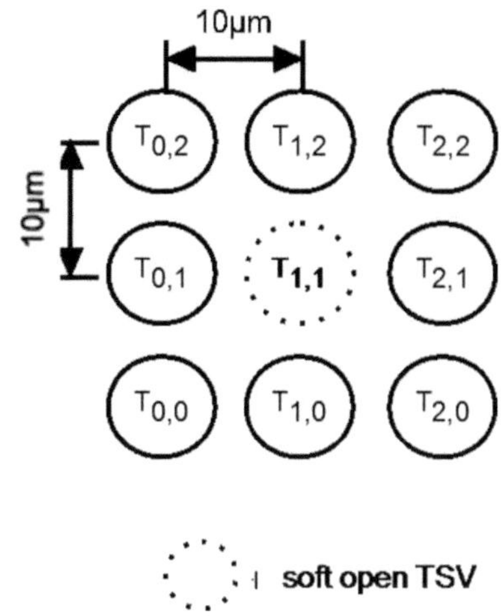


**Fig. 8.18** TSV array layout. *TSV* through-silicon via

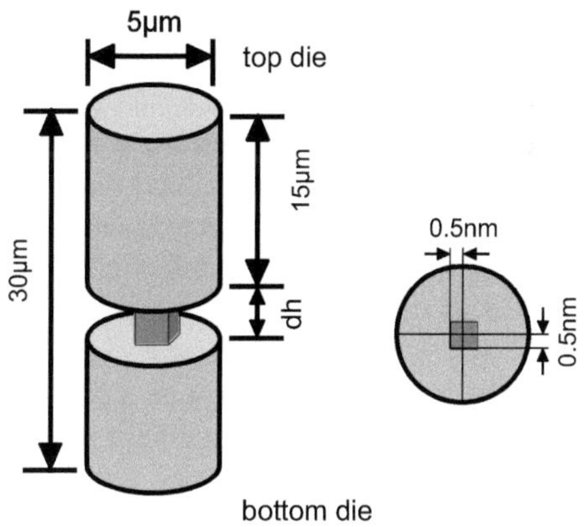


**Fig. 8.19** Soft-open model of a TSV

**Table 8.1** The experimental results obtained by the VDL

| *dh*(nm) | $T_{FF}$ | $N_{Sim}$ |
|---|---|---|
| normal | 7 | 18 |
| | 8 | 3 |
| 400 | 8 | 21 |
| 500 | 8 | 19 |
| | 9 | 2 |
| 600 | 8 | 7 |
| | 9 | 14 |
| 700 | 9 | 21 |
| 800 | 9 | 6 |
| | 10 | 15 |

21 Monte Carlo simulations. Since $T_{FF}=7.8$ is obtained for the fault-free circuit, the defect size dh≥700 nm is expected to be detected.

### 8.6.4 *Summary*

In this section, we have proposed two delay measurement circuits to detect excessive delay occurred at a TSV. One utilizes a TDC embedded in boundary-scan circuit. The other utilizes a VDL to measure the delay occurred at a TSV. Using electromagnetic simulation and circuit simulation, it is estimated that the delay caused by the resistive open in the TSV whose size is more than about 2.3 % of its depth, is expected to be detected using the proposed test circuit.

## 8.7 Electrical Interconnect Tests of Open Defects in a 3D-IC with a Built-In Supply Current Test Circuit

In this section, we describe a supply current test method and a built-in test circuit to detect open defects occurring at interconnects between dies in a 3D-IC. The test method and the test circuit have been proposed in [57]. Also, feasibility of tests with the test circuit has been examined by SPICE simulation. The simulation results show us that a hard-open defect and a soft-open one generating additional delay of 0.58 ns can be detected at a test speed of 100 MHz [57]. In order to examine feasibility of our tests by some experiments, we examined whether open defects could be detected in a circuit on a printed circuit board (PCB) made of a prototype IC in which the built-in test circuit was implemented. The experimental results are denoted in this section.

### 8.7.1 *Electrical Tests with a Built-In Supply Current Test Circuit*

It can be assumed that a 3D-IC is made of KGDs , since dies in the IC have been tested fully before stacking process. Open defects can occur at interconnects between KGDs inside a 3D-IC in the stacking process. Thus, we have developed how to detect open defects occurring at the interconnects after the stacking process [58].

Typically, IEEE 1149.1 test architecture is introduced in dies of 3D-IC so that interconnects between the dies can be tested by a boundary-scan test method. Thus, our test method utilizes the architecture for providing test input vectors in the tests.

Open defects can be classified into "hard-open defects" and "soft-open ones." In the case of a hard-open defect, an interconnect is divided into two parts completely and they are not connected each other. In the case of a soft-open defect, the parts are connected with each other in part electrically. A soft-open defect may be called a "weak open defect" [59]. The defect may be caused by a void or a crack in a TSV [60].

Open defects that generate logical errors will be able to be detected by a boundary-scan test method. However, soft-open defects that generate timing errors may not be detected by the test method.

Soft-open defects will reduce reliability of an IC. They will result in some increase of the propagation delay time. They will grow and can change into hard-open defects. Since a hard-open defect may generate a logical error [61] and also a timing error, a soft-open defect should be detected before it becomes a hard one. Thus, we have proposed an electrical test method with which hard-open defects and soft ones can be detected in [58]. We call the test approach "ET-Scan" (electrical test with scan FFs). The test method requests us to design ESD input protection circuits inside a die so as to be tested by the test method. We had prototyped an IC embedding in the ESD input protection circuits and shown feasibility of the tests with the prototype IC by some experiments [58].

Modification of ESD input protection circuits may not be accepted in real die design. Also, the test method proposed in [58] is applicable to tests of 3D-ICs, each of whose interconnects between dies is connected to only another one. If an interconnect is connected to more than one interconnect, an open defect occurring at one of the destinations may not be detected by the test method proposed in [58]. Thus, we developed a built-in test circuit for the supply current test method and examined the feasibility of our electrical tests with the circuit by SPICE simulation [57]. The results reveal us that hard-open defects and a resistive-open defect of 200 Ω occurring at an interconnect between two dies can be detected at a test speed of 100 MHz with our test circuit.

It is indispensable to examine feasibility of the tests with a real circuit. Thus, we prototyped a testable designed IC and built a circuit on a PCB made of the IC. We examined whether resistive-open defects and capacitive ones could be detected by some experiments with the circuit. We describe the experimental evaluation results in this section.

IEEE 1149.1 test architecture is utilized in our tests to provide test input signals to interconnects between dies in a targeted 3D-IC. A configuration of our test setup is shown in Fig. 8.20. As shown in Fig. 8.20, the control signals, *TDI, TCK, TMS,* and *TRST* are provided to a device under test (DUT) from a tester.

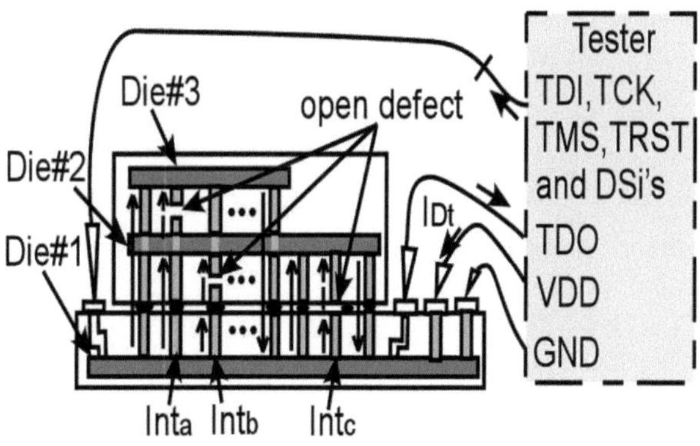


**Fig. 8.20** Test setup in our tests

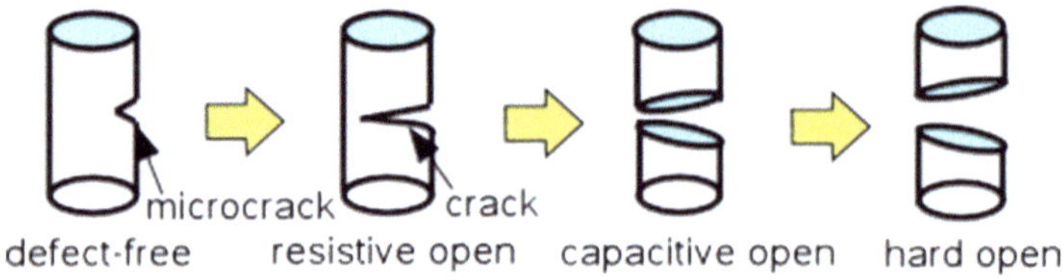


**Fig. 8.21** Generation process of open defects in a TSV

A generation process of a hard-open defect in a TSV is shown in Fig. 8.21. As shown in Fig. 8.21, a micro crack will grow to be a crack by electro migration or stress migration.

The TSV can be modeled as the one having a resistive-open defect. It will become a hard-open defect. The defective TSV is divided into two parts. Since the space between the parts is small, a high-speed logic signal will pass through the defective TSV [55]. The TSV can be modeled as the one having a capacitive-open defect. Finally, the defect will become a hard-open defect and a logic signal will not be propagated through the defective TSV. The TSV can be modeled as an interconnect having a hard-open defect.

Since capacitive and resistive-open defects and hard-open ones should be detected before shipping to a market, we select them as our targeted defects.

Principle of our test method proposed in [58] is shown in Fig. 8.22. The test circuit requests us to add a test circuit for the tests, which is depicted as an "Added Circuit Block" in Fig. 8.22. Interconnects are tested by measuring quiescent supply current, $I_{Dt}$, of the DUT.

When input interconnects connecting to a die are tested, a high-level (*H*) signal is provided to a test input terminal, *TST*, of the die to select a test mode from a tester, together with an *H* signal provided to a targeted input interconnect and a low-level (*L*) signal provided to input interconnects other than the targeted one. When the IC works in a normal mode, an *L* signal is provided to the *TST* terminal.

An interconnect *b* is targeted in Fig. 8.22. When the defect-free IC is tested by our test method, quiescent supply current, $I_{Dtn}$, will flow whose current path is shown in Fig. 8.22, since all of the nMOSs in the added circuit block turn on and an *H* signal is provided only to the targeted interconnect. On the other hand, when a hard-open defect at the targeted interconnect occurs, the quiescent current will not flow. Thus, if Eq. (8.3) is satisfied in our tests, it is determined that an open defect occurs at the targeted interconnect.

$$I_{Dtc} \leq I_{th}, \tag{8.3}$$

where $I_{Dtc}$ and $I_{th}$ are measured quiescent supply current in the DUT and a threshold value, respectively. $I_{th}$ will be specified from the variation of measured quiescent $I_{Dt}$'s of the defect-free ICs, since $I_{Dt}$ of defect-free IC depends on the variation of electrical characteristics of output protection circuits.

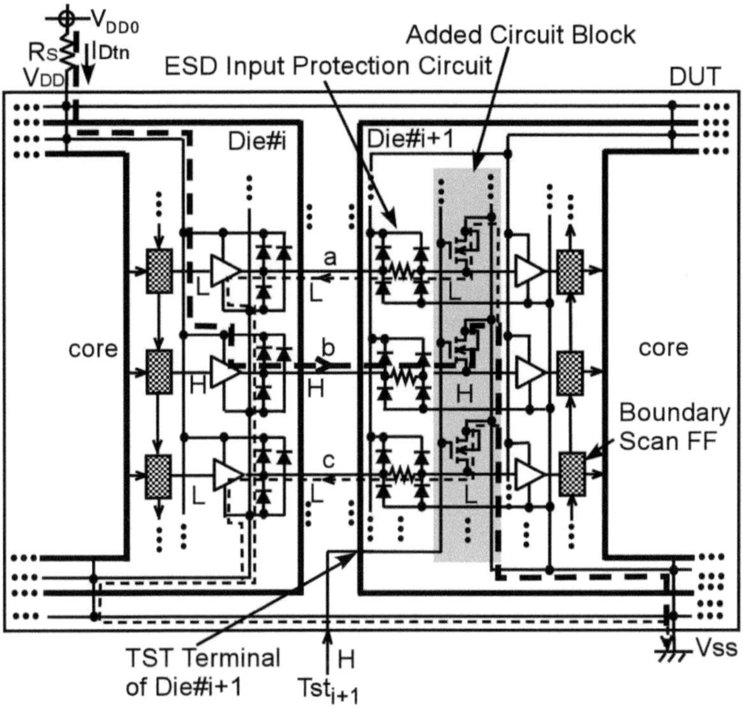


**Fig. 8.22** Principle of our electrical testing. *ESD* electrostatic discharge, *DUT* device under test, *FFs* flip flops

A resistor $R_S$ is added between the source voltage, $V_{DD}$, and the voltage source terminal, when the IC is tested. It is used to make $I_{Dtn}$ small in our tests. If $R_S$ is not added, large $I_{Dtn}$ will flow and the DUT may be destroyed in our tests.

When a soft-open defect, which is modeled as a resistive-open defect, occurs at the targeted interconnect, smaller supply current will flow than the defect-free ICs. It results in the fact that smaller $I_{Dtc}$ is measured than the defect-free ICs. When a soft-open defect, which is modeled as a capacitive-open defect, occurs at it, $I_{Dtc}$ of 0A will be measured. Thus, the soft-open defects are detected by Eq. (8.3).

In our tests, an H signal is provided to only one interconnect as a test input signal of the test method, and supply current flows through the interconnect in defect-free 3D-IC. Thus, if $I_{Dtc}$ is measured which is smaller than $I_{Dtn}$ in our tests, it is conclud-ed that an open defect occurs at the interconnect. That is, defective interconnects will be located by examining which interconnect an *H* signal is provided to when Eq. (8.3) is satisfied.

An interconnect between a primary input terminal of a 3D-IC and a die inside the IC can be tested in the same manner as in Fig. 8.22 by measuring supply current of the IC and logic gates in the tester that provides test input vectors to the IC. Also, an interconnect between a primary output terminal of a 3D-IC and a die inside the IC can be tested by connecting the 3D-IC to an IC in a tester whose configuration is the same as *Die#i*+1 in Fig. 8.22 and outputting our test input vectors from the primary output terminals.

### 8.7.2 Experimental Evaluation of Our Electrical Test Method

In order to evaluate whether open defects could be detected with our built-in test circuit, we designed a layout of a die that was designed by using our testable design method with a 0.18-μm complementary metal–oxide–semiconductor (CMOS) process of Rohm Co. Ltd. We prototyped the IC and built a circuit made of the IC on a PCB, whose circuit was shown in Fig. 8.23. The following are used as source voltages that are specified by the CMOS process used in our layout design: $V_{DD0}=3.3$ V, $V_{DD}=1.8$ V.

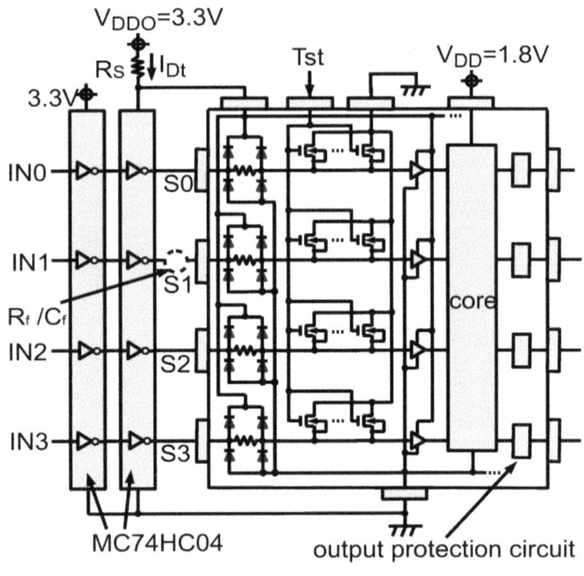


**Fig. 8.23** Experimental circuit

We inserted a hard-open defect at an interconnect *S1* by eliminating it from the defect-free circuit in order to examine the testability of a hard-open defect. Also, we inserted a soft-open defect at *S1* by adding either a resistor $R_f$ or a capacitor $C_f$ to the interconnect in the defect-free circuit. Test vectors for *IN0*, *IN1*, *IN2*, and *IN3* in our evaluations are shown in Fig. 8.24.

In our experiments, the test vectors are provided to the circuit per $T_s$ of 1 μs. An IEEE 1149.1 test circuit is not implemented in the IC. However, by providing the test vectors to *IN0*, *IN1*, *IN2*, and *IN3*, the same operation can be realized as the circuit in which an IEEE 1149.1 test circuit is implemented.

**Fig. 8.24** Test input signals

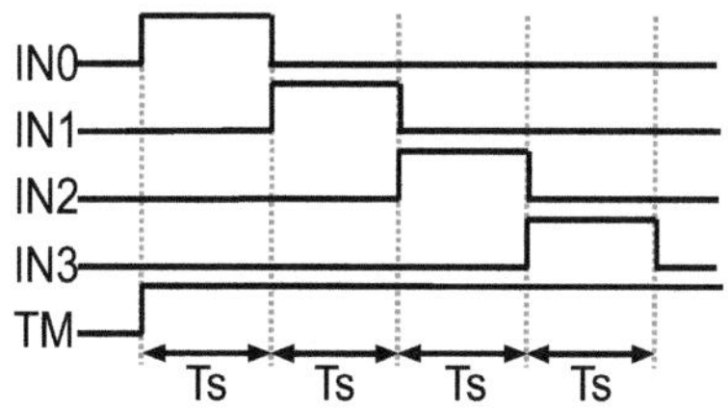


Our measured waveforms are shown in Figs. 8.25, 8.26, and 8.27. As shown in Fig. 8.25a, b, and c, the hard-open defect and the resistive-open one of 1 kΩ at *S1* are able to be detected by our test method, since smaller $I_{Dt}$ appears than in the defect-free circuit when *S1*=*H*. On the other hand, the resistive-open defect of 200 Ω

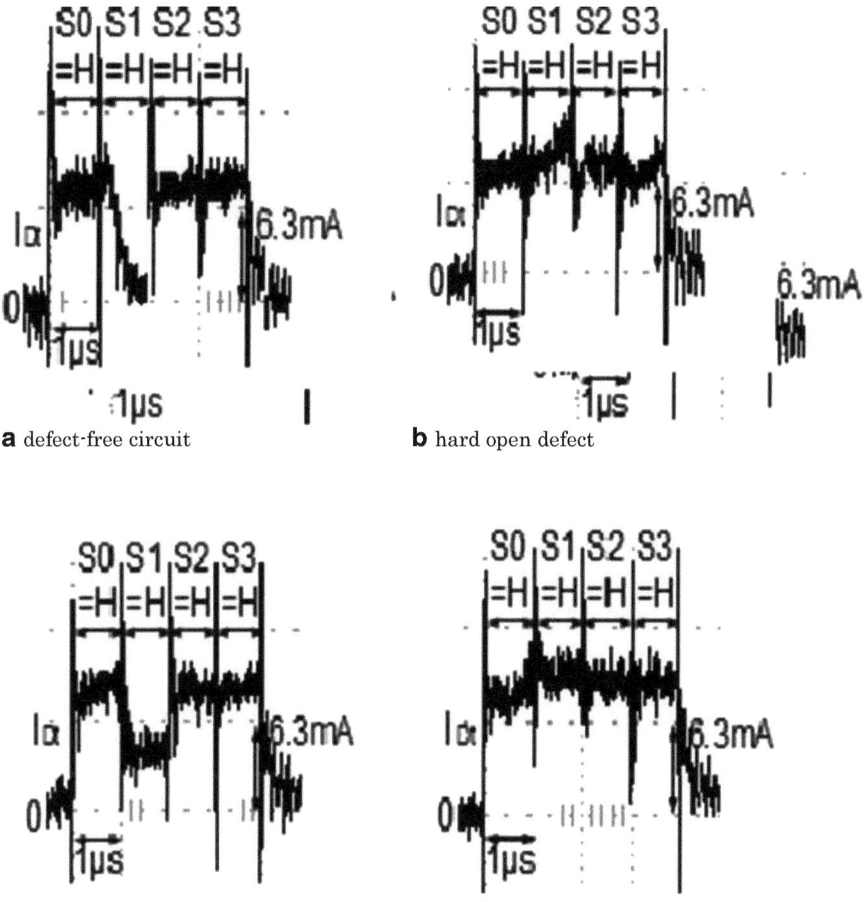


**a** defect-free circuit

**b** hard open defect

**c** resistive open defect of of $R_f$=1kΩ

**d** resistive open defect of $R_f$=200Ω

**Fig. 8.25** $I_{Dt}$ waveforms at test speed of 1MHz for hard-open and resistive-open defects

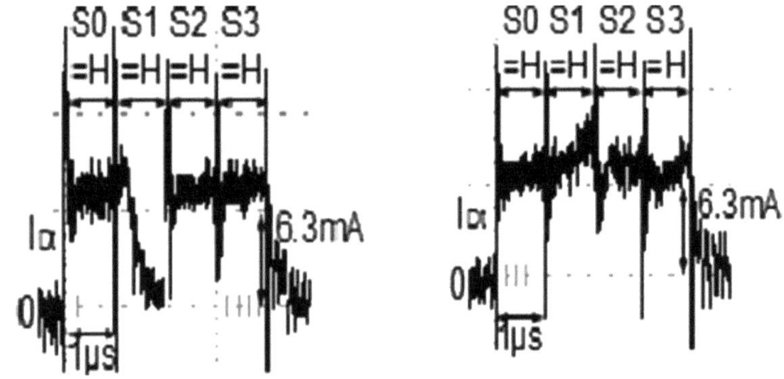


**a** defective IC of $C_f$=1nF **b** defective IC of $C_f$=10nF

**Fig. 8.26** $I_{Dt}$ waveforms at test speed of 1MHz for capacitive-open defects

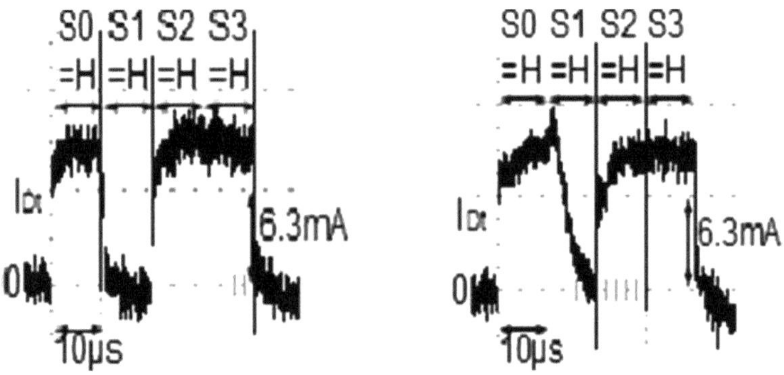


**a** defective IC of $C_f$=1nF **b** defective IC of $C_f$=10nF

**Fig. 8.27** $I_{Dt}$ waveforms at test speed of 100kHz

cannot be detected. It means that a hard-open defect and a resistive-open one whose resistance is greater than 1 kΩ can be detected by our test method.

Capacitive-open defects whose capacitance is smaller than 1 nF can be detected by our test method as shown in Fig. 8.26a. As the capacitance becomes large, $I_{Dt}$ cannot change quickly, and the capacitive-open defect of 10 nF cannot be detected by our test method as shown in Fig. 8.26b. However, it will be detected at a test speed of 100 kHz, as shown in Fig. 8.27. Thus, it is found out from Figs. 8.26 8.27 that testability of capacitive-open defects depends on test speed.

Test speed of our test method based on our built-in test circuit is specified as the time until quiescent $I_{Dt}$ appears. In our experiments, we used a circuit on a PCB as a targeted circuit. If 3D-ICs are used, faster test speeds will be realized. Simulation results denoted in [57] reveal us that open defects can be detected at a test speed of 100 MHz. Thus, it is expected that the open defects can be detected at a test speed that is faster than 1 MHz in a 3D-IC.

### 8.7.3 Summary

We have examined whether open defects occurring at an interconnect between dies in a 3D-IC can be detected by our supply current test method with a circuit made of our prototype IC on a PCB. The experimental results show us that resistive-open defects of greater than 1 kΩ and capacitive ones of smaller than 1 nF can be detected at a test speed of 1 MHz.

Feasibility of tests can be examined with a circuit on a PCB. However, test speed will not be examined in a 3D-IC. Thus, it is a future work to examine the test speed with a 3D-IC.

**Acknowledgments** The authors would like to thank Mr. T. Konishi and Mr. S. Umezu, The University of Tokushima, for their experimental evaluation. This work was supported by VLSI Design and Education Center (VDEC), University of Tokyo, in collaboration with Synopsys, Inc, Cadence Design Systems, Inc, and Mentor Graphics, Inc. The IC chip in this study has been designed in the chip fabrication program of VDEC, University of Tokyo, in collaboration with Rohm Corporation and Toppan Printing Corporation.

## References

1. Noia B, Chakrabarty K (2013) Pre-bond probing of through-silicon vias in 3-D stacked ICs. IEEE Trans Comput Aided Des Integr Circuits Syst 32(4):547–558
2. Noia B, Panth S, Chakrabarty K, Lim SK (2012) Scan test of die logic in 3D ICs using TSV probing. In: Proceedings of IEEE international test conference (ITC), pp 1–8
3. Chen PY, Wu CW, Kwai DM (2010) On-chip testing of blind and open-sleeve TSVs for 3D IC before bonding. In: Proceedings of IEEE VLSI test symposium (VTS), pp 263–268
4. Cho MK, Liu C, Kim DH, Lim SK, Mukhopadhyay S (2010) Design method and test structure to characterize and repair TSV defect induced signal degradation in 3D system. In: Proceedings of IEEE/ACM international conference on computer-aided design (ICCAD), pp 694–697
5. Cho MK, Liu C, Kim DH, Lim SK, Mukhopadhyay S (2011) Pre-bond and post-bond test and signal recovery structure to characterize and repair TSV defect induced signal degradation in 3-D system. IEEE Trans Compon Packag Manuf Technol 1(11):1718–1727
6. Huang LR, Huang SY, Sunter S, Tsai KH, Cheng WT (2013) Oscillation-based pre-bond TSV test. IEEE Trans Comput Aided Des Integr Circuits Syst 32(9):1440–1444
7. Deutsch S, Chakrabarty K (2013) Non-invasive pre-bond TSV test using ring oscillators and multiple voltage levels. In: Proceedings of design, automation & test in Europe conference & exhibition (DATE), pp 1065–1070
8. Deutsch S, Chakrabarty K (2013) Contactless pre-bond TSV test and diagnosis using ring oscillators and multiple voltage levels. IEEE Trans Comput Aided Des Integr Circuits Syst 33(5):774–785
9. You JW, Huang SY, Kwai DM, Chou YF, Wu CW (2010) Performance characterization of TSV in 3D IC via sensitivity analysis. In: Proceedings of 19th IEEE Asian test symposium (ATS), pp 389–394
10. Kandalaft N, Rashidzadeh R, Ahmadi M (2013) Testing 3-D IC through-silicon-vias (TSVs) by direct probing. IEEE Trans Comput Aided Des Integr Circuits Syst 32(4):538–546

11. Marinissen EJ, Verbree J, Konijnenburg M (2010) A structured and scalable test access architecture for TSV-based 3D stacked ICs. In: Proceedings of IEEE VLSI test symposium (VTS), pp 269–274
12. Marinissen EJ, Chi CC, Verbree J, Konijnenburg M (2010) 3D DfT architecture for pre-bond and post-bond testing. In: Proceedings of IEEE international 3D systems integration conference (3DIC), pp 1–8
13. Noia B, Goel SK, Chakrabarty K, Marinissen EJ, Verbree J (2010) Test-architecture optimization for TSV-based 3D stacked ICs. In: Proceedings of IEEE European test symposium (ETS), pp 24–29
14. Noia B, Chakrabarty K, Marinissen EJ (2010) Optimization methods for post-bond die-internal/external testing in 3D stacked ICs. In: Proceedings of IEEE international test conference (ITC), pp 1–9
15. Wu XX, Chen YB, Chakrabarty K, Xie Y (2008) Test-access solutions for three-dimensional SOCs. In: Proceedings of IEEE international test conference (ITC), p 1
16. Giri C, Roy SK, Banerjee B, Rahaman H (2009) Scan chain design targeting dual power and delay optimization for 3D integrated circuit. In: Proceedings of international conference on advances in computing, control and telecommunication technologies (ACT), pp 845–849
17. Marinissen EJ (2010) Testing TSV-based three-dimensional stacked ICs. In: Proceedings of design automation and test in Europe, pp 1689–1694
18. Deutsch S, Chickermane V, Keller B, Mukherjee S, Konjinenburg E, Marinissen EJ, Goel SK (2011) Automation of 3D-DfT insertion. In: Proceedings of IEEE Asian test symposium, pp 395–400
19. Noia B, Chakrabarty K, Goel SK, Marinissen EJ (2011) Test-architecture optimization and test scheduling for TSV-based 3-D stacked ICs. IEEE Trans Comput Aided Des Integr Circuits Syst 30(11):1705–1718
20. Deutsch S, Chakrabarty K, Marinissen EJ (2013) Uncertainty-aware robust optimization of test-access architectures for 3D stacked ICs. In: Proceedings of IEEE international test conference (ITC), pp 1–10
21. Agrawal M, Chakrabarty K (2013) Test-cost optimization and test-flow selection for 3D-stacked ICs. In: Proceedings of IEEE VLSI test symposium (VTS), pp 1–6
22. Rawat I, Gupta MK, Singh V (2013) Scheduling tests for 3D SoCs with temperature constraints. In: Proceedings of east-west design & test symposium, pp 1–4
23. Marinissen EJ (2012) Challenges and emerging solutions in testing TSV-based 2.5D- and 3D-stacked ICs. In: Proceedings of design, automation & test in Europe conference & exhibition, pp 1277–1282
24. Wang SJ, Chen YS, Li KSM (2013) Low-cost testing of TSVs in 3D stacks with pre-bond testable dies. In: Proceedings of IEEE international symposium on VLSI design, automation and test, pp 1–4
25. Liao CC, Chen AW, Lin LY, Wen CH (2013) Fast scan-chain ordering for 3-D-IC designs under through-silicon-via (TSV) constraints. IEEE Trans VLSI Syst 21(6):1170–1174
26. Yeh TH, Wang SJ, Li KSM (2011) Interconnect test for core-based designs with known circuit characteristics and test patterns. In: Proceedings of international conference on IC design & technology, pp 1–4
27. Panth S, Lim SK (2012) Transition delay fault testing of 3D ICs with IR-drop study. In: Proceedings of IEEE 30th VLSI test symposium (VTS), pp 270–275
28. Deutsch S, Chakrabarty K, Panth S, Lim SK (2012) TSV stress-aware ATPG for 3D stacked ICs. In: Proceedings of IEEE Asian test symposium, pp 31–36
29. Srinivasan S, Kundu S (2012) Functional test pattern generation for maximizing temperature in 3D IC Chip Stack. In: Proceedings of international symposium on quality electronic design, pp 109–116
30. Rajski J, Tyszer J (2013) Fault diagnosis of TSV-based interconnects in 3-D stacked designs. In: Proceedings of IEEE international test conference (ITC), pp 1–9
31. Shih CJ, Hsieh SA, Lu YC, Li JCM, Wu T, Chakrabarty K (2013) Test generation of path delay faults induced by defects in power TSV. In: Proceedings of IEEE Asian test symposium, pp 43–48

32. Lin YH, Huang SY, Tsai KH, Cheng WT, Sunter S, Chou YF, Kwai DM (2013) Parametric delay test of post-bond through-silicon vias in 3-D ICs via variable output thresholding analysis. IEEE Trans Comput Aided Des Integr Circuits Syst 32(5):737–747
33. You JW, Huang SY, Lin YH, Tsai MH, Kwai DM, Chou YF, Wu CW (2013) In-situ method for TSV delay testing and characterization using input sensitivity analysis. IEEE Trans Very Larg Scale Integr (VLSI) Syst 21(3):443–453
34. Gulbins M, Hopsch F, Schneider P, Straube B, Vermeiren W (2010) Developing digital test sequences for through-silicon vias within 3D structures. In: Proceedings of international 3D systems integration conference, pp 1–6
35. Noia B, Chakrabarty K (2013) Post-DfT-insertion retiming for delay recovery on inter-die paths in 3D ICs. In: Proceedings of IEEE VLSI test symposium (VTS), pp 1–6
36. Noia B, Chakrabarty K (2014) Retiming for delay recovery after DfT insertion on interdie paths in 3-D ICs. IEEE Trans Comput Aided Des Integr Circuits Syst 32(3):464–475
37. Huang SY, Huang LR, Tsai KH, Cheng WT (2013) Delay testing and characterization of post-bond interposer wires in 2.5-D ICs. In: Proceedings of IEEE international test conference (ITC), pp 1–8
38. Li J, Ye FM, Xu Q, Chakrabarty K, Eklow B (2013) On effective and efficient in-field TSV repair for stacked 3D ICs. 50th ACM/EDAC/IEEE design automation conference (DAC), pp 1–6
39. Ye FM, Chakrabarty K (2012) TSV open defects in 3D integrated circuits: characterization, test, and optimal spare allocation. In: Proceedings of 49th ACM/EDAC/IEEE design automation conference (DAC), pp 1024–1030
40. Millican SK, Saluja KK (2014) A test partitioning technique for scheduling tests for thermally constrained 3D integrated circuits. In: Proceedings of IEEE international conference on VLSI design, pp 20–25
41. Millican SK, Saluja KK (2012) Linear programming formulations for thermal-aware test scheduling of 3D-stacked integrated circuits. In: Proceedings of IEEE Asian test symposium, pp 37–42
42. Sen Gupta B, Ingelsson U, Larsson E (2011) Scheduling tests for 3D stacked chips under power constraints. In: Proceedings of 6th IEEE international symposium on electronic design, test and application (DELTA), pp 72–77
43. Kameyama S, Baba M, Higami Y, Takahashi H (2014) Precision resistance measurement method of TSVs in a 3D-IC by analog boundary-scan. IEICE Trans Inf Syst J97-D(4):887–890 (in Japanese)
44. Kameyama S, Baba M, Higami Y, Takahashi H (2014) Accurate resistance measuring method for high density post-bond TSVs in 3D-IC with electrical probes. In: Proceedings of international conference on electronics packaging (ICEP2014), pp 117–121
45. Kameyama S, Baba M, Higami Y, Takahashi H (2014) Measuring method for TSV-based interconnect resistance in 3D-SIC by embedded analog boundary-scan circuit. Trans Jpn Inst Electron Packag 7(1):140–146
46. Chung H, Ni CY, Tu CM, Chang YY, Haung YT, Chen WM, Lou BY, Tseng KF, Lee CY, Lwo BJ (2010) The advanced pattern designs with electrical test methodologies on through silicon via for CMOS image sensor. In: Proceedings of electronic components and technology conference (ECTC), pp 297–302
47. Stucchi M, Perry D, Katti G, Dehaene W (2010) Test structures for characterization of through silicon vias. In: Proceedings of IEEE international conference on microelectronic test structures (ICMTS), pp 130–134
48. IEEE Std. 1149.4-1999 (2000) IEEE standard for a mixed-signal test bus. ISBN: 0-7381-1755-2 SH94761, Mar.
49. IEEE Std. 1149.1™-2001 (R2008) (2001) IEEE standard test access port and boundary-scan architecture. ISBN: 0-7381-2944-5 SH94949, July
50. Parker KP, Kameyama S (2012) The boundary-scan handbook, 3rd edn. SEIZANSHA, Japan (Japanese Version, ISBN: 978-4-88359-303-3)

51. Yotsuyanagi H, Sakurai H, Hashizume M (2014) Delay line embedded in boundary scan for testing TSVs. IEEE international workshop on testing three-dimensional stacked integrated circuits (3D-TEST), Oct
52. Yotsuyanagi H, Makimoto H, Nimiya T, Hashizume M (2013) On detecting delay faults using time-to-digital converter embedded in boundary scan. IEICE Trans Inf Syst E96-D(9):1986–1993
53. Parker KP (2003) The boundary scan handbook, 3rd edn. Kluwer Academic, Dordrecht (ISBN: 978-1-4615-0367-5)
54. Nakamura M, Yotsuyanagi H, Hashizume M (2013) On fault detection method considering adjacent TSVs for a delay fault in TSV. IEICE Technical Report, NAID:110009728044 (in Japanese)
55. Kondo S, Yotsuyanagi H, Hashizume M (2011) Propagation delay analysis of a soft open defect inside a TSV. Trans Jpn Inst Electron Packag 4(1)119–126
56. Datta R, Sebastine A, Raghunathan A, Carpenter G, Nowka K, Abraham JA (2010) On-chip delay measurement based response analysis for timing characterization. J Electron Test 26(6):599–619
57. Konishi T, Yotsuyanagi H, Hashizume M (2012) A built-in test circuit for supply current testing of open defects at interconnects in 3D ICs. In: Proceedings of 4th Electronics System Integration Technologies Conference (ESTC), pp 1–6
58. Konishi T, Yotsuyanagi H, Hashizume M (2012) Supply current testing of open defects at interconnects in 3D ICs with IEEE 1149.1 architecture. In: Proceedings of IEEE international 3D systems integration conference, pp 8.2.1–8.2.6
59. Marinissen EJ, Zorian Y (2009) Testing 3D chips containing through-silicon vias. In: Proceedings of IEEE international test conference (ITC), pp 1–11
60. Marinissen EJ (2010) Challenges in testing TSV-based 3D stacked ICs: test flows, test contents, and test access. In: Proceedings of IEEE Asia pacific conference on circuits and systems, pp 544–547
61. Hashizume M, Kondo S, Yotsuyanagi H (2010) Possibility of logical error caused by open defects in TSVs. In: Proceedings of international technical conference on circuits, computers and communications, pp 907–910

# Chapter 9
# Dream Chip Project at ASET

**Morihiro Kada, Harufumi Kobayashi, Fumiaki Yamada, Haruo Shimamoto, Shiro Uchiyama, Kenichi Takeda, Kenichi Osada, Tadashi Kamada and Fumihiko Nakazawa**

M. Kada (✉)
The National Institution of Advanced Industrial Science and Technology (AIST), Ibaraki, Japan
e-mail: m-kada@aist.go.jp

H. Shimamoto
Advanced Industrial Science and Technology (AIST), Ibaraki, Japan
e-mail: haruo.shimamoto@aist.go.jp

H. Kobayashi
Lapis Semiconductor, Kanagawa, Japan
e-mail: kbys@jiep.or.jp

F. Yamada
IBM Research, Tokyo, Japan
e-mail: frank2007yokohama@yahoo.co.jp

S. Uchiyama
Micron Memory Japan, Kanagawa, Japan
e-mail: shuchiyama@micron.com

K. Takeda · K. Osada
Hitachi Ltd., Tokyo, Japan
e-mail: kenichi.takeda.hm@hitachi.com

K. Osada
e-mail: kenichi.osada.aj@hitachi.com

T. Kamada
DENSO Corp., Aichi, Japan
e-mail: TADASHI_KAMADA@denso.co.jp

F. Nakazawa
FUJITSU LABORATORIES LTD., Kanagawa, Japan
e-mail: nakazawa.fumihi@jp.fujitsu.com



K. Kondo et al. (eds.), *Three-Dimensional Integration of Semiconductors*,
DOI 10.1007/978-3-319-18675-7_9

Following is the sentence from Sect. 1.3.3.2 in Chap. 1.

The second full-scale national research and development (R&D) initiative of 3D integration technology using through-silicon via (TSV) was implemented over the 5-year period from 2008 to 2012. Association of Super-Advanced Electronics Technologies (ASET) conducted the project "Development on Functionally Innovative 3D-Integrated Circuit (Dream Chip) Technology Project," and it was managed by the NEDO organization that is based on "IT Innovation Program" of Japanese government's Ministry of Economy, Trade and Industry (METI). After the 2010 interim assessment, the two focus areas became 3D integration process basic technologies and application technologies using TSV. The former consisted of thermal management/chip-stacking technology, thin wafer technology, and 3D integration technology, while the latter's focus area comprised ultrawide bus 3D-SiP, mixed signal (digital–analog) 3D, and heterogeneous 3D (see Fig. 1.10. (Chap. 1) [1, 2])

In this chapter, R&D subjects and results are described in detail.

Re-posting Fig. 1.10, R&D subject of the Dream Chip Project (with permission from the Electrochemical Society, ECS) [2]

## 9.1 Overview of Japanese 3D Integration Technology Research and Development Project (Dream Chip)

ASET organized six teams to conduct these six R&D subjects. Following is the summary of the R&D subjects and the results.

### *9.1.1 3D Integration Process Basic Technologies*

#### 9.1.1.1 Thermal Management and Chip-Stacking Technology

The team was responsible for R&D of "Thermal Management and Chip-Stacking Technology." They conducted thermal design and thermal evaluation indispensable for highly integrated 3D structure using TSV. Fine pitch bump chip-stacking technology (chip-to-chip, C2C) development which realizes high-density 3D integration structure was performed.

1. High-power dissipation (efficient cooling) technology of 3D stacking chips under high-temperature environment for car driving:
   a) Defined necessary thermal bump density for the improvement of thermal performance within 3D chip stack
   b) Developed cooling structure from the peripheral of Si interposer, using heat pipe structure.
2. C2C micro-bonding technology subject:
   a) High-precision and in situ alignment method, utilizing infrared (IR) transmissive observation was developed.

b) Up to 11,520 series bumps were successfully connected in which the resistance was around 70 mΩ/bump.

#### 9.1.1.2 Thin-Wafer Technology

It was the team's responsibility to conduct and develop "thin wafer technology." They studied ultrathin wafer process important for a 3D process integration. They also carefully studied the fluctuation of electrical characteristics of thinned device.

1. Thin wafer process technology
   a) ±1 μm total thickness variation (TTV) of 10 μm thickness with micro-bumping wafer was achieved.
   b) New process underfill before dicing (UBD) was proposed for cost reduction.
   c) New measuring method of flexural bending strength was proposed to Semiconductor Equipment and Materials International (SEMI) Standardization.
2. Study on characteristic change of thinned device:
   a) Found that degradation of dynamic random access memory (DRAM) retention characteristics reduced to half was mainly induced by wafer-thinning process because of mechanical strain
   b) Proposed the possibility of new gettering technique

#### 9.1.1.3 3D Integration Technology

"3D Integration Technology" of Via-last TSV formation process was developed. The process includes photolithography, Si etching, plating, and wafer-to-wafer bonding (W2W) and so on. Development of elemental 3D-circuit technology and characteristic evaluation using TSV were performed.

1. 3D integration process technology
   a) Successfully developed via-last TSV formation compatible with Cu/low-k interconnect structure
   b) Good electrical contact was obtained
   c) Both C2C and W2W bonding process were developed
2. Elemental 3D circuit technology with TSV
   a) Established TSV layout design rule
   b) Using optimized transmission circuit for 3D integration and low-capacitance TSV, the highest transmission performance in W2W stacking was obtained
   c) Electrical model of TSV was defined

## 9.1.2 Application Technologies

### 9.1.2.1 Ultrawide Bus 3D-SiP Integration Technology

"Ultrawide bus 3D-SiP Integration technology" which can study the feature of TSV efficiently was performed. They confirmed that the high bandwidth signal transmission with low power consumption between logic and memory devices was possible.

- Designed and build 4096 wide input/output (I/O) memory and logic three-dimensional system in package (3D-SiP).
- Successfully demonstrated 102 GB/s @200 Mbit/s bus speed with low power consumption (0.56 pJ/I/O).

### 9.1.2.2 Mixed-Signal (Digital and Analog) 3D Integration Technology for Automotive Application

"Mixed signal (digital–analog) 3D integration technology" was developed. The system includes elemental technology of high-speed image-processing module using CMOS image sensor (CIS) necessary for car-driving support.

- C2C TSV forming and chip-stacking process were developed.
- CIS, correlated double sampling (CDS), analog-to-digital converter (ADC), interface (IF) chip for 3D car-driving support imaging system were made and evaluated individually.
- Si interposer with TSV-type decoupling capacitor was developed and evaluated.

### 9.1.2.3 Heterogeneous 3D Integration Technology for Radio Frequency Microelectromechanical Systems

"Heterogeneous 3D Integration technology" which can create a new functional device by integrating semiconductor devices and other kinds of devices such as microelectromechanical system (MEMS) was developed. The new functionality is radiofrequency (RF)-MEMS device which is a 3D integration of complementary metal-oxide semiconductor (CMOS) and low-temperature co-fired ceramic (LTCC)-based MEMS for RF tunable device.

- MEMS tunable filter was demonstrated.
- Wafer-level packaging (WLP) RF-MEMS switch had good RF performance.
- CMOS control integrated circuit (IC) was optimized properly.

To conduct R&D subjects, a 90-nm process shuttle test element group (TEG) chip was designed and manufactured as shown in Fig. 9.1.

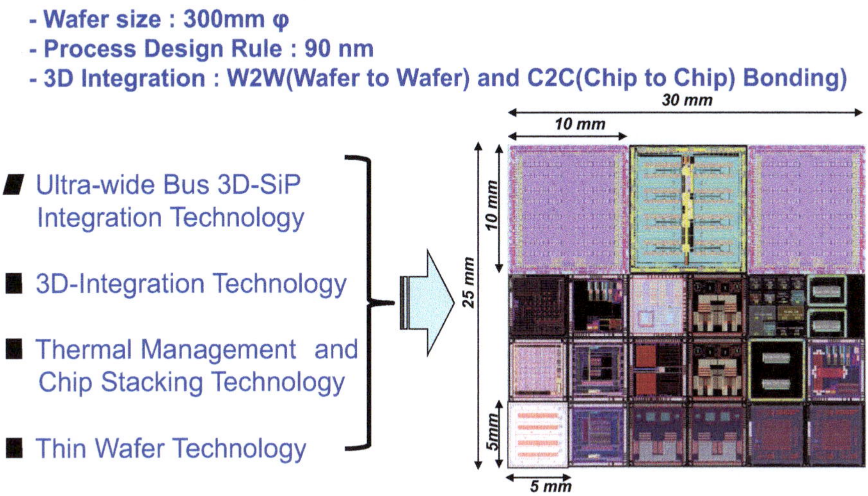


**Fig. 9.1** A 90-nm process shuttle TEG chip. *3D* three dimension, *SiP* system in package

## 9.2 Thermal Management and Chip-Stacking Technology

### *9.2.1 Background*

We carried out the R&D work for thermal management and chip-stacking technology in "Dream Chip Project." Major targets are as follows:

1. Chip-stacking/joining technology development and reliability study towards 10 μm connection bump pitch and more than 10,000 bumps per 1 $cm^2$ chip
2. Thermal management study for 3D stacked chip cooling
3. Chip stacking/joining and cooling system development of the 3D stereo image capture camera for automobile derive assistance

### *9.2.2 Chip-Stacking/Joining Technology*

#### 9.2.2.1 Metal Bump Materials and Structure

We have listed the cost-effective, well-recognized chip joining materials and structures, as shown in Table 9.1.

Among those candidates, we have finally chosen the structure and material as follows:

1. Cu or Ni pad
2. Cu pillar + Ni barrier + SnAg solder

**Table 9.1** List of chip joining materials and structures

← High Priority

| Chip (BEOL) | Interposer | Fine pitch | Stability | Low pressure process | Material cost | Bumping Takt time | Bonding Takt time | Tool |
|---|---|---|---|---|---|---|---|---|
| Au | Au | △ | ○ | ○ | ○ | ◎ | ◎ | ◎ |
| Au (Stud) | Solder | △ | × | ◎ | ○ | △ | ○ | ◎ |
| Au | ACF on Au pad | △ | △ | ◎ | △ | ◎ | ○ | ◎ |
| UBM+Solder (Bump) | Cu | △ | ○ | ◎ | ◎ | ◎ | ○ | ◎ |
| UBM+Cu-piller+Solder | Au + Ni + Cu | ○ | ○ | ◎ | ◎ | ◎ | ○ | ◎ |
| UBM+Cu-piller+Ni+Solder | Au + Ni + Cu | ○ | ○ | ◎ | ◎ | ◎ | ○ | ◎ |
| Cu | Cu | ◎ | ○ | × | ◎ | ◎ | × | × |

We have chosen solder to electrically and mechanically join between 3D chips because of the following reasons:

Chips/wafers have deformation caused by front-end-of-line (FEOL) and back-end-of-line (BEOF) processes. Since recent back-end-of-line (BEOL) consists of low-*k* dielectric material, it is preferable to apply lower mechanical pressure at the process in downstream. Because of that background, it is considered to be a better process to absorb the nonuniformity of the joint height. Solder joint requires lower pressure during the bonding process, and the joint height can be adjusted. In addition, the low melting point of the solder enables low bonding temperature which in turn helps in improving alignment accuracy between chips to be bonded. Low process temperature also helps to reduce residual mechanical stress of stacked chips.

We made a Cu pillar to stabilize the bump height. Without the pillar structure with higher melting point, the solder may cause bridging between neighboring joints, when the solder is melted under a certain pressure. Pillar structure is expected to keep proper distance between stacked chips. Cu is preferable for pillar material because of lower electrical resistance and higher thermal conductance.

Ni barrier helps in reducing intermetallic compound (IMC) formation. Chip/wafer with SnAg solder will pass through the reflow oven to melt the solder to form the round top shape prior to the bonding process. During this period, CuSn IMC will grow rather rapidly when the bump size is small. With a certain amount of IMC, it will be very difficult to perform bonding process, because the IMC has a much higher melting point than SnAg solder. The Ni barrier will reduce the reaction rate to grow the IMC by reducing the speed of metal diffusion.

#### 9.2.2.2 Reliability Study of Micro-bump

Temperature cycle accelerated reliability tests were carried out to study the mechanism of degradation caused by mechanical stress. The calculation by simulation model for deformation and mechanical stress by multilayer beam bending theory and finite element numerical calculations were also carried out in parallel for thermal cycle conditions. As a parameter for those calculations, hardness of the IMCs in the junction area were measured by nano-indentation. The calculated results for the stress of the micro-joint indicated that their dependency to the difference of hardness was negligibly small.

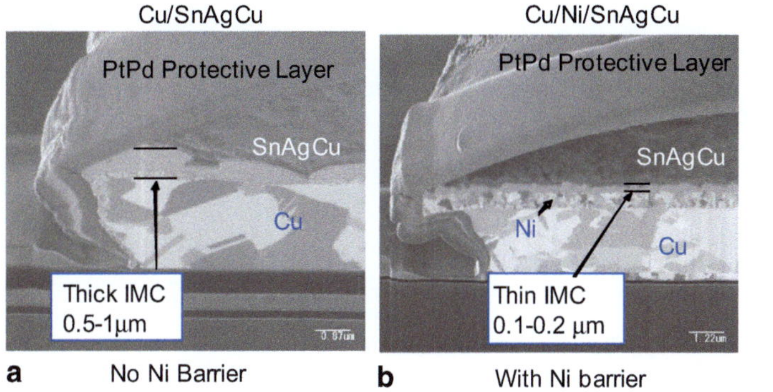


**Fig. 9.2** Cross-sectional SEM picture of the bump after reflow. *SEM* scanning electron microscope, *IMC* intermetallic compound

According to the high-temperature storage test, smaller number of voids were observed in the junction with Ni barrier structure than that without Ni barrier.

Figure 9.2 shows a cross section of the junction after the reflow; 0.5–1 -μm-thick IMC layer was observed in the junction without Ni barrier layer, while only 0.1–0.2-μm-thin IMC layer was observed in the junction with Ni barrier. Ni barrier layer is effective to improve the process stability.

As shown in Fig. 9.3, nano-indentation results of both hardness and Young's modulus are larger for NiSn IMC than that of CuSn IMC.

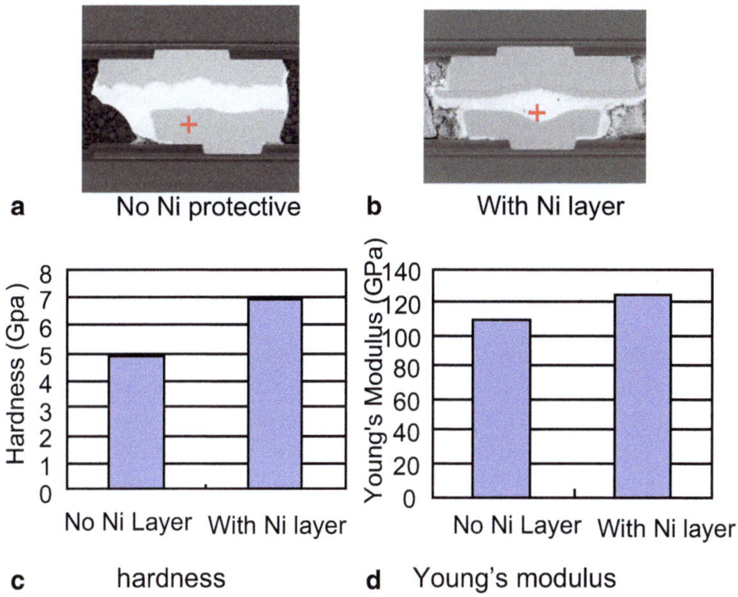


**Fig. 9.3** Hardness and Young's modulus of joint bump measured by nano-indentation

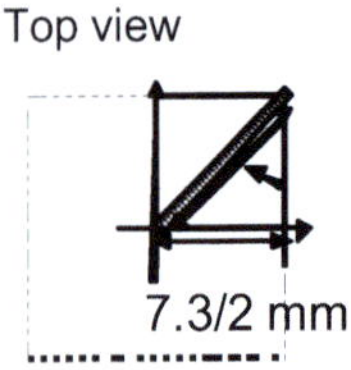


| Material | Young's Modulus | CTE ($10^{-6}$/K) | Poisson's ratio |
| --- | --- | --- | --- |
| CuSn IMC | 110 | 16 | 0.3 |
| NiSn IMC | 125 | 16 | 0.3 |

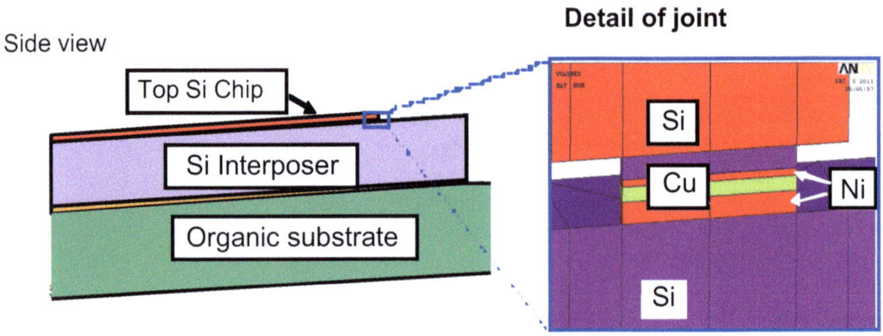


**Fig. 9.4** Model for analysis: one-fourth slice model. Simulation condition: 125C≥55C. *CTE* coefficient of thermal expansion

A model for the analysis is shown in Fig. 9.4, and the results are shown in Fig. 9.5. Mechanical stress with CuSn IMC and NiSn make no big difference [3, 4].

Prior to the bonding process, the solder on bumps is melted to form a round-shaped top of the bump by surface tension. However, if a large amount of SnCu IMC is formed during this melting process, it will become difficult to melt the bump again because the melting point of the IMC is much higher than that of SnAg solder. The Ni barrier layer between Cu stud and SnAg solder reduces the diffusion of Cu and Cu–Sn IMC. Ni and Sn also make the IMC ($Ni_3Sn_4$). The reaction rate constant of this reaction is much slower than that of the reaction between Cu and Sn. However, if NiSn IMC is formed, the mechanical stress might be larger, because $Ni_3Sn_4$ is harder than CuSn IMC.

#### 9.2.2.3 Electromigration Test to Understand the Current Density of Micro-bump Joint

EM test was carried out using ES#2 test chip as shown in Fig. 9.6. This test chip has daisy chains of six junctions with wiring for higher current and diode temperature sensors near junctions. For the reference, junctions with current flow of opposite directions and the junction without current flow are located next to each other. Their test structure is shown in Fig. 9.7 and Table 9.2. The SnAg solder junction is in between the Ni barrier and Ni stud. To observe the temperature dependency, an EM test with several temperature levels were carried out by monitoring those diode

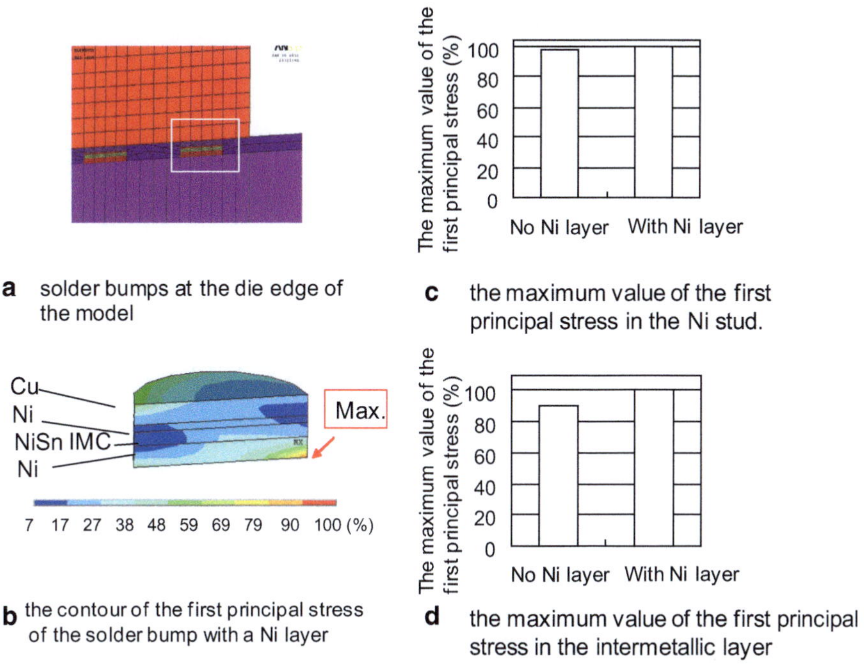


**Fig. 9.5** Mechanical analysis result. *IMC* intermetallic compound

sensors to control the temperature of the oven. Structure of junction is shown in Fig. 9.8. Figure 9.9 shows the test result with 50 and 25 KA/cm$^2$ under 100 °C. Energy dispersive X-ray spectroscopy (EDX) observation indicates that the solder in bonded junctions are mostly changed to NiSn IMC. There were no voids observed after 700 h of the test run; however, after a 1000-h test, electrical resistance for 50 KA/cm was increased by 10 %, while 25 KA/cm$^2$ test result showed no change, under 100 °C [5–7]

#### 9.2.2.4 Flip Chip Bonding Density Towards 10-μm Connection Bump Pitch

*Test Vehicles* We prepared two types of test vehicles (TVs) for joining study and one type of TV for cooling study. TV10 is to prove 10 μm pitch joining, TV40 is for joining study and reliability test, and TV200 is for thermal management study.

The manufacturing process of 10 μm pitch bump and TSV was not well established when we planned this project. TV40 was prepared for the study of joining process and reliability because the manufacturing process stability of 10 μm pitch TV was not mature when we started this project; however, the process stability of 40 μm was much better than those of 10 μm pitch.

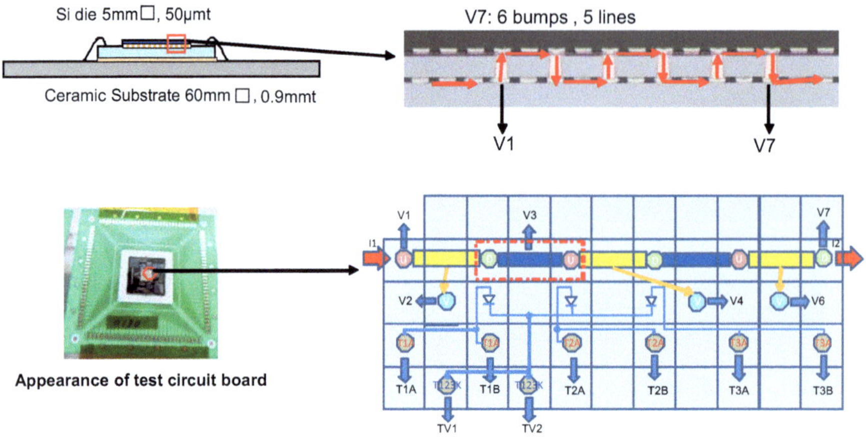


**Fig. 9.6** ES#2 test chip structure

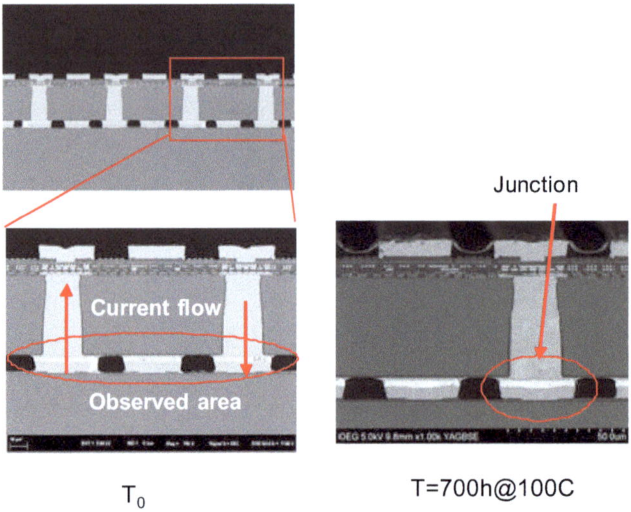


**Fig. 9.7** Test chip structure and cross section

**Table 9.2** Structure of joint

| Chip | Item | Specification |
|---|---|---|
| Chip | Substrate thickness | 50 μm |
| | Minimum pitch | 50.4 μm |
| | TSV diameter | 20 μm |
| | Bump structure | SnAg 3 μm/Ni 1 μm |
| | Bump layout | 50.4 μm Full grid |
| Si Interposer | Wiring rule | 1 layer, 2 μm min. |
| | Wire height | 0.5 μm |
| | Bump structure | Au 0.1 μm/Ni 6 μm |
| | Bump layout | 50.4 μm Full grid |

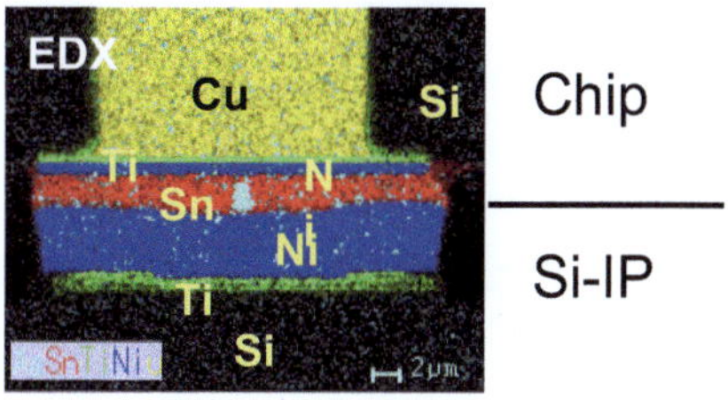


**Fig. 9.8** Structure of joint. *EDX* energy dispersive X-ray spectroscopy, Si-IP silicon interposer

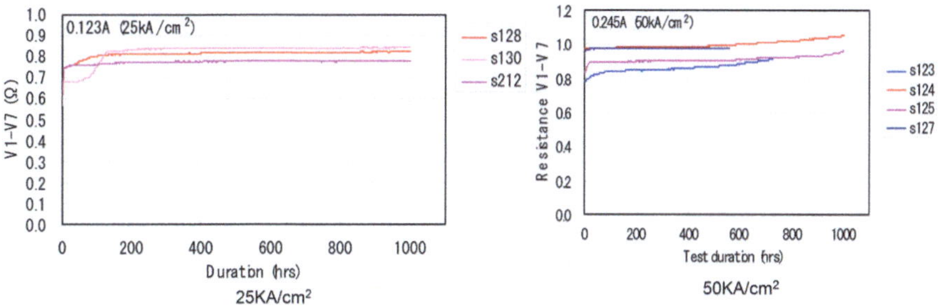


**Fig. 9.9** EM test result

a. TV10

The dimension of TV10 chip and interposer is shown in Fig. 9.10. Bump layout is shown in Fig. 9.11 that forms daisy chain of 11,520 connections. Structures of joint bumps on a chip and interposer are shown in Fig. 9.12a and 9.12b, respectively. Alignment marks for IR light are built in those chip and interposer, as shown in Fig. 9.13 [8].

b. TV40

TV40 is a set of three types of chips with 40 μm pitch connection consisting of TV40-1, TV40-2, and TV40-3 as shown in Fig. 9.14. Chip and joint bump structure are shown in Fig. 9.15. Daisy chain layout is shown in Fig. 9.16. The thickness and bump height of TV40-2 are 20–50 μm and 6 μm, including Cu-post+ SnAgCu-solder, respectively, similar dimensions to that of 10 μm pitch connection.

*10-μm Pitch Joining* Dynamic alignment scheme was employed as shown in Fig. 9.17. This scheme enables the correction of chip to chip/interposer misalignment induced by touching the bumps during the final approach of bonding process and stress release by solder melting on a bump. The position of alignment marks on the chip and interposer are monitored using IR light that transmit through those Si substrates.

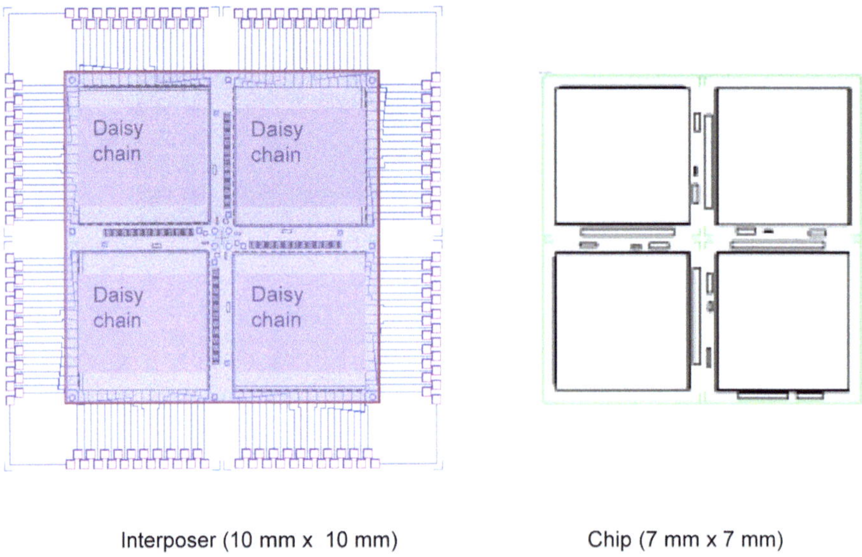


**Fig. 9.10** TV10 layout

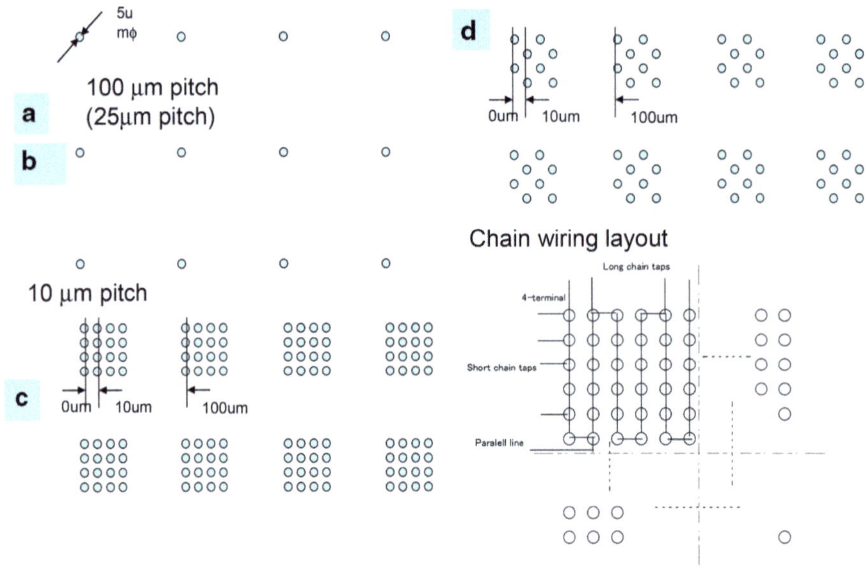


**Fig. 9.11** TV10 bump layout

Alignment accuracy in both *x* and *y* direction were within ±1 μm confirmed by IR microscope observation, as shown in Fig. 9.18.

The resistance of joint was 70 mΩ including wiring, measured by daisy chain of 11,520 connections, as shown in Fig. 9.19.

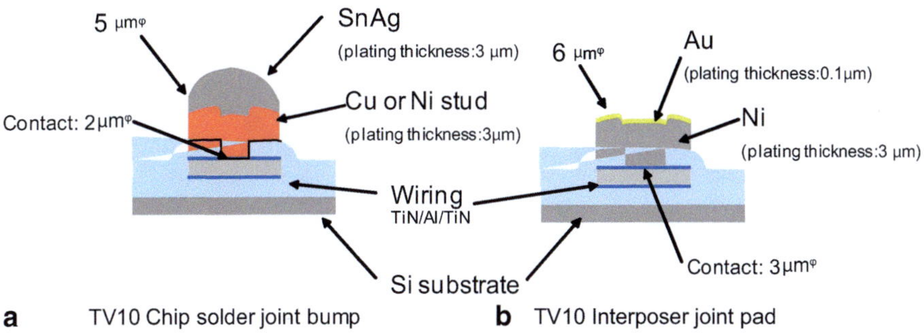


**Fig. 9.12** TV10 joint bump

**Fig. 9.13** TV10 alignment marks

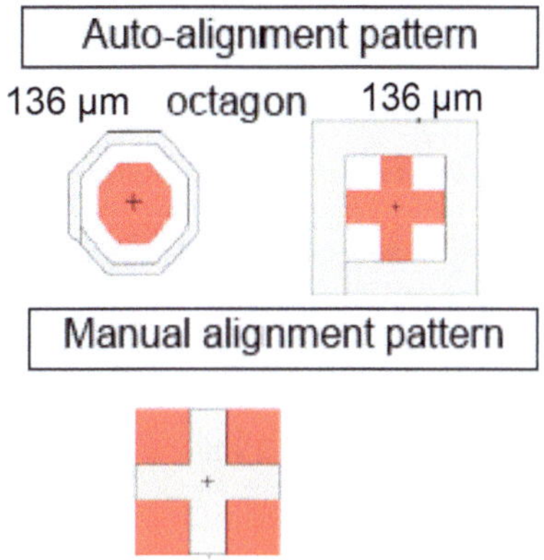


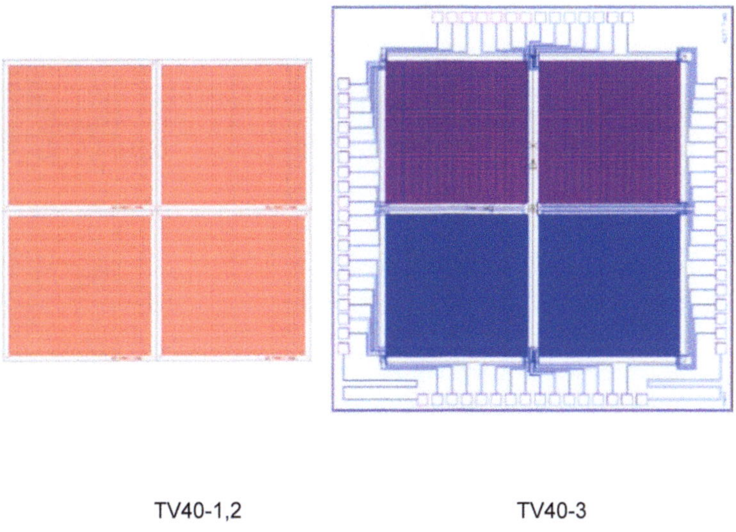


**Fig. 9.14** TV40 outline

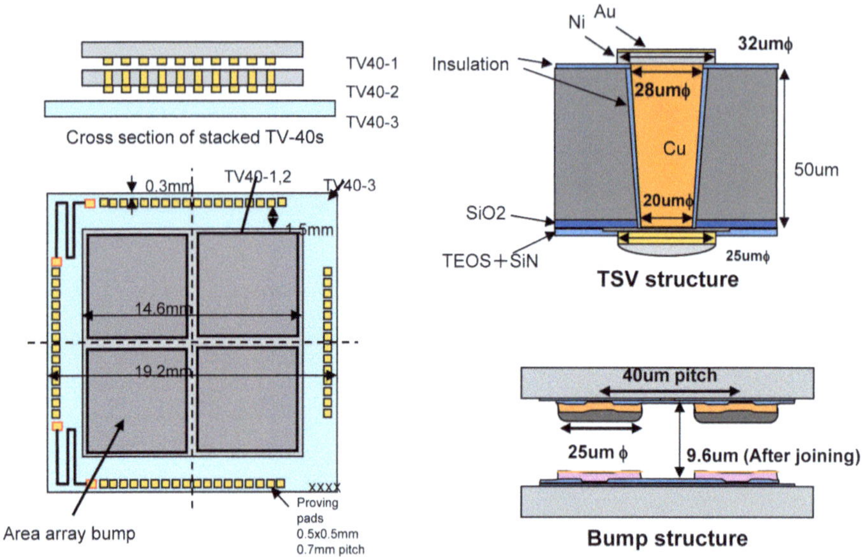


**Fig. 9.15** Structure of TV40. *TSV* through-silicon via

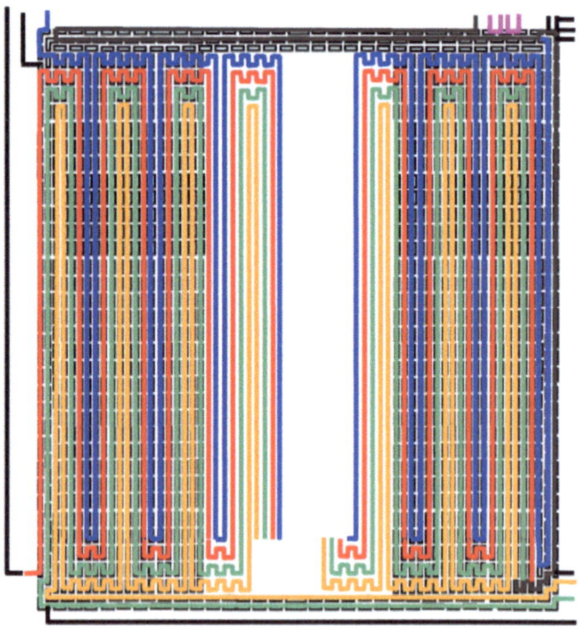

**Fig. 9.16** Wiring of TV40 daisy chain (shows one fourth of the chip, 7 × 7 mm)

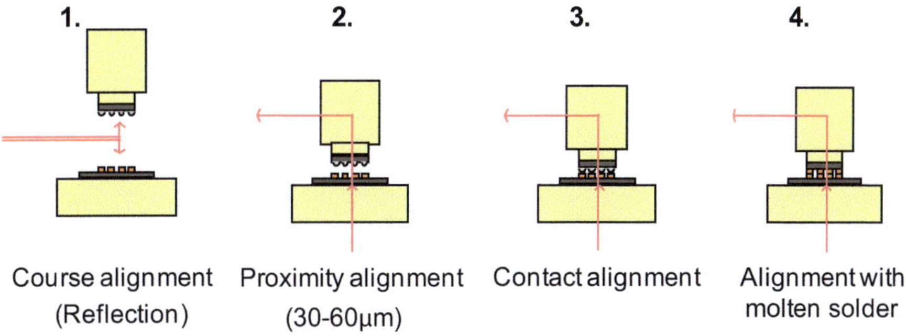


**Fig. 9.17** Dynamic alignment method

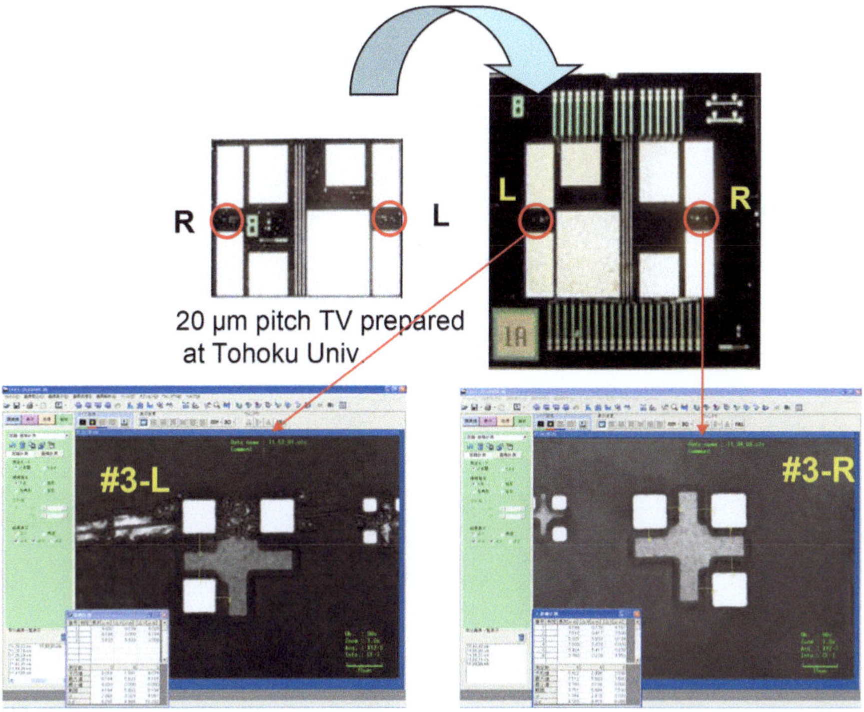


**Fig. 9.18** Observation of alignment marks by using confocal IR laser microscope. *TV* test vehicle

Measured results of three samples indicated the accuracy of < 0.6 mm.

*Thin Chip Joining* We observed waving or deformation of 1.2 $\mu m_{p\text{-}p}$ on the chip surface of 30 μm thin chip bonded with the substrate using regular bonding tool head made of aluminum nitride (AlN), as shown in Fig. 9.20b. This deformation was considerably large to cause stress on a chip and failure of joint contact. We assumed

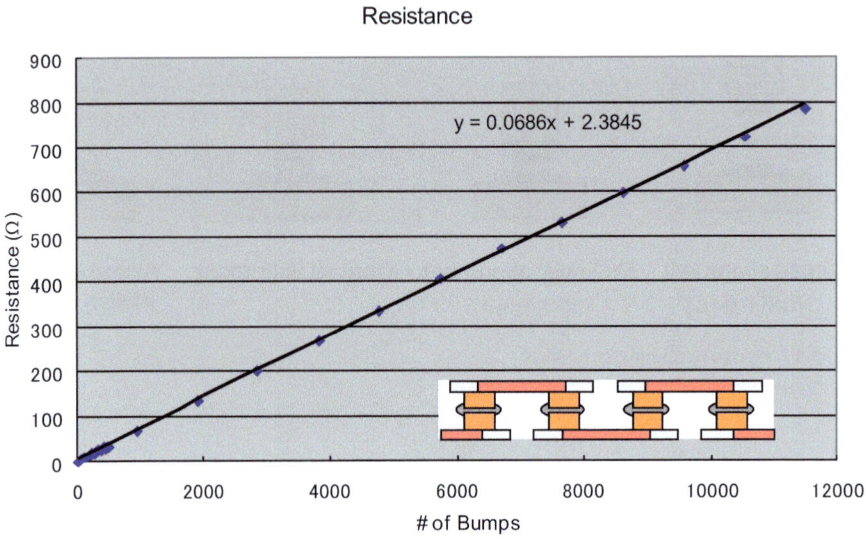


Fig. 9.19 Daisy chain joint resistance of TV10

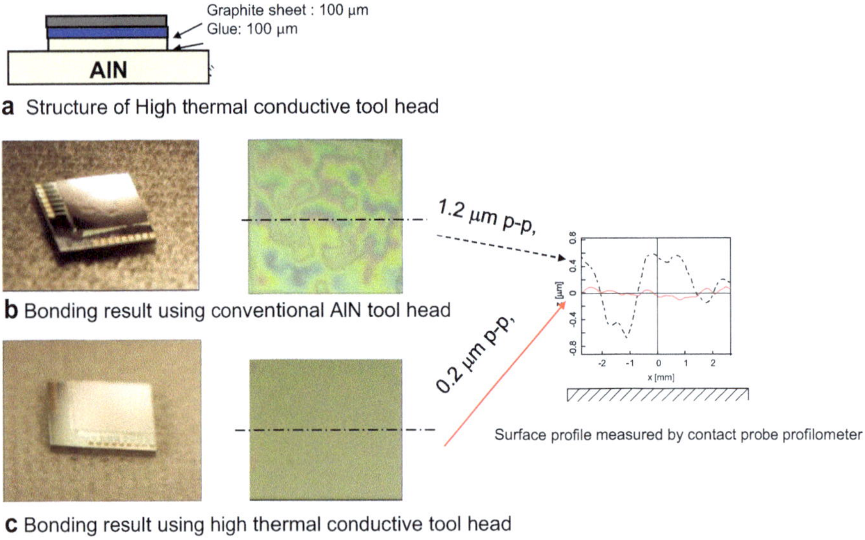


Fig. 9.20 Bonding result using high thermal conductive tool head

that this might be caused by nonuniformity of temperature over the thin chip during the cooling process of joining, resulting in solder freezing timing irregularity.

We developed new tool head as shown in Fig. 9.20a with high in plane thermal conductivity of 700 W/mK by employing carbon graphite sheet. By using this tool, we could obtain the result of 0.2 $\mu m_{p-p}$ surface flatness as shown in Fig. 9.20c [9].

### 9.2.2.5 Stack and Gang Bonding

In general, the 3D chip stack joining process is as follows:

1. Apply flux to the chip
2. Align the chip to the substrate (bottom chip)
3. Press the chip onto the substrate, and elevate the chip temperature to melt the solder and weld the bumps.
4. Repeat 1 to 3 as required.
5. Clean the flux.
6. Fill the inter-chip-fill resin and cure the stack.

There are several issues in this process:

1. Chips, especially the chip near to the bottom, will be exposed to the heat cycle repeatedly. Metal joint for the bottom chip should not be melted again during the following chip-stacking process.
2. Cleaning of the flux becomes difficult when the C2C clearance becomes small.
3. Process to fill the inter-chip-fill resin becomes very difficult when the C2C clearance becomes small.
4. Repeated bonding process may increase the process cost.

We developed a new multi-chip bonding process (Gang bonding) to provide possible solution for those issues, as shown in Fig. 9.21, expecting significant reduction of 3D chip-stack assembly cost.

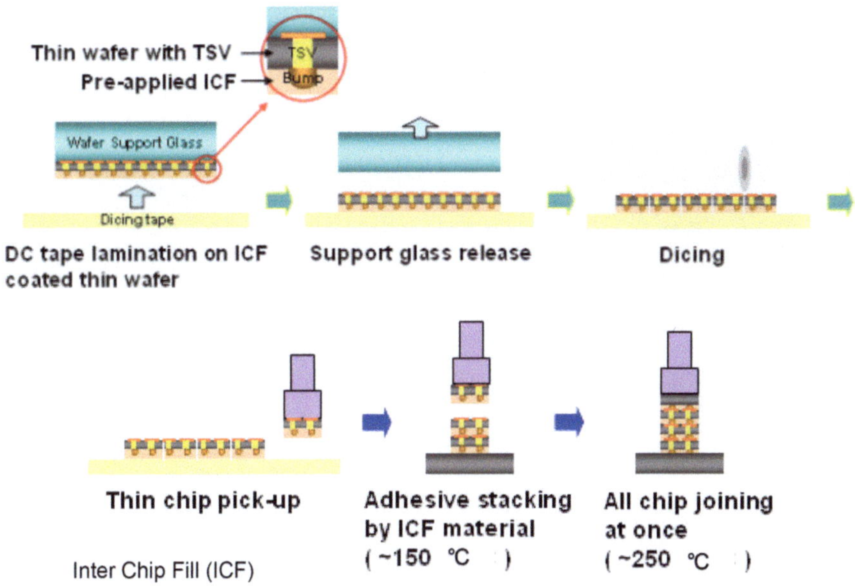


**Fig. 9.21** 3D C2C integration process using pre-applied inter-chip fill. *TSV* through-silicon via, *DC* double coated

Prior to the final gang bonding process, chips are aligned and temporarily bonded together using half-cured inter-chip-fill resin as a glue. More than two stacked chips and interposer are finally bonded together at once by melting solder bumps at an elevated temperature.

Inter-chip-fill resin for this process will be fully cured at the solder melting temperature. The functional group acts as a flux for soldering is built in the resin monomer, and this functional group will be incorporated into the polymer chain during the curing process. As a result, no flux cleaning is required.

*Pre-apply Inter-Chip-Fill resin* For this process development, the first step to be achieved was a pre-apply resin process with two chips. Resin is sometimes trapped between the metal bump and the counter pad and resulting to form a void to interrupt the electrical connection. This void may cause reliability issues.

We heated the bump with solder to be melted to form a round head by surface tension, by passing through the reflow oven prior to the joining process. Round head shape is effective to scrape the resin on top of the bump. We have to be careful not to grow the IMC during the reflow because the melting point of the IMC is significantly high in general. Joint materials and structure are optimized to satisfy this condition as shown in the previous section. We also have to control the curing temperature and viscosity of the resin as well as the temperature and pressure profile of the bonding tool.

Figure 9.22 is the result of this process, showing that the joint resistance is equivalent to the conventional process.

The second step of this process is a multi-chip joining at once. Figure 9.23 shows the successful result of three-chip stack [10].

We also developed a process using a film type pre-apply inter-chip-fill resin, as shown in Fig. 9.24 [11].

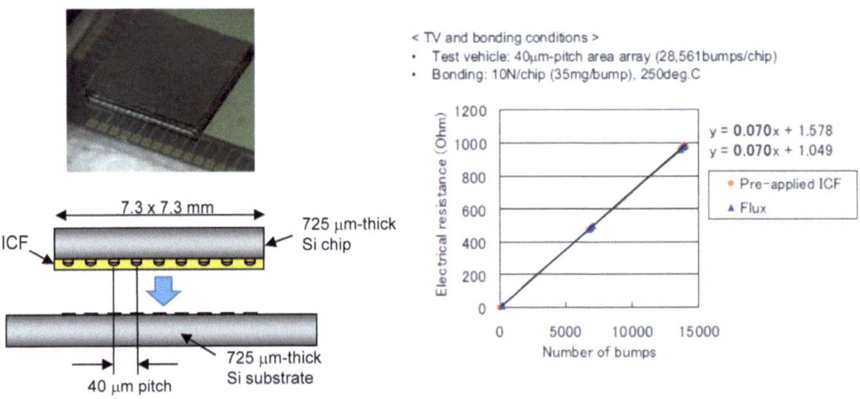


**Fig. 9.22** Bonding result of TV40 using pre-apply inter-chip-fill resin. *ICF* inter-chip fill

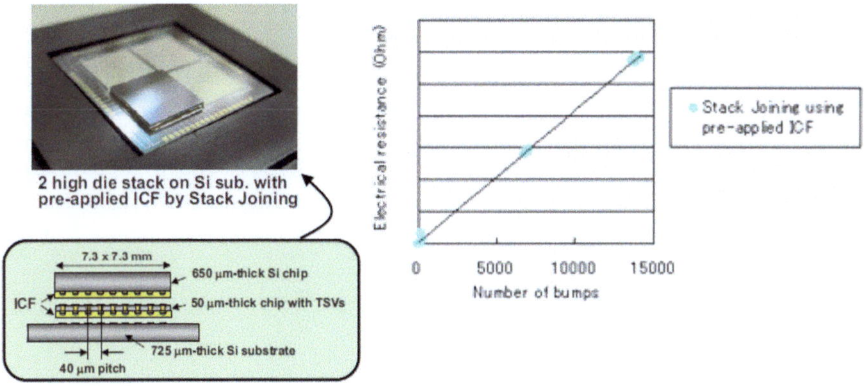


**Fig. 9.23** Gang bonding of TV40 using pre-apply ICF. *ICF* inter-chip fill

#### 9.2.2.6 Nondestructive Inspection Technologies of Micro-joint

We investigated nondestructive inspection technologies of micro-joint towards the size of 10 μm pitch. Connection pitch for the chip to the package was about 250 μm and smaller. We could use ultrasonic and X-ray for nondestructive inspection of the joint with this size. However, when the size becomes smaller down to less than 100 μm, it becomes very difficult to observe because of their limited resolution. When the chips are stacked, it becomes more difficult.

*Wafer-Level Nondestructive Void Detection Using X-ray Radiographic Apparatus* We observed voids in TSV of TV40 using X-ray radiographic apparatus. Since the bump size of TV40 is 20 μm, we have used X-ray light source with a spot size of 0.6 μm. For the test using TV200, we have used an X-ray light source with 1.0 μm spot. The observation method and the jig we designed for this observation is shown in Fig. 9.25. The picture of TV200 cross-section and transmission X-ray observation is shown in Fig. 9.26a and b. The result of X-ray data analysis is shown in Fig. 9.27. The observation result of TV40 is shown in Fig. 9.28 [12].

*Observation by X-ray CT* Nondestructive observation of micro-bump and TSV using X-ray CT.

We tried to observe voids in TSV using new equipment as shown in Fig. 9.29 for the test chip shown in Fig. 9.30. This new equipment uses normal X-ray source but equipped with new data processing unit to reduce measurement time, so that we could get a better picture by suppressing the influence of vibration, operating temperature variation, and other source of fluctuations. Figures 9.31a and b show the observation results. Figure 9.32 shows the results with two different measurement time. Left picture took 8.5 h (40X) while the right one took 1.5 h. The picture with 8.5-h measurement looks better than the one with 1.5 h; however, the difference was not very obvious. The 1.5-h measurement is more practical because the S to N ratio will be improved in proportion to the square root of the measurement time in theory. On the other hand, the influence of vibration, operating temperature variation, and

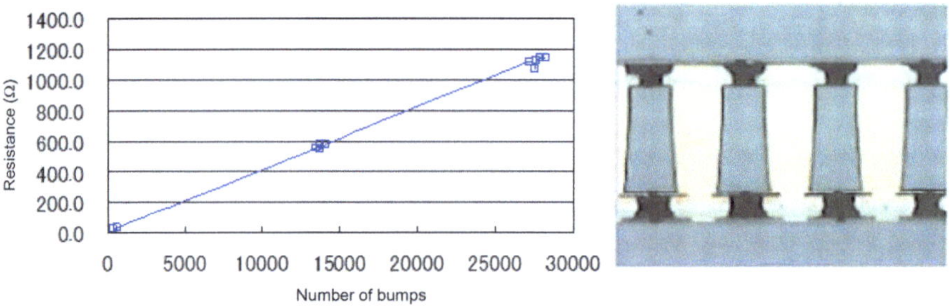


**Fig. 9.24** Bonding result of TV40 using film type ICF

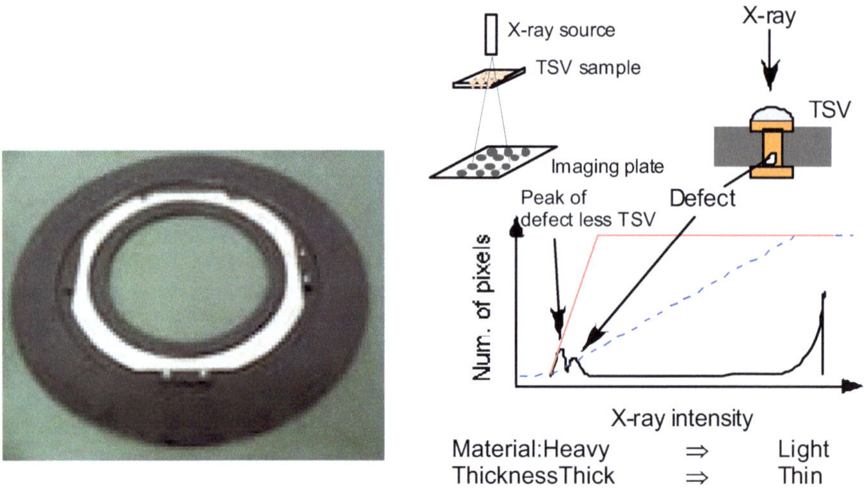


**Fig. 9.25** Transmission X-ray observation tool for 8′ wafer. *TSV* through-silicon via

**Fig. 9.26** Cross-sectional and transmission X-ray image. *TSV* through-silicon via

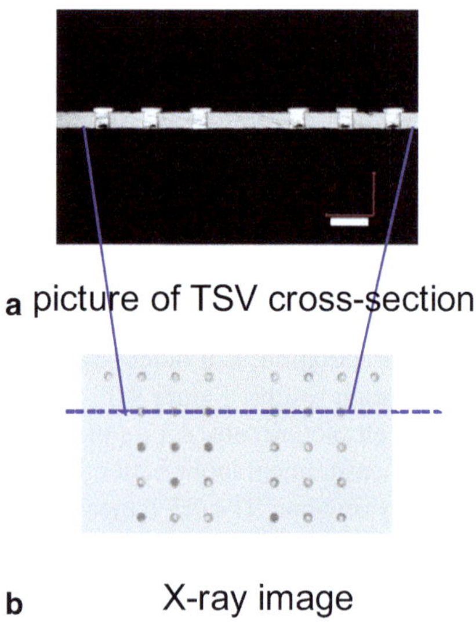

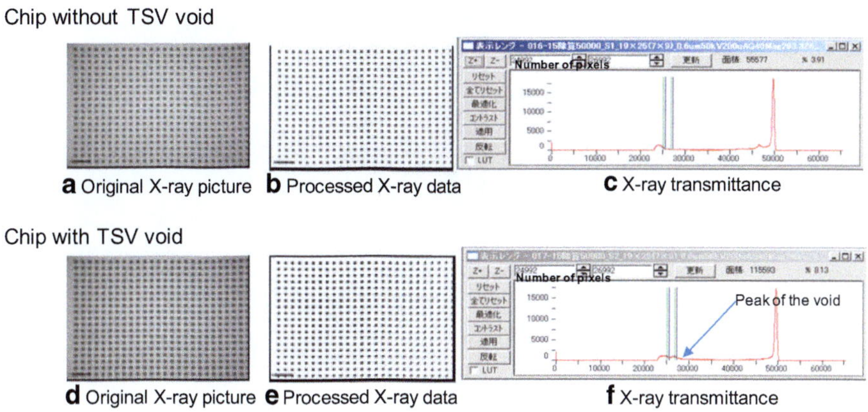


**Fig. 9.27** Results of transmission X-ray analysis. *TSV* through-silicon via

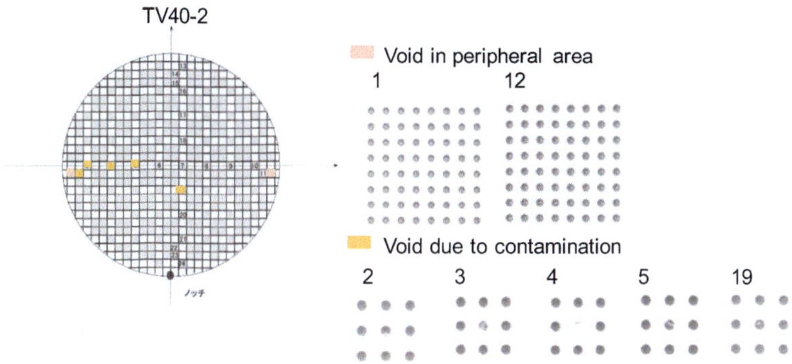


**Fig. 9.28** Observation result of voids in TV40 wafer

**Fig. 9.29** X-ray CT equipment used for this test

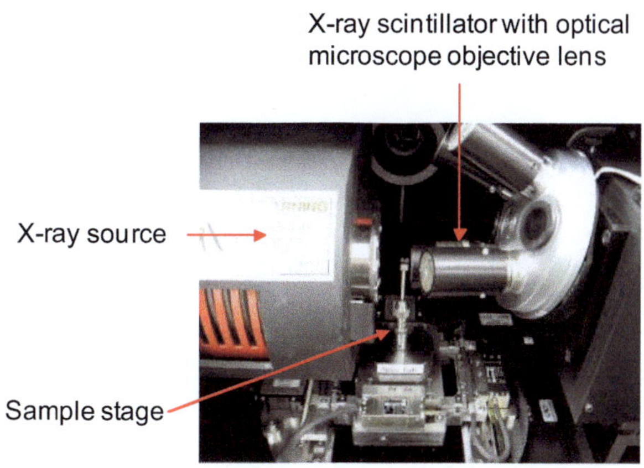

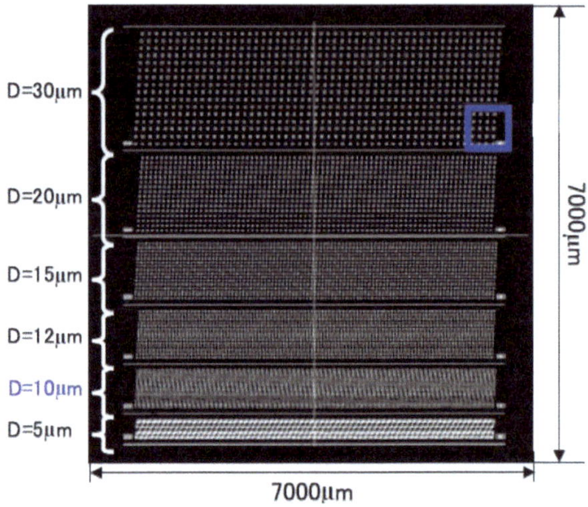


**Fig. 9.30** Test chip with variety of via sizes

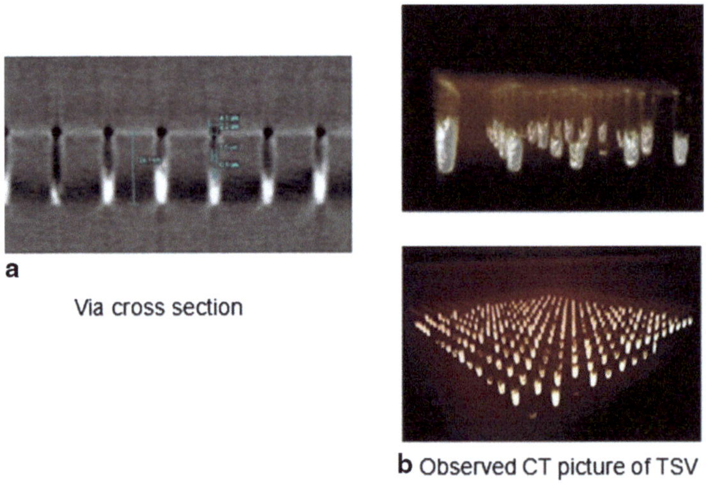


**Fig. 9.31** Results of X-ray CT observation. *CT* computed tomography, *TSV* through-silicon via

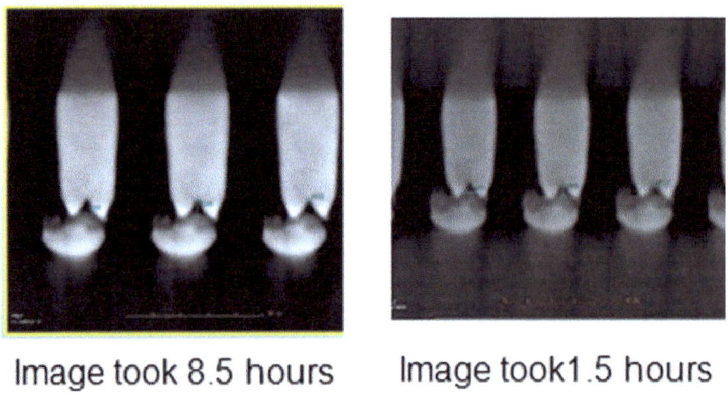


**Fig. 9.32** Results with different observation times

**Table 9.3** Conditions of observation

| | 1st trail | 2nd trail |
|---|---|---|
| Objective | 20X | 40X |
| Source settings (kV/W) | 80/7 | 80/7 |
| Pixel size (μm) | 0.751 | 0.374 |
| Start and end angle | 180 | 180 |
| Number of views | 800 | 4000 |
| Time per view (sec) | 4 | 5 |
| Total acquisition time (h) | 1.5* | 8.5* |

*We succeeded to reduce the measurement time significantly, and minimize the degradation of the observation quality, by the optimization of measurement conditions.

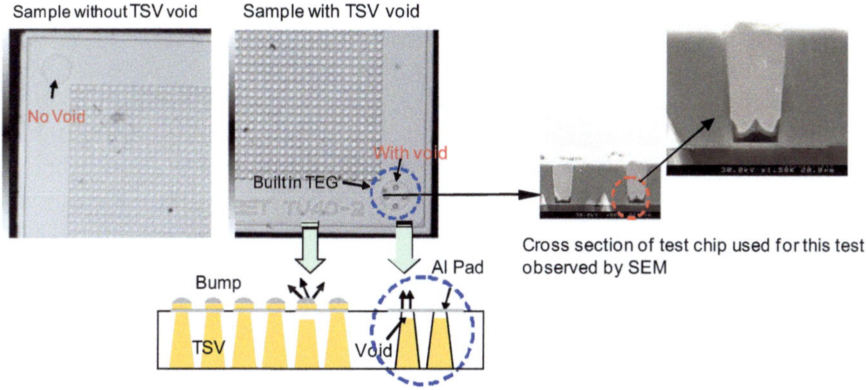


**Fig. 9.33** Principle and TEG structure of TSV void using ultrasonic tool. *TEG* test element group, *TSV* through-silicon via, *SEM* scanning electron microscope

other source of fluctuations will increase by taking time for observation. Table 9.3 shows the summary of experimental conditions.

*Observation of TSV Voids by Using Ultrasonic Apparatus* We tried to detect the TSV void of TV40-2. It is not possible to detect the void of TSV underneath the bump usually. We prepared a built-in TEG structure in chip design as shown in Fig. 9.33, and we found that it is possible to detect the void in TEG area. Although, it is a sample detection, it was possible to find voids in 20-μm-size TSV.

Al layer was prepared at the bottom of the TSV, the absorption of ultrasonic wave at this layer was different depending on the existence of the void. By this method, we could detect the TSV void caused by, for example, degradation of the TSV plating chemicals, without using the X-ray. This method might be useful when the device is sensitive to the exposure to X-ray. Specification of the ultrasonic detector is summarized in Table 9.4.

**Table 9.4** Specification of ultrasonic transmitter/detector

| Tool | CSAM images: Nippon BARNES |
|---|---|
| Frequency | 230 MHz, SW |
| Field of view | 1.02 × 1.02 mm |
| Resolution | 1024 × 1024 |

## 9.2.3 *Thermal Management Study*

### 9.2.3.1 Evaluation Technology of 3D Integrated Chip Stack

We evaluated the thermal conductivity of each component of 3D chip-stack structure.

Thermal conductivity evaluations of each 3D chip-stack structure component.

Steady Method Thermal Conductivity Measurement system was constructed and used for this study. The system is shown in Fig. 9.34. We evaluated the thermal conductivity of the joint structure to be 37–41 W/K, as shown in Table 9.5 based on the comparison of those experimental result and calculated thermal conductivity based on the structure and the values of bulk material from literatures. Measurements of the thermal conductivities were carried out using test chips with different sizes, as shown in Fig. 9.35. TV200 is a 9-mm$^2$ 3D stacked structure thermal characteristics evaluation chip set with 200-μm connection pitch, as shown in Fig. 9.36. Chip has several sets of a diode temperature sensor and a diffused resistor heater made in 1 mm square area. Sets of heaters and diodes are located as equally spaced $4 \times 4$ matrix, center, and a corner of the chip as shown in Fig. 9.37 [13]. Three temperature-sensing diodes were connected to monitor the local temperature. TV200s has four types (three types and spacer) of different on chip wiring to make it possible to lead connections of diodes and heaters from the different layers of stacked chips, as shown in Fig. 9.38 [14]. Equivalent thermal conductivity of the C2C joint is estimated to be 1.6 W/mK.

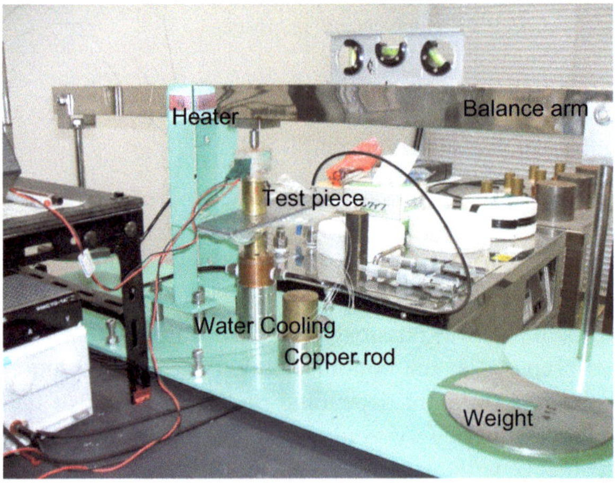


**Fig. 9.34** Steady method thermal conductivity measurement system

**Table 9.5** Evaluated thermal conductivity of the metal joint (W/mK)

| | |
|---|---|
| 250 μm pitch test chip | 35–48 |
| 500 μm pitch test chip | 37–41 |

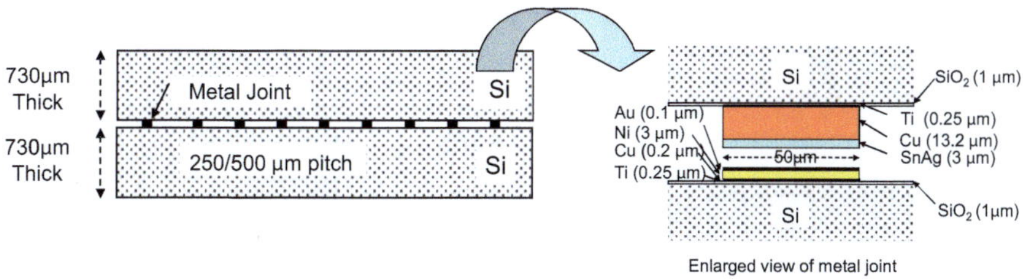


**Fig. 9.35** Test chip for measurement

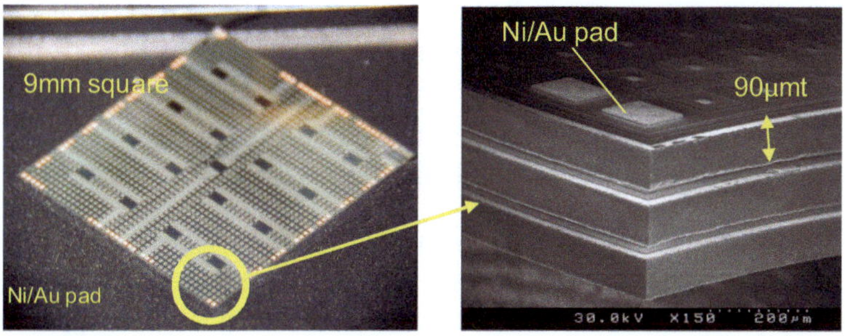


**Fig. 9.36** TV200

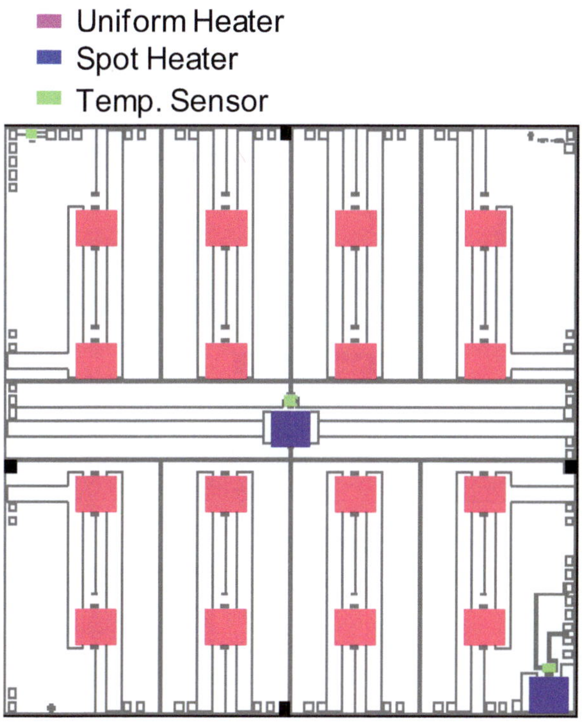


**Fig. 9.37** TV200 chip layout

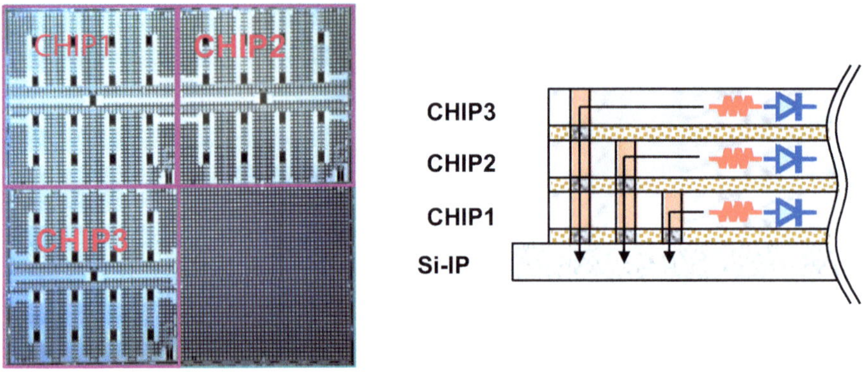


**Fig. 9.38** TV200 chip wiring

#### 9.2.3.2 TV200 Measurement Result and Correlation with Simulation

We constructed the simulation model to the TV200 three-layer 3D stack structure.

Figure 9.39a1, b1 show the top chip surface temperature profile observed by infrared temperature measurement system, when 2 W of power was applied to one heater on the top chip at the center, and 3 W of total power was applied to all the 16 heaters on a top chip uniformly, respectively. Figure 9.39a2, a3 are temperature measurement results using diode sensors on each chip, and the result of simulation,

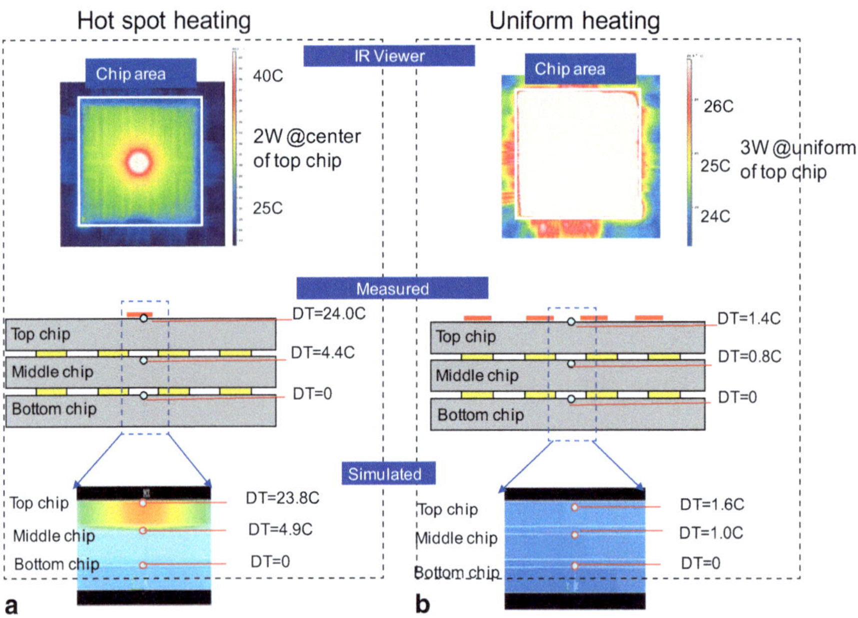


**Fig. 9.39** Temperature distribution depend on spot and uniform heating

Table 9.6 Parameters used for thermal simulation

| Material | Thermal conductivity (W/mK) |
|---|---|
| Si | 148 |
| Cu | 398 |
| $SiO_2$ | 1.3 |
| Interface (equivalent thermal conductivity) | 1.6 |
| Air | 0.0026 |

respectively, when 2 W was applied to the heater of the top chip at the center. Figure 9.39b2, b3 are the similar results when 3 W of power was applied uniformly to the all 16 heaters at the top chip. As shown in these figures, measured results and the simulation results are in reasonably good agreement. Parameters used for this simulation are shown in Table 9.6 [14].

#### 9.2.3.3 Thermal Conductivity Anisotropy Induced by Cu TSV

Thermal conductivity of Si substrate with Cu TSV has an anisotropy, because the in-plane thermal flow is interrupted by low thermal conductive dielectric material insulator ($SiO_2$(1.3 W/mK)), while the thermal conductivity of through thickness direction is increased, in general, because the thermal conductivity of Cu (398 W/mK) is larger than that of Si (148 W/mK).

We estimated the thermal conductivity anisotropy by the calculation based on one pitch model shown in Fig. 9.40 using parameters in Table 9.6. Result of this calculation is shown in Fig. 9.41 [14].

### *9.2.4 Development of Automobile Drive Assistance Camera*

#### 9.2.4.1 Development of Integration Process

CMOS image sensor is on top of the integrated 3D image sensor stack. Color filters made of organic dye or pigment and optical condensing lens array are fabricated on top of the CMOS image sensor. The temperature for the 3D stack assembly process should be less than 180 °C, because those organic optical materials are sensitive to the temperature.

We have chosen low-temperature SnBi solder for the chip joining of the pair of 3D stacks to the Si interposer. Optical components on top of 3D image sensor stack were fabricated at the last step of the 3D stack assembly, so that the process temperature can be higher than 180 °C during the 3D chip-stacking process. We developed the process to mount low-temperature SnBi solder (melting point is 139 °C on the bumps for joining. For the solder bump formation, ball mount, plating, solder paste printing, molten solder molding are typical examples of the process. There are

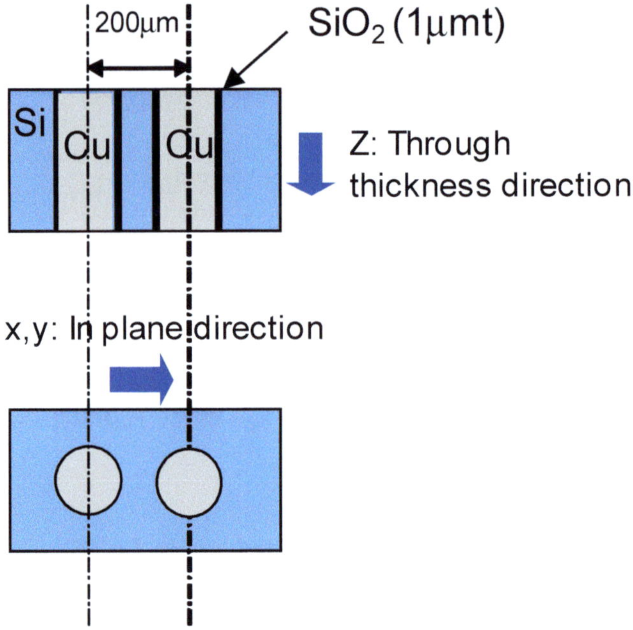


**Fig. 9.40** Model for thermal conductivity anisotropy caused by TSV

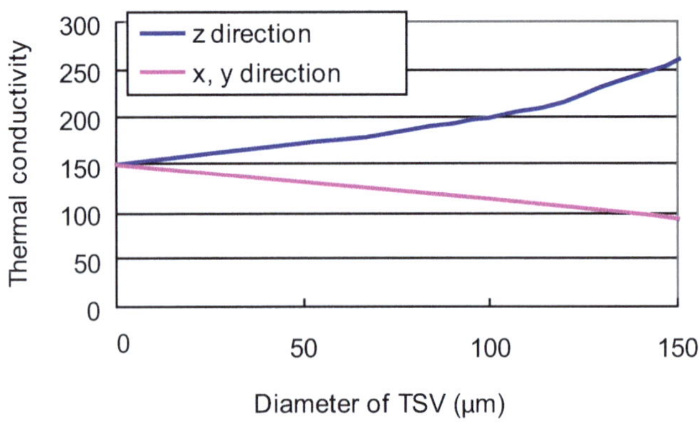


**Fig. 9.41** Thermal conductivity anisotropy depend on TSV diameter

two kinds of pitch/sizes on a chip to be assembled. Larger one is 200 μm/100 μm, and the other is 80 μm/50 μm. Solder paste printing method was not adequate for the smaller pitch/size. Low-temperature SnBi solder plating process was premature to apply. Molten solder molding process of the smaller size was still under development. Only the state of the art ball mounting technology had a potential to deal with this material and size. Minimum ball size for this latest technology was 50 μm. We have to cut out the excess amount of solder from the bump, since the bump height

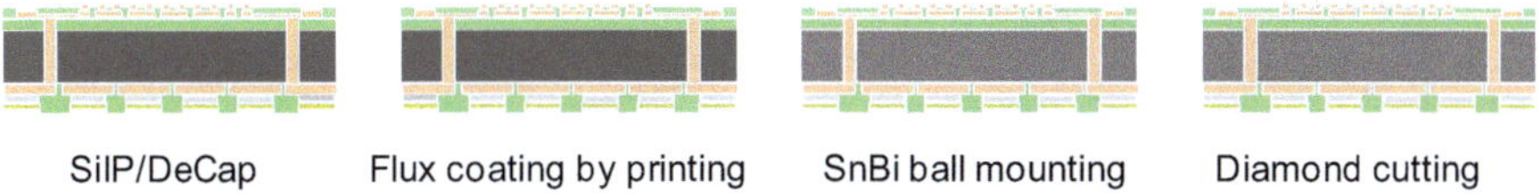


**Fig. 9.42** SnBi bumping by ball mount and diamond cutting

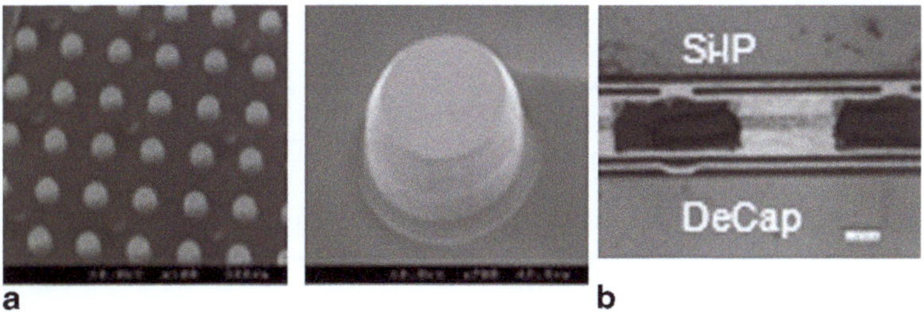


**Fig. 9.43** SnBi solder bump **a** and **b** a result of bonding. *Si-IP* silicon interposer, *DeCap* decoupling capacitor

of the smaller bump using this ball was too high as 40 μm to have a risk of bridging between the neighboring bumps during stacking process. We succeeded to trim the bump to the height of 10 μm using diamond cutting.

Figure 9.42 shows the process to make solder bumps. We printed flux on Cu post of Si Interposer, then mount 50 μmϕ eutectic SnBi solder ball. Si interposer wafer was put into the reflow oven to melt the solder once to form solder bumps. Then the wafer was put into the diamond cutting process to reduce the bump height of the small bump from 40 to 10 μm. Scanning electron microscope (SEM) picture of the solder bump is shown in Fig. 9.43a. Figure 9.43b is the cutout view of the bonded chips showing good result.

Figure 9.44 shows the image capture unit assembly process flow. 3D image sensor was bonded at 160 °C with a pressure of 5 N after Si interposer and decoupling capacitor (DeCap) chip were bonded.

#### 9.2.4.2 Development of Cooling System for Automobile Drive Assistance Camera

We developed cooling system for 3D stereo image sensor for automobile drive assistance. To assist the drivers of the car, stereo image sensor enables to measure the distance to the obstacles in front of the car.

Two sets of 3D image sensor stacks consists of CMOS image sensor chip, CDS chip, ADC chip, and interface chip are assembled on to Si-interposer/DeCap substrates, as shown in Figs. 9.45 and 9.46. Power consumption of each sensor stack is estimated to be about 2 W.

Cooling at high ambient temperature as 65 °C is required.

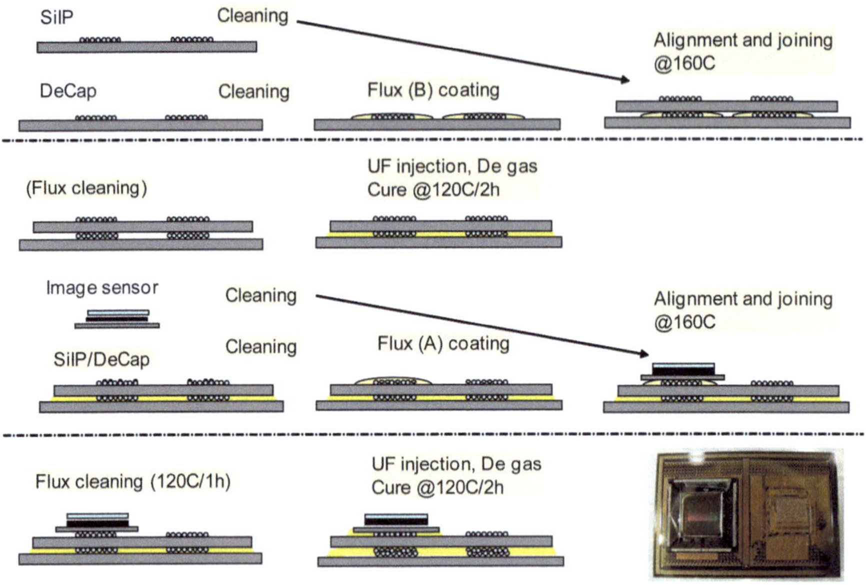


**Fig. 9.44** Image sensor assembly process. *Si-IP* silicon interposer, *UF* ultrafiltration, *DeCap* decoupling capacitor

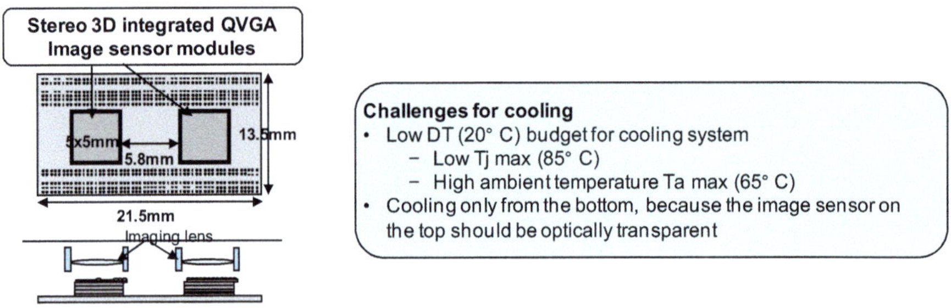


**Fig. 9.45** Structure overview of 3D stereo image sensor

Tj (junction temperature) should be less than 85 °C to keep low enough nose level.

Cooling from the top (Image sensor side) is not appropriate, because at the surface of the CMOS image sensor chip is cover by micro-lens array to optically improve the aperture ratio

We decided to divide the total temperature gradient budget of 20 °C. (85–65 °C) to two portions; 10 °C for top chip (Image sensor) to the surface of interposer, and 10 °C from the surface of the interposer to the ambient air, as shown in Fig. 9.47.

We considered the following three strategies to remove the heat generated within the 3D chip stack.

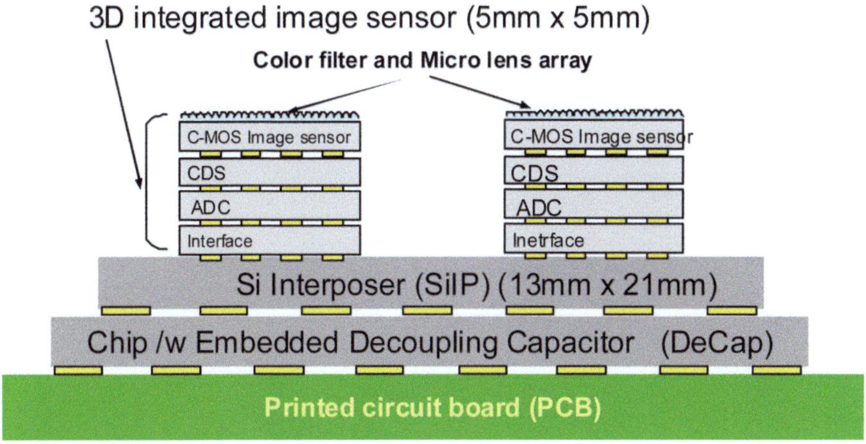


**Fig. 9.46** 3D chip-stack structure of 3D stereo image sensor

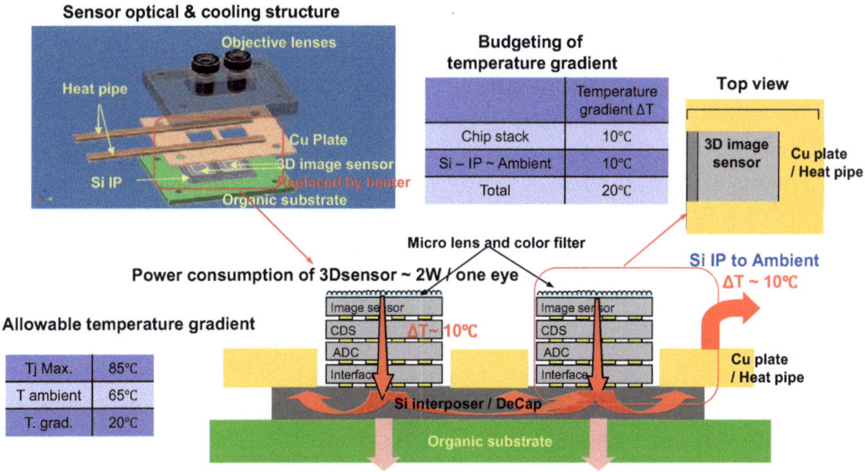


**Fig. 9.47** Strategy of cooling design under high ambient temperature

1. Lower thermal conductivity of the chip in-plane direction and the C2C interface.
2. Cooling from the bottom of the 3D integrated chip stack to the peripheral of the interposer. Then remove heat from the surface of the peripheral of the Si interposer to ambient through heat pipe heat conveyer.
3. Cooling from the bottom of the 3D integrated chip stack through the organic circuit board with a buried Cu heat conductor to the bottom of the organic board to the heat sink.

Third idea turned out to be not adequate since the organic circuit board under the Si interposer is extremely crowded with the wiring and ground plane to handle high-speed signals and power supply. We developed the cooling system by investigating 1 and 2 to satisfy the requirement.

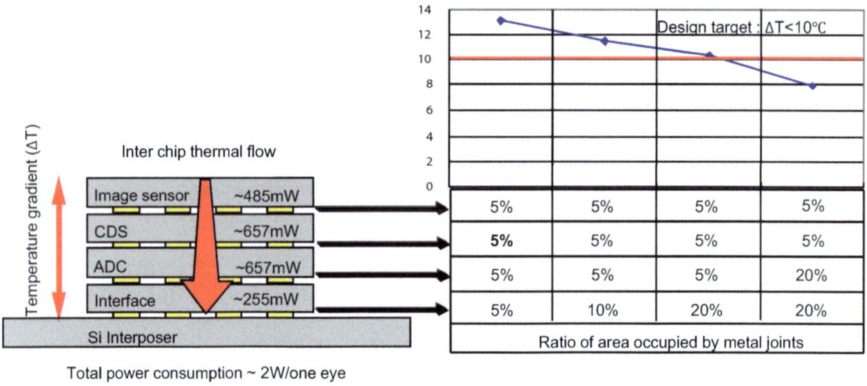


**Fig. 9.48** Cooling efficiency dependency on the area of thermal metal

Thermal conductivity of the metal joint and the inter-chip-fill resin are assumed to be 39 and 0.4 W/mK, respectively. Thermal conduction between stacked chips highly depends on the thermal conduction of the metal bump. Assuming the shape of the of metal joint bumps are circular and its diameter is a half of the connection pitch, the area occupancy of the metal joints is about 20 % with full array configuration. Figure 9.48 shows the $\Delta T$ between CMOS sensor chip and interface chip surface, as a function of metal joint areal occupancy ratio, considering the power consumption of each chip into account.

Because the thermal flow is from the CMOS sensor to the interface chip, increase of joint density between ADC, that is a second chip from the bottom, and interface, that is the most bottom chip is most effective to reduce the temperature gradient. It turned out that when the metal joint is fabricated as full array between ADC and interface, temperature gradient between the CMOS sensor chip and the interface could be about 10 °C, so that the cooling system can satisfy our target as shown in Fig. 9.49 [15]. For this calculation, thermal conductivities are assumed to be 0.4 W/mK for the inter-chip fill, and 39 W/mK for metal joint.

### *9.2.5 Summary*

We accomplished the research and development for thermal management and chip-stacking technology 5 years target of dream chip project. We reported in this chapter as follows.

1. We developed an evaluation and measurement method of 3D micro-joint for the cooling system design and simulation.
2. We developed test vehicles for the evaluations of 3Dstacked structure, and verified the consistency between the measurement result and thermal simulation.
3. We developed cooling system for the instruments used in a high-temperature car cabin.

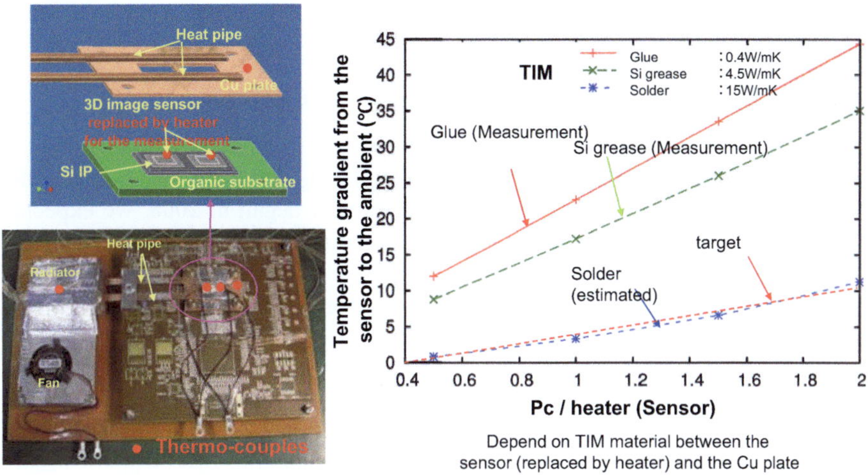


**Fig. 9.49** Cooling structure and measured performance

4. We made up design guide of thermal bump layout to keep enough thermal conductivity between 3D stacked chips [16, 17].
5. We studied and determined optimum joint material and structure. We also confirmed its reliability.
6. We developed pre-apply inter-chip-fill resin for 3D C2C integration process. We verified the junction performance of this new resin and process, as good as the resin for conventional capillary fill process.
7. We developed gang bonding up to three chips stack at one time using pre-apply inter-chip fill resin suitable for C2C 3D integration process, and confirmed its joint performance.
8. We developed high precision alignment method for 10-μm pitch joint.
9. We developed new bonding tool head made of very high thermal conductivity that enables flat surface topology after bonding even when the substrate is very thin. We often observed irregular and large surface variation relative to the ideal flat surface, using conventional tool head made of AlN, when the substrate is very thin.
10. We developed nondestructive observation technique using TEG structure on a chip by ultrasonic tool. We confirmed the resolution of this scheme capable to 20 μm bump size.
11. We developed wafer level X-ray observation scheme and tool possible to observe void in TSV cable to 20 μm bump size.

Those accomplishment is useful for integration process and cooling system design of next generation 3D integration circuit (3D-IC)

## 9.3 Thin Wafer Technology

### 9.3.1 Background of Wafer Thinning Technology

Figure 9.50 shows the historical trend of diminishing wafer thickness since 2003 and predictions until 2020. When the ASET project began in 2008, we expected the reduction of wafer thickness to follow the predictions indicated by the *blue line* in the figure. However, devices without TSV have already reached thicknesses of 30 μm in high-volume manufacturing, for example, negative AND (NAND) flash memory, and in special usages, such as backside illumination CMOS image sensors (BSI-CIS), wafers are slimming down to around 10 μm thickness [18]. High-volume manufacturing of middle-to-low voltage power devices has started using wafers with a thickness of 70 μm and a diameter of 300 mm. For system on chip (SoC) within SiP devices, such as smartphones, a 90-μm thickness wafer is usual.

Regarding TSV-based technologies, the thickness of a 3D integrated circuit (3DIC), the SoC die for smartphones is still 40–50 μm in 2013 (*red line* in Fig. 9.50) because of ultrathin die fragility [19]. Hybrid memory cube (HMC) and high bandwidth memory (HBM), which are types of high-performance DRAM, are also manufactured to this thickness range [20, 21]. The die thickness is determined by the easiness of handling and, the TSV diameter is determined by the circuit design and the via-processing technology. For example, when the Si dry etch aspect ratio (AR) is 10, a 50-μm die thickness is sufficient for 5 μm diameter via openings.

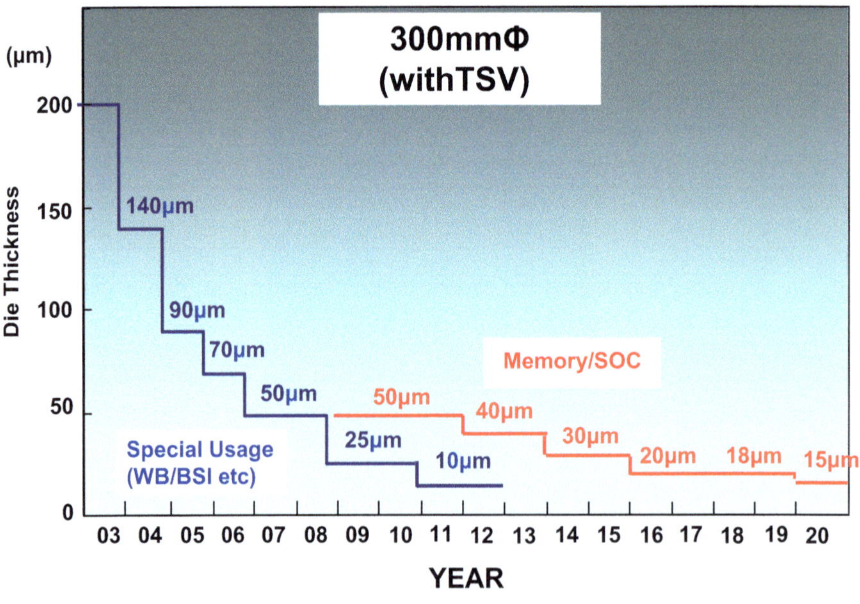


**Fig. 9.50** Roadmap of wafer thickness. *TSV* through-silicon via. *SOC* system on chip

To realize future capability for 1 μm diameter via openings, our final wafer thickness was determined to 10 μm in the ASET project.

### 9.3.2 Issues of Wafer Thinning

There are two main device-related issues in regard to wafer thinning. One is the wafer's capacity to withstand the higher stress concentrations. The other is the difficulty in eliminating the intrinsic gettering (IG) that forms from the bulk micro-defect (BMD) layer in the Si wafer. The operation of the device may alter depending on the wafer thickness. For example, in DRAM, retention time and error bit rate change [22].

Wafer thickness also influences externally derived metal contamination. In front-end thermal-treatment processes, the IG layer in thick Si substrates retains the micro-defect domain, which benefits from the gettering effect. Reducing the wafer thickness to the 50–10-μm range will eliminate the gettering effect because of the absence of BMD; the non-defect denuded zone (DZ) cannot trap pollutants. The Si backside grinding method that induces a micro-damage layer is a countermeasure to produce the gettering effect, but this also results in reduced wafer strength. The dry polish method is one solution to this problem [23].

In industry, there are several TSV formation processes, most prominently via first, via middle, and via last. These involve TSV etching, insulation liner deposition inside the TSV, and a viafilling process using a metallic material. Several kinds of wafer support system (WSS) have been proposed to handle the fragile ultrathin wafer during these formation processes [24–26].

Figure 9.51 describes several known issues in TSV 3DIC development. These manifest in the wafer backside thinning processing, WSS technology to laminate a wafer support material to a thick wafer temporary, de-bonding technology that separates the thin wafer from the support material, metal contamination countermeasures such as Cu-induced under thinning processing, and manufacturing-induced stresses.

### 9.3.3 Ultrathin Wafer Thinning Process

#### 9.3.3.1 Wafer Support System

It is common to use an Si or glass support carrier to process and handle the ultrathin wafer, for example, ASET use a glass carrier.

In ultrathin wafer processing, the total thickness variation (TTV) must be tightly controlled to ±1 μm after backside grinding to ensure process stability at the die-stacking solder joint, which is integral to C2C and W2W technologies. To achieve this precise TTV, there must be even more restrictive tolerances for the processing of the component materials (Fig. 9.52). That is, in the precision of wafer thinning by backside grinding equipment, in the material thickness variation of the Si wafer and

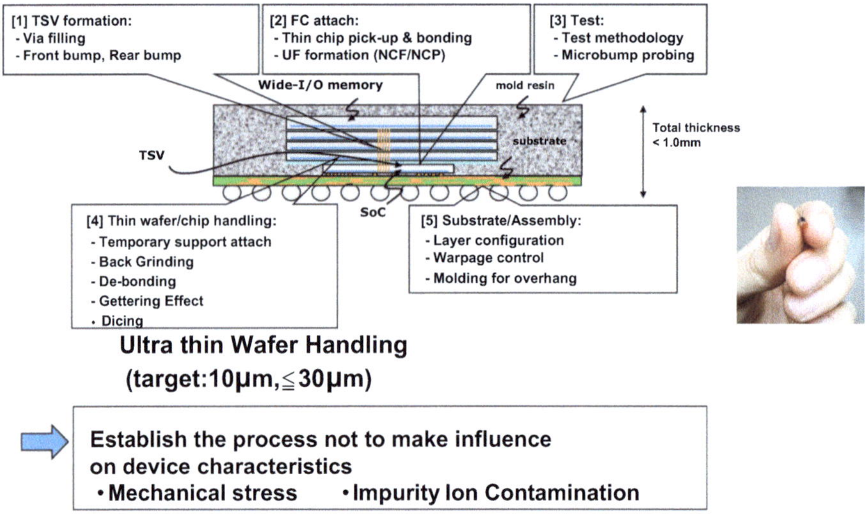


**Fig. 9.51** Issues in TSV 3DIC development. *TSV* through-silicon via, *UF* underfill resin, *NCF* nonconductive film, *NCP* nonconductive paste, *SoC* system on chip

Target thickness accuracy of 10 μm±1 μm wafer (TTV)

| Item | Current accuracy | Target |
|---|---|---|
| A: glass | 1~2 μm | 0.8 μm |
| B: wafer | 1 μm | 0.8 μm |
| C: adhesive | 2.5 μm | 0.8 μm |
| Total thickness (A+B+C) | 3~4 μm | D: 1.4 μm |
| BG | 3.~5 μm | E: 1.4 μm |

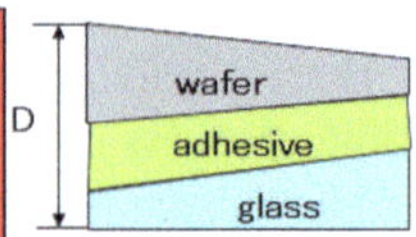


TTV (Total Thickness Variation)
Difference between max and min thickness in wafer

$$\sqrt{A^2+B^2+C^2}=\sqrt{0.64+0.64+0.64}=\sqrt{1.92}=1.4\,\mu\text{m}\,(D)$$

$$\sqrt{D^2+E^2}=\sqrt{1.4^2+1.4^2}=\sqrt{1.96+1.96}=2\,\mu\text{m}\,(\text{Final TTV})$$

**Fig. 9.52** Target accuracy in each component

the glass carrier, in the thickness variation of temporary adhesive, and in the precision of bonding parallelism between the Si and the carrier. In the wafer-stacking process, less than 2-μm overall variation is permitted in the whole positions. To achieve this, the TTV of, for example, the glass carrier, should be less than 0.8 μm (see Table 9.7).

Fig. 9.53 shows the WSS process flow. We will now describe the method developed by the 3M Company. The features of this method are bonding the Si wafer with the glass carrier by UV-hardening resin then, after the TSV process, the boundary portion between the glass and temporary adhesion is broken using laser irradiation. The wafer is de-bonded from the carrier at a low temperature and with minimal stress. To keep the bonding TTV tolerance within 0.8 μm, a flattening disk (FD) is often used to improve the coplanarity between the wafer and glass. The FD is a pad with a cushioning effect that uniformly distributes the load on the glass. To control

Table 9.7 ASET glass carrier specifications

| Outline (mm) | Material | Thickness (μm) | TTV(μm) | V Notch | Edge |
|---|---|---|---|---|---|
| Φ300.2±0.1 | Tempax | 700±1 | ≦0.8 | SEMI STD. | C-Cut (0.15 mm) |
| Φ200.2±0.1 | ↑ | ↑ | ↑ | ↑ | ↑ |

(700.0±5 mm)

(300.0±0.1 or 300.4±0.2 mm)

(Ra<1.0 nm)

(200.0±0.05 or 200.4±0.1 mm)

(Reference STD.SEMI 3DS-IC: 5173D, 5588)

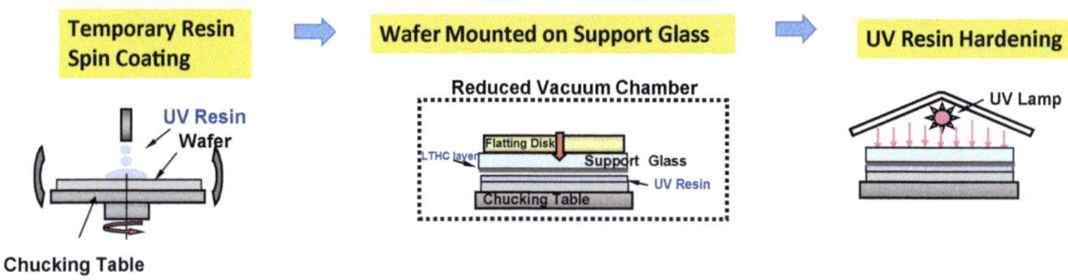


**Fig. 9.53** WSS process flow (e.g., 3M). *UV* ultraviolet, *LTHC* light to heat conversion

the thickness variance at the wafer periphery and notch area, the temporary adhesive upsurge (edge beat) that occurs during the adhesive spin coating process must be minimized. By optimizing the FD force and loading time, and by using a particular type of glass, a TTV after glass bonding of less than 1.4 μm can be achieved.

To control the wafer TTV after de-bonding to be less than 2 μm, the wafer backside grinding process under WSS is also important. The wafer thickness during the thinning process should be monitored in real time. Using an in-line process gauge (IPG) to monitor wafer thickness by direct contact is the standard method employed in the conventional back grinding (BG) machine. However, as shown in Fig. 9.54, a noncontact gauge (NCG) using an IR sensor has been recently introduced as a more accurate replacement. In addition, it is inevitable that an automated TTV control method (auto-TTV) will be adopted that can fine-tune the relative angle between the grinding whetstone and the wafer chucking stage to control the thickness distribution in wafers with a glass support [27].

In Fig. 9.54, we present some wafer-thinning results with measurements made using different thickness monitoring gauges. In these evaluations, we tested two types of wafers that had different bump densities (see Fig. 9.55). Sample W1 had the same bump density (=total bumps area/wafer area) as a commercial product with a bump density of 0.007, and bump diameter and height each of 5 μm, resulting in just two or three lines of bumps. In comparison, the W2 had a full matrix bump

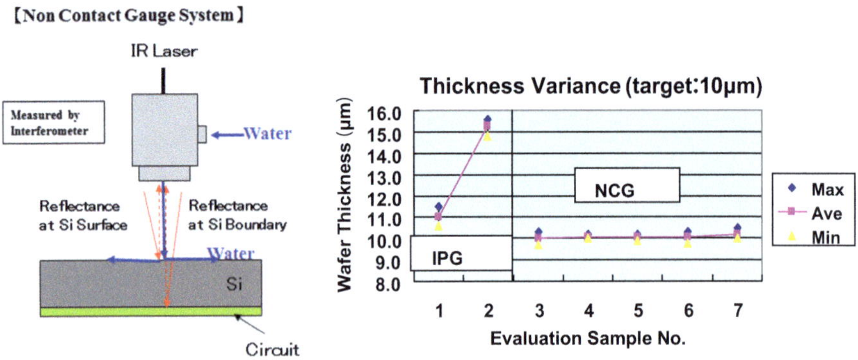


**Fig. 9.54** In situ measurement method of wafer thinning and evaluation results. *IR* infrared, *NCG* noncontact gauge, *IPG* in-line process gauge

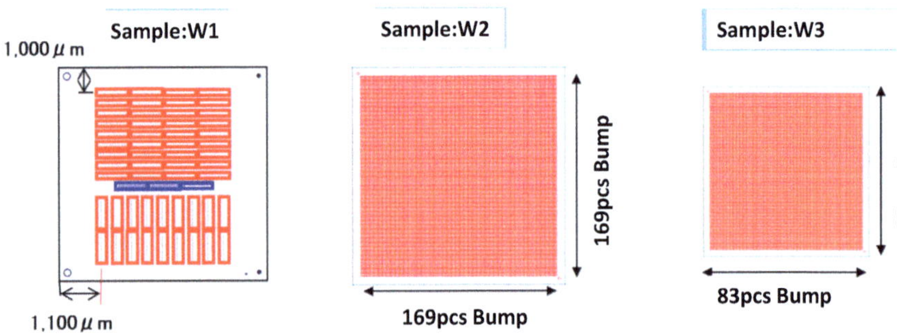


| | TEG | | |
|---|---|---|---|
| | W1 | W2 | W3 |
| Die Size | 7mm□ | 7.3mm□ | 5mm□ |
| Bump Pitch | 10μm | 40μm | 50μm |
| Bump Diameter | 5μmΦ | 33μmΦ | 35μmΦ |
| Bump Height | 5μm | 8μm | 45μm |
| DieThickness | 50μm | 50μm | 50μm |
| Bump Density | 0.007 | 0.218 | 0.024 |

**Fig. 9.55** The bump layout of three types of TEG samples. *TEG* test element group

array with bumps of diameter 32 μm and density of 0.218. Both wafers were thinned down to 50 μm using a C2C process.

The results after WSS mounting, BG/chemical mechanical polishing (CMP) are shown in Fig. 9.56. The TTV of W1 and W2 were 1.35 and 1.98 μm, respectively, which achieved the target value of 2 μm (±1 μm). Compared with the bare wafer, the initial thickness variation of the bumped wafer had an influence on the total thickness after the carrier glass was attached. For example, in the case of W1, TTV was 3.1 μm after the carrier glass was attached because the initial wafer TTV was around 2.5 μm. However, by using an auto-TTV function in the BG equipment, it was possible to achieve a final wafer TTV after CMP of 1.35 μm. Similarly, the TTV of sample W2 was reduced from 3.0 to 1.98 μm.

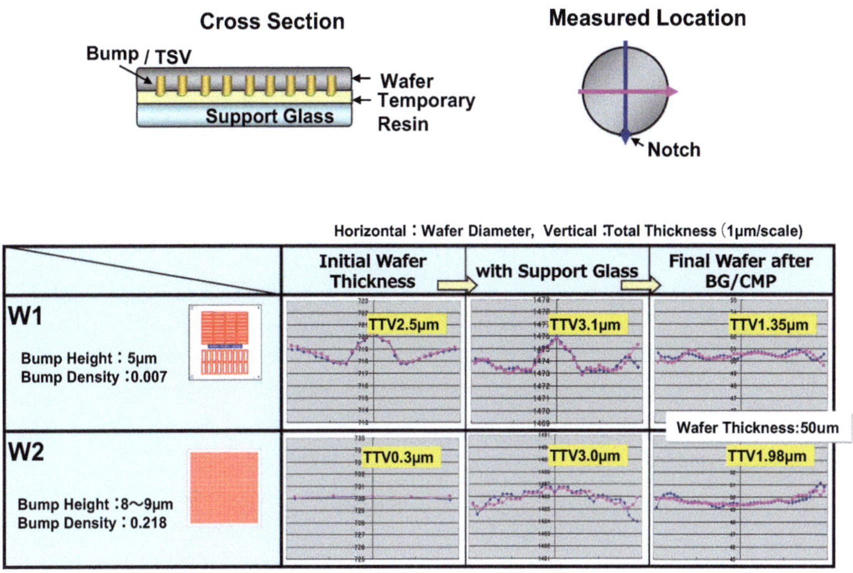


**Fig. 9.56** Experiment results of wafer thinning. *TTV* total thickness variation, *BG* back grind, *CMP* chemical–mechanical polishing

The main contributors to TTV after the glass carrier was attached were the thickness nonuniformity of the temporary adhesive coating, the height and density of the bump array, and the wafer notch shape.

### 9.3.3.2 Heat Resistance of the Resin Used for WSS Temporary Bonding

To improve the quality of the oxidation layer ($SiO_2$), which determines the insulation characteristics between the TSV and bulk Si, high-temperature chemical vapor deposition (CVD) is needed. Thus, a high-temperature resistance of the adhesive for the WSS is important. We evaluated the heat resistance of materials that induce wafer delamination from the carrier. The temporary adhesive properties of the WSS are shown in Table 9.8 and the heat resistance evaluation method is described in Fig. 9.57.

The study parameters were the wafer bump height and bump density (three kinds) and the type of temporary resin (two kinds). The wafer thickness was 50 μm as per the joint electron device engineering council (JEDEC) standard for Wide I/O memory usage. Resin A was the conventional resin, and resin B was the heat resistance improved version. The heat treatment consisted of applying the plasma CVD process at 220 °C for a period of 15 min. Results are reported in Table 9.9 and Fig. 9.58. The mirror wafer and W3 wafer that had a different bump density (see Fig. 9.55) are included for comparison. The applicable range of each resin in terms of wafer bump height and density is described in Fig. 9.59. Resin A experienced no problems for wafers having low bump height and low bump density but led to the

**Table 9.8** Physical characteristics of WSS temporary material. *TGA* Thermo Gravimetry Analyzer

| | | Resin A | Resin B |
|---|---|---|---|
| Viscosity (25℃) | cps | 2,500∼3,500 | 1,800∼2,000 |
| Modulus (25℃) | MPa | 160∼260 | 1,190∼1,270 |
| Breaking Elongation | % | 20≧ | 30∼40 |
| TGA (250℃ × 1hr) | % | -5.5∼-6.0 | -2.0∼-2.5 |
| Heat Resistance (Max) | ℃ × min | 200℃ × 60min | 250℃ × 60min |

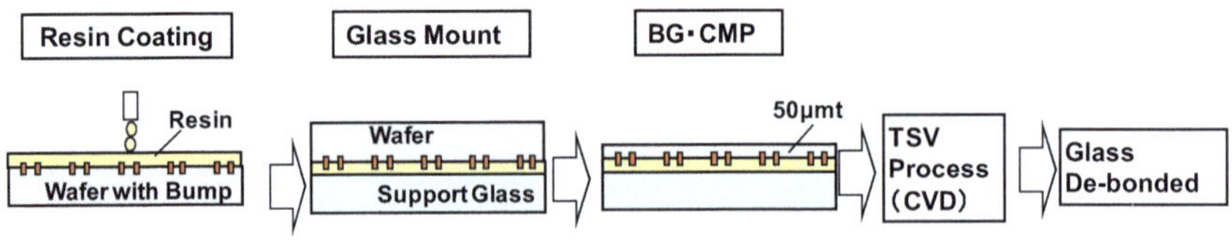


【Bump Condition】

| | Bump Height | Bump Density |
|---|---|---|
| 1 | 5 μm | 0.007 |
| 2 | 8∼9 μm | 0.218 |
| 3 | 45 μm | 0.024 |

【CVD Temperature】

| | Setting Temp. | Actual Temp. |
|---|---|---|
| Standard Condition | 80℃ × 15min | 150℃ |
| Evaluation Condition | 200℃ × 15min | 210∼220℃ |

**Fig. 9.57** Heat resistance evaluation method of temporary resin. *CVD* chemical vapor deposition, *BG* back grind, *CMP* chemical mechanical polishing, *TSV* through-silicon via

**Table 9.9** Heat resistance evaluation results. *WSS* wafer support system, *CVD* chemical vapor deposition

| Wafer Specification | | | | | WSS | | | CVD Condition | | Result | | | | Judgement |
|---|---|---|---|---|---|---|---|---|---|---|---|---|---|---|
| Wafer | Thickness | Bump Diameter | Bump Height | Bump Density | Resin | Thickness | Atmosphere | Temp. | Time | Floating | | Delamination | | |
| Mirror | 50μm | 0 | 0 | 0 | A | 75μm | Air | 200℃ | 20min | OK | ○ | - | - | (○) |
| W1 | ↑ | 5μm | 5μm | 0.007 | A | 50μm | ↑ | ↑ | 15min | OK | ○ | OK | ○ | ○ |
| | | | | | B | 65μm | Vacuum | ↑ | ↑ | OK | ○ | OK | ○ | ○ |
| W2 | ↑ | 33μm | 8∼9μm | 0.218 | A | 50μm | Air | ↑ | ↑ | Broken Wafer (a) | × | Cannot De-bond | × | × |
| | | | | | B | 65μm | Vacuum | ↑ | ↑ | Interference Fringe (b) | △ | Hard Removal | △ | △ |
| W3 | ↑ | 35μm | 45μm | 0.024 | A | 75μm | Air | ↑ | ↑ | Resin Floating & Burned Resin (c),(d) | × | - | - | × |
| | | | | | B | 70μm | Vacuum | ↑ | ↑ | OK | ○ | Hard Removal & Bump Removal | △ | △ |

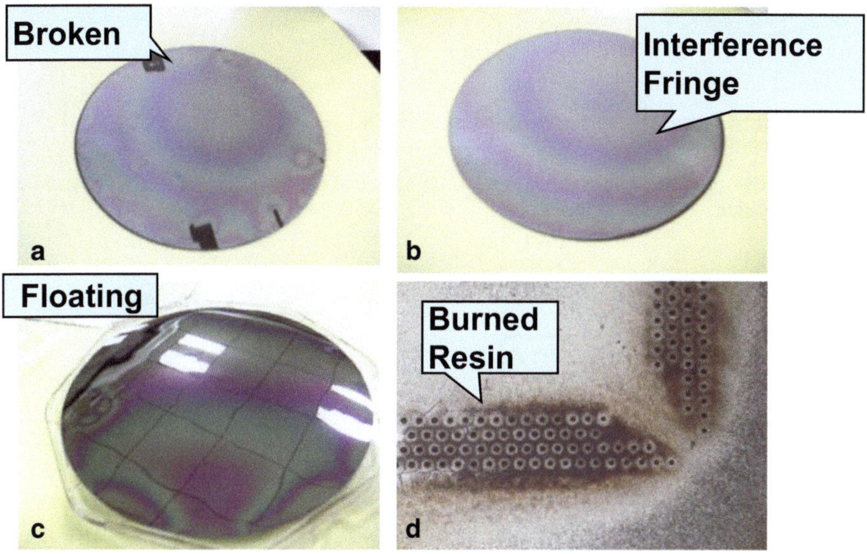


**Fig. 9.58** Example of failure modes

breaking of a wafer with a high bump density and the delamination of another that had tall bumps. In the high-density region, the carrier glass and adhesive resin could not be separated, and there was resin residue among the bump array. The cause of wafer breaking was the generation of air voids around the bump and the thermal expansion of these voids. The strong adhesive nature of the resin residue, which is dependent on how it hardens and sets at high temperatures, prevented it detaching from the bump array.

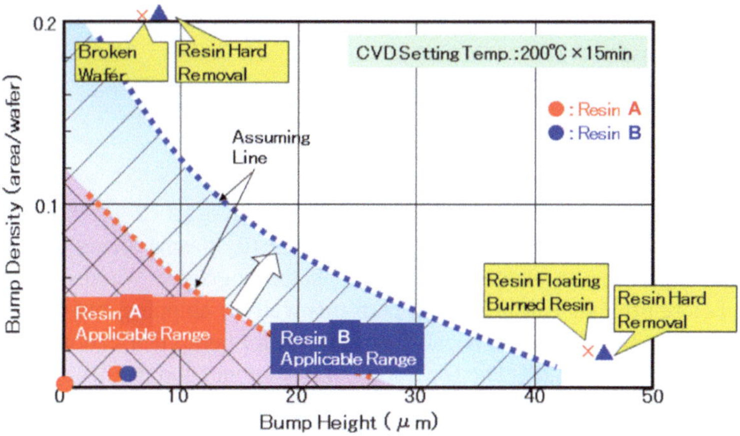


**Fig. 9.59** The usage range of two temporary adhesives, resin A and resin B. *CVD* chemical vapor deposition

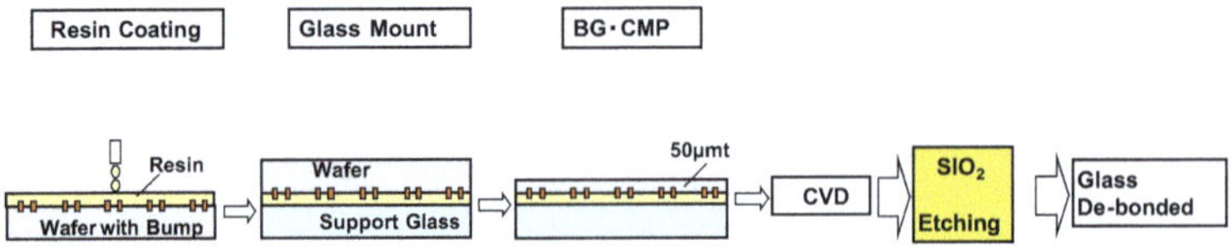


**Fig. 9.60** TSV process flow. *BF* back grind, *CMP* chemical mechanical polishing, *CVD* chemical vapor deposition

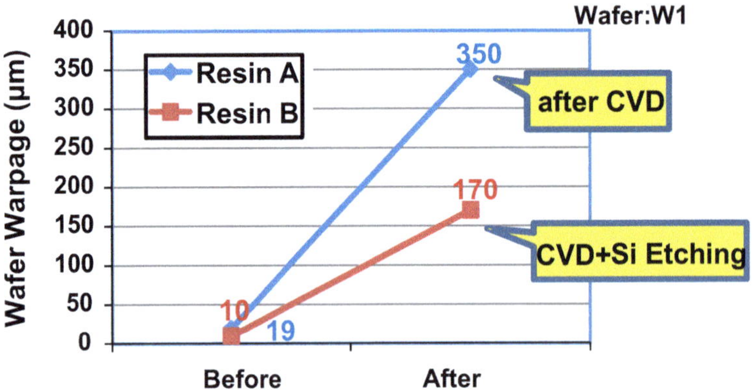


**Fig. 9.61** The effect of thermal-treatment on wafer warpage. *CVD* chemical vapor deposition

To better understand the thermal-treatment process in TSV fabrication, we examined the effect of including an $SiO_2$ etching process after the low-temperature CVD (see Fig. 9.60) on the W1 wafer. The post-fabrication wafer warpage was measured to determine whether a mask aligner would be required. The warpage of resin B was 170 μm, half the value of resin A, and satisfactory (less than 300 μm) to pass the photolithography process. Therefore, a resin heat resistance of at least 250 °C is necessary.

#### 9.3.3.3 Dicing Technology of Thin Chips

To evaluate the dicing process of ultrathin wafers, we chose a 10-μm-thick bare wafer. Several dicing methods are detailed in Table 9.10. Plasma dicing was rejected as a candidate because of the necessity of photolithography. We also considered dicing before grinding (DBG), which performs well for chipping, but could not be employed in 3DIC fabrication because of the need for bump fabrication before dicing and after grinding [28]. Die separation, die chipping, and die bending strength were evaluated for the blade dicing, stealth dicing, nano-Laser and femto-Laser dicing methods. The dicing results are summarized in Table 9.11. Figure 9.62 plots die bending strength for the five dicing methods, and Fig. 9.63 presents the associated side-view inspection photographs. In this evaluation, 10- and 30-μm-thick wafers

**Table 9.10** Various dicing methods. *DBG* dicing before grinding

| | Blade dicing | DBG | Stealth dicing | Nano-sec laser | Femto-sec laser |
|---|---|---|---|---|---|
| Method | Mechanical | Mechanical | Cracking | Fusion | Fusion |
| Tool | Diamond blade | Diamond blade | Pulse laser 1064 nm | Pulse laser 355 nm | Pulse laser 400 nm |
| Velocity | 20 mm/s | 50 mm/s | 180 mm/s | 200 mm/s | 5 mm/s |
| Kerfwidth | 40 µm(30 µm width blade) | 40 µm (30 µm width blade) | 0 µm | 15 µm | 25 µm |
| Chipping | ○ | ○ | ◎ | ○ | ○ |
| Productivity | ○ | ○ | ◎ | ◎ | △ |
| Initial cost | ◎ | ◎ | △ | △ | △ |
| Running cost | ◎ | ◎ | △ | △ | △ |

◎ Best, ○ Good, △ Fair, × Fail

were used. The blade dicing method had the best combined performance of die chipping (≈10-µm cross section) and bending strength (>1000 MPa on average), which is presently the state-of-the-art technology.

In contrast, bending strength decreased in the laser cleaving and laser ablation methods. Laser cleaving reduced bending strength because of the formation of a damaged layer on the side of the dicing street. The damaged layer was formed on the backside of the die so the bending strength measured from BG side decreased dramatically. Therefore, the control of the position of the damaged layer in the die cross section becomes very important.

Recently, improvements in laser equipment have made such precise control possible, leading to widespread adoption of stealth dicing. However, there remains the issue of designing the TEG layout on the dicing street so as not to disturb the optical path of the laser [29]. Irradiation from the backside of the wafer is one solution to this problem. Ablation laser induces the heat damage in the cutting zone and decreases bending strength. A common goal of all dicing processes is to design the structure of the dicing street to improve the chipping quality.

#### 9.3.3.4 Die Pickup Technology of Thin Chips

The wafer for C2C needs to be thin and have bumps on both sides. The following issues are known concerning the pickup process of the die with backside bumps:

1. The die moves and chipping increases during the dicing process because of the wafer backside bumps.
2. The cutting water flows around the backside of the wafer during the dicing process and mixes with the Si cutting powder because of the backside bumps' existence.
3. The adhesive of the dicing tape buries the bumps, which makes the die pickup process difficult as it becomes hard to separate the die from the dicing tape.

**Table 9.11** Dicing evaluation results. *DBG* dicing before grinding

| | Quality | | | | | | | | | | | | | | | Judge[b] |
|---|---|---|---|---|---|---|---|---|---|---|---|---|---|---|---|---|
| | Top chipping (μm) | | | | Backchipping (μm) | | | | Die strength (MPa)[a] | | | Remarks | | | | |
| Wafer thickness | 10 μm | | 30 μm | | 10 μm | | 30 μm | | 10 μm | 30 μm | | 10 μm | | 30 μm | | |
| Blade | 4~7 | ○ | 13 | △ | 9 | ○ | 9 | ○ | | 969~1507 | ○ | | ○ | | ○ | △ |
| | | | | | | | | | | 686~1471 | | | | | | |
| DBG | 24 | × | 6 | ○ | 4 | ○ | 0 | ◎ | | 740~1038 | ○ | Crack in BG | (×) | | ○ | ○ |
| | | | | | | | | | | 991~1344 | | | | | | |
| Stealth | 0 | ◎ | 0 | ◎ | 0 | ◎ | 0 | ◎ | | 1240~3587 | × | Unstable damagelayer | × | | ○ | × |
| | | | | | | | | | | 313~774 | | | | | | |
| nano second (ns)-laser | 2 | ○ | 3 | ○ | 0 | ◎ | 0 | ◎ | | 322~982 | × | | ○ | Adhesive residue | × | × |
| | | | | | | | | | | 347~1115 | | | | | | |
| femto second (fs)-laser | Cannot measure by debris | | | | Cannot pick up by strong adhesion | | 14~16 | △ | | 770~996 | △ | Cannot pick up by strong adhesion | | | ○ | × |
| | | | | | | | | | | 507~715 | | | | | | |
| Spec. | Less than 10 μm | | | | | | | | | c | | | | | | |

◎ Best, ○ Good, △ Fair, × Fail
[a]Upper: Mirror side evaluation, Lower Polished side evaluation
[b]Judged by 30μm thick (Die strength priority)
[c]Min. is more than 500 MPa and Max. is more than 1000 MPa→ ○

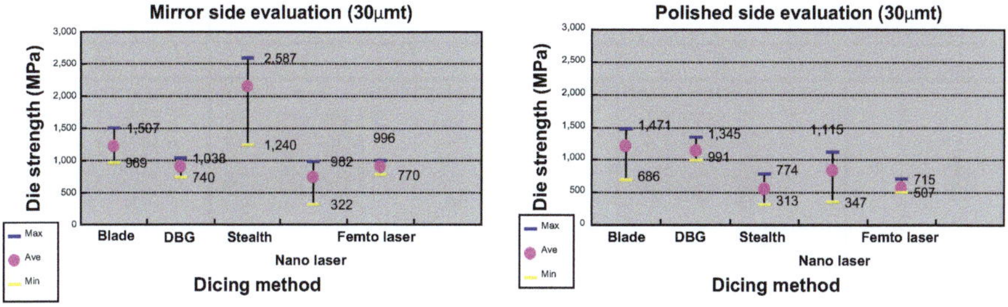


Fig. 9.62 The dependence of die bending strength on dicing method. *DBG* dicing before grinding

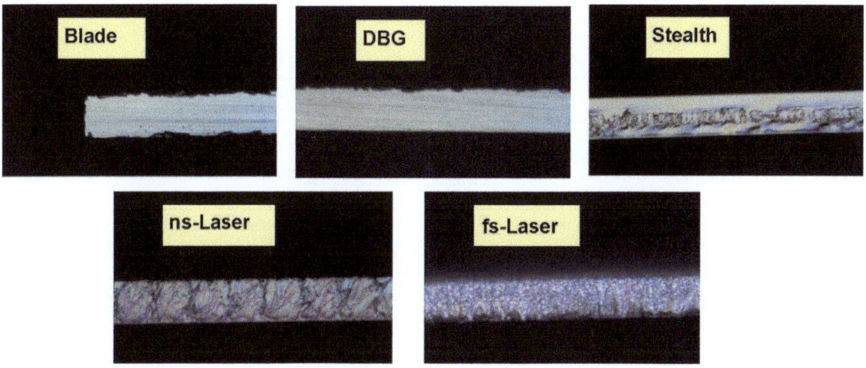


Fig. 9.63 Die cross-sectional view at the dicing street. *DBG* dicing before grinding

Optimization of the dicing tape is important to minimize these problems. For items 1 and 2, it is necessary to improve bump embedding. However, better embedding is more likely to cause problems 3. We thus decided to evaluate different dicing tapes for backside bumps and identify the relationship between the tape properties and the die pickup potential.

### (a) Evaluation of Die Pickup Without Backside Bumps

There are several die pickup methods which includes the push-up needle pin method, the die detachment method by UV bubble foaming of UV type dicing tape, and the slider operation method. In all approaches, the aim is to pick up a 10-μm-thick die and de-bond the die from the dicing tape in the lowest stress manner possible, beginning from the corner of the chip.

In the slider operation method, which can separate the dicing tape from the die edge, we evaluated the die overhang value, dicing tape expansion value, and needle push-up height. The dicing tape was a UV hardening type. The die pickup results are presented in Fig. 9.64, and some die broken modes are shown in Fig. 9.65. The two main factors contributing to die breaking in the pickup process were overhang value and expansion value. An overhang value of 0.3 mm was sustainable, but, at

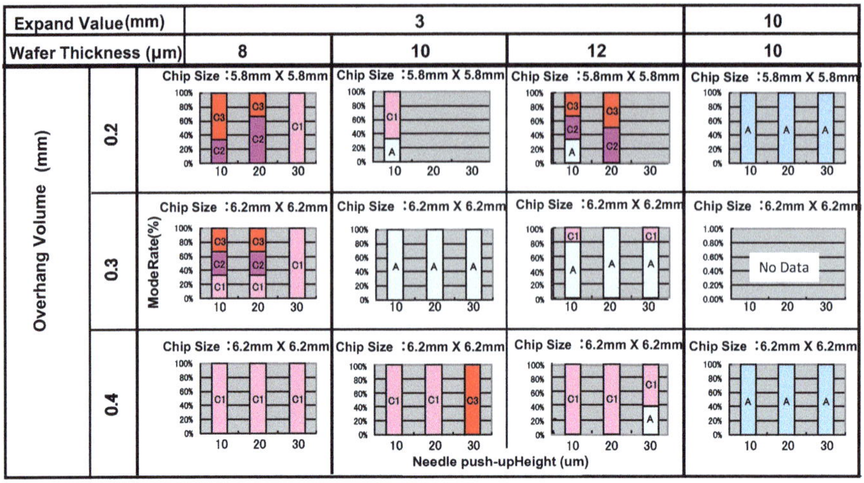


Fig. 9.64 Die pickup evaluation results

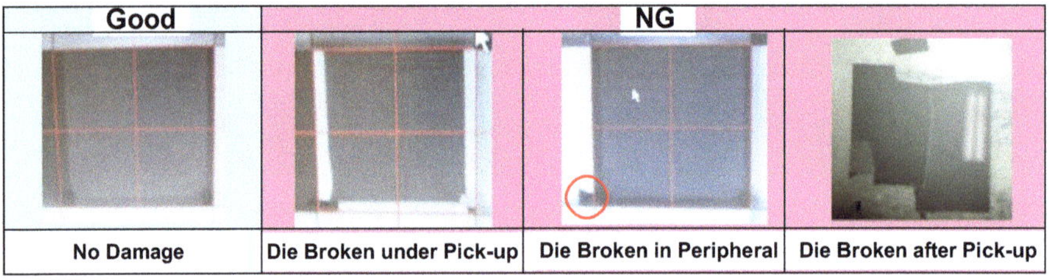


Fig. 9.65 Modes of die breakage after pickup. *NG* not good (or failure)

lower values, the neighbor die broke and, at higher values, the pickup die itself broke because the sticking tape's adhesion force (indicated by a wider die bending area) became too great.

To increase processing margins, wider expansion values are used to make it easier to pick up the die. In addition, it is necessary to ensure that the push-up amount of the push-up pins is not too large to avoid damaging the die by collision with the die collet tool.

We verified that an overhang value of 0.3 mm is adequate by analyzing the pickup mechanism. The pickup evaluation equipment is shown in Fig. 9.66. Details of the machine mechanism are given in Fig. 9.67.

While controlling the push-up height, we measured the die edge bending length with a laser displacement sensor. We also measured the loading force at the corner of the die edge by dicing tape tension with a load cell. The inflection points of the load cell data and the laser displacement data corresponded to the moment of de-bonding between the die and dicing tape. From these curves, the de-bonding point can be determined. The results for the 0.3-mm overhang are shown in Fig. 9.68. De-bonding commenced at 778 ms.

**Fig. 9.66** Equipment for measuring die pickup strength

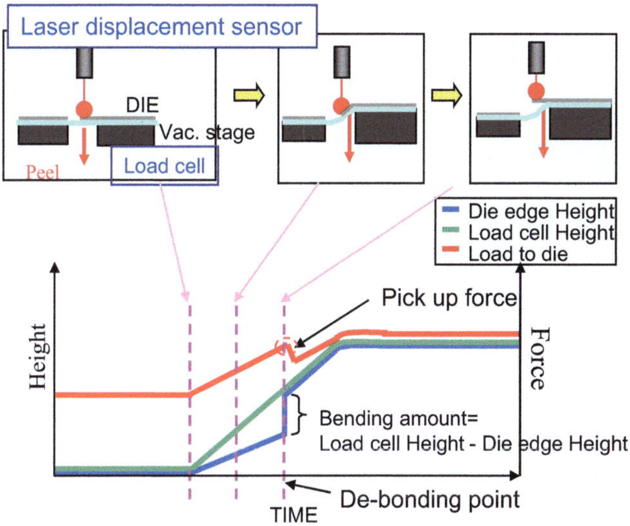


**Fig. 9.67** Analysis of the die pickup mechanism

Stress simulation at the die corner is shown in Fig. 9.69. The chip was triangular with one side fixed and the other two sides free. A downward force was applied at the center of the chip. The maximum stress occurred in the fixed edge.

Figure 9.70 shows the relationship between maximum stress in the simulated die pickup and peeling time in the actual die pickup experiment. It is clear that the smaller the overhang value becomes, the larger the stress on the chip. Furthermore, when the overhang value is 0.4 mm, it takes much longer to pick up the die compared with the 0.3-mm case. This means that if the pickup speed increases, the sticking die will be broken by a collision with the collet tool. We find that 0.3 mm is the most suitable overhang value for die pickup.

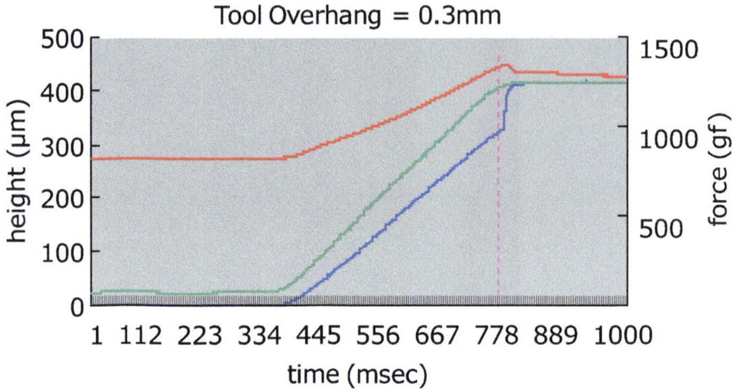


**Fig. 9.68** Experimental results of the die pickup action

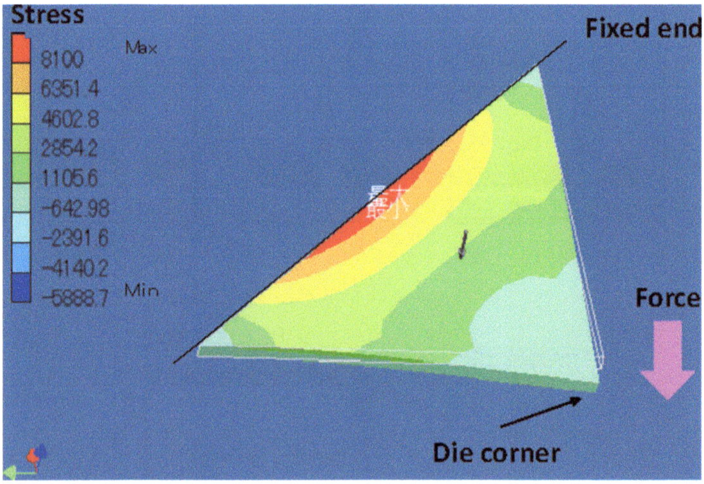


**Fig. 9.69** 3D simulation of applying a bending stress to the die edge

(b) Evaluation of Wafer Pickup with Backside Bumps

Next, we evaluated the pickup of the wafer with backside bumps using two types of dicing tape. These tapes have thick adhesive to absorb the bump height. Tape A is softer than normal to improve bump coverage, and tape B is the conventional tape for usage with bumps. We used the W1 and W2 wafers (Fig. 9.55). Tape A had good bump coverage and small backside chipping of around 10 μm after dicing, although it was not easy to pick up high-bump-density wafers like the W2. In contrast, with tape B, it was possible to pick up the W2 wafer, but the dicing did not perform as well due to larger backside chipping sizes and the inefficiency of slow dicing speeds.

In conclusion, tape A is suitable for average density bump arrays, and tape B is suitable for high-density full matrix bump arrays (see Fig. 9.71). The ease of dicing and pickup are competing properties, so an application-specific priority must be made.

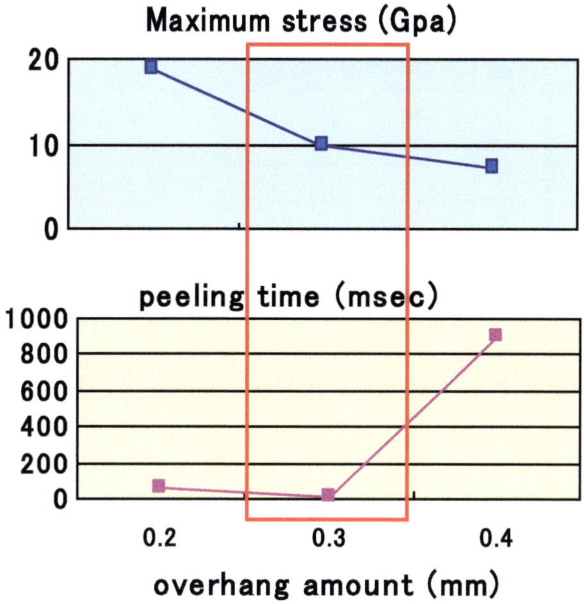


**Fig. 9.70** Results and analysis of the simulated die pickup

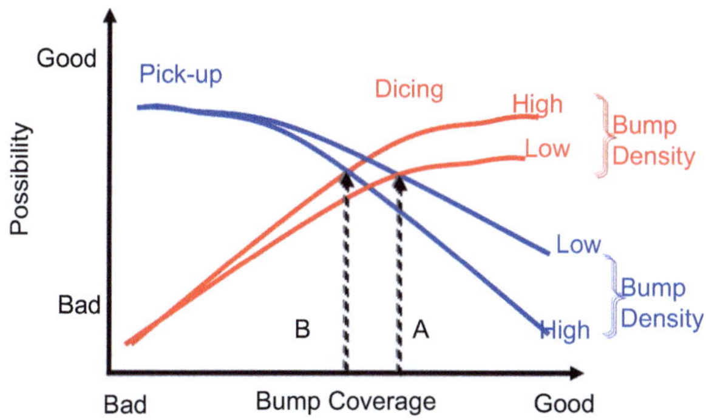


**Fig. 9.71** Relationship between die dicing and pickup

#### 9.3.3.5 Thin Wafer Processing Technique in the Wafer-Stacking Process

We now explain the backside grinding and dicing processes in the case of the stacking of three wafers. The third-level wafer is stacked on the W2W structure.

(a) BG Evaluation Results of the Wafer-Stacking Structure

In the case of the third wafer stacking, the TTV accumulates from the preceding wafers with the additional adhesive (see Fig. 9.72). The same methods as for a single

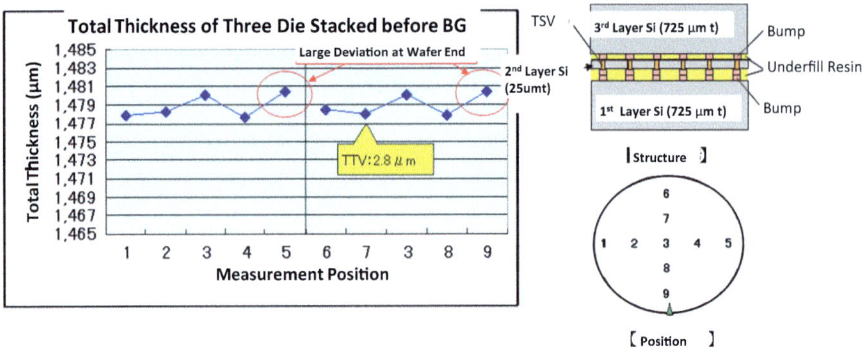


**Fig. 9.72** W2W wafer total thickness distribution. *TTV* total thickness variation, *BG* back grind

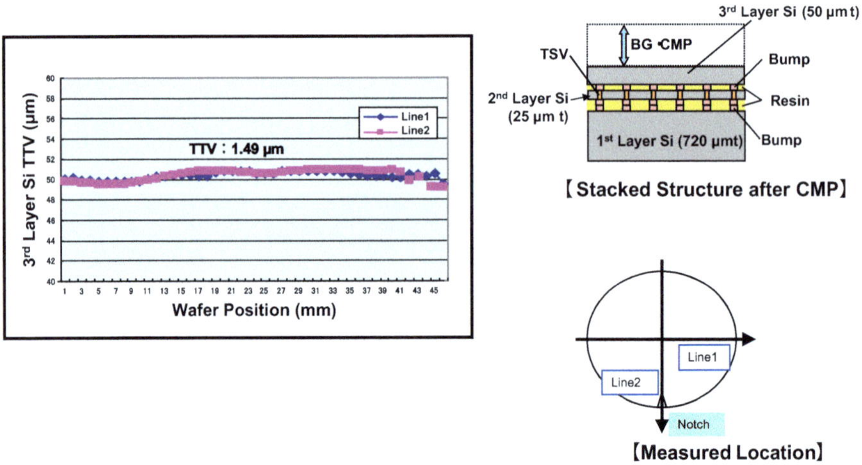


**Fig. 9.73** The thickness distribution of the third Si layer after CMP. *CMP* chemical mechanical polishing

wafer WSS case were used on the W2W stacked wafer, resulting in a small TTV size of 1.49 µm (Fig. 9.73).

(b) Dicing Evaluation Results of Wafer Stacking

Next, we evaluated the dicing process using a two-layer W2W structure bonded by three different adhesives: p-phenylene benzobisoxazole (PBO), epoxy, and polyimide (PI). The dicing results are presented in Figs. 9.74 and 9.75. The epoxy resin performed best in terms of chipping size, which is indicated by the higher modulus and lower elongation values than the others, making it easier to cut with a dicing blade.

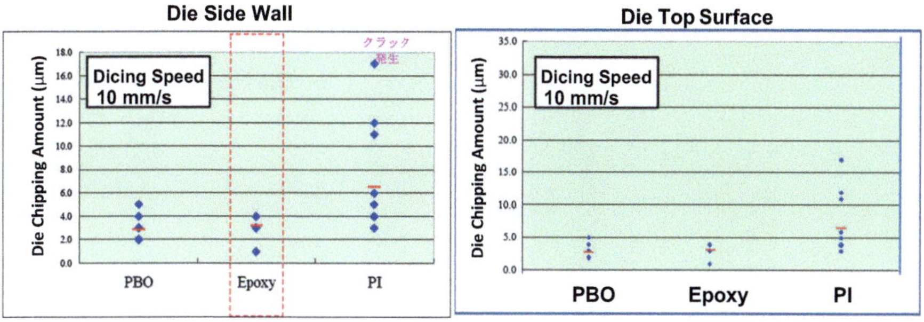


**Fig. 9.74** The amount of chipping from the dicing process depends on the underfill adhesive. *PBO* polybenzoxazole, *PI* polyimide

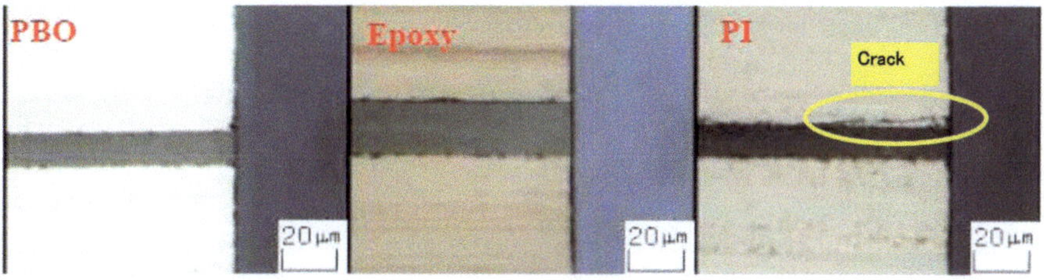


**Fig. 9.75** Side view of stacked die after dicing. *PBO* polybenzoxazole, *PI* polyimide

The PI adhesive had the largest chipping size, which is due to its lower bonding strength to the wafer, thus making it is easy to delaminate in the joining interface and easy to remove the die. In dicing W2W, the bonding strength of the adhesive directly influences die chipping.

To study the worst-case scenario in the dicing street, accelerated dicing evaluation was performed on a W2W sample (Fig. 9.76) with a large Cu area in the scribe line. The results are shown in Fig. 9.77; the upper wafer has a large die chipping.

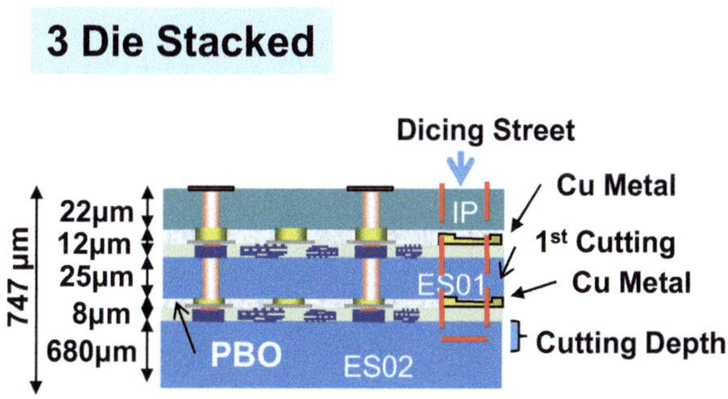


**Fig. 9.76** Cross-sectional structure of stacked die. *PBO* polybenzoxazole, *IP* Interposer

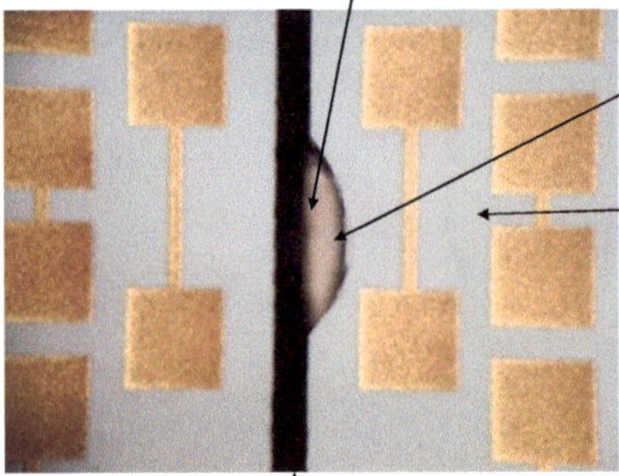


**Fig. 9.77** Top view of die chipping after dicing

The root causes of chipping are a Cu film attaching to the blade during the dicing process, which reduces the blade cutting life, and the stress concentrations in the wafer that lead to the die being displaced. We therefore looked at optimizing the dicing conditions.

When the cutting depth on the bottom wafer was 50 μm, the dicing blade could not remove the Cu metal scrap from the dicing line. However, when the cutting depth became deeper than 250 μm, Cu scrap was removed because the self-shaping effect of the dicing blade is proportional to the Si cutting depth. When the adhesive strength between wafers is not stable, Si chipping occurs. Setting the blade rotation speed lower is useful to prevent chipping. Lower rotation speed also improves the self-shaping effect.

To achieve the reliable dicing performance, the following two items are important:
1) The stability of the bonding strength between adhesive and wafer
2) The proper design of the dicing line to minimize the remaining Cu metal

(c) Backside Grinding of Wafer Stacking

In the case of the W2W process, we do not use WSS. The bottom wafer plays a role as the carrier. After the top wafer is bonded to the bottom wafer, the backside of the top wafer is ground and polished. Next, a middle-end process, such as TSV formation, redistribution layer fabrication, or bumping, is performed.

Therefore, the details of the W2W method and the material used between wafers can affect the wafer damage and TTV precision after backside grinding. Even if the TTV after two-layer stacking is around 4.6–6.2 μm, it is possible to achieve the target TTV value of 2.0 μm (± 1.0 μm) using the auto-TTV function. However, it is necessary to remove the fragile specific pattern or void that causes the wafer cracking.

### 9.3.4 *Issues in the Prevention of Device Characteristic Changes and Metal Contamination During Wafer Thinning*

Figure 9.78 shows the typical locations of IG and extrinsic gettering (EG) in the cross section of a thin wafer. Contaminating metal ions, such as from Cu, are captured by the IG effect that occurs because of wafer distortion fields from BMD in the bulk Si. Similarly, the distortion of the backside surface of the Si resulting from the finishing process leads to the capture of metal contamination ions, which defines the EG effect.

IG technology depends on the oxygen density in the crystal and the thermal history of the wafer front-end process. Using the standard Czochralski (CZ) crystal growth process, the wafer captures oxygen from the pot of melting Si quartz and the oversaturated oxygen is trapped in the interstitial Si lattice at room temperature. This oxygen reacts with Si and forms an $SiO_2$ deposition under a heat-treatment process, which results in a BMD zone inside the wafer.

However, during the thermal treatment process greater than 1150 °C, oxygen molecules which are trapped within the Si lattice near the surface, diffuse outward and form a defect-free DZ in the device fabrication area. The location of the boundary between the DZ and BMD zone is determined by the interstitial oxygen concentration and thermal history in the front-end process. So in the case of a device using a thinner CZ wafer, optimization of the two parameters described above are important to yield an IG effect.

On the contrary, EG technology does not depend on the wafer specifications. In fact, it is easy to add the extrinsic gettering potential through a backend process, such as a backside finishing method, which can be incorporated into the normal device cleanliness protocol (e.g., controlling the in-line impurity ion concentration) for a given manufacturing cost.

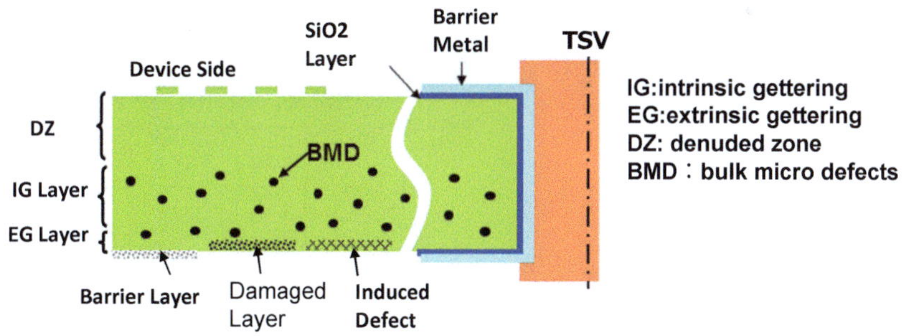


1 ) IG Layer : BMD induced inside of Si Wafer under crystal growth

2 ) EG Layer : Damaged Layer occurred by Backside grinding

3 ) Barrier Layer : Insulator/Metal Layer deposition on Backside

**Fig. 9.78** Gettering position in the thin wafer. *TSV* through-silicon via

However, when the device becomes ultrathin, the EG method should be avoided because backside grinding reduces the wafer bending strength. To demonstrate such gettering effects, we measured the concentration of Cu contamination on the wafer surface that had thermally diffused from the backside of the wafer; the quantity of evaporated Cu source material was fixed. The quantity of surface Cu was measured by the MOS capacitance time (MOS C-t) and a surface contamination analysis. The laser micro-Raman spectroscopy (μRS) measurement was effective in evaluating the residual distortion inside the crystal.

The measurement of DRAM retention time and metal–oxide–silicon field-effect transistor (MOSFET) keep-out zone (KOZ) value [30, 31] will be described in a later section.

#### 9.3.4.1 Evaluation of Crystal Defects and Metal Pollution in the Thin Wafer

Figure 9.79 shows the principle of the MOS C-t measurement techniquefor measuring IG potential [32]. Applying the pulse voltage changes the device from a normal charging state to an inversion state, which causes the depletion layer to form. The C-t curve demonstrates the presence of a minority carrier generated in the depletion layer through the change in capacity (C). We can calculate the lifetime of the generated minority carrier, $\tau_g$, by converting the C-t curve to a Zerbst curve and then finding the $y$-intercept of the linear portion. When a heavy metal impurity such as Cu exists in the depletion layer, there will be a rapid reduction of $\tau_g$, which we can readily observe.

#### 9.3.4.2 Backside Grinding Methods and Their EG Effect

The damaged layer formed by backside grinding processes has an EG effect that prevents the electrical characteristics of the device from being affected by heavy

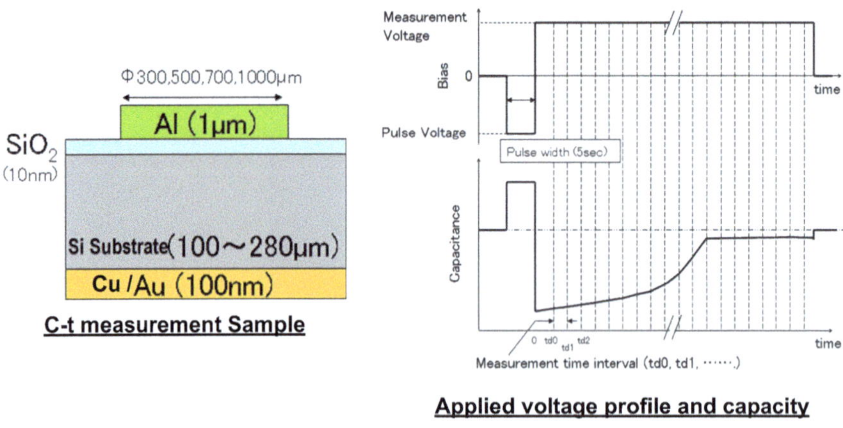


**Fig. 9.79** C-t method for evaluating the gettering effect. *C-t* capacitance time

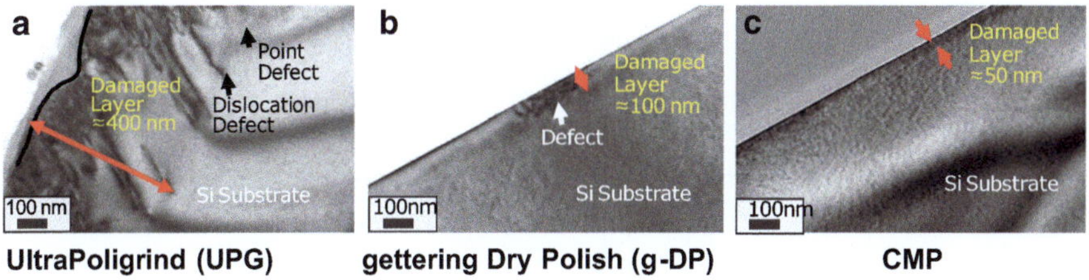


| BG | Damge after backgrinding | | Roughness | Damaged Layer |
|---|---|---|---|---|
| | Thickness 10 μm | 30 μm | | |
| UPG | BAD | GOOD | 6.1 nm | 400 nm |
| g-DP | GOOD | GOOD | 1.6 nm | 100 nm |
| CMP | GOOD | GOOD | 0.17 nm | 50 nm |

**Fig. 9.80** Cross-sectional TEM photos and results for three backside-finishing techniques. *CMP* chemical mechanical polishing, *BG* back grind

metal contamination. However, a stress-free backside grinding process has the advantage of preventing thin wafer breaking and warpage. We will now evaluate the ultra poligrind (UPG), gettering dry polish (g-DP) and CMP grinding methods [33].

Figure 9.80 shows the damaged Si layer in cross-sectional transmission electron microscope (TEM) images. We observe the crystal defects, which are mainly composed from dislocations; these become the gettering source. The larger the damaged layer is, the more effective the gettering. However, a larger damaged layer in a thinner wafer makes the wafer fragile and more easily broken.

We measured the wafer backside stress using μRS for the three different backside-finishing techniques (Fig. 9.81). The residual stress value in the processed side of the wafer was smallest for CMP and progressively larger for g-DP and UPG. Therefore, UPG is not suitable for 3DICs that use TSV technology.

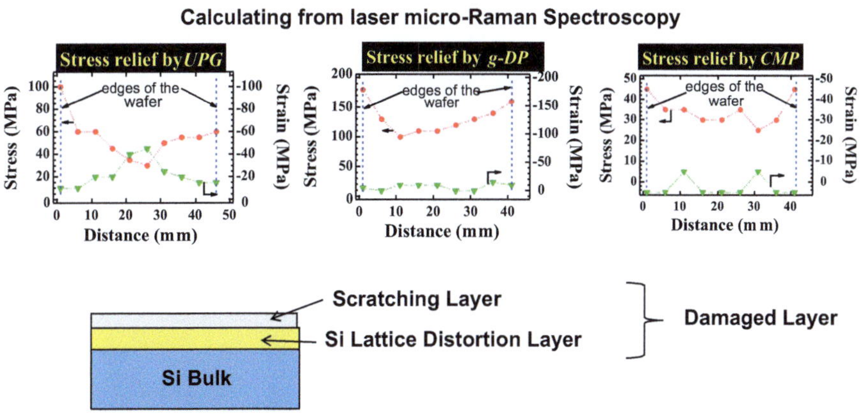


**Fig. 9.81** Stress relief of various backside finishes

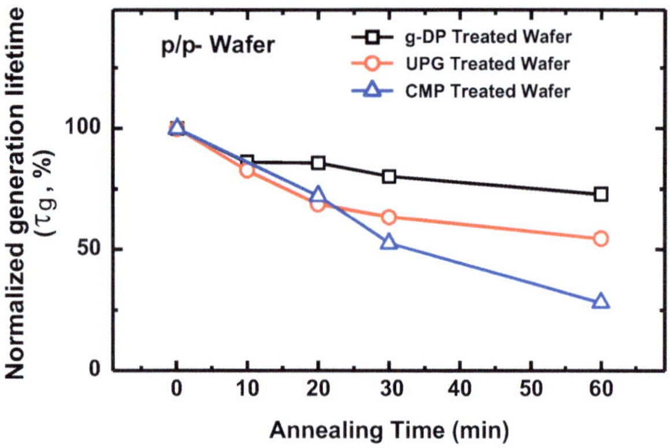


**Fig. 9.82** Evaluation of the gettering effect by C-t method. g-*DP* gettering dry polish, *UPG* ultrapoligrind, *CMP* chemical mechanical polishing

Figure 9.82 shows the comparison of minority carrier lifetime ($\tau_g$) for three kinds of an EG-derived damage layer.

To make the EG effect more obvious, a P/P–substrate wafer was used which has a relatively low resistance to gettering effects. A Cu film was evaporated from the damaged surface of the wafer on the assumption of there being Cu contamination present. The wafer was subjected to annealing for 60 min at 300 °C. The EG was least effective in the CMP-treated wafer and most effective when g-DP was applied.

The reason that the gettering effect is smaller in the wafer treated with UPG than with g-DP is because of the formation of a deeper and nonuniform damaged layer, which can be gleaned from cross-sectional TEM images (Fig. 9.80). When defects are generated deep within the material, they act as generation and recombination sources for the minority carriers, which decreases minority carrier life. In this regard, g-DP is a good solution.

Figure 9.83 shows the g-DP-processed cross-sectional SEM images before and after 30 min of thermal treatment at 300 °C. The surrounding zone indicated by

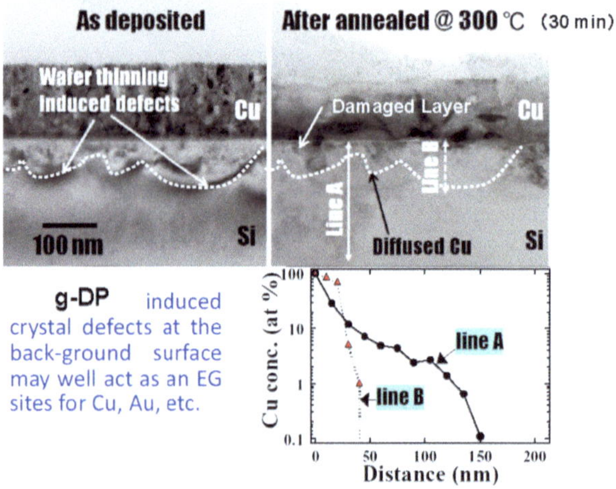


**Fig. 9.83** Cu EG effect by g-DP backside grinding

white dotted lines is the EG-affected area. It is evident that the Cu thermal diffusion length depends on the irregularity of the damaged zone boundary. This example demonstrates how the defects in the damaged area control the Cu diffusion. Therefore, to get a stable gettering effect, it is desirable not only to apply backside grinding but also use this in combination with another method, such as the barrier film depositions of substances like SiN.

#### 9.3.4.3 Variation of Electrical Characteristics with Mechanical Stress

(1) Investigation of Thin Wafer Basic Mechanical Characteristics

We evaluated the influence of wafer thinning on the mechanical properties of silicon using a nano-indenter.

The wafer surface was subjected to a 24-mN load by the tip of a pyramidal indenter and the change in Young's modulus of the thin wafer was measured. The test specimens were Si wafers with thicknesses of 30, 50, and 100 μm [34]. The backside was finished with CMP. The change of the Young's modulus of the thin wafer is shown in Fig. 9.84. The Young's modulus of the 30-μm wafer was much lower than for the two thicker specimens. The damaged layer, consisting of crystal defects and dislocations, penetrated several hundred nanometers into the backside treatment surface. Furthermore, crystal distortions may be present several tens of micrometers from the Si surface of the affected area. The reduction of the Young's modulus of Si was affected by distortions of the Si crystal structure that reached the surface when the Si thickness was 30 μm or less.

In 3DIC products, if the wafer is thinned down to less than 50 μm, then there are several problems in addition to their weak mechanical strength.

(2) Stress Analysis

Using the simple silicon bump structure shown in Fig. 9.85a, we investigated the relationship between the warpage of the thin Si wafer and the change in the device properties [35].

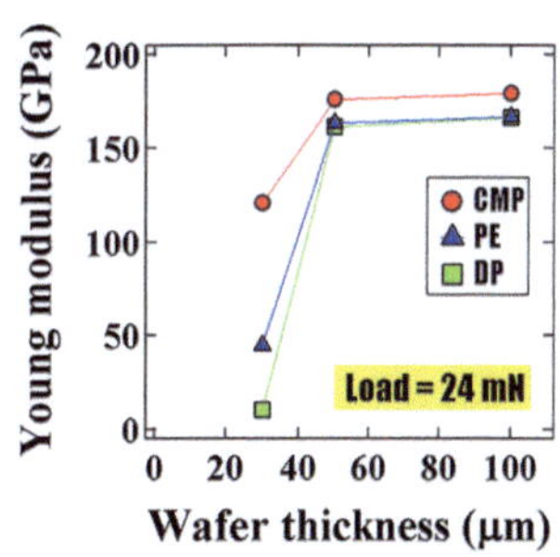


**Fig. 9.84** Relationship between wafer thickness and Young's modulus. *CMP* chemical mechanical polishing, *DP* dry polish *PE* plasma etching

**a** **Stress induced by dummy Si bump** **b** **Die warpage distribution by stress**

**Fig. 9.85** The influence of Si bump and Si wafer bonding on stress and strain

The bump array was fabricated on the thick Si substrate to which the thin wafer was bonded by epoxy resin to induce stress on the device. The bumps were $20 \times 20 \times 20$ μm. Epoxy resin was cured under 150 °C heat for 1 hr. Mechanical stress was applied to the Si substrate, and periodic bending occurred during the resin curing process because the coefficient of thermal expansion (CTE) was greatly different from the bulk Si and epoxy resin; see Fig. 9.85b.

After the Si substrate had bonded to the bump array, yielding a 35-μm-thick Si wafer, the total height distribution at the backside of the wafer was measured by laser microscope. As seen in Fig. 9.85b, the peaks and valleys of the Si bumps became more pronounced. The continuity of the stress field means that the bending stress profile of the Si bump array distributes within the thin Si substrate. The magnitude of this bending stress is dependent on the thickness and bump pitch of the Si.

Figure 9.86 presents the relationship between bending deflection and bump pitch. The thinner the Si thickness is, the larger the bending deflection becomes. Using the μRS laser spectrum, we also evaluated the mechanical stress of thinning the Si substrate (Fig. 9.87). A large tensile stress occurred on the surface of the thinned section of the Si substrate right above the Si bump when the bump pitch was wider than 300 μm. There was also a reciprocal compression stress on the Si surface between the bumps. It can also be seen from the figure that the mechanical stress applied by the Si bump decreased when the bump pitch was reduced to 100 μm.

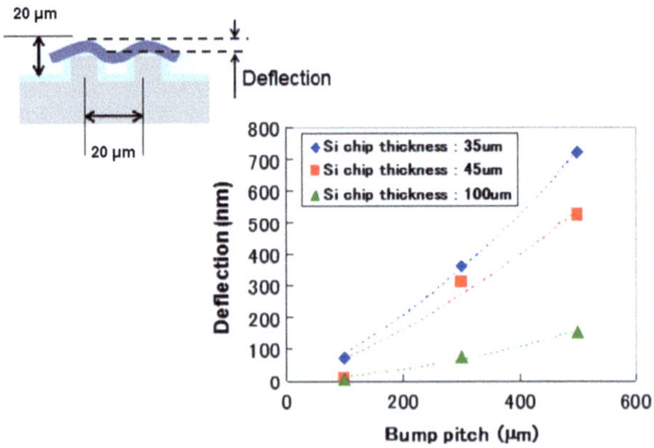


**Fig. 9.86** Relationship between the thin Si wafer bending value and dummy Si bump pitch

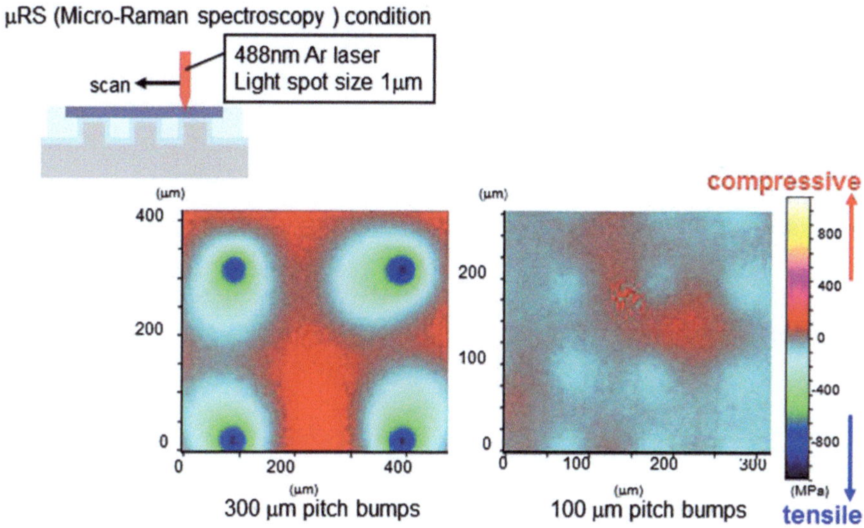


**Fig. 9.87** Mechanical stress mapping induced by dummy Si bump array

The change of the device's current–voltage characteristics due to mechanical stress was evaluated using an MOS transistor having 180-nm nodes. The Si was thinned down to 30 μm. The results are shown in Fig. 9.88.

The n-channel MOS transistor was positioned at an offset of 15 μm from the center of the Si bump; the bump pitch was 50 μm. Electric current flow was affected in the direction parallel to the tensile stress. The currents of the n-channel MOS transistor increased with the bending stress in the warped thin Si substrate. The current change was as much as 7 %, as seen in Fig. 9.88. Figure 9.89 shows the relationship between electron mobility (capacitance–voltage) and mechanical stress, which is indicated by the effective electrical field of the Si surface. After applying the tensile stress, the electron mobility increased by 14 %.

Returning to Fig. 9.88, we consider the change of the current–voltage when the n-channel MOS transistor was 22 μm from the Si bump center; see Fig. 9.88d. In this position, there was compressive stress parallel to the direction of the MOS transistor current flow, which slightly decreased (4 %) its electric current. In summary, the electron mobility increases with tensile stress in the neighborhood of the Si bump and decreases with compression stress in the central section between Si bumps (Fig. 9.88b).

In the experiments, Si dummy bumps were fabricated to introduce stress in the C2C bonding area. However, the bending stress in a thinning silicon substrate in the real 3DIC would be smaller because the bumps would be fabricated from Cu or SnAg, which have slightly higher CTE than Si. However, the stress-based electron mobility phenomena demonstrated above also apply.

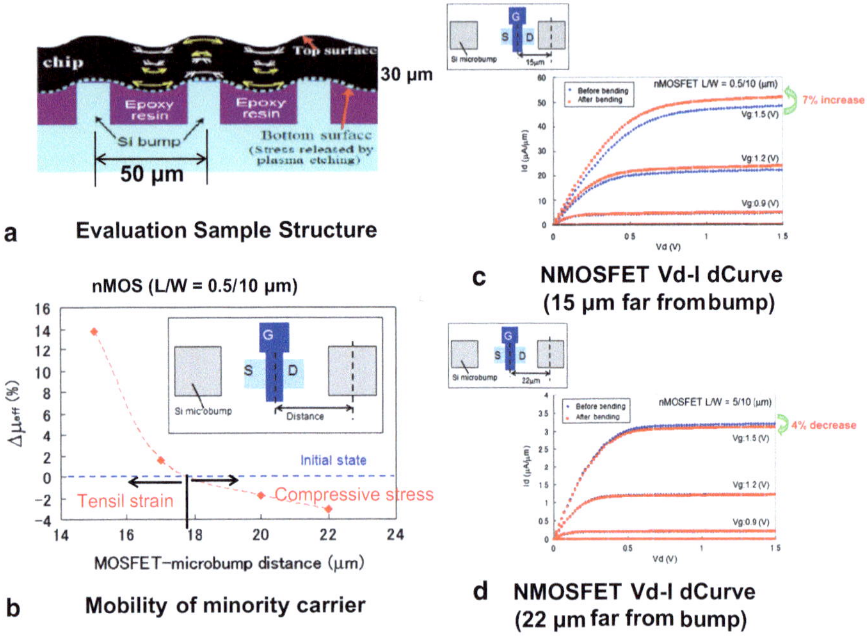


**Fig. 9.88** The influence of Si bump and Si wafer bonding on stress and strain. *MOSFET* metal–oxide–silicon field-effect transistor, *nMOSFET* n-type metal–oxide–silicon field-effect transistor, nMOS n-type metal–oxide semiconductor

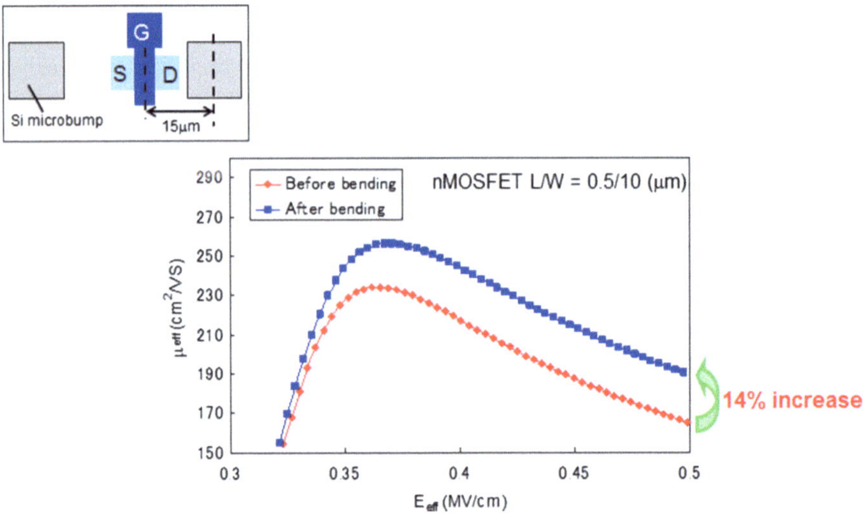


**Fig. 9.89** Variation of electron mobility with mechanical stress. *nMOSFET* n-type metal–oxide–silicon field-effect transistor

### 9.3.5 Standardization

The establishment of a 3DIC standard, including a supply chain, is important for the development of high-volume manufacturing of 3DICs. Within Semiconductor Equipment and Materials International (SEMI), a global industry association of companies, the Three-Dimensional Stacked Integrated Circuits (3DS-IC) Committee of North America was founded in the autumn of 2010 to promote precisely this activity. Under the committee, three task forces were organized: Thin Wafer Handling TF, Wafer Bonded Stacks TF, and Inspections & Metrology TF. One year later, 3DS-IC Taiwan was also founded, with two task forces: Middle End Process TF and Testing TF. In Japan, the 3DIC study group as a subunit of the SEMI Package Committee was organized in 2012. This group established the "Thin die bending strength measurement" task force. The purpose of this task force was to design a simpler measurement tool and to simplify the bending strength measurement method for ultrathin dies (<30 μm thick) with small deviations. In cases when the die becomes very flexible, the bending strength cannot be measured accurately using conventional methods. In SEMI standard SEMI G86-0303, the conventional method "Three-point bending test method" is defined. This standard also describes the method for die thicknesses under 100 μm, but it has been found impracticable. ASET subsequently proposed the "cantilever method," which is depicted in Fig. 9.90. This new method was standardized (with the support of the SEMI 3DIC Study Group) in May 2014 as SEMI G96-1014.

Figure 9.91 presents the results of a comparison between the three-point bending method and the cantilever bending method. The two approaches produced similar average values, but the cantilever method had a narrower range of maximum and

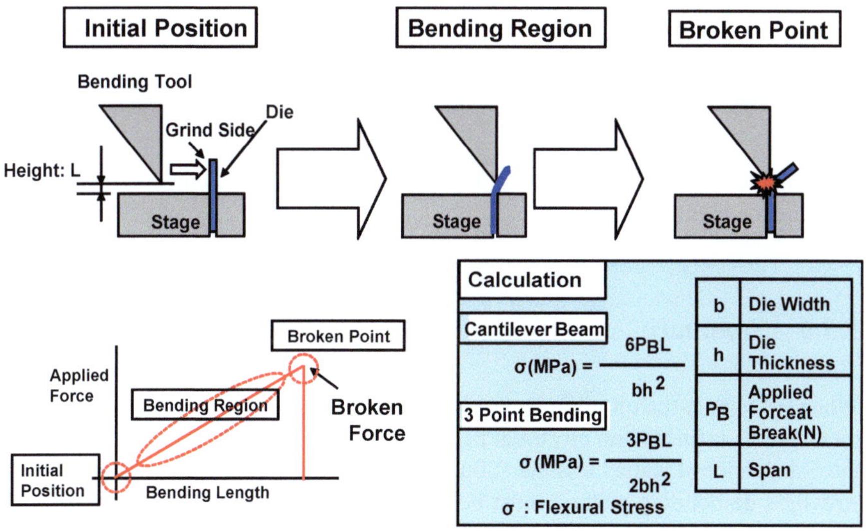


**Fig. 9.90** Schematics of the cantilever bending method

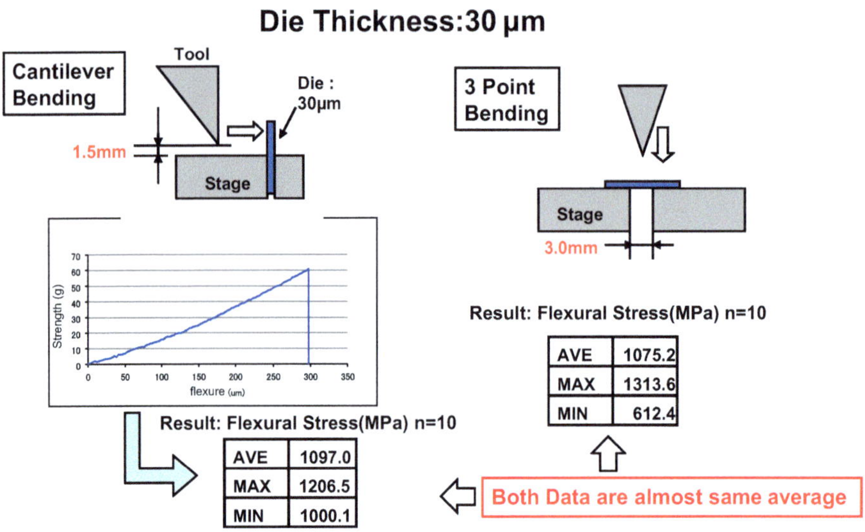


**Fig. 9.91** Comparison between the three-point bending method and the cantilever bending method

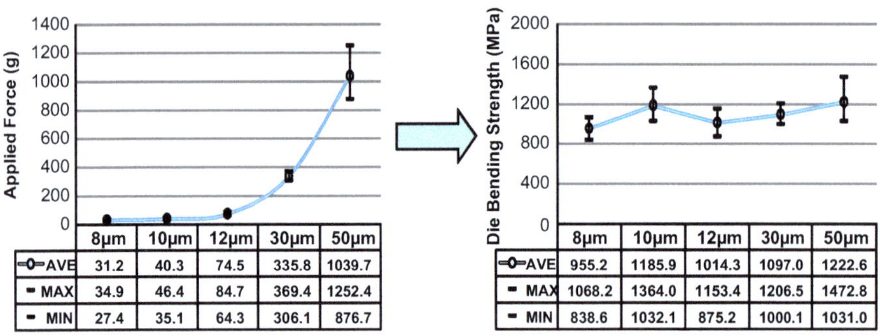


| | 8μm | 10μm | 12μm | 30μm | 50μm |
|---|---|---|---|---|---|
| AVE | 31.2 | 40.3 | 74.5 | 335.8 | 1039.7 |
| MAX | 34.9 | 46.4 | 84.7 | 369.4 | 1252.4 |
| MIN | 27.4 | 35.1 | 64.3 | 306.1 | 876.7 |

| | 8μm | 10μm | 12μm | 30μm | 50μm |
|---|---|---|---|---|---|
| AVE | 955.2 | 1185.9 | 1014.3 | 1097.0 | 1222.6 |
| MAX | 1068.2 | 1364.0 | 1153.4 | 1206.5 | 1472.8 |
| MIN | 838.6 | 1032.1 | 875.2 | 1000.1 | 1031.0 |

**Fig. 9.92** Test results of the cantilever bending method

minimum values; see also Fig. 9.92. Therefore, the cantilever method is superior for ultrathin die measurement [36].

### *9.3.6 Summary*

When we handle the ultrathin wafer, it is necessary to consider the stress environment and its influence on electrical performance. This is particularly important for the Si behavior of wafers less than 30μm thick. For the WSS, the room temperature stressless de-bonding method must be used. The cost of the WSS technology has also now become more important. Furthermore, the temporary adhesive that is used in the TSV formation process must be heat resistant.

In summary, it is necessary to prevent electrical characteristic changes not only from the mechanical stress but also from metal contamination. Successively more sensitive countermeasures will need to be made to keep up with device scaling.

Section 9.3 includes the research and development results of a co-research program conducted by ASET and Tohoku University.

## 9.4 3D Integration Technology

### *9.4.1 Background and Scope*

Next-generation 3D-SiP requires electrical connections between stacked ICs for electrical communications and power delivery, and these are achieved through bumps and TSV. For further performance improvement of such next-generation 3D-SiP, wider bandwidth communication between large-scale integrations (LSIs), lower power consumption of the 3D-SiP, and more stable power supply network are inevitable. Thus, performance improvements of bumps and TSVs are also necessary. Especially, TSV has definitely different structure from ordinal IC interconnect, that is, metal interconnect with surrounding dielectric layer embedded with silicon substrate. Therefore, optimizations of its structure and communication circuit are required for the performance improvement. Moreover, layout designs should consider MOSFET performance change by TSV placement because mechanical stress induced by TSV probably affects MOSFET characteristics. These circuit designs and layout designs should be determined for each 3D integration scheme because these design rules are strongly affected by the process flow, the structure, and the materials of TSV and the bump.

Many types of 3D integration scheme have been proposed, and we have not had “standard material/structure” or “standard process” for 3D integration (Table 9.12).

This is because each application requires different specifications for 3D integration. Scope of this chapter includes C2C process, stacking a chip on the other

**Table 9.12** Process choices for 3D integration. *TSV* through-silicon via, *3D* three dimension

| Category | Items | Major choices |
|---|---|---|
| TSV | Timing | Via-first, via-middle, via-last |
| | Direction | Backside, frontside |
| | Material | Cu, W, poly-Si |
| Stacking | Form | W2W, C2W, C2C |
| | Direction | F2F, B2F |
| | Timing of thinnign | Thinning before bonding, thinning after bonding |
| Bonding | Material | Cu, Solder(SnAg,Sn), Au |
| | Underfilling | Post-applied(CUF), Pre-applied |

*W2W* wafer-to-wafer, *C2W* chip-to-wafer, *C2C* chip-to-chip, *F2F* face-to-face, *B2F* back-to-face, *CUF* capillary underfill

chip, and W2W process, and stacking in a form of wafer then dicing. C2C process enables chip stacking using known good die (KGD) chips and is also applicable for stacking of dies with different size. Since one-by-one chip stacking is required for C2C process, this process is suitable for small volume production or prototyping rather than high-volume production because of its less production throughput. On the other hand, W2W process is suitable for high-volume production because 3D stacking is achieved in wafer form, and this process has an advantage for chip thinning which is required for smaller TSV diameter because thinned chip is difficult to handle. On the other hand, size of stacked die must be same, and it is difficult to stack only using KGD chips. As a TSV formation process, various types of process are proposed (Table 9.12). Here, we focus on "backside via-last TSV process" which has advantages of no modification in BEOL process and maximizing wiring resources. The following sections explain typical process flow, its design rules, and integration results for both C2C and W2W processes.

## 9.4.2 C2C Process

### 9.4.2.1 C2C Integration Overview

This section describes the overview of C2C process integration. Figure 9.93 illustrates a schematic cross section of three-layer stacked chips (one Si-IP chip and two device chips) and its target specifications.

Here, Si-IP chip means a silicon substrate with interconnects for redistribution wiring and usually has no active devices. Two device chips are stacked in "Face-up"

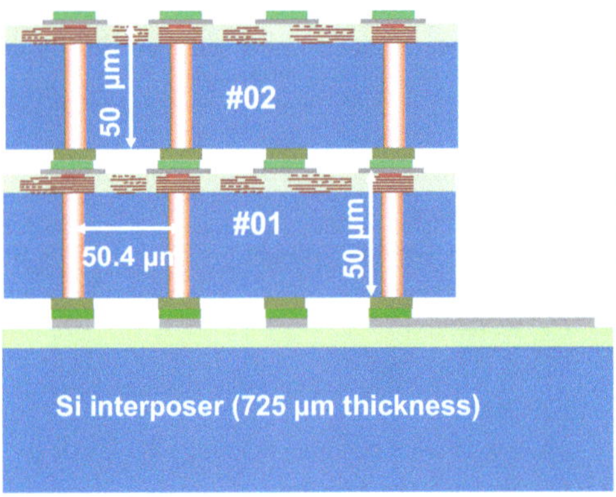


| Structure | One silicon interposer chip and two device chips |
|---|---|
| TSV resistance | 0.2 Ω |
| TSV capacitance | 0.5 pF |
| Bump density | 320 /mm$^2$ |
| Bump diameter | 30 μm |
| TSV pitch | 50.4 μm |
| TSV diameter | 20 μm |
| TSV length | 50 μm |
| KOZ | 2 μm |

**Fig. 9.93** Schematic cross section of C2C stacked chip and its specifications. *TSV* through-silicon via, *KOZ* keep-out zone

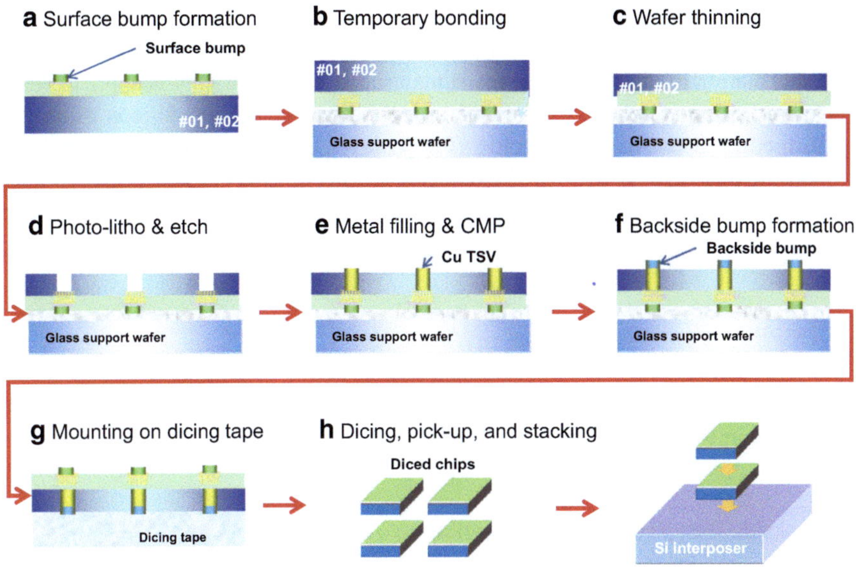


**Fig. 9.94** MEOL (middle-end-of-line) process flow for C2C integration. *TSV* through-silicon via, *CMP* chemical mechanical polishing

configuration on a Si-IP chip with aluminum interconnect. Thickness of the device chip is about 50 μm and pitches of bumps and TSVs are 50.4 μm. TSV diameter is 20 μm and bump diameter is 30 μm. Figure 9.94 shows middle-end-of-line (MEOL) process flow for C2C integration. Temporally bonding method using glass support wafer was used for thinned wafer handling. The details are as follows:

(a) Surface Bump (Au/Ni) Formation

Surface bumps are formed on a prepared wafer. After Cu/Ti seed layer formation by sputtering for electroplating, surface bump of Au (100 nm)/Ni (6 μm) is electroplated using semi-additive method.

(b) Temporary Bonding to Glass Support Wafer

The wafer is bonded in "face-down" configuration to a glass support wafer with an adhesive material.

(c) Wafer Thinning

Backside of the bonded wafer is subjected to wafer thinning using BG and CMP. Target thickness is 50 μm in total and its TTV is around 2 μm.

(d) TSV Lithography and Etching

After formation of photosensitive resist material, TSV lithography is applied. Here, backside IR alignment is used for TSV pattern alignment. Then, silicon etching is performed until pre-metal dielectric (PMD) layer appears. After silicon etching, silicon dioxide layer is deposited using low-temperature CVD and is etched back by Reactive Ion Etching (RIE) in order to reveal TSV contact wiring at the bottom of the TSV.

(e) TSV Filling and Planarization

Metal seed layer of Cu/Ti is formed by sputtering. Then, copper electroplating is performed to fill the TSV and excessive copper film is removed by CMP.

(f) Backside Bump (SnAg/Ni) Formation

Metal seed layer of Cu/Ti is sputtered again. Backside bump of SnAg (3 μm)/Ni (1 μm) is formed by semi-additive process.

(g) Mounting on Dicing Tape

The wafer is de-bonded from the glass support substrate and is mounted on a dicing tape for blade dicing.

(h) Dicing, Pickingup, and Chip stacking

For blade dicing, blade thickness of 25 μm and cutting speed of 5 mm/s is employed to minimize chipping. Flexural strength of the diced chip is about 400 MPa and no crack or breakage was observed during pickup and chip stacking. After dicing, the diced device chips are mounted on the Si-IP chip using a flip chip bonder.

#### 9.4.2.2 C2C Integration Results

(a) Stacked Chip

Figure 9.95a–c shows a bird's eye view and cross-sectional views of a stacked chip. Four device chips with TSVs are stacked on a Si-IP chip. Four-chip stacking is successfully obtained.

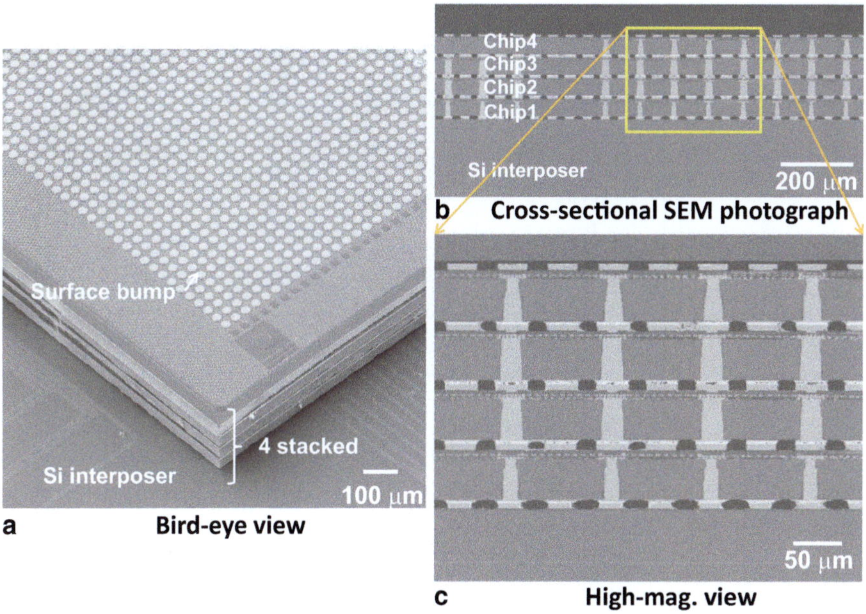


**Fig. 9.95** SEM photos of a stacked chip. *SEM* scanning electron microscope

(b) Resistance, Capacitance

This section describes an evaluation method of electrical resistance of a daisy-chain test structure containing bumps and TSVs. Total resistance ($R_{total}$) of the daisy chain can be expressed as follows:

$$R_{total} = M(N \times R_{bump\text{-}TSV} + R_{wire}), \tag{9.1}$$

where $M$ is the number of bumps or TSVs, $N$ is the number of stacked chips, $R_{bump-TSV}$ is a series resistance of bump–TSV per chip, $R_{wire}$ is the resistance of interconnections between the bumps or TSVs on terminal chips where bumps or TSVs of the chain are electrically shunted. When $R_{bump-TSV}$ is relatively small (e.g., below 100 mΩ, ), contribution of $R_{wire}$ in $R_{total}$ cannot be ignored. As shown in Eq. (9.1), precise value of $R_{bump-TSV}$ can be derived from stacking number *(N)* dependence of total resistance ($R_{total}$). Figure 9.96 shows the relationship between the number of stacked chips *(N)* and unit resistance. Unit resistance is the total resistance divided by the number of bumps or TSVs ($R_{total}/M$). As shown in this figure, unit resistance depends linearly on the number of stacked chips. We can obtain $R_{bump-TSV}$ of 40 mΩ and $R_{wire}$ of 80 mΩ from linear regression analysis.

Figure 9.97 shows relationship between TSV voltage and TSV capacitance. TSV voltage means TSV potential to P-type WELL.

As shown in this figure, TSV capacitance decreases with increasing TSV voltage. By increasing frequency from 100 kHz to 1 MHz, TSV capacitance decreases.

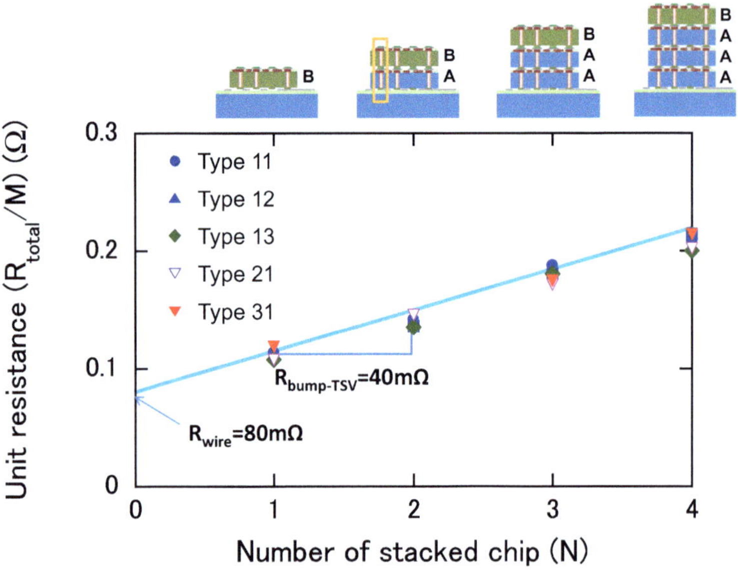


Fig. 9.96 Relationship between the number of stacked chips *(N)* and unit resistance ($R_{total}/M$)

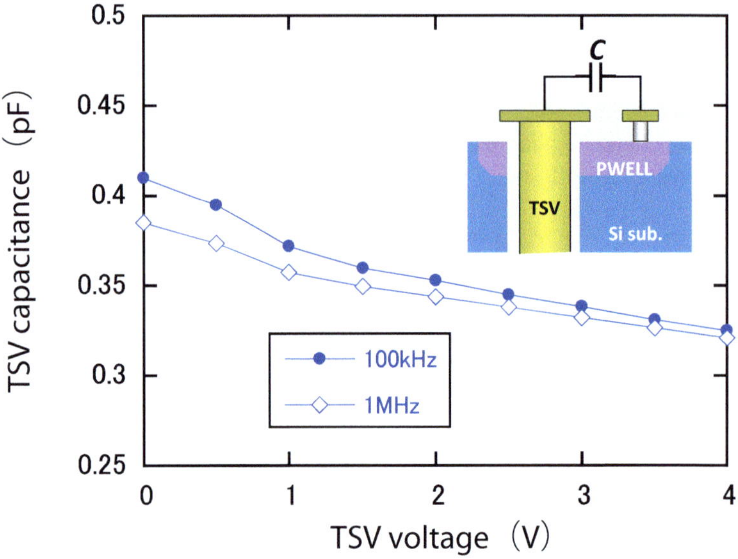


**Fig. 9.97** TSV voltage dependence of TSV capacitance. *TSV* through-silicon via, *Si sub*. silicon substrate

Such kind of voltage and frequency dependencies of TSV capacitance is similar to that of an MOS-type capacitor which can be modeled as series capacitor of insulator dielectric and a depletion layer. The measured TSV capacitance at 1 V is about 0.4 pF and capacitance of TSV liner of 0.8 μm thickness is calculated as about 0.7 pF. If uniform depletion layer is generated around the TSV liner, depletion layer capacitance is calculated as 0.3 pF. Based on a depletion layer model for MOS-type capacitor, a calculated depletion layer capacitance is around 0.5 pF at 1–4 V and roughly matched to the estimated capacitance from the measured TSV capacitance.

### (c) Keep-Out Zone

TSV process may affect drain current characteristics or threshold characteristics of MOSFET and parasitic capacitance of IC interconnect. Thus, design rules considering these influences must be established for 3D-IC using TSVs. These influences are evaluated using MOSFET test structures, interconnect test structures, and standard libraries.

First, KOZ for via-last TSV process was evaluated by MOSFET placed in the vicinity of TSVs. KOZ is usually defined as the region where a specific characteristic of MOSFET changes in specific ratio from its original value (e.g., 5 % changes in on current of MOSFET). As shown in Fig. 9.98a, a fundamental $I_d$–$V_g$ characteristic was obtained for p-/n-type MOSFETs with different distance from TSV edge (2 μm, 5 μm), different channel direction (vertical, horizontal), and relative alignment to TSV (0 °C, 90 °C from [100] direction). Figure 9.98b summarizes on current obtained from these MOSFETs. This figure clearly shows almost same (within 1$\sigma$) on currents were obtained independent on the distance from TSV edge, the channel direction, and the relative alignment of MOSFET. Moreover, the on current of MOSFET near TSV is the same as that of MOSFET without TSV.

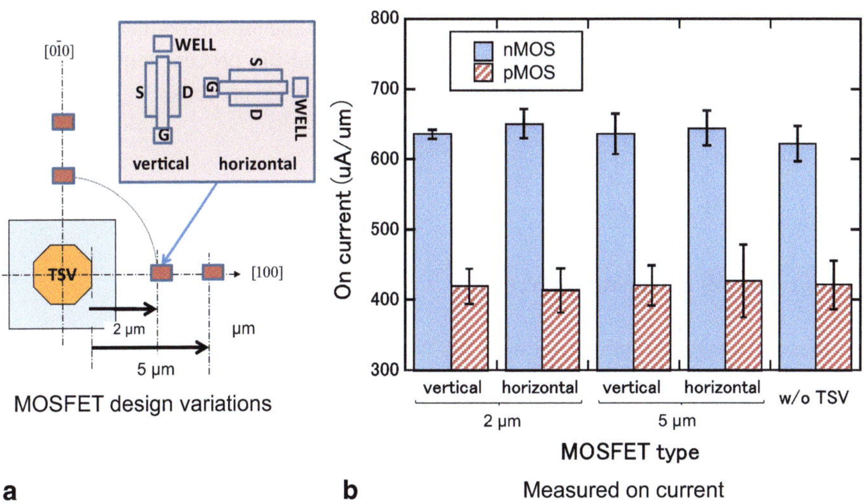


**Fig. 9.98** MOSFET characteristic near TSV. *MOSFET* metal–oxide–silicon field-effect transistor, *nMOS* n-type MOS, *p-MOS* p-type MOS

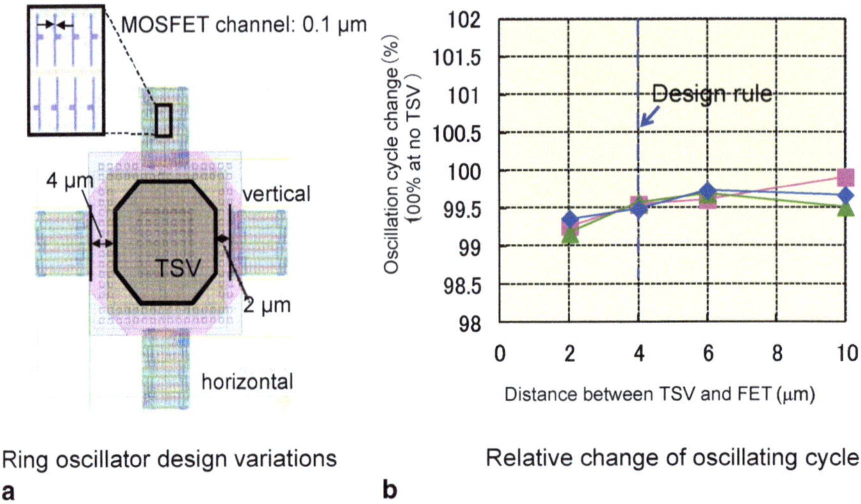


**Fig. 9.99** Ring oscillator characteristic near TSV. *MOSFET* metal–oxide–silicon field-effect transistor, *TSV* through-silicon via, *FET* field-effect transistor

Next, KOZ for via-last TSV process was also evaluated by oscillating cycle change of ring oscillator placed in the vicinity of the TSVs. Figure 9.99a shows alignment of the ring oscillator near the TSV. The oscillating cycles of the ring oscillator were measured as a function of distance from TSV edge and of the channel direction (Fig. 9.99b). According to this figure, oscillating-cycle change of a ring oscillator is within 1 % at 2 μm from a TSV edge. This result suggests that a MOSFET can be placed as close as 2 μm from a TSV, namely, KOZ < 2 μm. A possible cause for this relatively small KOZ is relatively small residual stress in silicon substrate because copper residual stress which causes the silicon residual stress can be lowered by lowering the process temperature, and the process temperature of via-last TSV process is lower than that of via-middle TSV process.

## *9.4.3 W2W Process*

### 9.4.3.1 W2W Integration Overview

This section explains an overview of W2W integration process. A 3D-IC composed of a Si-IP and two active chips is shown schematically in Fig. 9.100.

The target specifications of the 3D-IC are also summarized in the table in the figure. Based on electrical requirements, structural parameters of TSVs were determined. The pitches of bumps and TSVs were set to 25.2 μm in accordance with the connection–density requirement (over 1000/mm$^2$). TSV dimensions (i.e., 8–12 μm in diameter and 25 μm in length) were determined in view of a practical balance between TSV formation difficulties and TSV parasitic capacitance. In order to suppress IR drop of power supply line, target resistance of TSV was set to 0.2 Ω. And

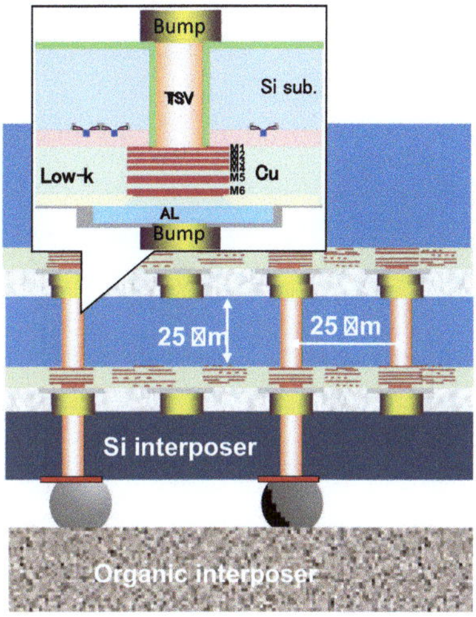


Target specifications

| TSV resistance | < 0.2 Ω |
|---|---|
| TSV capacitance | < 50 fF |
| Bump density | 4–10 x10$^5$ /cm$^2$ |
| TSV pitch | 25.2 μm |
| TSV diameter | 8–12 μm |
| TSV length | 25.2 μm |
| Liner thickness | 4–6 % of diam. |
| KOZ | 2 μm |

**Fig. 9.100** Schematic cross section of 3D-LSI and its specifications. *TSV* through-silicon via, *KOZ* keep-out zone

to achieve TSV capacitance below 50 fF, required TSV liner thickness was set to 4– 6 % of TSV diameter (i.e., around 500 nm). Target KOZ for W2W process is also below 2 μm.

W2W process flow employed here is determined from the following viewpoints:

- The process must be applicable to 25-μm-thickness substrate. The substrate thickness must be 25 μm because the target TSV length is 25 μm as shown in Fig. 9.100. A thinned silicon wafer is easy to bend and easy to crack, thus the thinned wafer is necessary to bond to a support wafer. Wafer bonding to a support wafer is categorized into two methods. One is temporary bonding and the other is permanent bonding. Because the support wafer must be de-bonded from the thinned wafer for further integration, wafer breakage at de-bonding process is of concern. Therefore, permanent bonding process was employed. Among the permanent bonding process, thinning after bonding process, thinning the silicon wafer down to 25-μm thickness after permanent bonding, was employed.
- The process must realize (bump) connection pitch of 25 μm. Although solder material like SnAg base material, which is widely used in flip-chip bonding is a prime candidate, one big concern is its low melting point (e.g., 221 °C for Sn-3,5Ag). If more than three wafers are stacked, solder bumps which is bonded the wafer stack may remelt during successive wafer bonding, causing a de-bonding of misalignment of the solder bumps. Moreover, degradation of electromigration reliability is another concern for finer bumps. To overcome these challenges, copper which has high melting temperature (1085 °C) is employed as the bump material of the W2W process.
- Backside via-last TSV process in combination with thinning-after-bonding method requires protecting bonded bumps from process damage during wafer thin-

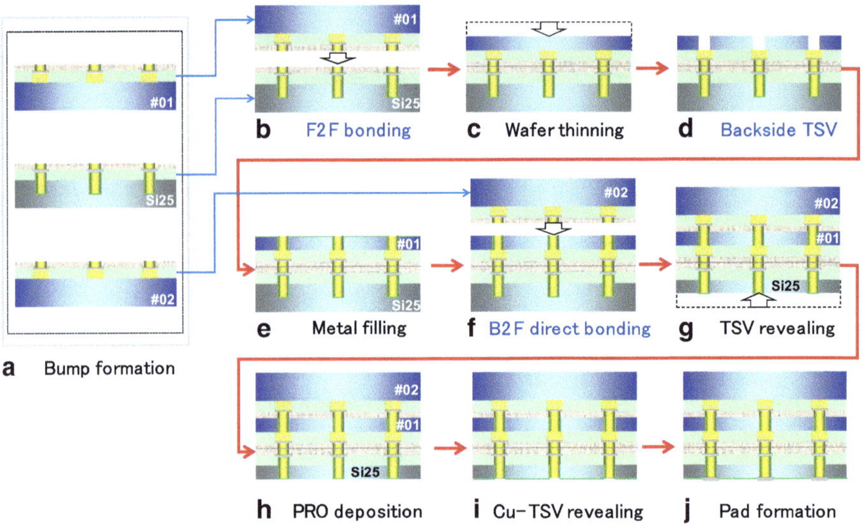


**Fig. 9.101** MEOL (middle-end-of-line) process flow for three-layer wafer stacking. *F2F* face to face

ning and TSV formation. As a bump protection method for C2C process, Capillary Underfill (CUF) method, filling a gap between chips with an encapsulation polymer (e.g., epoxy-based material) by capillary phenomenon. CUF method is difficult to apply in the W2W process because small the gap between the wafers is hard to be completely filled with the polymer which is injected from the edge of the bonded wafers. Pre-applied underfill (PAUF) is also known as a polymer underfilling method for C2C process. In this method, an underfilling polymer is formed before C2C stacking and both curing of the polymer and bonding of the solder bumps are achieved simultaneously. This method is also not applicable to W2W process because the pre-applied polymer remains between copper bumps and cause increase in the bump resistance. To solve this problem, hybrid wafer bonding method, a modified method of PAUF, is employed. The detailed process is explained later in this section.

Based on the abovementioned decisions, a W2W process flow was determined. The detailed process flow for three-layer wafer stacking (one Si-PI and two device wafers) is shown in Fig. 9.101.

(a) Bump Formation on Device Wafer

First, polymer layer is formed on the wafers (Si-IP and device wafers) and openings for bumps are formed into the polymer layer. Since bumps are electrically connected to pad electrodes on the wafer, the surface of the electrode is exposed at the bottom of the opening. Then, copper seed layer is deposited by sputtering method over the wafer, and thick copper layer is electroplated to fill the opening of the polymer. After that, CMP is applied to remove copper film on the polymer, resulting in flat copper/polymer surface.

(b) Wafer Bonding (F2F Bonding)

Wafers with copper bumps (silicon interposer and first device wafer) are subjected to thermo-compression bonding in face-to-face configuration (F2F bonding). Thermo-compression bonding is a bonding technique where force and heat is simultaneously applied to the stack to gain bonding strength.

(c) Silicon Substrate Thinning

First, the device wafer is thinned down to 25 μm thickness using BG and CMP. Via-last TSV process requires precise wafer thickness and its uniformity control because thickness deviation causes increase in TSV over etching, resulting in degradation in TSV shape (notching). To avoid this problem, target TTV (total thickness variation) is set to 2 μm for target silicon thickness of 25 μm. Another concern about this process is wafer chipping during the wafer thinning process. "Knife edge" shape is created at the wafer edge during grinding and the subsequent thinning step, and this edge shape easily causes wafer cracking. To remove this shape, edge trimming (reshaping wafer edge cross section) is necessary. For the edge trimming scheme, trimming after bonding was employed.

(d) Backside Lithography and TSV Etching

First, photolithography for TSV is performed. Conventional mask alignment (using visible light) cannot be applied because there is no alignment mark on the thinned silicon surface. Thus, mask alignment using IR which is almost transparent for silicon is useful. Next, silicon etching is performed with photo-resist mask and is stopped at PMD or the device passivation layer. After resist stripping, silicon oxide as a TSV liner dielectric which electrically isolates silicon substrate from the TSV metal is formed using plasma-enhanced physical vapor deposition (PECVD). Because step coverage of PECVD is around 20–25 % at target dimension, deposition thickness of 2–3 μm is required for TSV liner thickness of 0.5 μm. Then, etch back of silicon oxide is performed to reveal the surface of IC interconnect at the bottom of the TSV.

(e) Metal Filling and CMP

After cleaning of the IC interconnect surface, Cu/Ti seed layer is sputtered and copper is electroplated. To avoid electroplating failure (e.g., void), bottom-up copper plating is preferred. After that, copper and seed layer is removed except inside TSV by CMP.

(f) Wafer Bonding (B2F Bonding)

Copper bumps on the second device wafer are bonded to backside (TSV side) of the bonded wafer stack (Si-IP and first device wafer) in Back-to-Face (B2F) configuration. Bonding condition is the same for F2F bonding.

(g) Silicon Substrate Thinning

The Si-IP substrate is thinned in order to reveal embedded TSVs which formed using via-middle process (TSV revealing). BG and subsequent silicon RIE is employed to reveal the TSV.

(h) Backside Protection Dielectric and (i) TSV Revealing

Silicon oxide film is deposited over the TSV-revealed surface using PECVD. Silicon oxide film on TSV top is removed by CMP.

(j) Backside Pad Formation

Semi-additive method is employed for Au/Ni backside pad. After Cu/Ti seed and photo-resist formation, Au/Ni layer is electroplated for the region where photo resist has the opening. Then, photo resist and Cu/Ti seed layer is removed.

The following section describes the detail of wafer bonding process.

#### 9.4.3.2 Wafer Bonding Technology

Backside via-last TSV process in combination with thinning after bonding method requires bump protection to prevent process damage during wafer thinning and TSV formation. To protect the bumps, hybrid bonding method was employed because the bumps were simultaneously covered with polymer during the bonding. Key process technologies for this are (a) wafer bonding without bonding void and (b) surface cleaning prior to wafer bonding. The following section explains the details.

The biggest challenge concerning hybrid wafer bonding is the formation of voids in bonded wafers. The void distribution of a bonded wafer after wafer bonding is shown in Fig. 9.102a. Since the bright area represents void area, large voids can be seen in the wafer. As a result of dicing, a significant amount of the chips on the wafer are delaminated (Fig. 9.102b), and the area of the delaminated chips corresponds well to the void area.

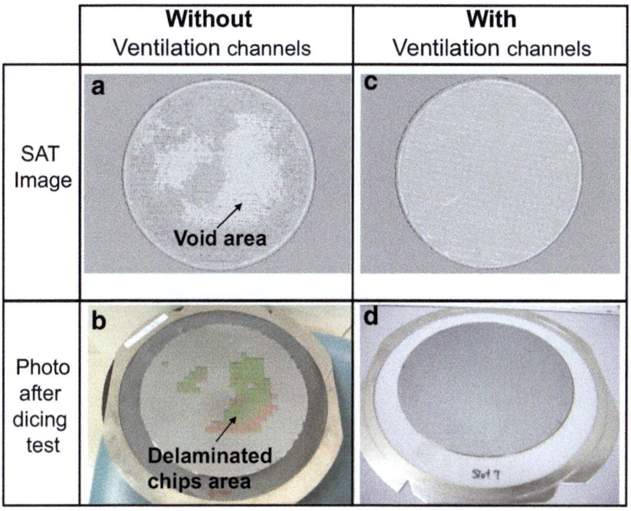


**Fig. 9.102** Void distribution of bonded wafers and chip delamination after dicing

This result indicates that the void formation caused chip delamination in the dicing process. A possible cause of these voids is captured gas between the wafers during the wafer bonding. To prevent capturing the gas, ventilation channel structure is introduced to the wafers. A process flow to fabricate the ventilation channel is shown in Fig. 9.103. First, polymer patterns, for both copper pads and ventilation channels, were formed by photolithography. Next, copper and a barrier metal were deposited on the polymer patterns. Since the filling ratio of electroplating depends on pattern size, it is possible to create patterns of ventilation channels that are not completely filled with copper and patterns of bumps that are completely filled.

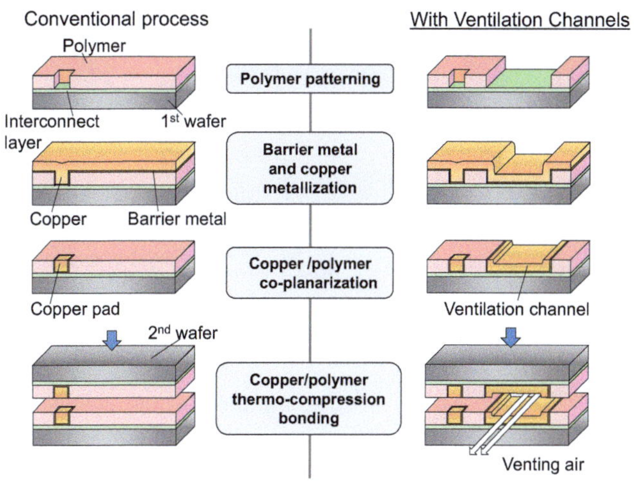


**Fig 9.103** Process flow for fabricating ventilation channel structure

The excess copper and barrier metal are removed by CMP. Finally, the wafers are bonded. As shown in Fig. 9.103, no additional process is necessary for creating the ventilation channels. When ventilation channels were formed in the copper bump layer, the formation of voids was reduced substantially, as shown in Fig. 9.102c. Moreover, dicing yield was 100 % (Fig. 9.103d). These results suggest that the formation of ventilation channels can reduce void formation, and this can help to prevent chip delamination in dicing.

This section explains the necessity of wafer surface cleaning prior to wafer bonding for reliable bump connection. Figure 9.104a is a cross-sectional SEM photograph of bonding interface of copper bumps after hybrid bonding. Interfacial layer can be clearly seen between copper bumps, suggesting incomplete copper bonding (no grain growth across initial bonding interface). Secondary ion mass spectrometry (SIMS) depth profile analysis of the copper bump surface revealed that copper sulfide exists on the bump surface ("without cleaning" in Fig. 9.104c). In order to reduce the copper sulfide, two cleaning process: Ar (argon) plasma cleaning, H (hydrogen) radical cleaning, were evaluated. As shown in Fig. 9.104c, both processes can effectively reduce copper sulfide on the bump surface. To clarify the effect of the cleaning, dicing tests of the bonded wafer were achieved.

For the wafer with Ar plasma cleaning, all chips were delaminated during dicing. On the other hand, no delamination occurred for the wafer with H radical cleaning. SIMS analysis of polymer surface after Ar plasma cleaning suggests that the plasma treatment causes a surface composition change and results in degradation of bonding strength between the polymers. From a cross-sectional SEM photograph of copper–copper bonding interface with H radical cleaning (Fig. 9.104b), good bonding interface (with no interfacial layer) was obtained. These results indicate H radical

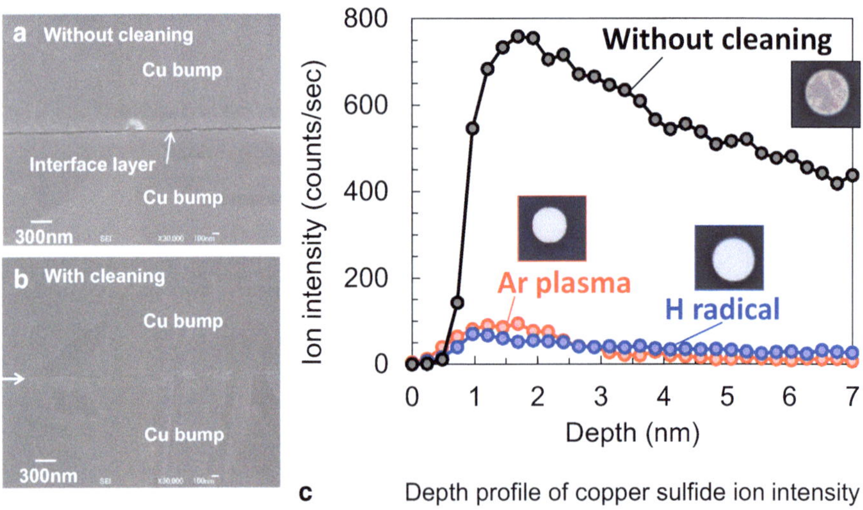


**Fig. 9.104** Cross-sectional SEM of bonded interface and SIMS profile of copper bump surface. *Ar* argon *H* hydrogen

cleaning prior to Cu hybrid bonding provides both good copper–copper bonding interface and sufficient polymer bonding strength.

As explained in MEOL process flow for wafer stacking (Fig. 9.101), stacking of more than two-wafer layer requires B2F bonding. To reduce the process step, direct B2F bonding was employed. Here, direct B2F bonding means bonding wafer surface with copper bumps to wafer backside where Cu TSVs and backside dielectric are exposed. This means that copper bumps are bonded to copper TSV, and the polymer on the wafer surface is bonded to the backside dielectric (SiO in this case). From dicing test, enough bonding strength between polymer and SiO was confirmed. To reduce bonding void, ventilation channel structure was also applied for B2F bonding.

#### 9.4.3.3 W2W Integration Results

This section explains fundamental performances of stacked chips fabricated using the abovementioned W2W integration process flow. The basic concept of test structures is almost the same as that for C2C process.

(a) Stacked Chip

Figure 9.105a shows a cross-sectional SEM image of packaged 3D chip. The 3D chip was mounted on an organic substrate. Gold (Au) stud bumps of 200-μm pitch were used for electrical connection between a Si-IP of the 3D chip and the organic

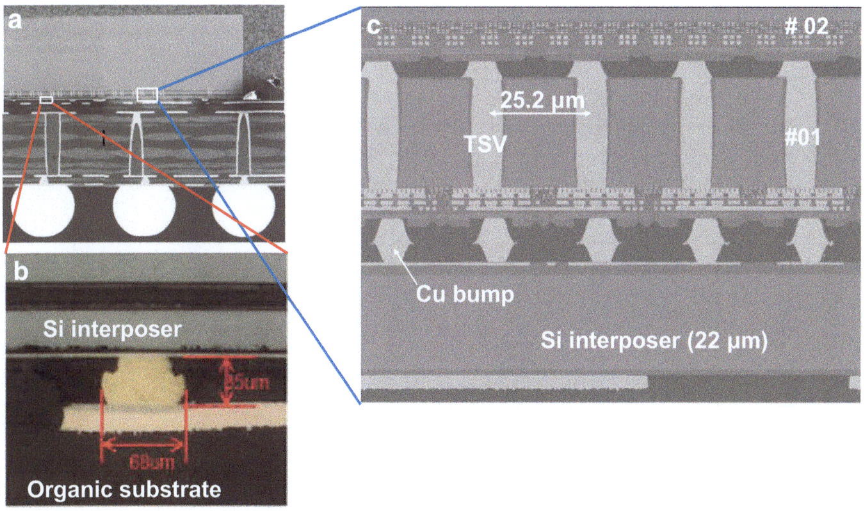


**Fig. 9.105** **a** Cross-sectional SEM image of a packaged 3D chip on an organic substrate, **b** cross-sectional photo around bonding interface between the 3D chip and the organic substrate, and **c** high-magnification view of the 3D chip. *TSV* through-silicon via

substrate (Fig. 9.105b). Within the 3D chip, the chip is connected to each other with 25-μm-pitch copper bumps and TSVs (Fig. 9.105c).

(b) Resistance, Capacitance

Voltage–current characteristic of F2F bumps were measured with Kelvin test structure (Fig. 9.106). Measured resistance consists of wiring resistances of both the lower chip and the upper chip, resistance of copper bumps bonded in F2F configuration, and their contact resistances. As shown in the figure, good liner relationship is observed even in the low-voltage region and obtained resistance which is calculated from the slope of the plot is very low (0.028 Ω).

Figure 9.107a shows TSV capacitance as a function of TSV voltage. Here, TSV voltage means voltage applied to the TSV. In this measurement, capacitance com-

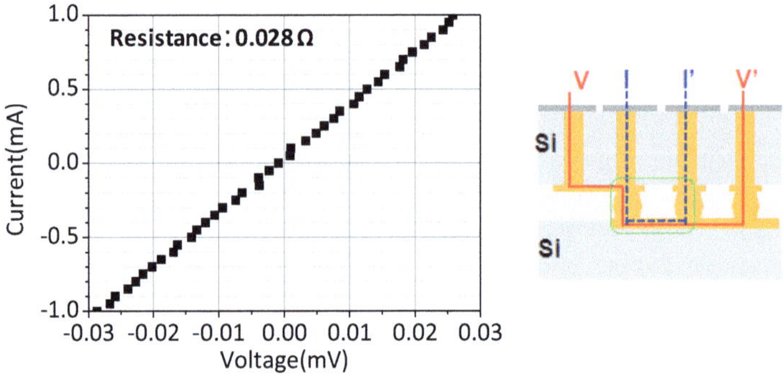


**Fig. 9.106** Voltage–current characteristic of F2F bumps. Measured value is a series resistance of wirings, copper bumps, and their interfaces

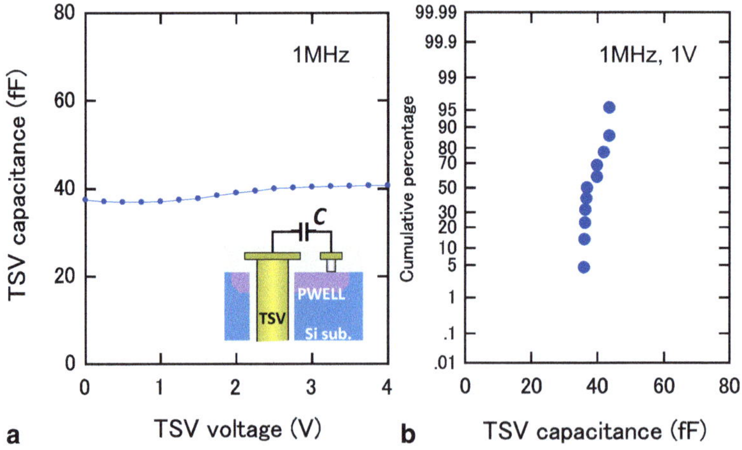


**Fig. 9.107** **a** Capacitance—voltage characteristic of TSV. **b** Variability of TSV capacitance (1 MHz, 1 V). *TSV* through-silicon via

ponents except for TSV capacitance were excluded from the measured capacitance value by using a test structure for capacitance calibration. As shown in this figure, TSV capacitance is around 40 fF and almost constant in this voltage range. Variability of TSV capacitance across a wafer (Fig. 9.107b) is found to be relatively small (coefficient of variation is within 10 %). Electrostatic capacitance of TSV with 7 μm in diameter and 25 μm in length and oxide liner thickness of 600 nm is estimated to be 160 fF, and its depletion layer capacitance is around 50 fF at 1 V. Thus, series capacitance of the TSV liner oxide and the depletion layer becomes about 40 fF, which agrees well with measured TSV capacitance. It is confirmed that TSV capacitance can be modeled as a series capacitance of TSV liner oxide layer and depletion layer.

One concern in regard to the via-last TSV process is degradation of copper/low-*k* interconnects. In the via-last TSV scheme, metal interconnect layers connecting to TSVs act as etch-stop layers during TSV etching as well as electrical contact electrodes for TSVs. Thus, process damage to surrounding low-*k* dielectric caused by TSV etching and subsequent wet processing must be clarified because low-*k* damage leads to increased parasitic capacitance and reliability degradation. Low-*k* damage caused by the via-last TSV process was evaluated by capacitance change of comb-type capacitors placed near TSV. Figure 9.108a shows interline capacitance of comb-type capacitors as a function of measurement frequency. Space between the comb lines is 190 nm, and distance from TSV edge is 2 μm. As shown in the figure, the capacitance is almost the same within this frequency range with or without TSV. Figure 9.108b is a cumulative percentage distribution of capacitance change across a wafer. Here, capacitance change is capacitance difference between capacitors with and without TSV divided by capacitance of capacitor without TSV. Because the capacitance change is within 1 %, the via-last TSV process has negligible impact on interline capacitance of copper/low-*k* interconnect.

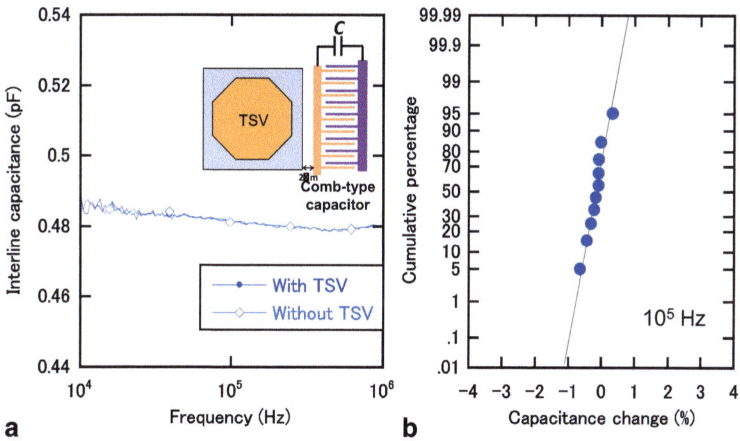


**Fig. 9.108** **a** Frequency dependence of interline capacitance of comb-type capacitor place around TSV. **b** Variability of capacitance change of interline capacitance across a wafer. *TSV* through-silicon-via

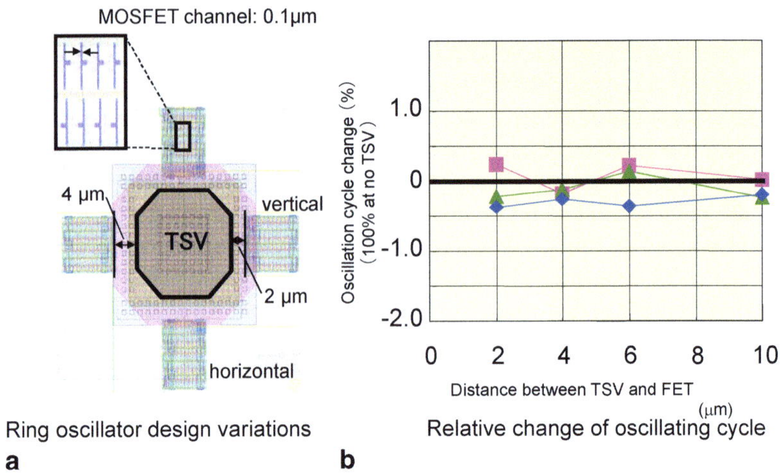


**Fig. 9.109** Ring oscillator characteristic near TSV. *TSV* through-silicon via, *MOSFET* metal–oxide–silicon field-effect transistor, *FET* field-effect transistor

(c) Keep-Out Zone

Same as for C2C process, KOZ for W2W process was evaluated by oscillating cycle change of ring oscillator placed in the vicinity of the TSVs. Figure 9.109a shows alignment of the ring oscillator near the TSV and measured results are shown in Fig. 9.109b. This figure clearly indicates that oscillating cycle change is below 1 % and a MOSFET can be placed as close as 2 μm from a TSV edge, namely, KOZ < 2 μm. This KOZ value is relatively small as compared with reported values [37, 38].

To clarify this reason, residual stress in silicon substrate in the vicinity of TSV was measured by electron backscatter diffraction (EBSD) (Fig. 9.110). The stress is under 50 MPa at about 2 μm from the TSV edge. It is known that on-current change in a MOSFET is below 2 % when silicon residual stress is below 50 MPa. Thus, this extremely small KOZ (i.e., 2 μm from the TSV) is mainly attributed to low residual stress in silicon. It is therefore concluded that a major advantage of a via-last TSV is its small KOZ resulting from the low residual stress in silicon.

(d) TSV Transmission Performance

TSV is expected to have drastically small interconnect delay time and power consumption as compared with wire bonding or package interconnect because TSV has very small coupling capacitance, thanks to its short length (e.g., 25 μm in W2W process). However, TSV also has coupling capacitance to silicon substrate, and this coupling capacitance may degrade TSV transmission performance. Measured power consumption of an optimized transmission circuit for TSV communication is shown in Fig. 9.111 as a function of TSV pad. From the slope of this plot, over-

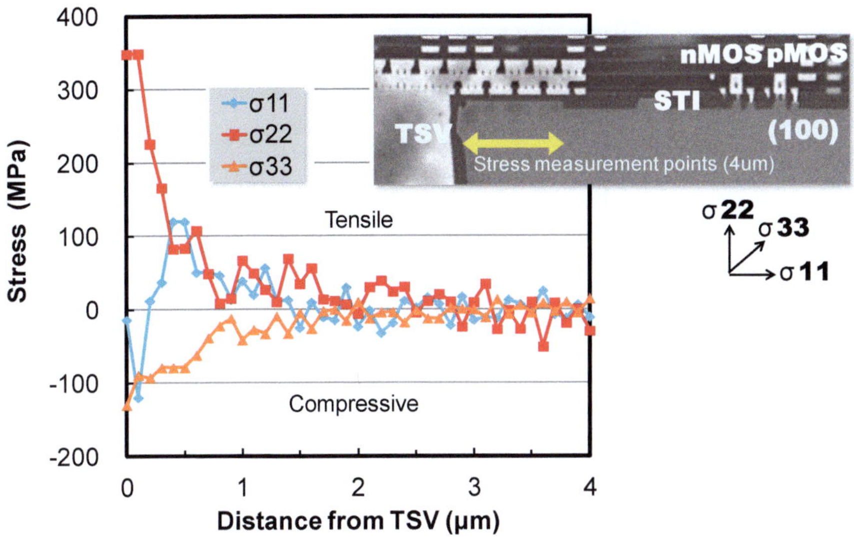


**Fig. 9.110** Residual stress distribution in silicon substrate near TSV. The stresses were measured by EBSD. *TSV* through-silicon via, *nMOS* n-type MOS, *pMOS* p-type MOS, *STI* shallow trench isolation

all coupling capacitance (TSV capacitance, wiring capacitance, bump capacitance) was obtained to be 84 fF. The parasitic capacitance of TSV, 48 fF, was estimated from the power consumption and parasitic extractions of a layout of the TSV pad. Thanks to this low TSV capacitance, the highest-level transmission performance of 17 Tbps/W and 3.3 Tbps/mm$^2$ was achieved (Fig. 9.112).

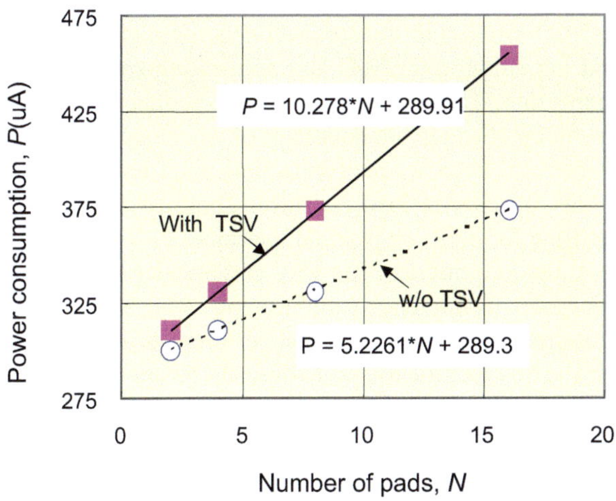


**Fig. 9.111** Power consumption of an optimized transmission circuits for 3D integration. *TSV* through-silicon via

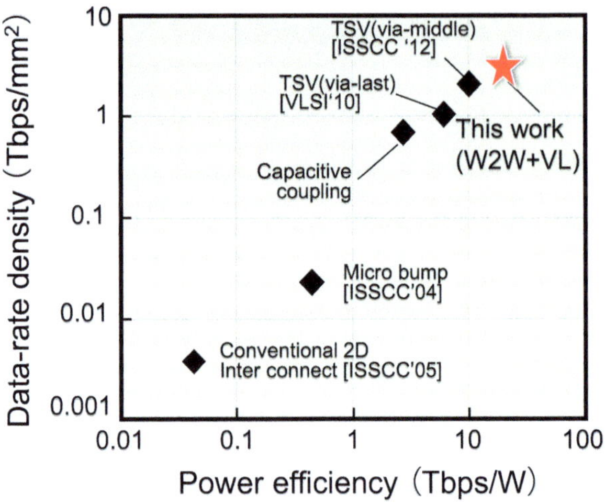


**Fig. 9.112** Transmission performance comparison. *TSV* through-silicon via, W2W wafer-to-wafer bonding, *VLSI* very-large-scale integration, *2D* two dimension

## *9.4.4 Standardization*

### 9.4.4.1 Reference Model of TSV Electrical Characteristics and Guideline of Test Conditions

This section describes standardization content we expect to propose to an international standardization organization. Reference model of TSV electrical characteristics required for 3D-LSI design and guideline of test conditions to specify TSV electrical characteristics were studied.

The conventional multi-chip interconnect specifications were used for limited applications and not suitable for SoC and application-specific integration circuit (ASIC) interconnection. The wider range of interconnect technology realized by TSV and micro-bump changes methodology of interconnect. Wide range I/O enables to extract on-chip bus to outside connection, and small size of TSV and micro-bump enables low-capacitance load interface. These two technologies permits to use on-chip signaling for multi-chip signal interface.

This specification does not describe TSV and micro-bump technology and also does not describe how to implement multi-chip modules as a level of packaging technology or interposer materials. It just shows them as a reference.

The basic idea of this specification is to describe reference model of TSV electrical characteristics required for 3D-LSI design and a guideline of test conditions to specify TSV characteristics.

(a) Reference Model of TSV Electrical Characteristics

Reference model of TSV electrical characteristics is composed of a capacitance ($C_v$) and a resistance ($R_v$). $C_v$ is the total capacitance and consists of $C_{ox}$, $C_{dep}$, and $C_{fr}$ shown in the left-hand side of Fig. 9.113. $C_{ox}$ and $C_{dep}$ are capacitances between the TSV and the semiconductor substrate and dependent on the TSV voltage ($V_{cc}$) since depletion layer exists. $C_{fr}$ is a fringe capacitance between a bump and the semiconductor substrate. Total capacitance $C_v$ is defined as a function of $V_{cc}$ and as a function of frequency shown on the right-hand side of Fig. 9.113. The node of $C_{dep}$ is connected to the ground thorough the substrate. When the interval of TSV is very near, coupling model is recommended.

Table 9.13 shows a policy for model standardization. Circuit definition explains the view of the proposed model. Device structure shows the device structure which will be the requisite in the case of using the proposed model. Test condition explains the device structure and operating conditions for the test.

(b) Test Conditions to Specify TSV Electrical Characteristics Resistance Measurement:

TSV resistance ($R_v$) consisting of bulk resistances of LSI interconnect, TSV, and bumps and their contact resistances is obtained by four-point measurement. A constant current *(I)* is supplied via current supply wirings connecting to terminal 1A and terminal 1B and the generated voltage *(V)* between wirings connecting to the terminals is measured by voltmeter as shown in Fig. 9.114. According to Ohm's law, a pair resistance for the first chip ($R_1$) is defined as *V*/*I*. Based on same the setup, a resistance for the first chip and second chip ($R_2$) is obtained. TSV resistance ($R_v$) is defined as $(R_2–R_1)/2$. This method is only valid when the second chip is es-

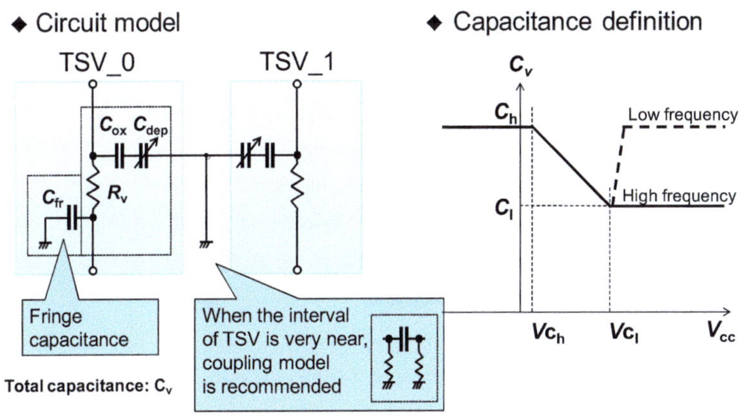


**Fig 9.113** TSV electrical characteristic model. TSV through-silicon via

**Table 9.13** Policy for model standardization. *TSV* through-silicon via

| Item | Subitem | Policy |
|---|---|---|
| Circuit definition | Parameter | Resistance is defined with average value |
| | | Capacitance is defined by a waveform<br>- Fringe capacitance is removable |
| | | Inductance isn't included because of little influence |
| | Model for every application | No definition |
| Device structure | Surrounding TSV | No definition |
| | | When the interval of TSV is very near, coupling model is recommended |
| | Semiconductor substrate power supply | Substrate is connected to the ground |
| | | The defined TSV circuit model is inapplicable when substrate isn't connected to power supply |
| | Semiconductor substrate | P type |
| Test condition | Structure | TSV array |
| | | There is a difference by single TSV and array TSVs |
| | Frequency | Frequency dependence of capacitance is measured |
| | Voltage | Voltage dependence of capacitance is measured |

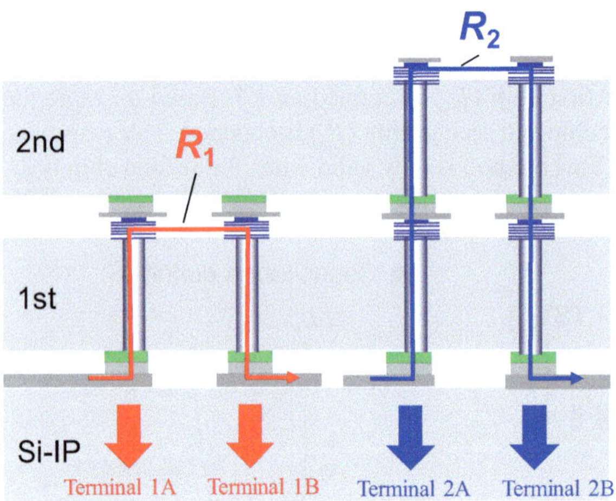


$$R_v = \frac{R_2 - R_1}{2}$$

**Fig. 9.114** Resistance measurement method

sentially the same as the first chip. To minimize the measurement error, the voltage measurement wiring should be connected as close as possible.

**Capacitance Measurement:** TSV capacitance ($C_v$) is obtained by electrical impedance measurement. Overall capacitance ($C_1$) between terminal 1A and terminal 1B is measured by impedance meter as a function of signal frequency *(f)* and

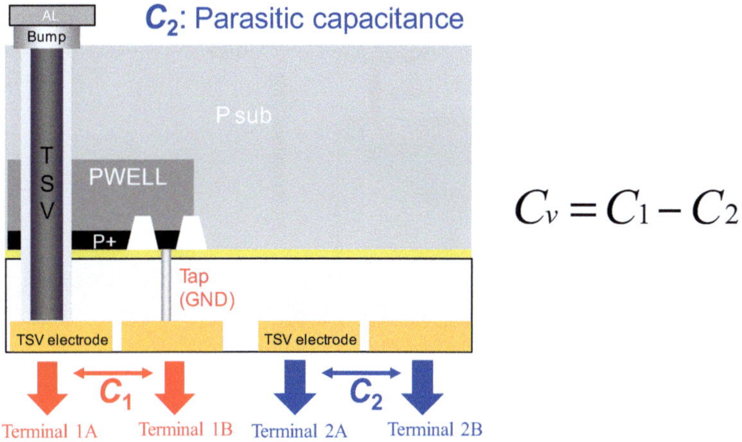


**Fig. 9.115** Capacitance measurement method. *TSV* through-silicon via, *GND* Ground level

direct current (DC) voltage ($V_{dc}$) as shown in Fig. 9.115. This capacitance ($C_1$) is composed of TSV capacitance ($C_v$) and the parasitic capacitance ($C_2$) caused by the measurement wiring and experimental setup. Therefore, TSV capacitance ($C_v$) is given by $C_1$−$C_2$. Parasitic capacitance ($C_2$) between terminal 2A and terminal 2B is obtained by impedance measurement for a test structure having neither TSV nor bump. Using TSV array is recommended to reduce parasitic capacitance.

**High-Frequency Characteristic** As a TSV high-frequency characteristic, S parameters of TSV are measured using a vector network analyzer (VNA). Every two TSVs short circuited on the backside form a pair of a grand line and a signal line using four TSVs as shown in the left 3D drawing of Fig. 9.116. Two ground-signal (GS) contact microwave probes with the contact pin pitch corresponding to the TSV interval pitch (200 um) are used in the measurement. After calibration operation for Open, Short, and Load using a calibration substrate, S parameters' measurement of TSV is carried out. Typical measurement results of $S_{21}$, $S_{11}$ are shown in the right chart of Fig. 9.116.

**Reference Dimension of the TSV Model** This section will describe reference dimension of the TSV model. Figure 9.117 shows a device structure and reference values.

(c) Keep-Out Zone

This chapter will describe evaluation standard for KOZ. TSV fabrication causes mechanical stress around TSVs. This stress causes transistor current variation. KOZ is the area surrounding each TSV from which transistors must keep out so that they are not influenced by the TSV-induced stress. The parameters affecting KOZ are defined in Table 9.14 and Fig. 9.118. The distance from the edge of TSV to the edge

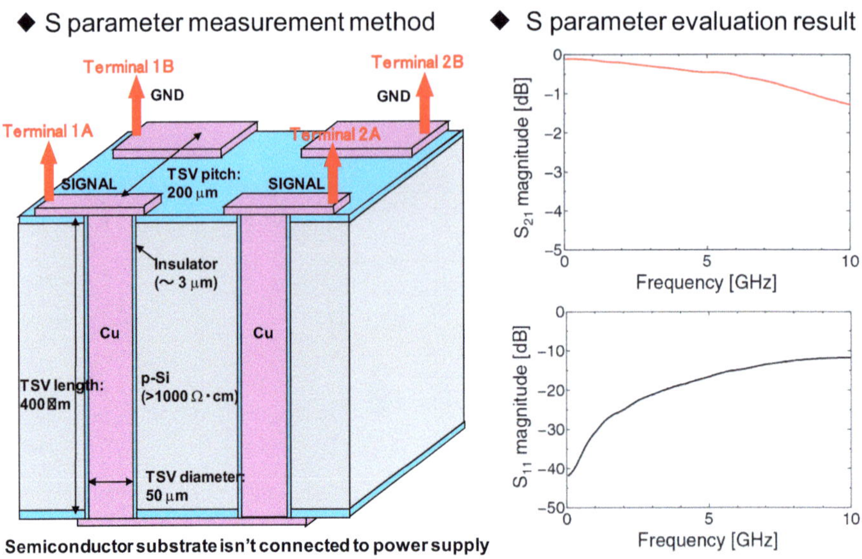


**Fig. 9.116** Test conditions specify TSV electrical characteristics when substrate is not connected to the power supply. *TSV* through-silicon via

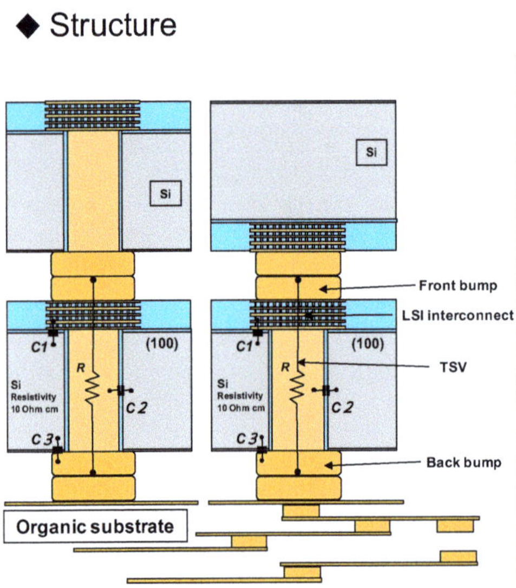


◆ Parameter

| | | Parameter | Reference Value |
|---|---|---|---|
| C1 | LSI interconnect | Size | 20/30um□ |
| | Pre-metal dielectric | Dielectric | ≈ 4 |
| | | Thickness | ≈ 0.3um |
| C2 | TSV | Diameter | 10/20um |
| | | Length | 25/50um |
| | TSV liner | Dielectric | 4.5 |
| | | Thickness | 0.5um |
| C3 | Back bump | Size | Φ10/30um |
| | Backside ILD | Dielectric | 4.5 |
| | | Thickness | TBD |
| R | Front bump-LSI interconnect | Contact resistance | |
| | LSI interconnect-TSV | Contact resistance | |
| | TSV | Resistivity | 2.0 uOhm cm |
| | | Diameter | 10/20um |
| | | Length | 25/50um |
| | TSV-Back bump | Contact resistance | |
| | Front bump/Back bump | Resistivity | |
| | | Thickness | 5/5um |
| | | Size | 10/30um |
| | LSI interconnect | Resistivity | |
| | | Structure | |

**Fig. 9.117** Reference dimension of the TSV model. *TSV* through-silicon via, *LSI* large-scale integrations, *ILD* interlayer dielectric

of gate on diffusion area of a transistor is defined as the dimension (D). When the dimension *(D)* is the specific value and the current of a transistor is changed by *n*% at most, the dimension *(D)* is defined as KOZn.

**Table 9.14** Parameters affecting KOZ. *TSV* through-silicon via

| Item | Subitem | | Parameter | Reference value |
|---|---|---|---|---|
| Chip stack structure | Front bump | | Size | Φ10/30 μm |
| | | | Material | Cu/SnAg-Cu |
| | Back bump | | Size | Φ10/30 μm |
| | | | Material | Cu/Au-Ni |
| | TSV | | Size | 10/20 μm |
| | | | Material | Cu |
| | | | Single/array | Array |
| | Underfill resin | | Material | PI/Epoxy series |
| Transistor structure | Process | | Generation | 90 nm |
| | Channel direction | | Parallel/vertical/ diagonal | Parallel/vertical/ diagonal |
| | Gate width | W | Size | 0.25 μm(n), 0.5 μm(p) |
| | Gate length | L | Size | ≈0.1 μm (n,p) |
| | Gate oxide | T | Thickness | |
| | MOS type | | N/P | N/P |

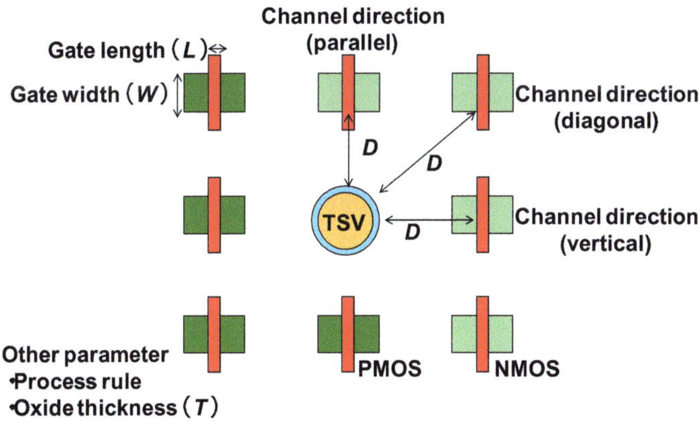


**Fig. 9.118** KOZ definition

## 9.4.5 Summary

Next-generation 3D-SiP requires electrical connections between stacked ICs for electrical communications and power delivery, and these are achieved through bumps and TSVs. Two types of three-dimensional (3D) integration schemes (C2C and W2W) were introduced in this chapter. A wafer-level 3D integration scheme using via-last TSV formation was explained in detail. As a key of this integration scheme, bonding of a copper/polymer hybrid wafer with a ventilation channel structure provides good copper-to-copper bonding as well as good polymer-to-polymer bonding without producing any large bonding voids. Via-last TSVs (7 μm in diameter and 25 μm in length) were successfully formed in the bonded wafer, indicating the effectiveness of the proposed 3D integration scheme. Sulfur-contained impurity degrades bondability between copper bumps during copper hybrid bonding and hydrogen radical treatment effectively eliminates copper sulfide from copper bump surface.

Effectiveness of proposed integration scheme was confirmed by its electrical characteristics. Thanks to the via-last processes, TSV contact wiring needs only two interconnect levels. The estimated KOZ is up to 2 μm from a TSV because of low-silicon stress (less than 50 MPa). Copper/low-*k* damage is negligible after via-last TSV formation (i.e., interconnect capacitance change is within 1 %). Low TSV capacitance (about 40 fF) was achieved for W2W process. As a result, TSV transmission performance was the highest level of 17 Tbps/W and 3.3 Tbps/mm$^2$.

## 9.5 Ultrawide Bus 3D System in Package Technology

### 9.5.1 Background

Since the wide bus memory such as Wide I/O2 and high bandwidth memory (HBM)began to appear in the market, it has led to drastic improvements based on 3D technologies including that of TSV [39–41]. These trends are shown in Fig. 9.119. This figure indicates that the market requires wider and faster bus to improve the bandwidth between memory and logic. Based on the current trend, this development was designed the 3D structure including ultrawide bus with Si-IP which would be necessary in the future because it is assumed that this is the direction the market is heading in the future. In the current environment, the supply chain is the problem in

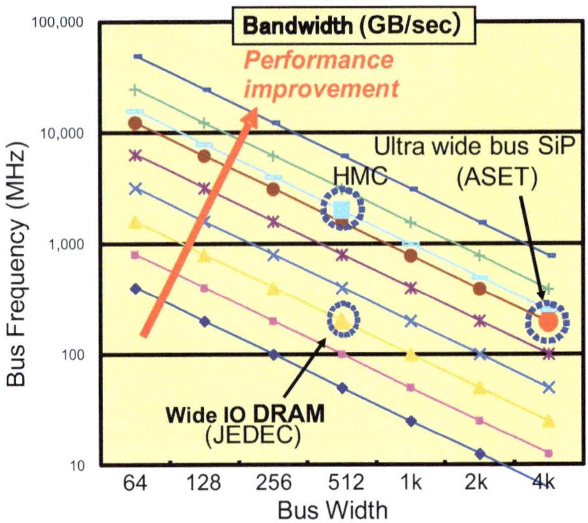


**Fig. 9.119** Memory bus width and transmission ability (bandwidth). *IO* input/output, *SiP* system in package, *ASET* super-advanced electronics technologies, *HMC* hybrid memory cube, *DRAM* dynamic-random access memory

3D development due to the complexity of the manufacturing process. Fortunately, in our development activity, this problem was solved by using a wafer foundry and Outsourced Semiconductor Assembly and Test (OSAT). In other words, a supply chain was built successfully by combining outsourcing with design rule adjustment throughout the manufacturing process.

After die preparation, device characteristics including TSV are tested by joint test action group (JTAG) and scan; these tests consist of a simple circuit and are modified based on 3D cube performance. KGD provided by these tests perform 3D device evaluation.

The advantage of 3D structure is the wide signal transmission width, but this might cause operation simultaneous switching output (SSO) noise and large transmission energy. Therefore, this development was performed with high attention to basic evaluation such as proof of the ultrawide bus performance, possibility of energy reduction, and noise reduction. The circuit which monitors the quality of the signal and the power supply in Si-IP including TSV was performed in situ to evaluate transmission capability including TSV itself.

### *9.5.2 The Test Vehicle Fabrication*

The memory such as HBM and Wide I/O2 with 512 or 1024 I/O began to appear on the market. Therefore, the TVs for this evaluation were designed as 3D-SiP consisting of memory die and logic die of 4096 I/O [42]. The 3D-SiP also has inserted Si-IP between the dies and is equipped with a monitor circuit to measure the synchronous power supply noise during operation of the ultrawide bus. In other words, this Si-IP is designed as an active interposer which does not affect the design of both logic die and memory. In addition, this monitor circuit would make use of the redundancy of the ultrawide bus to the maximum.

The reduction of power consumption is required for more than 4000 I/O systems. The following examinations were performed.

1. Reduction of the TSV capacitance: To achieve less than 100 fF, reduction of the device die thickness down to 50 μm and optimization of sidewall insulation film thickness was performed.
2. The I/O circuit with minimum control logic: The transmission speed between memory and logic was lowered to 200 Mbps and I/O drivability was optimized.
3. Minimization of electrostatic discharge (ESD) circuits: Because TSV between memory and logic are not outside pins, it is possible to minimize ESD circuits.
4. I/O power supply separated from Vdd: Because power supply fluctuation caused deterioration of the signal quality in 4096 I/O. The block diagram is shown in Fig. 9.120.

The device wafer was designed using 90-nm-shuttle TEG based on the foundry design rule and design kit with 7Cu/1AL process. However, the TSV formation is not supported by the FEOL foundry processes of 90-nm-shuttle TEG, therefore forming

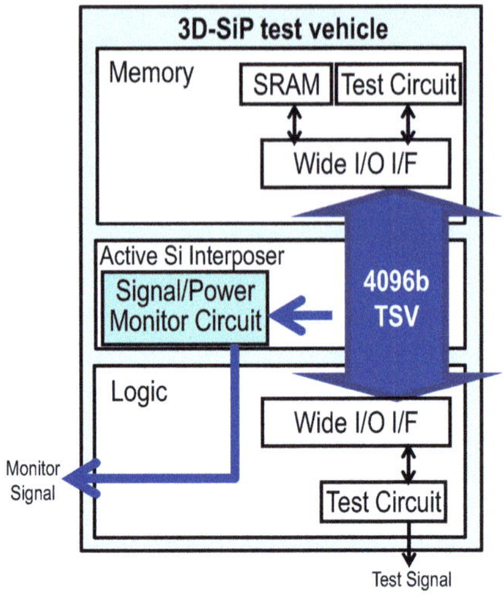


**Fig. 9.120** Block diagram of test vehicle. *TSV* through-silicon via, *I/O* input/output, *SRAM* Static random access memory, *I/F* Interface

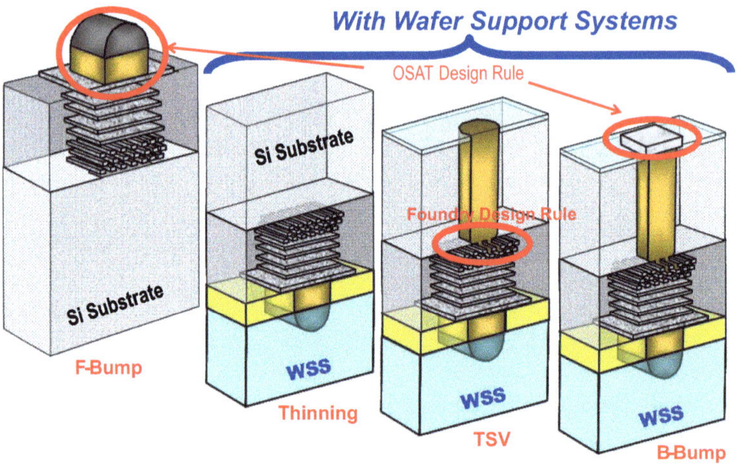


**Fig. 9.121** Bump and TSV fabrication flow. *OSAT* outsourced assembly and test, *TSV* through-silicon via, *WSS* wafer support system

the TSV via last TSV process was selected. The TSVs were formed at R&D process of ASET. The bump and TSV formation flow of the logic and active interposer die are shown in Fig. 9.121.

TSVs of these dies are assumed to connect to the 1Cu (the first metal) layer to make an appropriate structure of via-last TSV. Via-last TSVs of these dies require a pad consisting of 1Cu and 2Cu (the second metal layer) line pattern based on the wiring design rules of the FEOL for optimum interconnectivity between the TSV and the pad. 1Cu and 2Cu were formed with maximum line width which was allowed by design rule and had the same value of wiring prohibition area which means the KOZ of wiring.

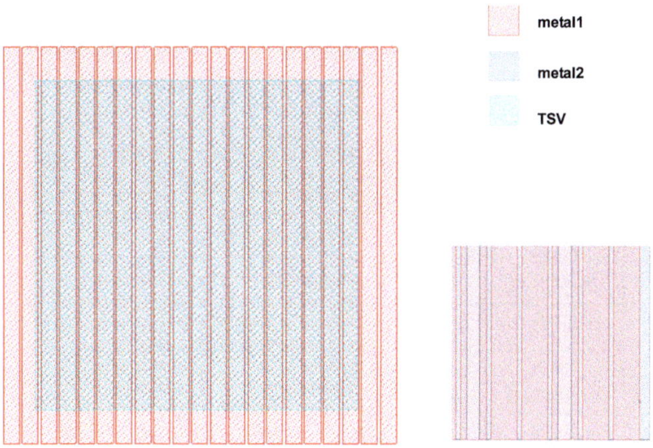


**Fig. 9.122** TSV layout for Logic die. *TSV* through-silicon via

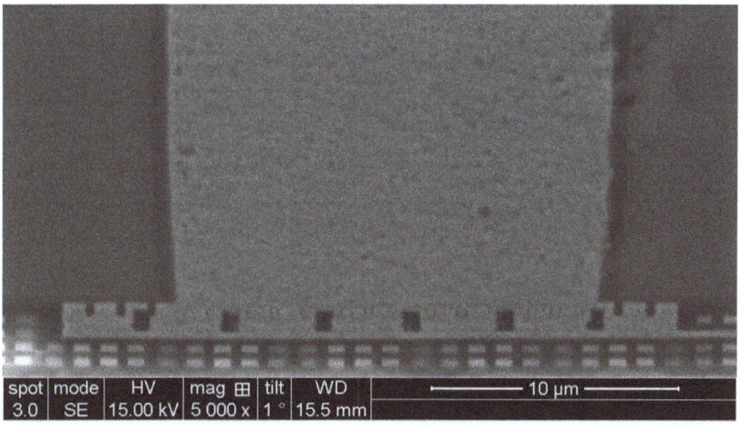


**Fig. 9.123** TSV vertical SEM image

2Cu cannot be used as the wiring layer in the TSV area because 1Cu pad formed by maximum wiring width restricted by the design rule is not able to cover TSV completely, so TSV reaches to 2Cu. The TSV layout of the logic die and vertical structure are shown in Figs. 9.122 and 9.123. In addition, TSVs were completely covered by 1Cu+2Cu composed of the wiring pattern so that we have no more prohibition area in the upper layer than 3Cu.

The top aluminum layer and passivation layer were also designed using the foundry BEOL design rule to coordinate the BEOL design rule and the bump design rules. This meant that the bump pad's design was different from the normal probe pad rule and generated original contact structure including verification rules.

Also, the integration was carried out to implement new design rule for this test vehicle and the adjustment between the bump formation and the assembly process

<table>
<tr><td></td><td colspan="3">Test vehicle component</td></tr>
<tr><td></td><td>Logic</td><td>Active Si Interposer</td><td>Memory</td></tr>
<tr><td>Chip Size</td><td colspan="3">9.93mmx9.93mm</td></tr>
<tr><td>Wafer</td><td colspan="3">90nm Shuttle TEG</td></tr>
<tr><td>TSV Depth</td><td colspan="2">50um</td><td>-</td></tr>
<tr><td>TSV Diameter</td><td colspan="2">20um</td><td>-</td></tr>
<tr><td>TSV Pitch</td><td>200um</td><td>50um</td><td>-</td></tr>
<tr><td>TSV Fabrication</td><td colspan="2">Via-Last</td><td>-</td></tr>
<tr><td>Assembly</td><td colspan="3">OSAT</td></tr>
<tr><td>Stacking</td><td>Face up</td><td>Face up</td><td>Face down</td></tr>
<tr><td>Connection</td><td></td><td>Solder</td><td>Solder</td><td></td></tr>
<tr><td>Underfill</td><td colspan="3">CUF</td></tr>
</table>

**Fig. 9.124** The outline of test vehicle package. *TSV* through-silicon via, *TEG* test element group, *OSAT* outsourced semiconductor assembly and test, *CUF* capillary underfill

at the same time. The assembly design rule had the prohibition of bump pattern shifts less than bump pitch and interconnection structure with micro-bumps.

Backside 3D process was performed using WSS. The bump fabrication process in ASET and assembly requirement of OSAT were organized together to make the ultrawide bus 3D-SiP.

The chips are connected with 50-μm pitch solder micro-bumps of 7332 pins and TSV with backside bumps which were complete center adjustment placement layout without the backside redistribution layer (RDL). The connections between the dies were performed with solder bumps. The logic die was stacked on the Si-IP die with face to back (F2B) connection, and the SI-IP die was stacked on memory die with face to face (F2F). The outline of the package is shown in Fig. 9.124, and the 3D-SiP structure after assembly is shown in Fig. 9.125.

### *9.5.3 Evaluation*

In case of evaluation using test system, the test circuit including redundancy and detectors of interconnected defectiveness on each die were confirmed to be effective as the 3D-SiP structure which has ultra many pins (4k I/O). The test architecture and test block diagram are shown in Fig. 9.126. The test items for judgment of KGD which were evaluated in both wafer state and 3D-SiP by a semiconductor tester are shown in Fig. 9.127. The test circuits were connected to not only the probe pad but also the bump to judge the KGD by the test technique that was usually used such

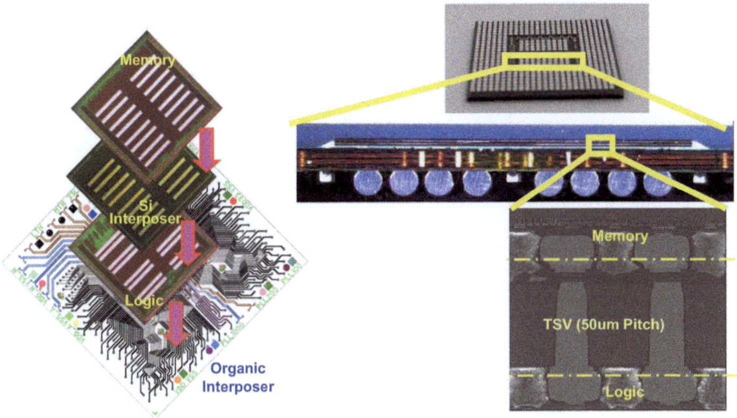


**Fig. 9.125** 3D-SiP structure. *TSV* through-silicon via

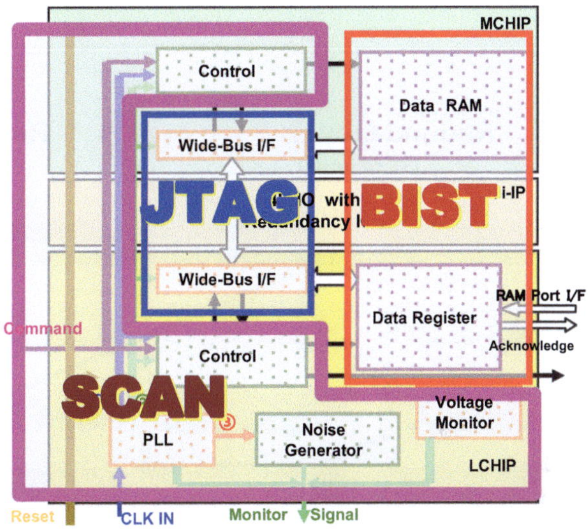


**Fig. 9.126** Test architecture and test block diagram. *RAM* random access memory, *JTAG* joint test action group, *BIST* built-in self-test, *MCHIP* Memory chip, *LCHIP* Logic chip, *CLK* clock, *PLL* phase locked loop, *I/F* Interface

as SCAN and built-in self test (BIST) in the wafer state and 3D-SiP. Furthermore, 3D-SiP which became the KGD were evaluated by the transmission performance between the dies. The evaluation of interconnection between the dies was carried out by JTAG. Read and write operations were performed using control registers which had inter-integrated circuit (I2C) interface.

These were able to confirm the effectiveness of the circuit according to the design.

JTAG test was performed to detect the bad interconnect of test access port (TAP) at first, then confirmed the flip-flop (F/F) chain connection, and transmission between logic die and memory die. In addition, this test confirmed that it was effective for the identification of the place by converting poor I/O into the physical map (Fig. 9.128).

**Fig. 9.127** The test items including both wafer and 3D-SiP. *3D-SiP* three-dimensional system in package, *TSV* through-silicon via, *JTAG* joint test action group, *PLL* phase-locked loops, *DLL* delay-locked loops, *MBIST* memory built-in self test

| Test Item | Test Items | | | |
|---|---|---|---|---|
| | Wafer Level | | 3D SiP | |
| | Logic | Memory | Logic | Memory |
| Power Lane Short | ● | ● | ● | ● |
| Pin Leak | ● | ● | ● | – |
| IO Open/Short | ● | ● | ● | – |
| Interconnect IO Loop | ● | ● | – | – |
| SCAN | ● | ● | ● | ● |
| MBIST | – | ● | – | ● |
| JTAG | – | – | ● | ● |
| Connection | – | – | ● | ● |
| PLL/DLL | ● | ● | ● | ● |
| TSV Path | – | – | ● | ● |
| Function | – | – | ● | ● |

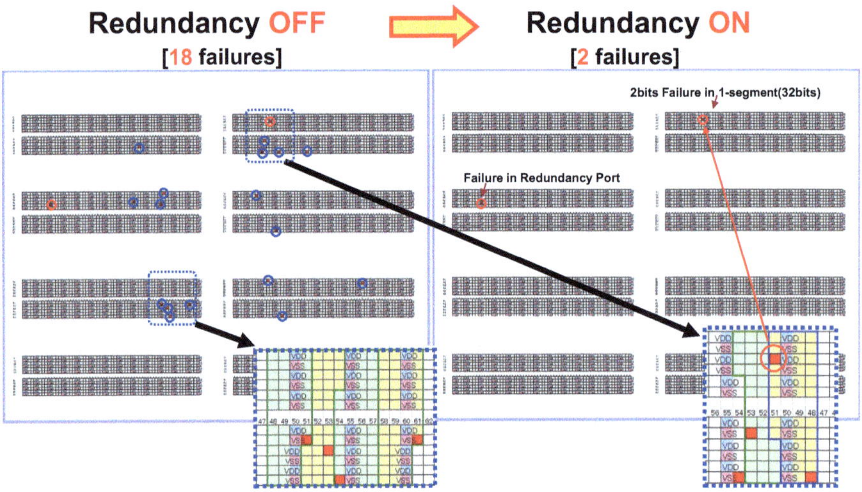


**Fig. 9.128** Redundancy (interconnect fail recovery)

In case of ultrawide bus structure, it is thought that power supply noise becomes the problem due to SSO. This SSO noise was predicted by the simulation and compared with actual measurement. The correlation between power distribution network (PDN) frequency properties and power supply noise due to structure and a layout of 3D-SiP was investigated on this development. The design and fabrication of the evaluation board were carried out for the purpose of power supply noise measurement of 3D-SiP and the voltage-dependent evaluation of the functional frequency.

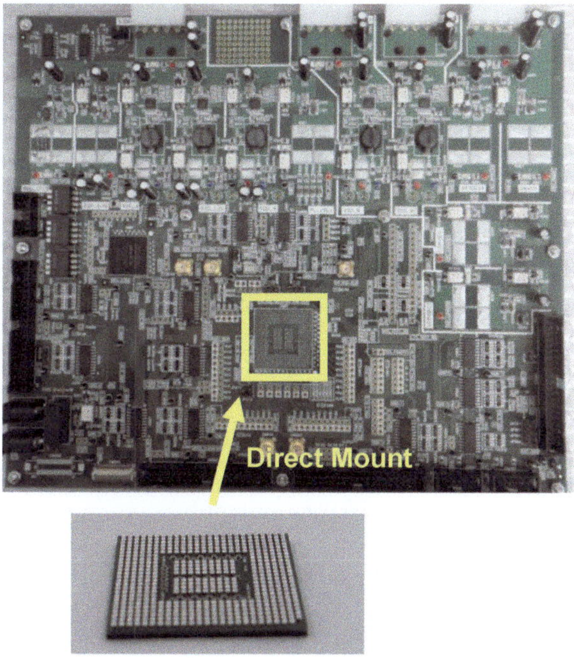


**Fig. 9.129** The photograph of board for PDN evaluation

The evaluation system with graphical windows user interface (GUI) was made. The system was made by six wiring layers Flame Retardant type 4 (FR-4) board. The photograph of the board which implemented the parts and 3D-SiP is shown in Fig. 9.129.

The modeling simulation procedure extracts the individual PDN model of each component from a layout and structure information by a simulation tool based on electromagnetic field analysis at first. Then each component connects everything between a power supply and the GND pins of the individual treatment PDN model. In the next step, these make a unification model to express the whole PDN of 3D-SiP implemented on an evaluation board.

Finally, a unification model was analyzed by a circuit simulator, then frequency properties and a transient response were obtained. In addition, the decoupling capacitance which derived from on die elements extracted by a cell unit using SPICE analysis of the transistor level beforehand were implemented on die PDN model. The current behavior of the ultrawide bus operation that is necessary for time domain analysis was made according to each operation modes and each frequency based on test vector and circuit delay information.

As for the comparison of the frequency domain analysis and measurement data, verification of individual PDN model was done first. Impedance of each die in the 3D-SiP was measured by connecting the probe to the bonding pad directly, then frequency domain analysis was done.

Measurement method of on-chip PDN impedance is shown in Fig. 9.130.

The measure was performed by vector network analyzer (VNA). Then, the comparison between an actual measurement and the simulation result of the PDN

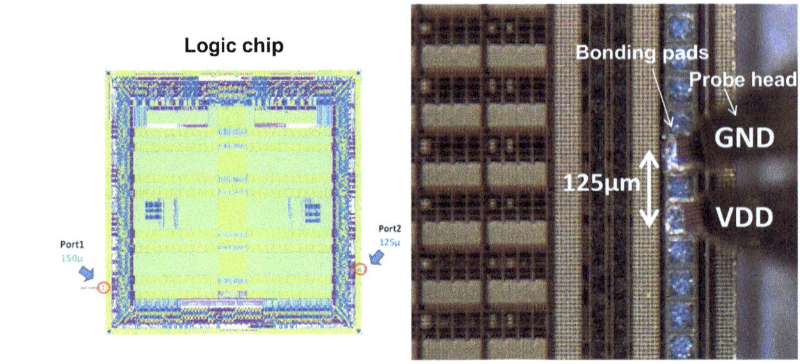


**Fig. 9.130** Measurement method of on-chip PDN. *GND* ground, VDD Positive Power supply Voltage (FET) Drains

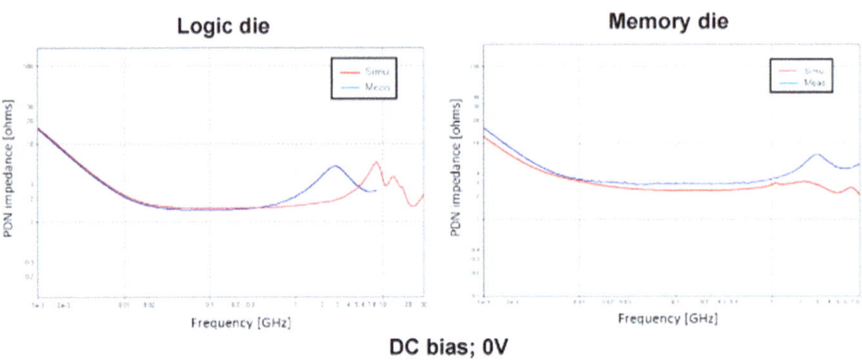


**Fig. 9.131** PDN impedances for memory die and logic die. *DC* direct current

impedance with the logic die and memory chip were done. Both data were well compatible to around 1 GHz as shown in Fig. 9.131. The comparison between the PDN impedance simulation analysis and the actual measurement in the Si-IP level is shown in Fig. 9.132. With an equivalent capacitance, around 2 nF difference was seen.

Using the same method, the comparison between the simulation analysis and the actual PDN impedance value in the package board itself was done. The result is shown in Fig. 9.133. In both cases, with/without decoupling capacitor, these have good correlation. From this evaluation result, the validity of the individual PDN modeling would be confirmed. Finally, in the unification model of 3D-SiP on the evaluation board, PDN impedance simulation analysis was performed. The observation points were implemented on each die, and the comparison result is shown in Fig. 9.134. The simulation analysis result shows that the strong anti-resonance peak appears at around 80 MHz due to the inductance of the package and the capacitance on die.

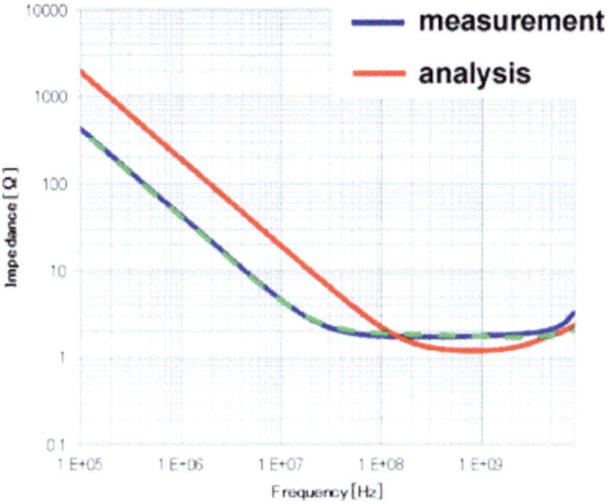


Fig. 9.132 PDN impedance measurement and analysis in the Si-IP

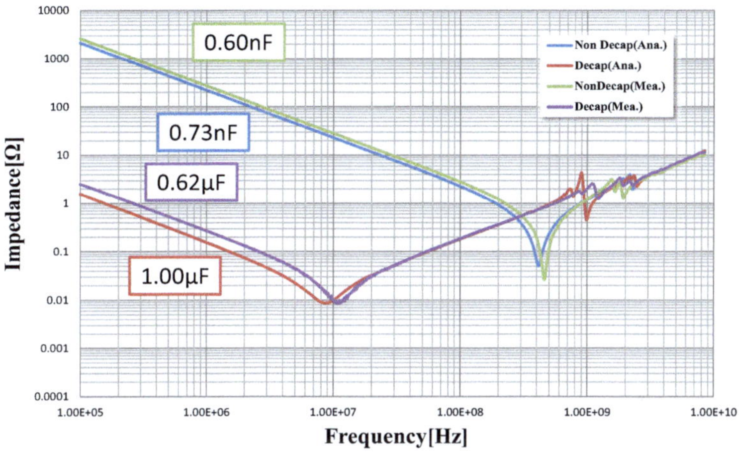


Fig. 9.133 PDN impedance measurement and analysis in the package. *DeCap* decoupling capacitance

From the individual die evaluation results mentioned above, the comparison with the time domain analysis of 3D-SiP was performed. To analyze SSO noise, this test vehicle was operated with the condition below:

1. Each dies current behavior as a drive source
2. Write operation with four kind of clock frequency (50/75/100/200 MHz)
3. A total of 4096 I/O operation with simultaneous or with/without phase shift

The actual measurement wave patterns of power supply noise obtained by a voltage monitor are shown in Fig. 9.135. The data were compared between with and without phase shift. The noise amplitude is the strongest at 75 MHz, and tendency is good

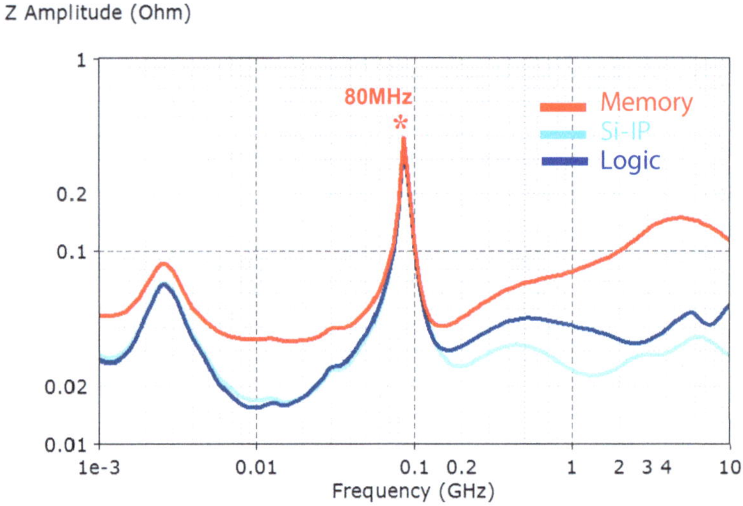


**Fig. 9.134** PDN impedance comparison result. *Si-IP* silicon interposer

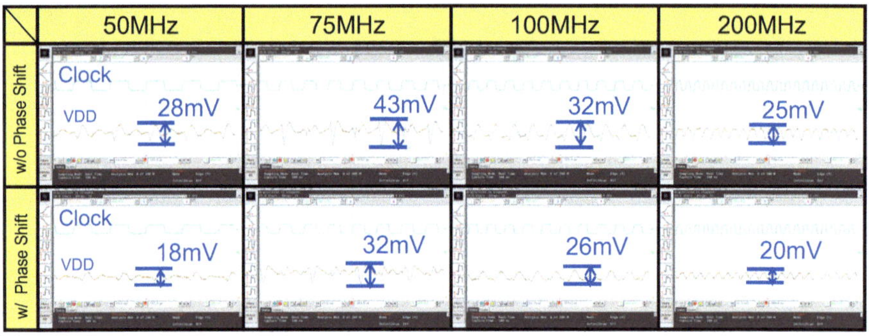


**Fig. 9.135** Measured SSO noise waveform @ 4k-bit Write operation. *W* with, *w/o* without

match with impedance properties of a provided unification model on frequency domain analysis. In addition, from an actual measured wave pattern, SSO noise reduction by the phase-shift circuit would be confirmed to be effective [43].

The evaluation of the function using ultrawide bus was carried out with phase-locked loop (PLL) circuit. I/O transfer functioned under the operation clock for ultrawide bus in 200/150/100/50 MHz using internal PLL with 100 MHz outside clock. The result showed that internal PLL frequency was equal to the output clock monitored terminal pin under the same power supply current. The pass judgment was also done using bit error rate (BER).

The measurement result of the I/O consumption current with/without PLL is shown in Fig. 9.136 and pass/fail map of the operation compared with/without PLL is shown in Fig. 9.137.

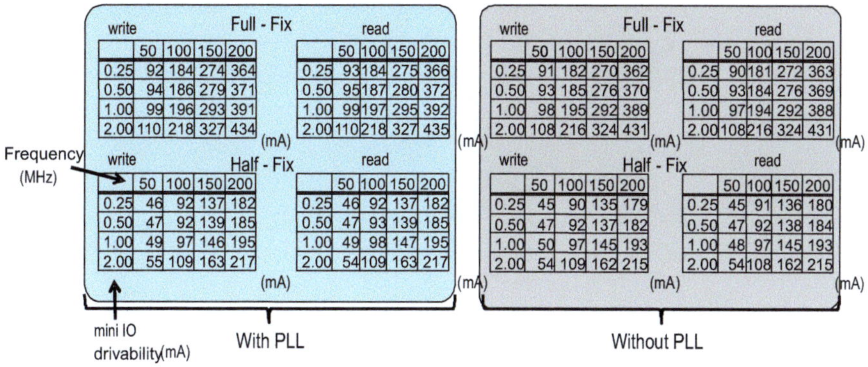


**Fig. 9.136** The measurement result of the I/O consumption current. *IO* input/output, *PLL* phase-locked loop

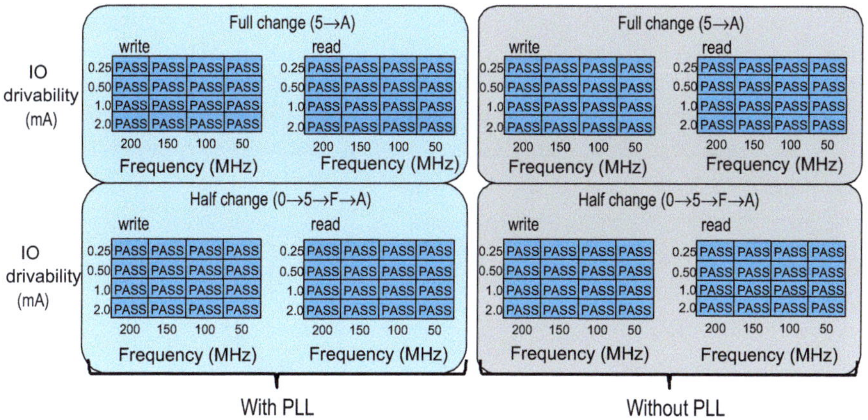


**Fig. 9.137** Pass/Fail map of the operation compared with/without PLL. *IO* input/output, *PLL* phase-locked loop

About signal input/output and power supply through TSV of 3D-SiP, it is very difficult to monitor from the outside, so indirect evaluation by situation is general under good quality. The direct routing was connected to a signal and a power supply passed through the TSV in this test vehicle and routing was connected to a monitor circuit. Then, an analog signal was converted into a digital code by a quality evaluation circuit and outputted to a package terminal pin to evaluate signal quality and power noise.

The result that measured an eye opening in the signal waveform when wide bus of 4096 bits was operated is shown in Fig. 9.138. Because wide bus I/O composes of a variable driver, the eye opening changes by regulating driver ability. For the signal input and output in 3D state, the driver ability was confirmed that around 0.5 mA was enough this time. In addition, the power supply waveform was mea-

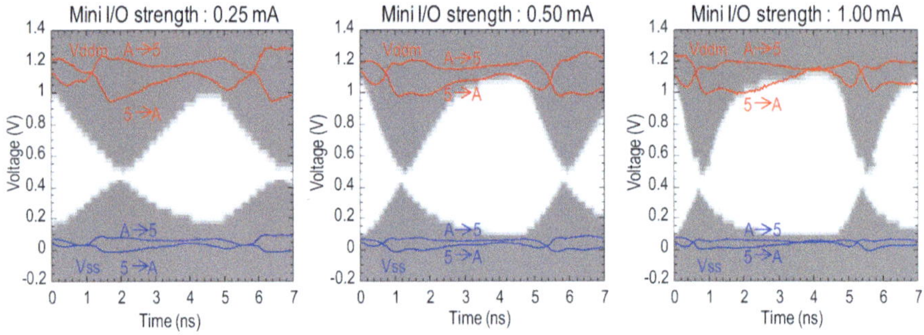


**Fig. 9.138** Eye opening in the signal waveform with power fluctuation. *I/O* input/output

sured at the same time of the eye-opening measurement, and it was confirmed that a power supply change is big when transition is in a state.

In addition, Shmoo between wide bus power supply and the clock was measured to evaluate frequency properties of the wide bus function. It was confirmed that a transfer speed was achieved at 200 Mbps. Then an evaluation of the transfer energy was carried out.

Maximum current was consumed by transferring 0101 → 1010 (inversion pattern). The result of transfer energy measurement at read/write operation is shown in Fig. 9.139. The transfer energy per 1 bit became 0.56 pJ (less than 1 pJ) [44].

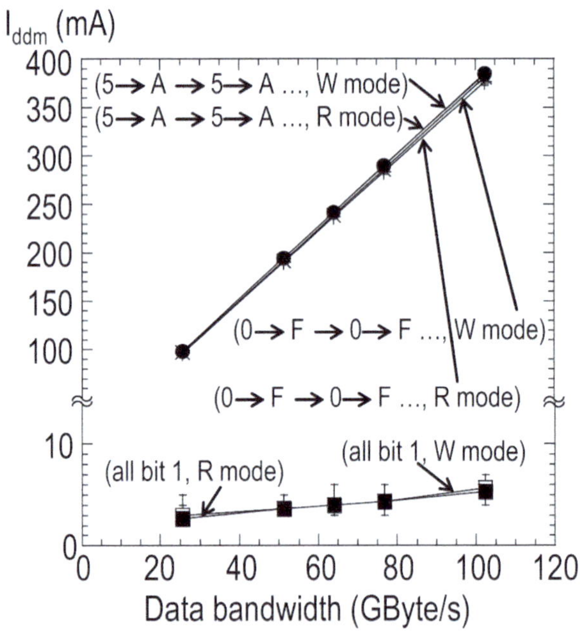


**Fig. 9.139** The transfer energy at read and write operation

### 9.5.4 Summary

3D-SiP which consisted of memory die and logic die with ultrawide bus (4096 bits) was designed, and the test vehicle was fabricated using shuttle TEG foundry and OSAT. This test vehicle implemented via-last TSV and solder bump connection to build 3D-SiP package consisting of the memory die and logic die and Si-IP. For KGD confirmation of 3D-SiP, the test circuit was implemented and would be confirmed by actual measurement with 3D-SiP operation. In addition, the interface that was used for this ultrawide bus was confirmed as follows:

1. Bus transmission ability of 102 GB/s in the operation under relatively low speed such as 200 Mbit/s.
2. The transmission energy was achieved around 0.562 pJ/bit, and load capacitance of the transmission lane confirmed. The capacitance value is less than 1/10 in comparison with conventional memory.

The reduction of the power supply noise in all frequency domains by operating the phase-shift circuit which reduced SSO noise with timing control was confirmed. The anti-resonance peak calculated by PDN impedance simulation analysis was almost accorded with an actual measurement and the validity was confirmed.

The effectiveness of the monitor circuit for evaluation of transmission properties was confirmed by monitoring the signal input and output of the ultrawide bus through TSV using active interposer. The designed I/O circuit would have enough transmission capability between the dies with 0.5 mA drivability. The interconnection performance between the dies would be close to the wiring in the single die.

## 9.6 Mixed Signal (Digital and Analog) 3D Integration Technology for Automotive Applications

### 9.6.1 Introduction

In recent years, the development of advanced vehicular driving support systems has made significant advances. It has been said that fully/semi-automatic driving systems will be put into practical use by 2020. One indicator of the trend towards this milestone is the growth rate of the integration of in-vehicle camera systems, which has been increasing by more than 40 % annually in new car models.

In this study, we developed a mixed-signal 3D integrated image sensor module that provides an image-processing system for automotive driving support. The issues specifically relating to 3D integration technology are discussed throughout this chapter. Table 9.15 summarizes our target product features. It is expected to overcome the principle challenges in manufacturing, design, and quality.

Figure 9.140 shows the main components of the computer development system used in the present research. These are the (i) 3DIC sensor module, (ii) the

**Table 9.15** Target features of the development products

| | Technology | Length of the ranging | Accuracy of the ranging | Night vision performance | Cost |
|---|---|---|---|---|---|
| Target | | OK | OK | OK | OK |
| Existing technology | Millimeter-wave radar | OK | Not enough | Not enough active | NG |
| Existing technology | Ultrasonic Sonar | × | Not enough | Not enough Active | OK |
| Developed products company A | 3D Camera | NG | NG | NG | OK |
| Developed products company B | Stereo camera | Not enough | Not enough | NG | OK |
| Developed products company C | IR camera | × No function | × No function | OK | NG |
| This study | 3D-integrated Camera | Not enough | Not enough | Not enough | Prepro-duction only—N/A |

*NG* No Good

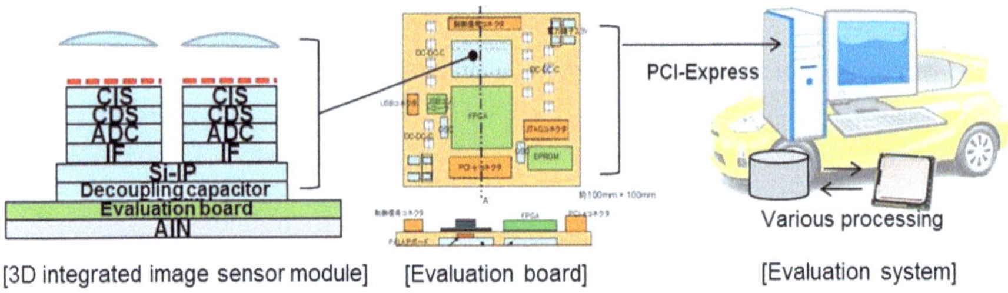


**Fig. 9.140** Components of the computer development system. *CIS* CMOS image sensor, *CDS* correlated double sampling, *ADC* analogue-to-digital converter, *Si-IP* silicon interposer

secondary mounting board required to perform 3DIC operations, and (iii) the personal computer evaluation system.

## 9.6.2 Challenges

There are five main challenges, which we now describe.

**(i) Establishment of a Heterogeneous C2C Integration Process** In practical devices, the chips have different wafer diameters, wafer thicknesses, interconnect wiring/passivation structures and materials, and manufacturing processes. It is necessary to establish a flexible C2C integration process.

**(ii) Clarification of the 3DIC Design** A consensus on the design of 3DICs has not yet been reached. One must take into consideration the complexity of each electrical property, the heat and stress environments, and the electromagnetic compatibility

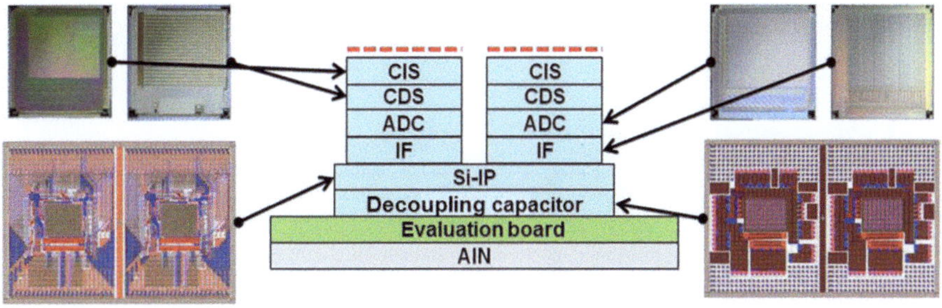


**Fig. 9.141** Conceptual outline of a mixed-signal 3D integrated imaging sensor module. *CIS* CMOS image sensor, *CDS* correlated double sampling, *ADC* analogue-to-digital converter, *Si-IP* silicon interposer

at design time. There is currently no standard procedure for the evaluation of trial products nor for the inspection of the manufacturing process.

**(iii) Design Manual for the Operational Stability and Methods of Evaluating 3DICs** The unstable operation of a 3DIC renders the product unserviceable. Therefore, appropriate guidelines need to be put in place to define acceptable design margins and methods that need to be established to ensure that these margins result in a stable device.

**(iv) Establishing a Heat Dissipation Mechanism** Even in a system with low power consumption, there may be some focal points of heat concentration, typically where the circuit density is highest. The heat current in TSV circuits will be different from that in a conventional thin silicon wafer because of an increase in copper (Cu) and micro-bump connections. Since 3D hot spots may also occur, establishing a heat dissipation mechanism is critical.

**(v) The Challenge of a Secondary Mounting Technology, and Clarification of Countermeasures** 3DICs are not necessarily assembled in the conventional package because of their heat dissipation and sensory requirements. This introduces the necessity for a secondary mounting technology, which may require low-temperature processing. Figure 9.141 shows a conceptual outline of a mixed-signal 3D integrated imaging sensor module.

## *9.6.3 Results of Development of Mixed-Signal 3D Integration Technology*

### 9.6.3.1 Basic Technology Development of a 3D Integrated Imaging Sensor Module for In-Vehicle Support Systems

Our 3D integrated image sensor module consists of 10 chips comprising six types for an in-vehicle electronic device. The module has small size-2 "eyes" that can range (measure the distance of) an obstacle from the vehicle. The system consists of

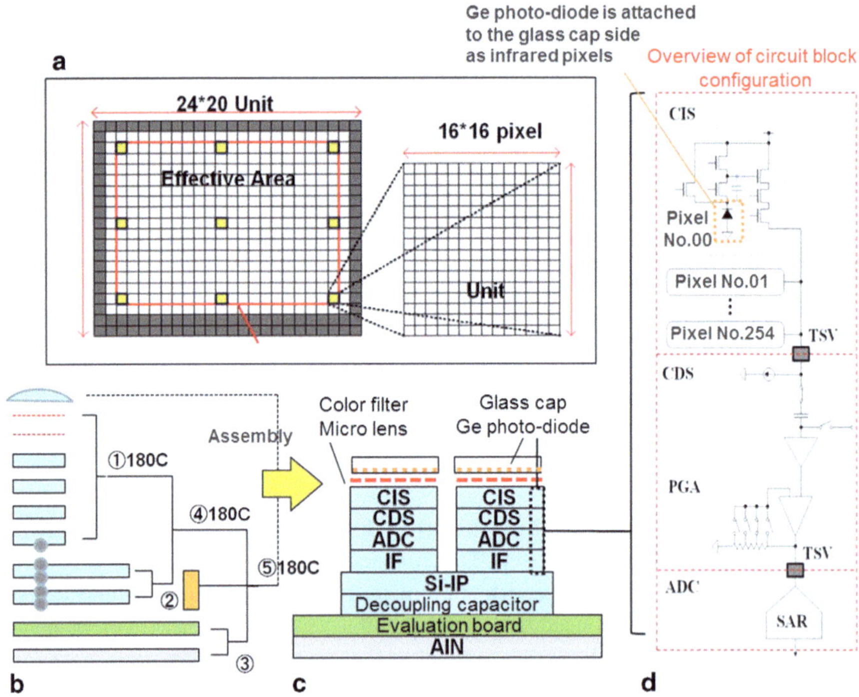


**Fig. 9.142** Overview of the **a** image sensor, **b** assembly process, **c** module structure, and **d** circuit configuration. *TSV* through-silicon via, *CIS* CMOS image sensor, *CDS* correlated double sampling, *ADC* analog-to-digital converter, *Si-IP* silicon interposer, *PGA* Programmable Gain Amp

a TSV connection integrated with an SiP structure comprising a CIS, a CDS chip, an ADC, and an interface circuit (IF). It also has night-operation sensitivity and a high-speed frame rate. Our image sensor module operates stably with an integrated Si-IP that contains a TSV-type decoupling capacitor. Its key features are:

1. Physically small: Ranging according to a corporal vision by micro-solid-state stereo structure.
2. Robustness: The interposer permits operation with poor power supplies.
3. Night operation: Infrared sensitivity and an expanded dynamic range made possible by the heterogeneous integration of 3D.
4. High-speed operation: 1,200 frames per second (fps) can be captured when operating in parallel processing mode with 256 pixels per frame.

Figure 9.142 gives an overview of the image sensor, the assembly process, the module structure, and the circuit configuration.

**Fig. 9.143** Photographs of the chips developed in this study. *CIS* CMOS image sensor, *CDS* correlated double sampling, *ADC* analog-to-digital conversion

(1) Design, Manufacture, and Verification of the CIS/CDS/ADC Chip for Sensor Modules

We designed and evaluated three chips that operate as an image sensor through 3D integration. The sensor has a germanium (Ge) photo-diode for infrared sensitivity, a novel feature. A capture rate of 1200 fps was realized by using a 480 block parallel operation with 255 pixels per block. The ADC rate was 4 MS/s with an accuracy of 7.28 effective number of bits (ENOB) at 85 °C .

Figure 9.143 shows photographs of the three chips that we developed. Their size was restricted to 5 × 5 mm in the trial commercial production. Operating characteristics were checked for conformance to the design specifications.

(2) Design, Manufacture, and Verification of the IF Chip for Sensor Modules

The image receiver is contained in the IF chip of the sensor module. The digital image signal is contained in 20 × 24 data blocks that are divided every 16 × 16 pixels for parallel passage by the TSVs. A parallel-serial conversion circuit collects eight signals and produces a series signal. There is a driver for high-speed exterior output; the clock generation circuit was designed in-house. The engineering sample 1 (ES1) IF chip had an I/O of 1200 fps and the I/O of engineering sample 2 (ES2) was 5000 fps. The layout pattern of the high-speed low-power consumption ES2 chip is shown in Fig. 9.144.

#### 9.6.3.2 The Mixed-Signal (CIS/CDS/ADC/IF) Integrated Structure with TSV Connection

In this project, we integrated a compound sensor, amplifier, and communication circuit, which form one of the required components of an in-vehicle support system and exemplifies the future of 3D integration applications. Specifically, we devel-

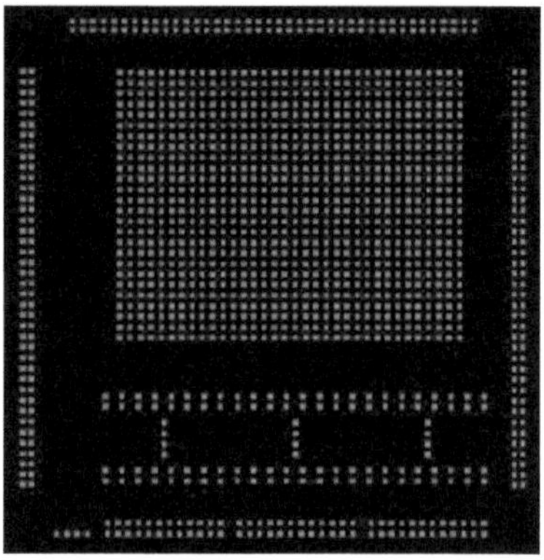

**Fig. 9.144** Layout pattern of the IF chip

oped a mixed-signal 3D imaging sensor module for driving assistance systems. The sensor module was an imaging sensor with a small range having an SiP structure of four chips (CIS, CDS, ADC, IF) integrated by TSV. Since it is necessary to be able to integrate chips having different wafer diameters, wafer thicknesses, interconnect wiring/passivation structure and materials, and manufacturing processes, we chose to use the flexible C2C integration process for construction; see Fig. 9.145. C2C integration processing involves wafer thinning, TSV, micro-bump formation, and the integration process with the Si-I0050 for assembly. The conditions for this all-encompassing process technology were optimized to our device in the production of the trial product. The germanium photodiode was formed when the glass end cap was in place (step 9 in Fig. 9.145).

(1) Formation of TSV

First, each chip was thinned while attached to a support wafer. Next, TSV and micro-bump formation were carried out. Before proceeding to trial production, we evaluated the technology involved in the following 13 procedures:

Planing the chip surface by CMP

Chip attachment to a support board evaluation of the strength of the adhesive

1. CMP of a Si chip
2. Deep via formation in Si from the backside of the chip
3. Oxide liner formation in the deep Si via
4. Barrier metal (Ta, Ti) formation in the deep Si via
5. Cu seed layer formation in the deep Si via
6. Cu plating of the deep Si via
7. Simultaneous formation of Cu-TSV and Cu/Sn micro-bumps by Cu plating
8. Surface planing of a lower layer chip, and Cu/Sn micro-bump formation
9. Bonding of the lower and upper layer chips
10. Support wafer de-bonding from the upper chip after bonding
11. Cu/Sn micro-bump formation on the chip surface

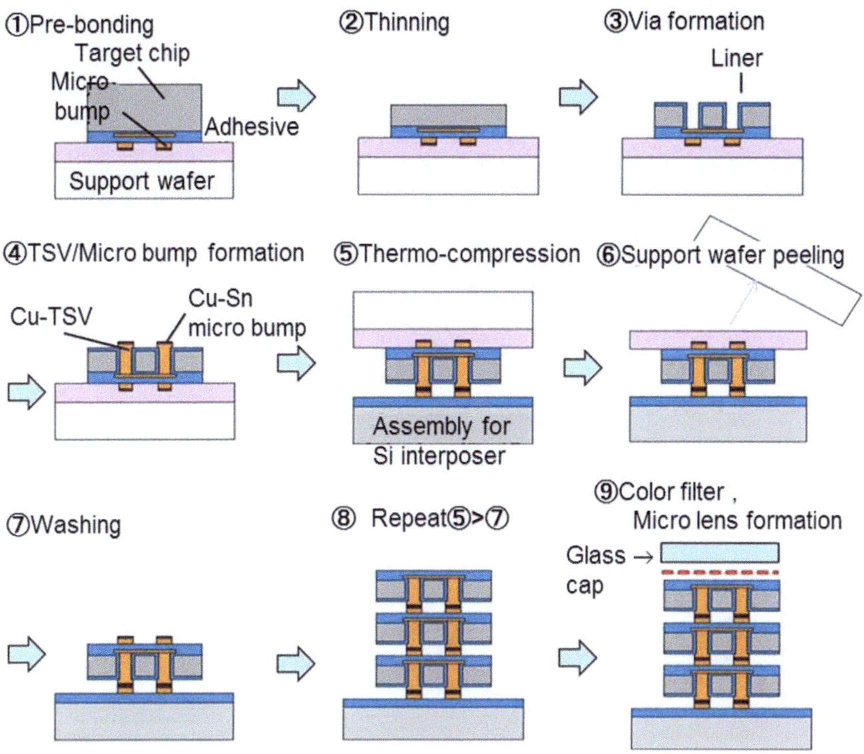


**Fig. 9.145** The C2C integration process. *TSV* through-silicon via

Figure 9.146 shows the technical details of procedures 9 and (10, and cross-sectional photographs of bottom-up plating and Cu CMP conditions. Figure 9.147 shows cross-sectional and plan view photographs of the CDS chip after procedure 10 was performed.

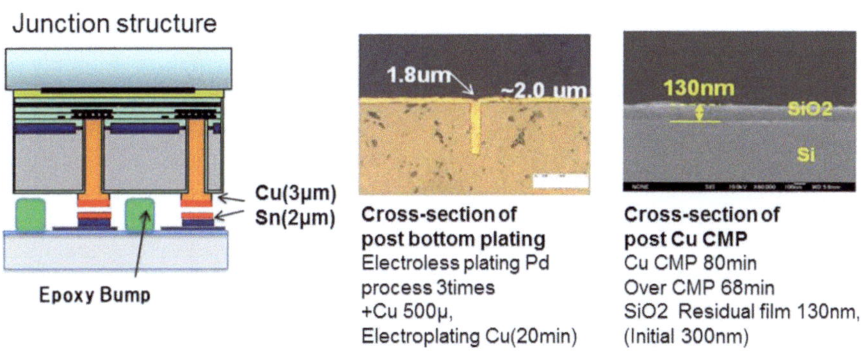


**Fig. 9.146** Formation of Cu-TSV and Cu/Sn micro-bumps. *CMP* chemical mechanical polishing

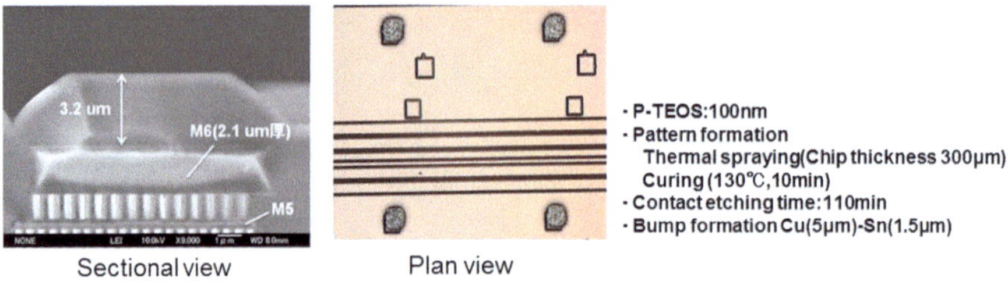


**Fig. 9.147** Micro-bump formation on an Al pad of a CDS chip

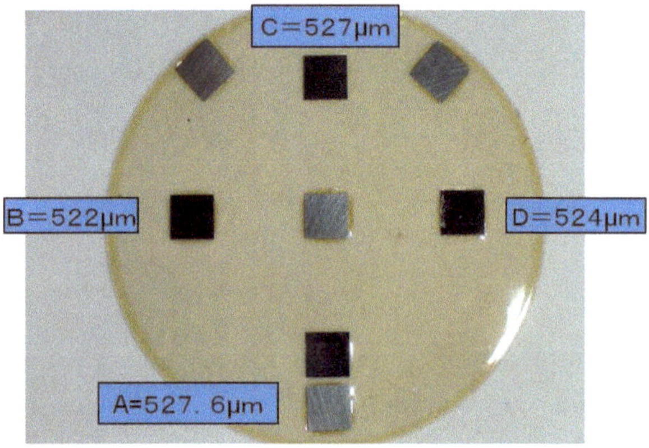


**Fig. 9.148** Photograph of chips bonded to a supporting glass substrate

(2) 3D Chip Integration

Figure 9.148 shows the photograph of a chip bonded to a supporting glass substrate before wafer thinning. The substrate can handle four chips at a time for a trial run. The variation in surface flatness after chip attachment was less than 5 μm. It has been suggested that many chips can be bonded simultaneously in high-volume manufacturing.

Figure 9.149 shows the chips as seen from the supporting glass side. Steps 5–8 of the process described in Fig. 9.145 were repeated. The surface of the IF, ADC, CDS, and CIS chips are shown. Chips need to be bonded firmly to the substrate for the TSV formation process, so the strength of the adhesive as well as its ability to separate easily without leaving residue are important.

Figure 9.150 shows the cross section of a multilayer-stacked chip for design verification. The SEM image (gray scale) shows the cross-sectional view of a Cu/Sn bump designed to connect 5-μm TSV and metal-1 of an ADC chip.

(3) Heterogeneous Stacking of Other Materials

After four chips (CIS, CDS, ADC, and IF) were stacked on the Si-IP, the color filter micro-lens was formed (Fig. 9.145, step 9). Figure 9.151 (left) shows a photograph

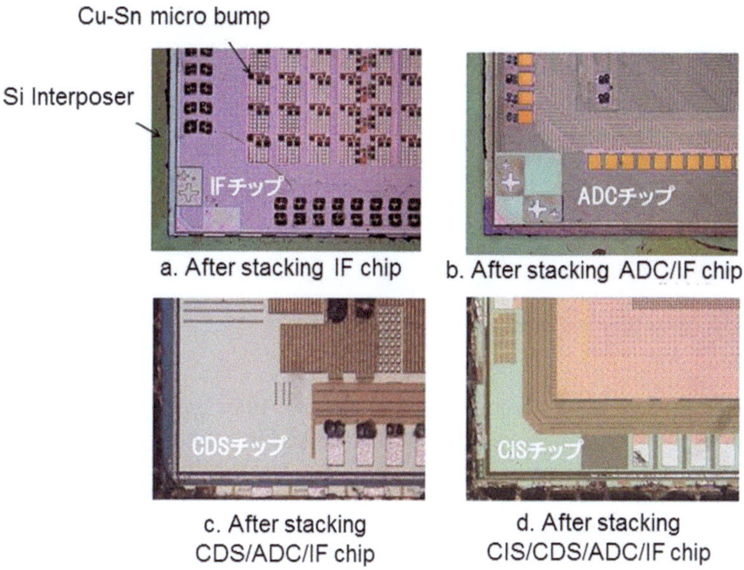


**Fig. 9.149** chips viewed from the supporting glass side. *CDS* correlated double sampling, *ADC* analog-to-digital conversion, *IF* interface

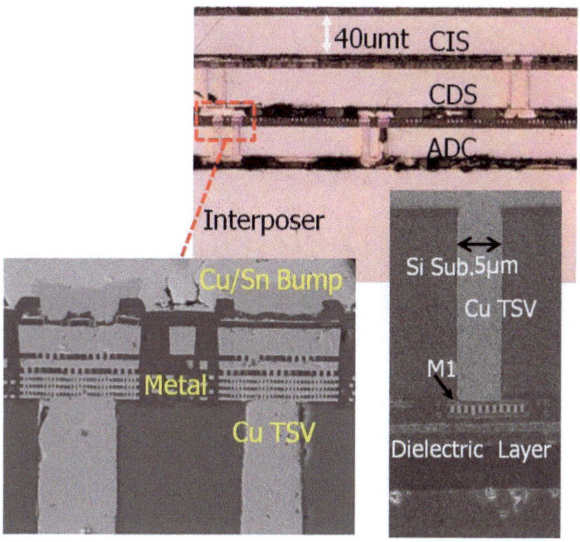


**Fig. 9.150** Cross section of a multilayer stacked chip. *CIS* CMOS image sensor, *CDS* correlated double sampling, *ADC* analog-to-digital conversion

of the lens, which is part of a red green blue (RGB) filter; the portion that does not have a colored filter (white part) enables near-IR sensitivity.

Figure 9.151 (right) is a photograph of the micro-lens and a plot of the convex lens profile. Because such materials have little heat resistance, a low-temperature process (under 180 °C) is needed.

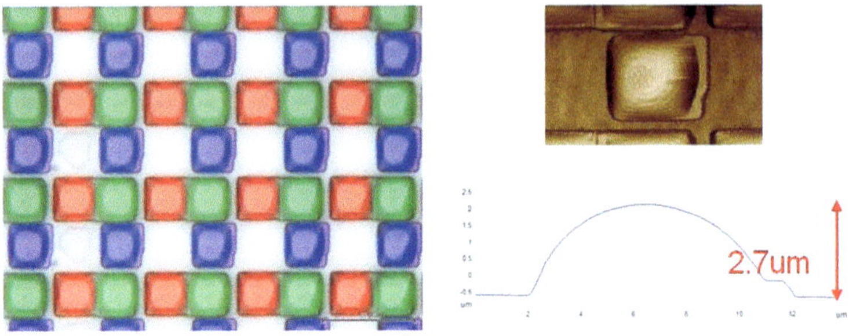


**Fig. 9.151** Color filter and micro-lens formation

Although the glass protection cap was attached after the color filter micro-lens was formed, to obtain infrared sensitivity, the Ge photodiode was formed on the backside of the cover glass. This structure was electrically connected to the CIS chip.

Figure 9.152 shows the production process of the Ge photodiode and the diode surface is shown in Fig. 9.153.

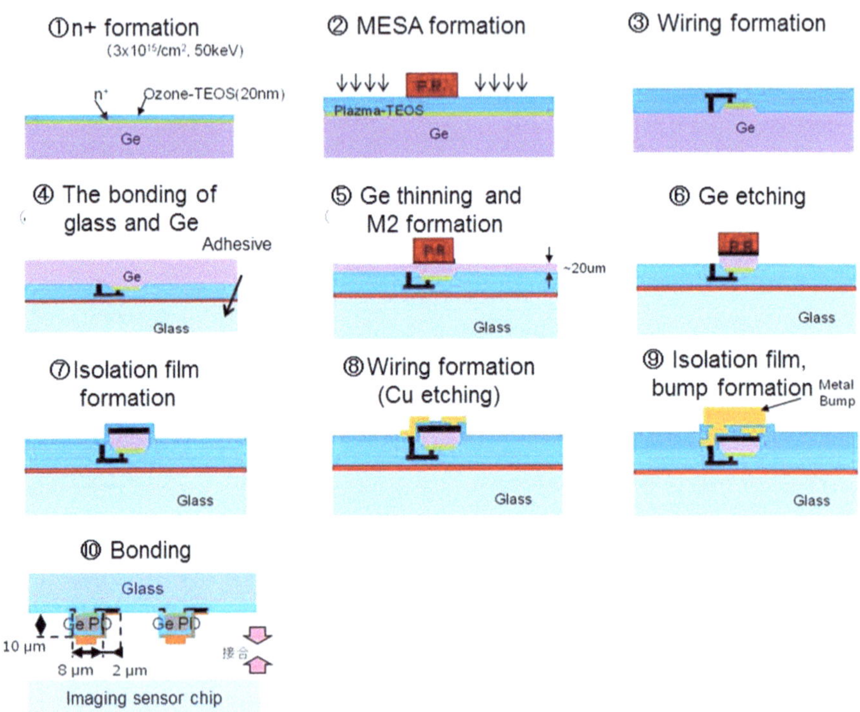


**Fig. 9.152** Ge photodiode fabrication *MESA* Mesa Transistor

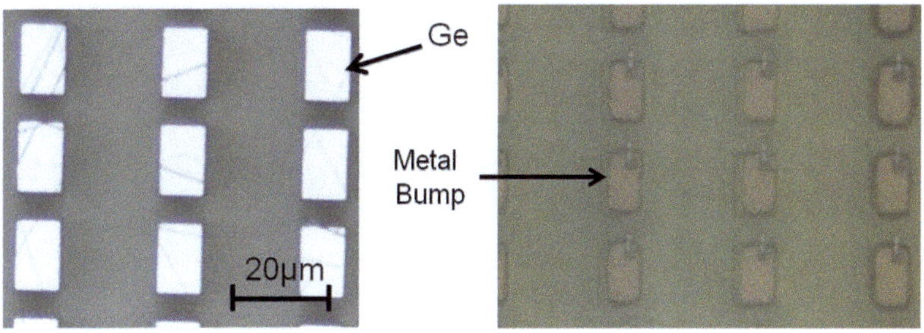


**Fig. 9.153** The Ge photodiode

3D Integration Technology can realize an unprecedented level of device multi-functionality as it is possible to mechanically and electrically connect a variety of non-Si chips.

#### 9.6.3.3 Development of an Si Interposer with TSV-Type Decoupling Capacitor

We now describe the development of an Si-IP with low electric impedance and high power dissipation for a 3DIC chip operating in a stable condition in the vehicle environment. For a mixed-signal 3D integrated image sensor module, multiple stable power supplies that do not have mutual interference are needed. The Si-IP we developed had a TSV-type decoupling capacitor, which is shown in Fig. 9.154. The bypass capacitor was formed in the TSV with a coaxial structure having a highly dielectric layer.

Prior to system design, a TEG experiment was performed. The TEG consisted of 16 parallel cells, shown in Fig. 9.155 *(center)*. A photograph of the trial production wafer is shown in Fig. 9.155 *(left)*, and a cross-sectional view is shown in Fig. 9.155 *(right)*. The wafer thickness was 400 μm, the diameter of the TSV was 50 μm, and the back and front side copper wiring layers (for interconnections) were 5 μm thick.

The breakdown voltage and capacitance data are shown in Fig. 9.156. The targets in the 16-capacitor parallel TEG structure were a capacitance of 52 pF and a breakdown voltage greater than 100 V; the results were 33 pF and 60 V.

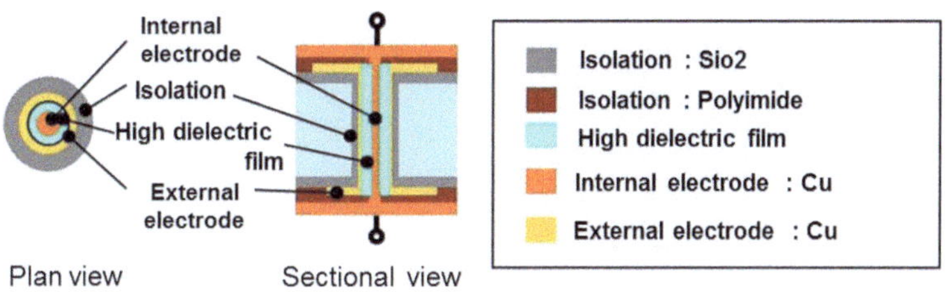


**Fig. 9.154** Structure of TSV-type decoupling capacitor

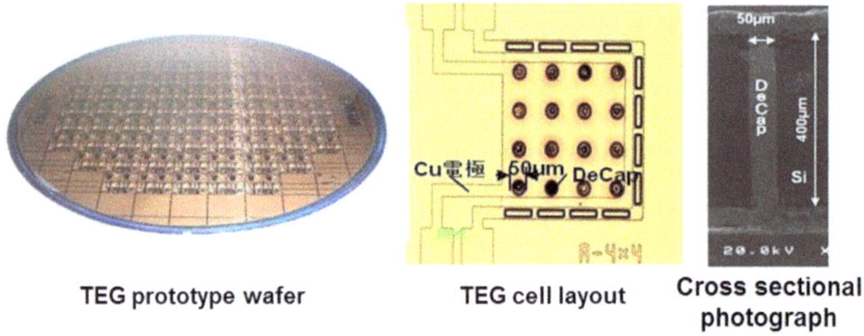


**Fig. 9.155** Test element group (TEG)

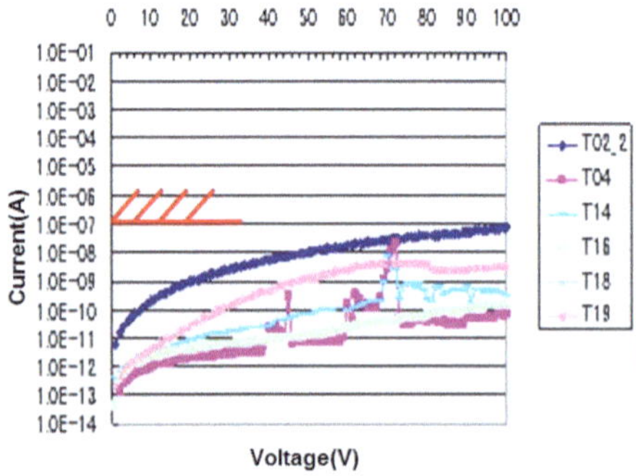


| Characteristic | Target | Results |
|---|---|---|
| Voltage[V] | >100 | 50 |
| Capacitance[pF/unit] | 53 | 33 |

**Fig. 9.156** TEG evaluation results

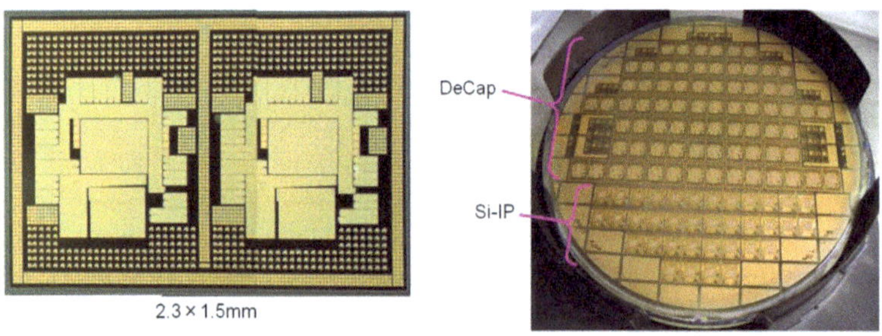


**Fig. 9.157** Photographs of the decoupling capacitor and the prototype wafer. *DeCap* decoupling capacitor, *Si-IP* silicon interposer

Although these values were less than the targets, we believed that they fell in the range that could be adjusted by optimization of the dielectric film thickness. So that a decoupling capacitor of approximately 70 nF could be built in the Si-IP, we did a layout design of the decoupling capacitor chip joined to an actual 3D imaging sensor module. We subsequently proceeded with wafer trial production.

Figure 9.157 shows photographs of the chip *(left)* and trial production wafer *(right)* of a decoupling capacitor. As the Si-IP was also made in the trial production,

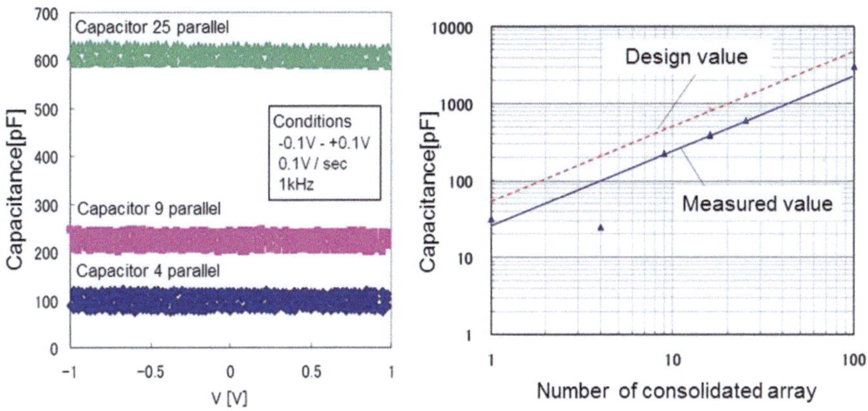


**Fig. 9.158** Capacitance characteristics of the decoupling capacitor chip

there are two kinds of chips on the same wafer. Although the Si-IP and the decoupling capacitor chip were made separately and stacked for the trial production, it is desirable both economically and for optimal performance to make the interposer and the decoupling capacitor in the same chip. Figure 9.158 shows the stable current–voltage characteristics of the decoupling capacitor chip *(left)* and its capacitance performance *(right)*. The absolute value of capacitance was only about half of the design value. This could be because of an unformed Ta layer (GND electrode) at the via-side wall, which would prevent the storage of electric charge.

Figure 9.159 shows the breakdown voltage measurements; 83 V compared with the target value of 100 V. However, 93 V was measured when we formed the dielectric film into the plate shape. There is likely some structure-based electric field concentration that prevents reaching the target value.

Figure 9.160 *(left)* plots the frequency characteristics of the decoupling capacitor chip in a TEG pattern. When there were few sequence connections, even a

**Fig. 9.159** Breakdown voltage characteristics of the decoupling capacitor chip

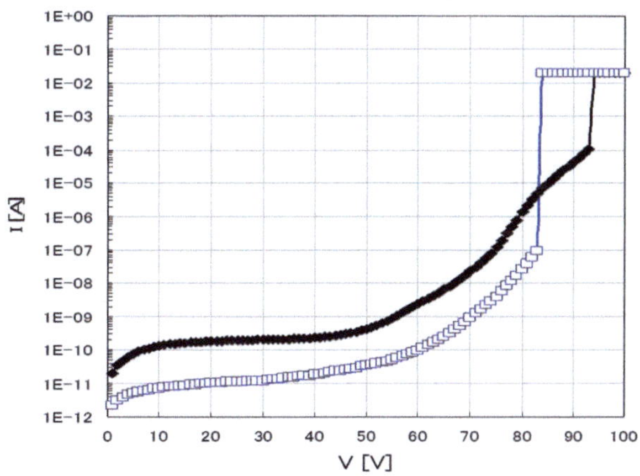

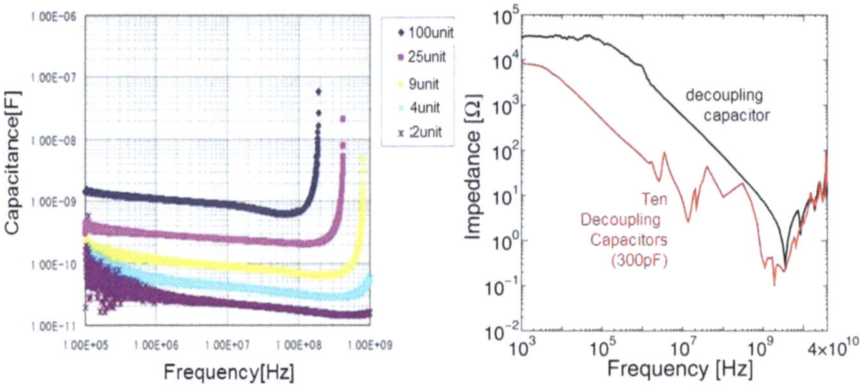


Fig. 9.160 Frequency characteristics of the decoupling capacitor chip

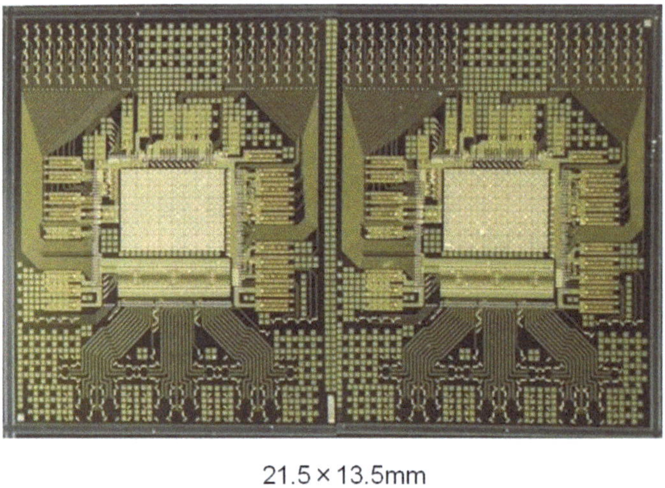


Fig. 9.161 Chip that integrates the functions of the decoupling capacitor and Si interposer

gigahertz-level belt did not have resonance. However, when there were many connections, resonance reached a level just a few hundred megahertz under that of the outgoing line. Therefore, an outgoing line was incorporated into the layout of the actual design. Compared with the chip that does not carry the decoupling capacitor, the power impedance was able to be reduced by introducing a decoupling capacitor in a large frequency domain (Fig. 9.160, right).

We designed our decoupling capacitor and wafer prototype to operate stably in unstable power supply environments to simulate the real-world in-vehicle conditions.

Fig. 9.161 shows the chip that integrates the functions of the Si-IP and the decoupling capacitor. The unified chip was manufactured in a process that shared formation process of a decoupling capacitor and TSV. In practical use, it is expected that process step can be reduced sharply.

#### 9.6.3.4 A Trial Production and Evaluation of the Driving-Assistance Image-Processing System for Automobiles

(1) Evaluation Board

An experiment was conducted to measure the performance and usefulness of the 3D integrated image sensor module. A photograph of the evaluation board is shown in Fig. 9.162. The sensor module has two 3D integrated image sensors at the top-center and has dimensions of $25 \times 13$ mm (325 mm$^2$), which is a small footprint.

With this trial production board (operating at 3.3 V), data from a sensor module were converted into PCI-Express format by a field-programmable gate array (FPGA), located in the central-lower section of the board, and then transmitted to the personal computer (PC) system.

In the board evaluation procedure, the object lens and two heat dissipation devices (which were made separately) were attached. One of the heat dissipation devices was an AlN heat sink covering the whole back surface of the evaluation board, while the other was a heat pipe system for the Si-IP. In the same way that the conversion circuit can be integrated with an interface chip in an actual product, the power supply circuit can be integrated with the interposer. The PC processing software only sees the data from a single stacked chip, that is, a 3D integrated circuit.

**Fig. 9.162** 3D integrated image sensor module evaluation board

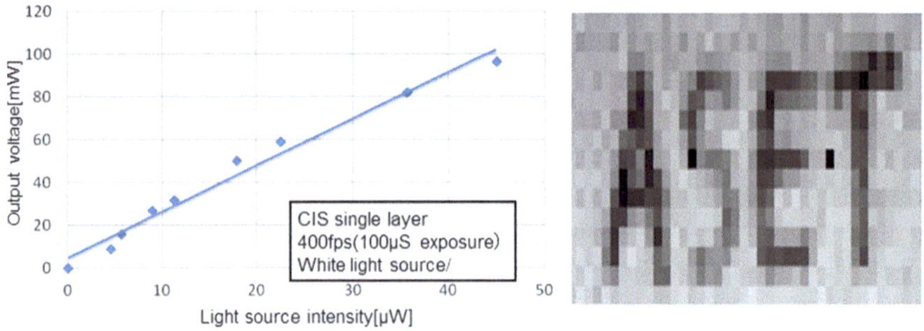


Input light intensity dependence of sensor output voltage

Output image example (96x16pixels)

**Fig. 9.163** CIS sensitivity results

(2) Evaluation Results

The trial board was evaluated in a suite of standard imaging tests: sensitivity, spectral characteristics, temperature characteristics, contrast response, spatial frequency, color reproducibility, gradation accuracy, noise, distortion of standard camera properties, night vision, ranging, and skin detection. The results of the CIS sensitivity evaluation are shown in Fig. 9.163.

We believe that our results presented here—applicable to real-world driving-assistance image-processing systems—establish a technological knowledge base for the heterogeneous C2C 3D stacking process, electric power supply, and high-speed signal transmission.

## *9.6.4 Conclusion*

1. We manufactured CIS, CDS, ADC, and IF chips as a basic technology for automotive driving-assistance image-processing systems. The design targets were achieved (operating speed greater than 1200 fps and power consumption is less than 2 W).
2. We developed a heterogeneous integrated C2C process. 3D integration of the abovementioned four chips was carried out by TSV connection. We constructed a sensor module stacked with color filter, micro-lens, and cover glass.
3. We confirmed the stable operation of our Si interposer, which had a 70-nf TSV-type decoupling capacitor, making it suitable for the automobile environment.
4. We fabricated two-eye image sensor modules (325 mm$^2$) with ranging functionality. These used a TSV connection 3D-SiP structure, which integrated 10 chips comprising six types of CIS/CDS/ADC/IF chips, and a Si interposer/decoupling capacitor chip.

5. We created the evaluation board from this module as a driving-assistance image-processing system for cars. It was evaluated using software for practical applications.
6. Our research results clarify the following aspects of the implementation of 3D integration in an in-vehicle electronic device:

   a. The structure, design manual, and the effect of a Si-IP, which also improves the environmental resistance of 3DICs
   b. The challenges in designing a 3D-IC using optimal semiconductor processes
   c. The effect of a secondary 3DIC packaging method.

7. Issues of reliability and cost, which remain to be solved.

## 9.7 Heterogeneous 3D Integration Technology for Radio Frequency Microelectromechanical Systems

### *9.7.1 Background and Issues*

The purpose of heterogeneous 3D integration is the creation of an unprecedented multifunctional device and a high-performance semiconductor device by integration of semiconductor device with other functional device. For example, the microelectromechanical system (MEMS) device is a promising device and will play an important role which COMS devices cannot play for new applications.

The MEMS device has so many expecting fields, such as sensor devices, RF devices, photo devices, BIO devices, and so on. Also, the MEMS device can handle a variety of signals, such as very small amplitude analog signals, high frequency signals for communication, high-voltage driving signals for actuators. 3D integration of the MEMS device which handles a variety of signals and CMOS devices which handle high-voltage driving signals for actuators, results in interference between chips spoils performance or deteriorate signals and at worst, malfunction.

The MEMS device has a tiny movable mechanical part that requires a hermetic shield to protect the tiny mechanical part from contamination during the 3D integration process [45].

We adopt 3D integration technology developed for the CMOS device and find issues and solve them. The MEMS device deals with many application fields and adopts not only silicon but also glass or ceramic as a wafer. 3D integration technology deals with deference of thermal expansion, while young's modulus between materials needs to be developed.

To develop heterogeneous 3D integration technology, device design technology and device fabrication technology and 3D integration technology of the device is being studied using MEMS on low-temperature co-fired ceramics (MEMS-on-LTCC) technology which has better RF characteristics. With MEMS-on-LTCC technology, LTCC substrate play a role not only as a device substrate but also as an interposer, and high-performance and high-density integration are expected to be realized.

Aiming for practical use for mobile phones, a fully 3D integrated MEMS tunable filter module has been designed and fabricated as a demonstration device. We developed LTCC substrate with a tunable filter circuit, MEMS switches and CMOS driving IC and 3D integration technology for these devices.

The developed 3D tunable filter module is necessary to replace the conventional RF front-end module consisting of numbers of fixed frequency circuit with a tunable frequency RF front-end module.

Construction of the 3D integration RF module, heterogeneous integration technology for high-density integration, simulation, design fabrication, and evaluation are being studied.

## *9.7.2 Development Result*

### 9.7.2.1 Structure of 3D Integration RF Module

We studied the miniaturization of the 3D integration RF module using heterogeneous 3D integration technology. The configuration of the proposed 3D integration RF module is shown in Fig. 9.164. The 3D integration RF module is based on a LTCC interposer with the filter circuit directly formed on the front surface. The MEMS switches and CMOS driving IC are mounted on the front and reverse sides of the interposer, respectively.

Pads are formed on the reverse side of the ring wall which is mounted on the same side of the LTCC wafer to provide in/out paths to the CMOS IC and to the outside. The inner wiring of the LTCC wafer can provide dense interconnects among

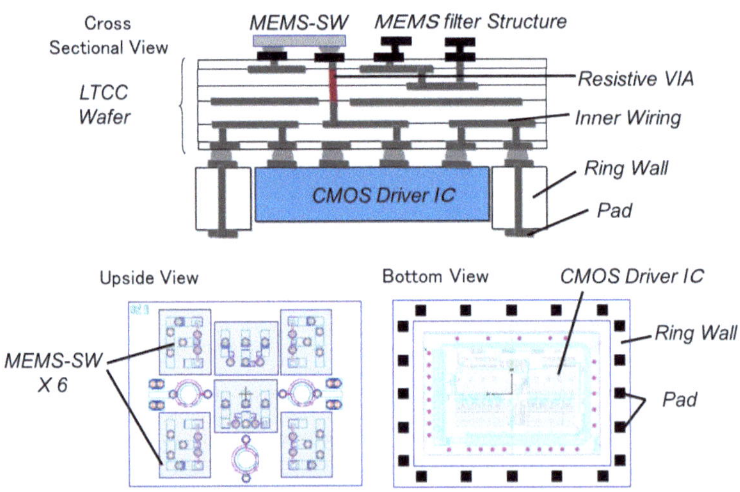


**Fig. 9.164** Configuration of 3D the integration RF module. *CMOS* complementary metal–oxide–semiconductor, *IC* integrated circuit, *MEMS* microelctromechanical systems

the integrated filter circuit, MEMS switches and CMOS driving IC. These 3D integration technologies realized a miniaturized module size of 3.6 mm by 4.7 mm.

#### 9.7.2.2 MEMS Tunable Filter

Miniaturized high performance digital MEMS tunable filter which is an important element of the 3D integration RF module is designed fabricated and significant performance of the fabricated filter is confirmed.

(1) Circuit Configuration of Tunable Filter

The proposed filter circuit is symmetric as shown in Fig. 9.165. The filter consists of three LC resonators that are connected to a T-network with a capacitive coupling (Cc) between the in and out terminal to form transmission zeros at both the upper and lower bandwidth limits. Resonator 1 and 3 consisting of inductor L1 and capacitor bank C1 have completely the same arrangement with each other. Resonator 2 having inductor L2 and capacitor bank C2 is shunted to the ground. Capacitor banks Cm are parallel connected at in/out terminal respectively to match the in/out impedance. SPDT MEMS switches are used in capacitor bank C1, C2, Cc, and Cm to discretely select capacitances. Each capacitor bank includes three capacitors. One of the three capacitors is arranged to be normally on and the other two are selected by a single pole double throw (SPDT) MEMS switch. Therefore, each capacitor bank

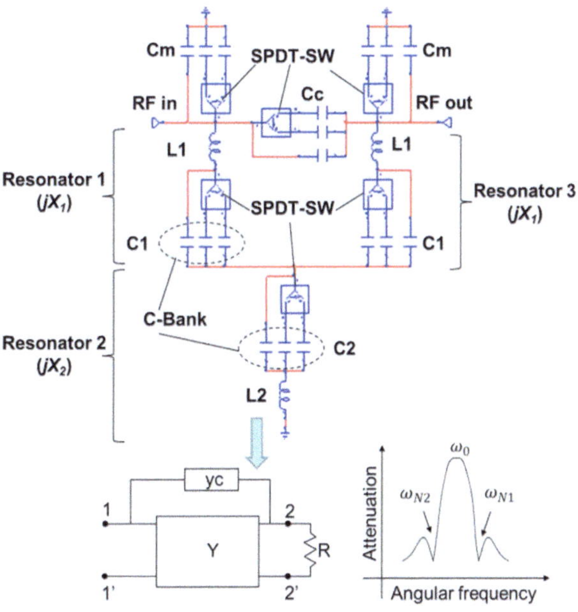


**Fig. 9.165** Configuration of proposed tunable filter circuit. *SPDT* single pole double throw

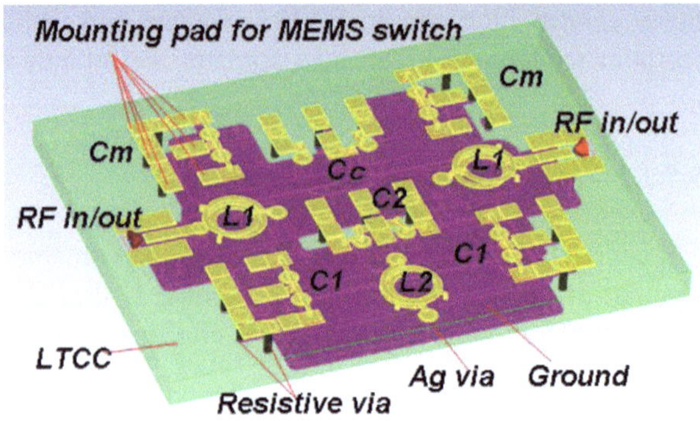


**Fig. 9.166** Electromagnetic simulation model. *MEMS* microelctromechanical systems, *LTCC* low-temperature co-fired ceramic

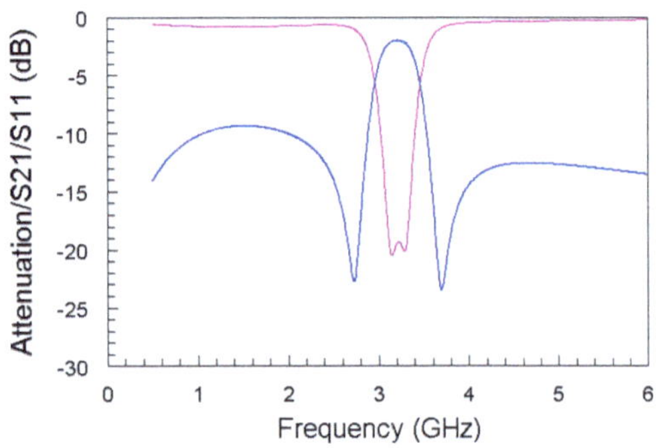


**Fig. 9.167** Simulated filter S-parameter

can provide 4 capacitance states equivalent to 2 bit tuning [46–48]. Eight channels of 20 v signals from the CMOS IC are used for driving six SPDT MEMS switches.

(2) Electromagnetic Simulation

The filter circuit shown in Fig. 9.165 is developed into an electromagnetic simulation model shown in Fig. 9.166 and electromagnetic simulation is carried out. A two-layered aerial spiral coil inductor and 3D interconnection in the air are formed to increase the quality factor and to reduce the parasitic capacitance.

The simulated filter S-parameters are shown in Fig. 9.167. These simulation results indicate the designed filter can achieve an insertion loss of 2 dB and provide transmission zeros at both the upper and lower bandwidth limits.

(3) Construction of LTCC Wafer

Figure 9.168 shows layer construction of LTCC wafer. Its thickness is 460 μm. Three upper layers are for the RF circuit and two bottom layers are for the DC circuit. The ground layer between the RF layer and DC layer isolate the RF circuit

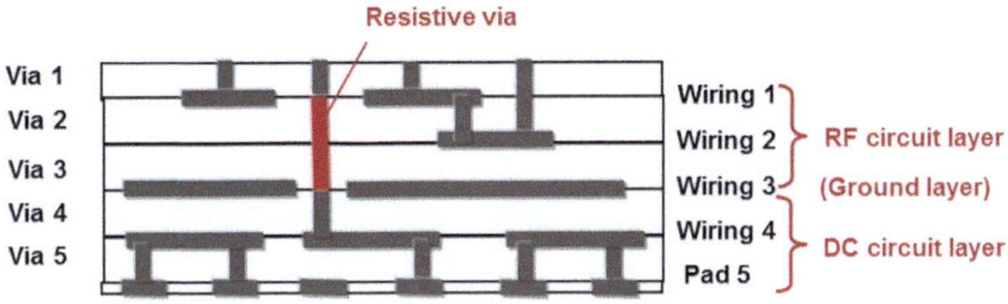


**Fig. 9.168** Layer construction of LTCC wafer. *DC* direct current

and DC circuit. Resistive vias are placed under the driving pad of the MEMS switch and provide high impedance of 10 kΩ to the driving path and isolate the RF signal from the DC driving path. Pads are formed directly on the surface of the wafer to provide in/out paths to the CMOS IC and also on the ring wall which is mounted on the same side of the LTCC wafer to provide in/out paths to the outside. LTCC wafer provide not only device wafer but also an interposer of the module.

To realize digital tunability, the filter is constructed with a lumped circuit element. LTCC wiring wafer having a dielectric constant of 7 is used to realize high Q factor and high self-resonance frequency of the lamped circuit element.

The LTCC wiring wafer has printed circuit on both sides of the surface. To prevent vacuum chucking error in the photolithography process on the top side, embedded pads were formed on the reverse side of LTCC wafer, to enable a flat reverse surface, to avoid chucking failure during manufacturing processes.

The RF and DC paths are separated by an inner ground to reduce the interference from DC paths. The driving signals from the reverse side mounted CMOS IC are connected to the front side mounted MEMS switch through the inner resistive vias to further improve the isolation between the RF and driving paths [49].

(4) Fabrication Process Technology

The filter circuit is directly formed on the LTCC substrate using MEMS-on-LTCC process technology consisted of photolithography and the thin film deposition shown in Fig. 9.169. The bottom electrode of the capacitor, dielectric layer, upper electrode of capacitor and first layer of aerial spiral coil inductor, in connecting pillar of two-layer coil inductor, and second layer of the two-layered aerial spiral coil are stacked sequentially. The two layered aerial inductor and 3D interconnection in the air are formed using a multistage plating technology based on a sacrifice layer to increase the quality factor and to reduce the parasitic capacitance. White wafer causes diffuse reflection because ultraviolet rays ingresses from the surface in the exposure process. To solve this, we introduced colored wafer to prevent diffuse reflection and it results in higher productivity in the exposure process. The fabricated LTCC filter chip is shown in Fig. 9.170. Also, an enlarged view of the two-layered coil and capacitor is shown in Fig. 9.170. A high-resolution 3D structure using the proposed process technology is confirmed.

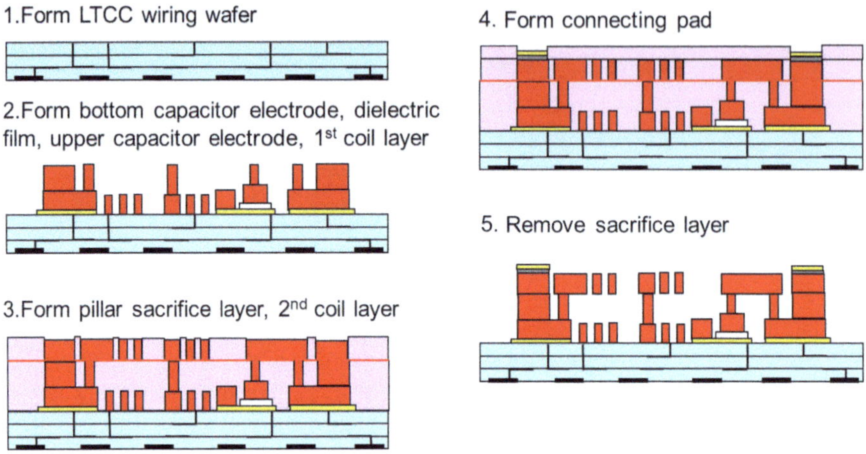


**Fig. 9.169** Fabrication process. *LTCC* low-temperature co-fired ceramic

#### 9.7.2.3 MEMS Switch

RF-MEMS switches are attractive due to their low insertion loss, high isolation, and excellent linearity. But none of the MEMS switches have been commercially assembled to mobile phones yet. This is because switches in mobile application are required for low-voltage operation (< 10–20 V) and long switch lifetime (> 109–1010 cycles). It is difficult to achieve these two demands at the same time. Although highly reliable MEMS switches using electrostatic actuation have been realized, their actuation voltages are 60–90 V. Because large contact force is necessary for sustaining low contact resistance through quite a lot of switching cycles, large restoring force is also required to avoid stiction problems, so reducing the stiffness of the movable part is not allowed [50–52].

The piezoelectric actuator is promising to generate large contact force with low driving voltage and low power consumption. Several piezoelectric actuated MEMS switches with low- voltage actuation (< 15 V) have been demonstrated, but the switch lifetime is referred to only a little. One of the reasons is that the fabrication process of the piezoelectric switch is more complex than the electrostatic type. PZT film needs the high-temperature process and has strong film stress. Therefore, formation of contact/signal electrodes and narrow air-gap is difficult [53, 54].

We use a stiff SCS beam to endure the high-temperature process and to reduce deflection originated from film stress. In addition, a single contact structure is adopted to achieve high reliability.

(1) Design of MEMS Switch

Figure 9.171 shows the structure of our metal-to-metal contact RF-MEMS switch that uses piezoelectric actuation. The switch has a single-crystal silicon (SCS)

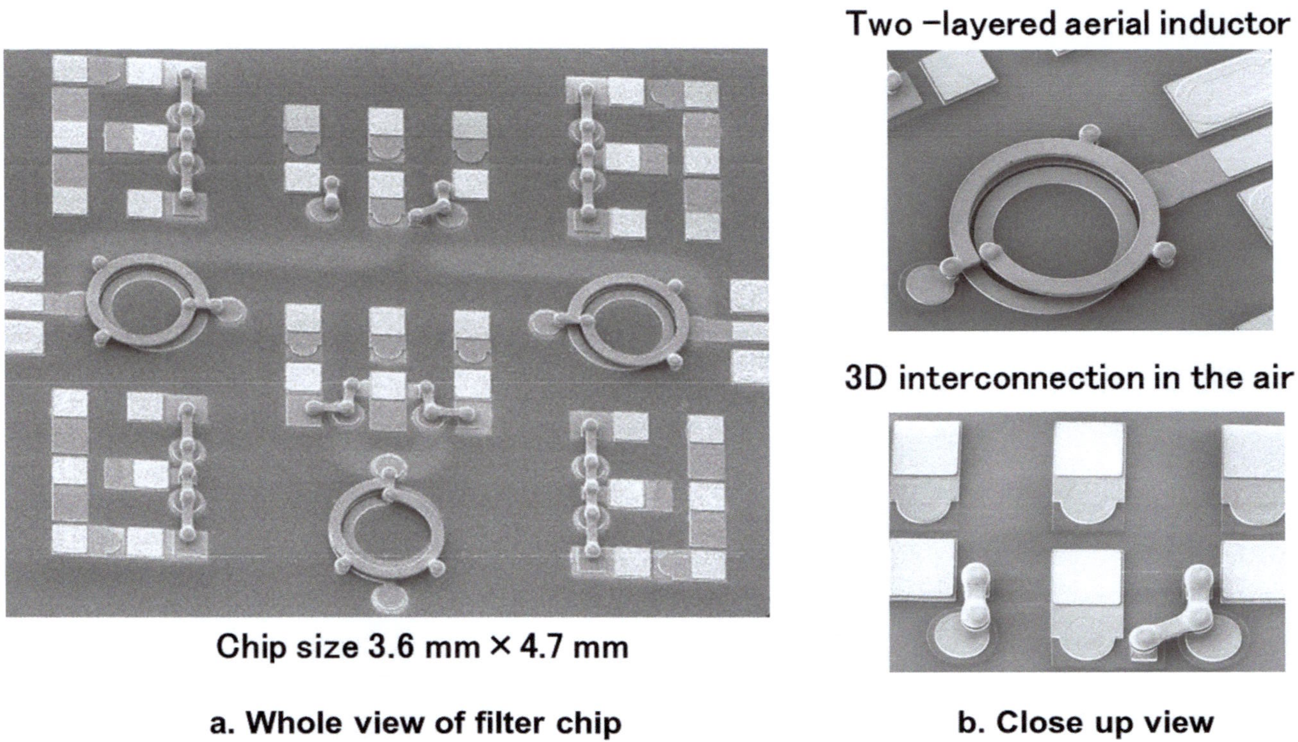


**Fig. 9.170** SEM photograph of fabricated filter chip. *3D* three dimension

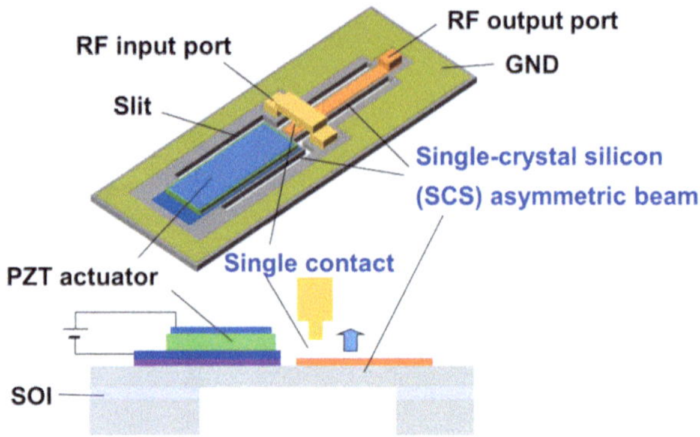


**Fig. 9.171** Schematic drawing of proposed RF-MEMS switch. GND ground

fixed-fixed beam, on which a lead zirconate titanat (PZT) unimorph actuator and a bottom RF signal electrode are patterned. The beam is separated from the fixed part by a slit. A bridge-shaped top RF signal line is formed above the beam by electroplating with narrow air gap. RF ground (GND) surrounds the switch to improve impedance matching.

When a DC bias voltage is applied to top and bottom electrodes sandwiching the PZT film, the PZT shrinks in a parallel direction to the substrate. Then SCS beam moves upwards and the switch turns ON-state. The fixed–fixed beam is a designed asymmetric shape. In order to reduce actuation voltage, the portion having the bottom electrode is narrower and longer than the other portion having the PZT actuator. Our beam is very stiff (>2000 N/m of spring constant) so as to reduce its undesirable deflection.

Another standout feature is that our switch has only one contact point, while the actuator is electrically isolated from the signal lines. The conventional switches have symmetric RF input/output signal lines and two corresponding contacts. These switches need low contact resistance for both series contacts, so achieving a long lifetime should be harder than a single contact structure. The isolated actuator allows us to simplify the switch driving circuits and to avoid self-actuation against high power input. Figure 9.172 shows SEM images of the fabricated switch [55].

(2) Evaluation of Fabricated MEMS Switch

The DC and RF characteristics of the fabricated single pole single throw (SPST) switches were measured. Figure 9.173 shows the relation between contact resistance and applied voltage for piezoelectric switch. Increasing the applied voltage, the switch closed to ON-state at 6 V. The contact resistance became lower than 2Ω at 13 V or less. Decreasing the voltage, it returned to OFF-state at 3 V.

Figure 9.174 presents the measured RF performance of the piezoelectric switch. Note that subtraction of the thru-line loss was not carried out. The insertion loss was −0.2 dB up to 2 GHz, and −0.3 dB up to 5 GHz. The isolation was −33 dB up to 2 GHz, and −25 dB up to 5 GHz. The main factor of the insertion loss is the resistance of the sputtered bottom signal line that has 0.9 Ω resistance [56, 57].

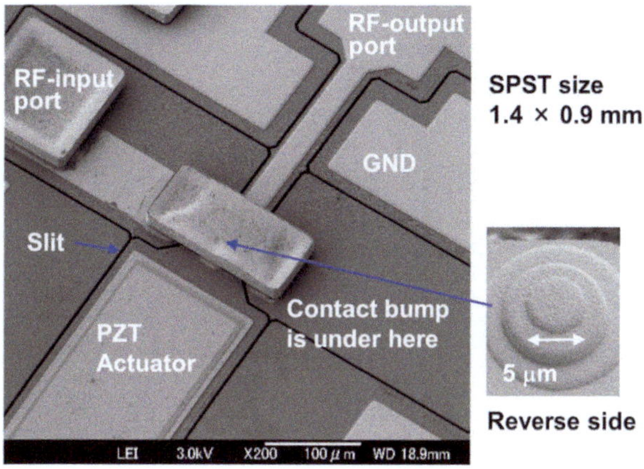


**Fig. 9.172** SEM images of piezoelectric switch. PZT Lead zirconate titanate

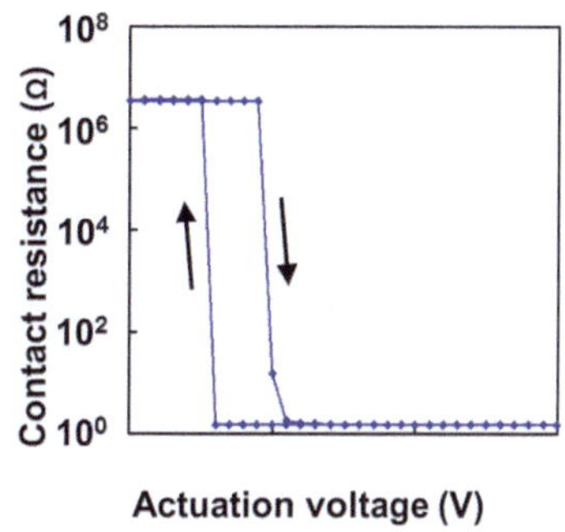


**Fig. 9.173** Measured ON–OFF characteristic of piezoelectric switch

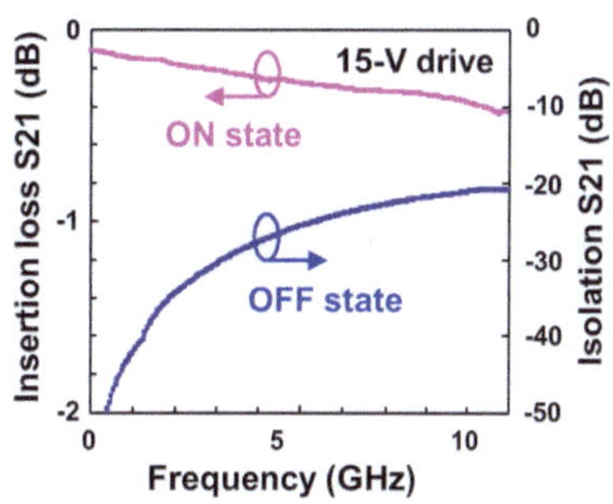


**Fig. 9.174** Measured RF performance of piezoelectric switch

We measured switch lifetime at a low power condition (DC 1 mW) using a 10 kHz switching rate in air. As shown in Fig. 9.175, piezoelectric switch operated with 15 V keeping low contact resistance (<2 Ω) until 10 × 107 cycles. Reducing the actuation voltage to 10 V in order to weaken the contact force, stiction did not occur as shown in Fig. 9.176. Instead, contact resistance increased to around 20 Ω, then saturated and kept operating up to 10 × 109 cycles. This indicates that more than 10 × 109 cycles durability of the PZT actuator can be expected.

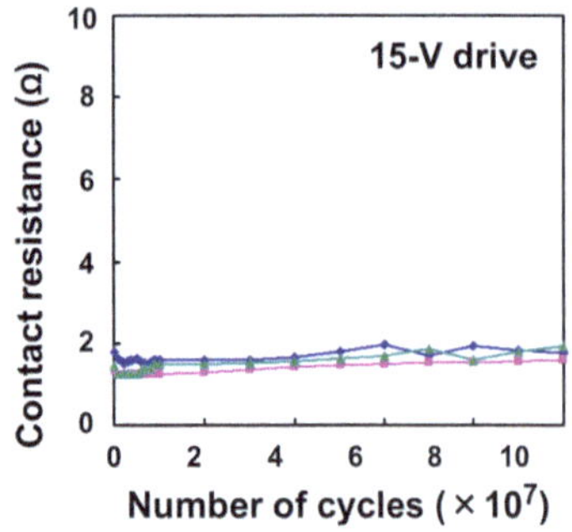


**Fig. 9.175** Measured contact resistance of piezoelectric switch with 15 V operation

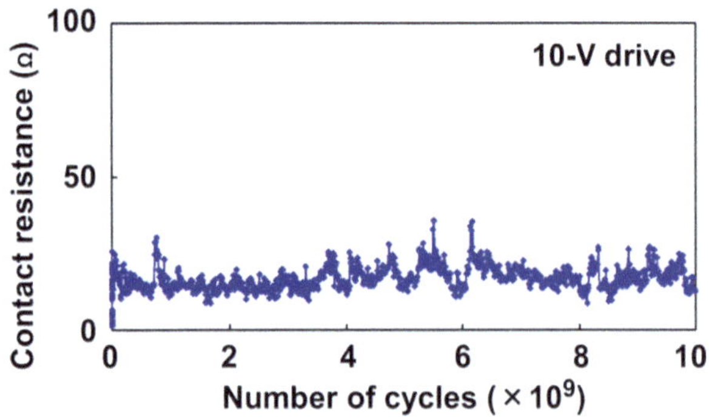


**Fig. 9.176** Measured contact resistance of piezoelectric switch with 10 V operation

#### 9.7.2.4 CMOS Driving IC

The CMOS driving IC is fabricated by a foundry service. The fabricated IC is shown in Fig. 9.177. The chip size is 2.4 mm by 3.5 mm. The driving IC includes three wire digital serial interface, bias, VCO, 2-stage charge pump and 8-channel output controlling circuit blocks. The external source is 3.3 V and the output voltages can have eight-step selection between a voltage of 20 and 40.

**Fig. 9.177** Fabricated CMOS driving IC

#### 9.7.2.5 3D Integration of Tunable Filter Module

To realize tunable 3D integration RF module shown in Fig. 9.164, we studied bonding technology and parameters in the bonding process for each chip.

The CMOS driving IC was mounted on the backside by solder bonding technology and underfill was used to enhance die share hardness. Figure 9.178a shows LTCC substrate after CMOS driving IC bonding. The size of LTCC substrate is five by four chip blocks and six inner blocks are used to provide chucking space on the perimeter of substrate. Figure 9.178b shows LTCC substrate diced into each chip. The MEMS switches were mounted on the front side of the LTCC filter substrate by Au-flip-chip-bonding technology. The module was mounted onto an evaluation board for measurement using printed solder paste on evaluation board and a heat process of 250 °C and underfill was used to enhance die share hardness. The MEMS switches were mounted on the front side of the LTCC filter substrate by Au flip-chip bonding technology. The fabricated fully integrated tunable filter module is shown in Fig. 9.179. The module size is 3.6 mm by 4.7 mm.

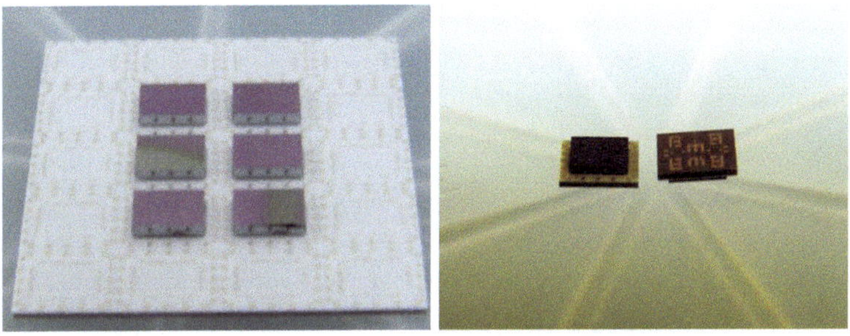

**a** 5 by 4 chip size LTCC substrate with CMOS **b** Diced chip

**Fig. 9.178** LTCC substrate with CMOS IC in work size and diced LTCC chip. *CMOS* complementary metal–oxide–semiconductor, *LTCC* low-temperature co-fired ceramic

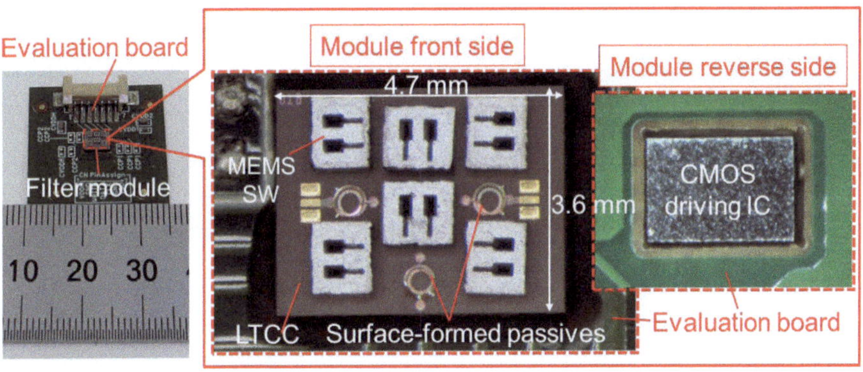


**Fig. 9.179** Fully integrated tunable filter module and valuation board. *CMOS* complementary metal–oxide–semiconductor, *IC* integrated circuit

In this project, the MEMS switch without a hermetic seal was adopted and mounted on the module in the last process to avoid contamination of contact during the soldering process. The WLP MEMS switch with hermetic sealed contact is already developed by our group. Adoption of the WLP MEMS switch will shorten the 3D integration process and realize a lower profile 3D construction.

#### 9.7.2.6 RF and Tuning Performances of the Fabricated 3D Tunable Filter Module

The RF and tuning performances of the fabricated 3D integrated tunable filter module were evaluated. The measured and simulated filter performances for the initial state [Cm (00) C1 (00) C2 (00) Cc (00)] without any MEMS SW being mounted are compared in Fig. 9.180. The measurement result roughly accords with the simulated ones.

With a 3.3 V outside voltage supplement and 8bit driving of 20 V, the integrated CMOS IC to control the MEMS switches which turns on around 10 V and have contact resistance of 0.5 Ω. Eight driving signals from the integrated CMOS IC control six SPDT MEMS switches and control six capacitor banks (Cm, Cc, C1, C2) shown in Fig. 9.165. The capacitor bank consists of three capacitors, one capacitor is connected directly to the signal line, two capacitors are switched with SPDT MEMS switches. This design results in only two driving signals controlling one capacitor bank.

Figure 9.181 presents the measured filter module tuning response. The fabricated tunable filter module achieves a 44 % discrete tuning from 1.8 GHz to 2.8 GHz. The measured insertion loss is 1.5–2.5 dB over the whole tuning range. Designing transmission 0 close to the pass band result in a shape factor less than 3 and the relative bandwidth is around 13 %. The out-of-band rejection is better than −25dB. In addition, bandwidth tuning performance also was confirmed. Figure 9.182 presents the bandwidth tuning around a center frequency of 2.2 GHz. The obtained bandwidth tuning range is 115 % from 260 MHz to 560 MHz. These results verify that the filter module can tune both the center frequency and bandwidth tuning with a steep frequency response.

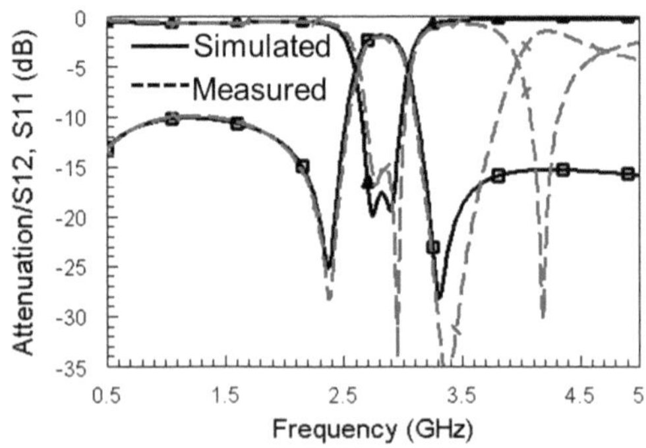


**Fig. 9.180** Simulated and measure filter performances

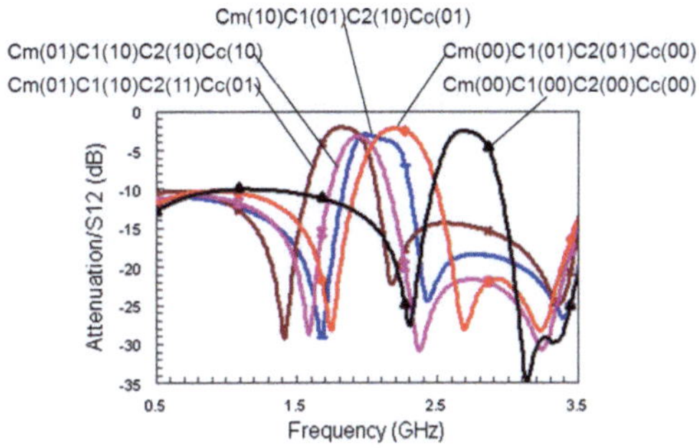


**Fig. 9.181** Measured tuning response of tunable filter module

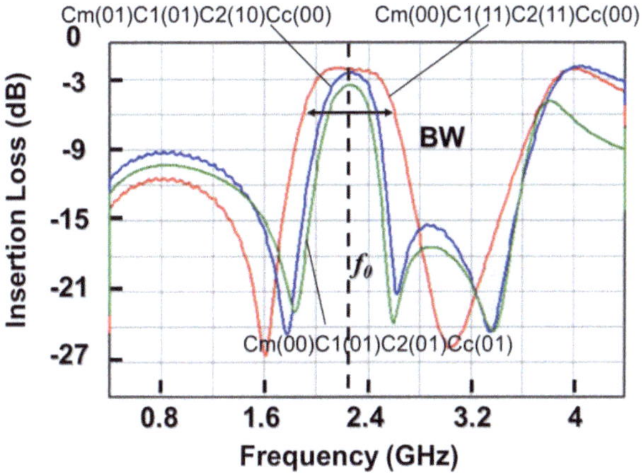


**Fig. 9.182** Measured bandwidth tuning response at 2.2 GHz

## *9.7.3 Summary*

To develop heterogeneous 3D integration technology, device design technology and device fabrication technology, and 3D integration technology of the device was studied using MEMS-on-LTCC technology which has better RF characteristics. A fully 3D integrated MEMS 8-bit digital tunable filter module based on LTCC filter substrate has been designed and fabricated as a motif. The fabricated module is constructed with LTCC substrate with filter circuits, MEMS switches mounted on one side of LTCC substrate, and CMOS-driving IC mounted on the other side of LTCC substrate. The developed 3D tunable filter module achieved a compact size of 3.6 mm by 4.7 mm and a wide discrete tuning range of 44 % from 1.8 GHz to 2.8 GHz with an insertion loss less than 2.5 dB over the whole tuning range. Furthermore, the bandwidth tuning range of 115 % has been confirmed. It has been verified that the proposed 3D filter module enables digital tuning of both center frequency and bandwidth and provides a steep frequency response.

**Acknowledgments** Part of this work was entrusted by NEDO "Development of Functionally Innovative 3D-Integrated Circuit (Dream Chip) Technology" project that is based on the Japanese government's METI "IT Innovation Program."

## References

1. ASET (2013) Presentation slide dream-chip project by ASET (final result). March 8, http://aset.la.coocan.jp/english/e-kenkyu/Dream_Chip_Pj_Final-Results_ASET.pdf. Accessed 12 Jul 2014
2. Kada M (2014) R & D overview of 3D integration technology using TSV worldwide and in Japan. 2014 ECS and SMEQ joint international meeting (October 5–10)
3. Kohara S et al (2012) Thermal stress analysis of die stacks with fine-pitch IMC interconnections for 3D integration. Presented at 3D system integration conference (3DIC), 31 January–02 February 2012) Osaka, Japan
4. Kohara S et al (2012) Thermal cycle tests and observation of fine-pitch IMC bonding in 3D chip stack. Presented at microjoining and assembly technology in electronics (MATE) conference, (31 January–1 February 2012), Yokohama, Japan (in Japanese)
5. Yoo JH et al (2010) Analysis of electromigration for cu pillar bump in flip chip package. Presented at electronics packaging technology conference (EPTC), (08–10 December 2010) Singapore
6. Wright SL et al (2006) Characterization of micro-bumpC4 interconnects for Si-carrier SOP applications. Presented at electronic components and technology conference (ECTC), (31 May–2 June 2006) San Diego
7. Lu M et al (2012) Effect of joule heating on electromigration reliability of Pb-free interconnect. Presented at electronic components and technology conference (ECTC), (29 May–01 Jun 2012), San Diego
8. Sueoka K et al (2013) High precision bonding for fine-pitch interconnection. Presented at international conference on electronics packaging (ICEP), (10–12 April 2013), Osaka, Japan
9. Sueoka K et al (2011) Joining method for thin si chips. Presented at micro electronics symposium (MES), (8–9 September 2011), Osaka, Japan (in Japanese)
10. Horibe A et al (2011) High density 3D TSV chip integration process. Presented at international conference on electronics packaging (ICEP), (13–15 April 2011), Nara, Japan
11. Kitaichi K et al (2013) 3D package assembly development with the use of the dicing tape having NCF layer. Presented at international conference on electronics packaging (ICEP), (10–12 April 2013), Osaka, Japan
12. Sueoka K et al (2011) TSV diagnostics by X-ray microscopy. Presented at electronics packaging technology conference (EPTC), (07–09 December 2011), Singapore
13. Matsumoto K et al (2011) Experimental thermal resistance evaluation of a three-dimensional (3D) chip stack. Presented at semiconductor thermal measurement and management symposium (SEMI-THERM) (20–24 March 2011), San Jose
14. Matsumoto K et al (2012) Experimental thermal resistance evaluation of a three-dimensional (3D) chip stack, including the transient measurements. Presented at semiconductor thermal measurement and management Symposium (SEMI-THERM), (18–22 March 2012), San Jose
15. Yamada F et al (2013) Cooling strategy and structure of 3D integrated image sensor for automobile application. Presented at international conference on electronics packaging (ICEP), (10–12 April 2013), Osaka, Japan
16. Matsumoto K et al (2013) Thermal design guidelines for a three-dimensional (3D) chip stack, including cooling solutions. Presented at semiconductor thermal measurement and management symposium (SEMI-THERM), (17–21 March 2013), San Jose

17. Matsumoto K et al (2013) Thermal design guideline and new cooling solution for a three-dimensional (3D) chip stack. Presented at 3D system integration conference (3DIC), (02–04 October 2013), San Francisco
18. Suekawa S et al (2013) A 1/4-inch 8Mpixel Back-Illuminated Stacked CMOS Image Sensor. ISSCC 27.4, 484–486
19. Kim D-W et al (2013) Development of 3D Through Silicon Stack (TSS) assembly for wide IO memory to logic devices integration. Electron Compon Technol Conf 77–80
20. Graham S (2014) Hybrid Memory Cube: your new standard for memory performance. ICEP 821–825
21. Lee DU et al (2014) A 1.2V 8 Gb 8-channel 128 GB/s High-Bandwidth Memory (HBM) stacked DRAM with effective microbump I/O test methods using 29 nm process and TSV. ISSCC 25.2, 432–433
22. Shimamoto H et al (2014) Study for CMOS device characteristics affected by ultra thin wafer thinning. IEEE ICSJ Symp 22–24
23. www.disco.co.jp/eg/solution/library/stresslief.html
24. John R-S-E et al (2013) Low cost, room temperature debondable spin-on temporary bonding solution: a key enable for 2.5/3D IC packaging. Electron Compon Technol Conf 107–112
25. Jourdain A et al (2013) Integration and manufacturing aspects of moving from WaferBOND HT-10.10 to ZoneBOND material in temporary wafer bonding and debonding for 3D applications. Electron Compon Technol Conf 113–117
26. Kubo A et al (2014) Development of new concept thermoplastic temporary adhesive for 3D-IC integration. Electron Compon Technol Conf 899–905
27. www.disco.co.jp/eg/apexp/grinder/ncg.html
28. www.disco.co.jp/eg/solution/library/dbg.html
29. www.disco.co.jp/eg/solution/library/stealth.html
30. Lee K et al (2013) Degradation of memory retention characteristics in DRAM chip by Si thinning for 3-D integration. IEEE Electron Device Lett 34(8):1038–1040
31. Kim YS et al (2014) Ultra thinning down to 4-um using 300-mm wafer proven by 40-nm node 2 Gb DRAM for 3D multi-stack WOW applications. VLSI Symp Technol Dig 3.2, 26–27
32. Bea J et al (2011) Evaluation of Cu contamination at backside surface of thinned wafer in 3-D integration by transient-capacitance measurement. IEEE Electron Device Lett 32(1):66–68
33. www.disco.co.jp/eg/solution/apexp/polisher/gettering.html
34. Murugesan M et al (2013) Mechanical characteristics of thin dies/wafers in three-dimensional large-scale integrated systems. 24th SEMI-Advanced Semiconductor Manufacturing Conference, 66–69
35. Murugesan M et al (2010) Wafer thinning, bonding, and interconnects induced local strain/stress in 3D-LSIs with fine-pitch high-density microbumps and through-Si vias. IEDM Technol Dig 2.3.1
36. Shimamoto H (2015) Thin wafer bending strength measurement standardization. 21th Mate 199–202
37. Ryu S et al (2012) Effect of thermal stresses on carrier mobility and keep-out zone around through-silicon vias for 3-D integration. IEEE Trans Device Mater Reliab 12:255–262
38. Mercha A et al (2010) Comprehensive analysis of the impact of single and arrays of through silicon vias induced stress on high-k/metal gate CMOS performance. International Electron Devices Meeting, P.2.2.1
39. Takahashi K et al (2001) Current status of research and development for three-dimensional chip stack technology. Jpn J Appl Phys 40:3032–3037
40. Lee DU et al (2014) An Exact Measurement and Repair Circuit of TSV Connection for 128 GB/s High-Bandwidth Memory(HBM) Stacked DRAM. In: Proceeding of VLSI-Circuit, pp 27–28
41. Jang DM et al (2007) Development and evaluation of 3-D SiP with vertically interconnected through silicon vias (TSV). In: Proceeding of 57th ECTC, pp 847–852

42. Ikeda H et al (2012) Development of TSV/3D Memory+ Logic SiP with 4k-IO Interconnects. In: Proceeding of ICEP-IAAC, pp 60–76
43. Takatani H et al (2012) PDN impedance and noise simulation of 3D SiP with a widebus structure. IEEE electronic components and technology conference, pp 673–677
44. Takaya Satoshi et al (2013) A 100 GB/s wide I/O with 4096b TSVs through an active silicon interposer with in-place waveform capturing. In: Proceeding of ISSCC, pp 434–436
45. Rebeiz M G, Entesari K, Reines I C, Park S, El-Tanani MA, Grichener A., and Brown R. A. (2009) Tuning into RF MEMS. IEEE Microwave Magazine 55–72, Oct
46. Entesari K, Rebeiz GM (2005) A differential 4-bit 6.5–10-GHz RF MEMS tunable filter. IEEE Trans Microw Theory Tech 53(3):1103–1110
47. Mi X, Kawano Y, Toyoda O, Suzuki T, Ueda S, Hirose T, Joshin K (2010) Miniaturized microwave tunable bandpass filters on high-k LTCC. In: Proceeding of APMC. pp 139–142
48. Inoue H, Mi X, Fujiwara T, Toyoda O, Ueda S, Nakazawa F (2012) A novel tunable filter enabling both center frequency and bandwidth tunability. In: Proceeding of 42nd EUMC, pp 269–272
49. Mi X, Takahashi T, Ueda S (2008) Integrated passives on LTCC for achieving chip-sized-modules. In: Proceeding of 38th EUMC. pp 607–610
50. Rebeiz GM (2003) RF MEMS theory, design, and technology. New Jersey: Wiley
51. Majumder S, Lampen J, Morrison R, Maciel J (2003) A packaged, high-lifetime ohmic MEMS RF switch. IEEE MTT-S International Microwave Symposium Digest 1935–1938 (Jun)
52. Newman HS, Ebel JL, Judy D, Maciel J (2008) Lifetime measurements on a high-reliability RF-MEMS contact switch. IEEE Microw Wirel Compon Lett 18(2):100–102 (Feb)
53. Guerre R, Drechsler U, Bhattacharyya D, Rantankari P, Stutz R, Wright RV, Milosavljevic ZD, Vaha-Heokkila T, Kirby PB, Despont M (2010) Wafer-level transfer technology for PZT-based RF MEMS switches. J Microelectromech Syst 19(3):548–560 (Jun)
54. Pulskamp JS, Judy DC, Polcawich RG, Kaul R, Chandrahalim H, Bhave SA (2006) PZT actuated seesaw SPDT RF MEMS switch. J Phys: Conf Ser 34(304):304–309
55. Nakatani T, Katsuki T, Okuda H, Toyoda O, Ueda S, Nakazawa F (2011) PZT actuated reliable RF-MEMS switch using single crystal silicon asymmetric beam. In: Proceeding of IEEE Asia pacific microwave conference
56. Nakazawa F, Shimanouchi T, Nakatani T, Katsuki T, Okuda H, Toyoda O, Ueda S (2011) Effect of frequency in the 3D integration of a PZT-actuated MEMS switch using a single crystal silicon asymmetric beam" In: Proceeding of 3DIC
57. Nakazawa F, Mi S, Shimanouchi T, Nakatani T, Katsuki T, Okuda H, Toyoda O, Ueda S (2012) 3D heterogeneous integration using MEMS devices for RF applications. In: Proceeding of MRS spring conference

# Index





K. Kondo et al. (eds.), *Three-Dimensional Integration of Semiconductors*,
DOI 10.1007/978-3-319-18675-7

## J



## K



## L



## M



## N

MIX
Papier aus verantwortungsvollen Quellen
Paper from responsible sources
FSC® C105338

If you have any concerns about our products,
you can contact us on
**ProductSafety@springernature.com**

In case Publisher is established outside the EU,
the EU authorized representative is:
**Springer Nature Customer Service Center GmbH**
**Europaplatz 3, 69115 Heidelberg, Germany**

Printed by Libri Plureos GmbH
in Hamburg, Germany